The Electric Power Engineering Handbook

ELECTRIC POWER SUBSTATIONS ENGINEERING

THIRD EDITION

The Electric Power Engineering Handbook
Third Edition

Edited by
Leonard L. Grigsby

Electric Power Generation, Transmission, and Distribution
Edited by Leonard L. Grigsby

Electric Power Transformer Engineering, Third Edition
Edited by James H. Harlow

Electric Power Substations Engineering, Third Edition
Edited by John D. McDonald

Power Systems, Third Edition
Edited by Leonard L. Grigsby

Power System Stability and Control
Edited by Leonard L. Grigsby

The Electric Power Engineering Handbook

ELECTRIC POWER SUBSTATIONS ENGINEERING

THIRD EDITION

EDITED BY
JOHN D. McDONALD

CRC Press
Taylor & Francis Group
Boca Raton   London   New York

CRC Press is an imprint of the
Taylor & Francis Group, an **informa** business

CRC Press
Taylor & Francis Group
6000 Broken Sound Parkway NW, Suite 300
Boca Raton, FL 33487-2742

© 2012 by Taylor & Francis Group, LLC
CRC Press is an imprint of Taylor & Francis Group, an Informa business

No claim to original U.S. Government works

Printed in the United States of America on acid-free paper
Version Date: 20111109

International Standard Book Number: 978-1-4398-5638-3 (Hardback)

This book contains information obtained from authentic and highly regarded sources. Reasonable efforts have been made to publish reliable data and information, but the author and publisher cannot assume responsibility for the validity of all materials or the consequences of their use. The authors and publishers have attempted to trace the copyright holders of all material reproduced in this publication and apologize to copyright holders if permission to publish in this form has not been obtained. If any copyright material has not been acknowledged please write and let us know so we may rectify in any future reprint.

Except as permitted under U.S. Copyright Law, no part of this book may be reprinted, reproduced, transmitted, or utilized in any form by any electronic, mechanical, or other means, now known or hereafter invented, including photocopying, microfilming, and recording, or in any information storage or retrieval system, without written permission from the publishers.

For permission to photocopy or use material electronically from this work, please access www.copyright.com (http://www.copyright.com/) or contact the Copyright Clearance Center, Inc. (CCC), 222 Rosewood Drive, Danvers, MA 01923, 978-750-8400. CCC is a not-for-profit organization that provides licenses and registration for a variety of users. For organizations that have been granted a photocopy license by the CCC, a separate system of payment has been arranged.

Trademark Notice: Product or corporate names may be trademarks or registered trademarks, and are used only for identification and explanation without intent to infringe.

Library of Congress Cataloging-in-Publication Data

Electric power substations engineering / editor, John D. McDonald. -- 3rd ed.
   p. cm. -- (Electrical engineering handbook)
  ISBN 978-1-4398-5638-3 (hardback)
  1. Electric substations. I. McDonald, John D. (John Douglas), 1951-

TK1751.E44 2013
621.31'26--dc23   2011044124

Visit the Taylor & Francis Web site at
http://www.taylorandfrancis.com

and the CRC Press Web site at
http://www.crcpress.com

Contents

Preface vii
Editor ix
Contributors xi

1   How a Substation Happens 1-1
    Jim Burke and Anne-Marie Sahazizian

2   Gas-Insulated Substations 2-1
    Phil Bolin

3   Air-Insulated Substations: Bus/Switching Configurations 3-1
    Michael J. Bio

4   High-Voltage Switching Equipment 4-1
    David L. Harris and David Childress

5   High-Voltage Power Electronic Substations 5-1
    Dietmar Retzmann and Asok Mukherjee

6   Interface between Automation and the Substation 6-1
    James W. Evans

7   Substation Integration and Automation 7-1
    Eric MacDonald

8   Oil Containment 8-1
    Thomas Meisner

9   Community Considerations 9-1
    James H. Sosinski

10   Animal Deterrents/Security 10-1
    Mike Stine

11   Substation Grounding 11-1
    Richard P. Keil

12   Direct Lightning Stroke Shielding of Substations 12-1
    Robert S. Nowell

13 Seismic Considerations 13-1
    Eric Fujisaki

14 Substation Fire Protection 14-1
    Don Delcourt

15 Substation Communications 15-1
    Daniel E. Nordell

16 Physical Security of Substations 16-1
    John Oglevie, W. Bruce Dietzman, and Cale Smith

17 Cyber Security of Substation Control and Diagnostic Systems 17-1
    Daniel Thanos

18 Gas-Insulated Transmission Line 18-1
    Hermann Koch

19 Substation Asset Management 19-1
    H. Lee Willis and Richard E. Brown

20 Station Commissioning and Project Closeout 20-1
    Jim Burke and Rick Clarke

21 Energy Storage 21-1
    Ralph Masiello

22 Role of Substations in Smart Grids 22-1
    Stuart Borlase, Marco C. Janssen, Michael Pesin, and Bartosz Wojszczyk

Index **Index**-1

Preface

The electric power substation, whether generating station or transmission and distribution, remains one of the most challenging and exciting fields of electric power engineering. Recent technological developments have had a tremendous impact on all aspects of substation design and operation. The objective of *Electric Power Substations Engineering* is to provide an extensive overview of substations, as well as a reference and guide for their study. The chapters are written for the electric power engineering professional for detailed design information as well as for other engineering professions (e.g., mechanical and civil) who want an overview or specific information in one particular area.

The book is organized into 22 chapters to provide comprehensive information on all aspects of substations, from the initial concept of a substation to design, automation, operation, physical and cyber security, commissioning, energy storage, and the role of substations in Smart Grid. The chapters are written as tutorials and provide references for further reading and study. A number of the chapter authors are members of the IEEE Power & Energy Society (PES) Substations Committee. They develop the standards that govern all aspects of substations. In this way, this book contains the most recent technological developments regarding industry practice as well as industry standards. This book is part of the Electrical Engineering Handbook Series published by Taylor & Francis Group/CRC Press. Since its inception in 1993, this series has been dedicated to the concept that when readers refer to a book on a particular topic, they should be able to find what they need to know about the subject at least 80% of the time. That has indeed been the goal of this book.

During my review of the individual chapters of this book, I was very pleased with the level of detail presented, but more importantly the tutorial style of writing and use of photographs and graphics to help the reader understand the material. I thank the tremendous efforts of the 28 authors who were dedicated to do the very best job they could in writing the 22 chapters. Fifteen of the twenty chapters were updated from the second edition, and there are two new chapters in the third edition. I also thank the personnel at Taylor & Francis Group who have been involved in the production of this book, with a special word of thanks to Nora Konopka and Jessica Vakili. They were a pleasure to work with and made this project a lot of fun for all of us.

John D. McDonald

Editor

John D. McDonald, PE, is the director of technical strategy and policy development for GE Digital Energy. In his 38 years of experience in the electric utility industry, he has developed power application software for both supervisory control and data acquisition (SCADA)/energy management system (EMS) and SCADA/distribution management system (DMS) applications, developed distribution automation and load management systems, managed SCADA/EMS and SCADA/DMS projects, and assisted intelligent electronic device (IED) suppliers in the automation of their IEDs.

John received his BSEE and MSEE in power engineering from Purdue University and an MBA in finance from the University of California, Berkeley. He is a member of Eta Kappa Nu (electrical engineering honorary) and Tau Beta Pi (engineering honorary); is a fellow of IEEE; and was awarded the IEEE Millennium Medal in 2000, the IEEE Power & Energy Society (PES) Excellence in Power Distribution Engineering Award in 2002, and the IEEE PES Substations Committee Distinguished Service Award in 2003.

In his 25 years of working group and subcommittee leadership with the IEEE PES Substations Committee, John led seven working groups and task forces who published standards/tutorials in the areas of distribution SCADA, master/remote terminal unit (RTU), and RTU/IED communications protocols. He was also on the board of governors of the IEEE-SA (Standards Association) in 2010–2011, focusing on long-term IEEE Smart Grid standards strategy. John was elected to chair the NIST Smart Grid Interoperability Panel (SGIP) Governing Board for 2010–2012.

John is past president of the IEEE PES, chair of the Smart Grid Consumer Collaborative (SGCC) Board, charter member of the IEEE Brand Ambassadors Program, member of the IEEE Medal of Honor Committee, member of the IEEE PES Region 3 Scholarship Committee, VP for Technical Activities for the US National Committee (USNC) of CIGRE, and past chair of the IEEE PES Substations Committee. He was also the director of IEEE Division VII in 2008–2009. He is a member of the advisory committee for the annual DistribuTECH Conference. He also received the 2009 Outstanding Electrical and Computer Engineer Award from Purdue University.

John teaches courses on Smart Grid at the Georgia Institute of Technology, for GE, and for various IEEE PES chapters as a distinguished lecturer of the IEEE PES. He has published 40 papers and articles in the areas of SCADA, SCADA/EMS, SCADA/DMS, and communications, and is a registered professional engineer (electrical) in California, Pennsylvania, and Georgia.

John is the coauthor of the book *Automating a Distribution Cooperative, from A to Z*, published by the National Rural Electric Cooperative Association Cooperative Research Network (CRN) in 1999. He is also the editor of the Substations chapter and a coauthor of the book *The Electric Power Engineering Handbook*—cosponsored by the IEEE PES and published by CRC Press in 2000. He is the editor in chief of the book *Electric Power Substations Engineering*, Second Edition, published by Taylor & Francis Group/CRC Press in 2007, as well as the author of the "Substation integration and automation" chapter.

Contributors

Michael J. Bio
Alstom Grid
Birmingham, Alabama

Phil Bolin
Mitsubishi Electric Power Products, Inc.
Warrendale, Pensylvania

Stuart Borlase
Siemens Energy, Inc.
Raleigh, North Carolina

Richard E. Brown
Quanta Technology
Raleigh, North Carolina

Jim Burke (retired)
Baltimore Gas & Electric Company
Baltimore, Maryland

David Childress
David Childress Enterprises
Griffin, Georgia

Rick Clarke
Baltimore Gas & Electric Company
Baltimore, Maryland

Don Delcourt
BC Hydro
Burnaby, British Columbia, Canada

and

Glotek Consultants Ltd.
Surrey, British Columbia, Canada

W. Bruce Dietzman
Oncor Electric Delivery Company
Fort Worth, Texas

James W. Evans
The St. Claire Group, LLC
Grosse Pointe Farms, Michigan

Eric Fujisaki
Pacific Gas and Electric Company
Oakland, California

David L. Harris
SPX Transformer Solutions
(Waukesha Electric Systems)
Waukesha, Wisconsin

Marco C. Janssen
UTInnovation
Duiven, the Netherlands

Richard P. Keil
Commonwealth Associates, Inc.
Dayton, Ohio

Hermann Koch
Siemens AG
Erlangen, Germany

Eric MacDonald
GE Energy–Digital Energy
Markham, Ontario, Canada

Ralph Masiello
KEMA, Inc.
Chalfont, Pennsylvania

Thomas Meisner
Hydro One Networks, Inc.
Toronto, Ontario, Canada

Asok Mukherjee
Siemens AG
Erlangen, Germany

Daniel E. Nordell
Xcel Energy
Minneapolis, Minnesota

Robert S. Nowell (retired)
Commonwealth Associates, Inc.
Jackson, Michigan

John Oglevie
POWER Engineers, Inc.
Boise, Idaho

Michael Pesin
Seattle City Light
Seattle, Washington

Dietmar Retzmann
Siemens AG
Erlangen, Germany

Cale Smith
Oncor Electric Delivery Company
Fort Worth, Texas

Anne-Marie Sahazizian
Hydro One Networks, Inc.
Toronto, Ontario, Canada

James H. Sosinski (retired)
Consumers Energy
Jackson, Michigan

Mike Stine
TE Energy
Fuquay Varina, North Carolina

Daniel Thanos
GE Energy–Digital Energy
Markham, Ontario, Canada

H. Lee Willis
Quanta Technology
Raleigh, North Carolina

Bartosz Wojszczyk
GE Energy–Digital Energy
Atlanta, Georgia

1
How a Substation Happens

Jim Burke (retired)
Baltimore Gas & Electric Company

Anne-Marie Sahazizian
Hydro One Networks, Inc.

1.1	Background...	1-1
1.2	Need Determination..	1-2
1.3	Budgeting...	1-2
1.4	Financing...	1-3
1.5	Traditional and Innovative Substation Design....................	1-3
1.6	Site Selection and Acquisition..	1-4
1.7	Design, Construction, and Commissioning Process..........	1-5
	Station Design • Station Construction • Station Commissioning	
	References..	1-8

1.1 Background

The construction of new substations and the expansion of existing facilities are commonplace projects in electric utilities. However, due to its complexity, very few utility employees are familiar with the complete process that allows these projects to be successfully completed. This chapter will attempt to highlight the major issues associated with these capital-intensive construction projects and provide a basic understanding of the types of issues that must be addressed during this process.

There are four major types of electric substations. The first type is the switchyard at a generating station. These facilities connect the generators to the utility grid and also provide off-site power to the plant. Generator switchyards tend to be large installations that are typically engineered and constructed by the power plant designers and are subject to planning, finance, and construction efforts different from those of routine substation projects. Because of their special nature, the creation of power plant switchyards will not be discussed here, but the expansion and modifications of these facilities generally follow the same processes as system stations.

The second type of substation, typically known as the customer substation, functions as the main source of electric power supply for one particular business customer. The technical requirements and the business case for this type of facility depend highly on the customer's requirements, more so than on utility needs; so this type of station will also not be the primary focus of this discussion.

The third type of substation involves the transfer of bulk power across the network and is referred to as a system station. Some of these stations provide only switching facilities (no power transformers) whereas others perform voltage conversion as well. These large stations typically serve as the end points for transmission lines originating from generating switchyards and provide the electrical power for circuits that feed transformer stations. They are integral to the long-term reliability and integrity of the electric system and enable large blocks of energy to be moved from the generators to the load centers. Since these system stations are strategic facilities and usually very expensive to construct and maintain, these substations will be one of the major focuses of this chapter.

The fourth type of substation is the distribution station. These are the most common facilities in power electric systems and provide the distribution circuits that directly supply most electric customers.

They are typically located close to the load centers, meaning that they are usually located in or near the neighborhoods that they supply, and are the stations most likely to be encountered by the customers. Due to the large number of such substations, these facilities will also be a focus of this chapter.

Depending on the type of equipment used, the substations could be

- Outdoor type with air-insulated equipment
- Indoor type with air-insulated equipment
- Outdoor type with gas-insulated equipment
- Indoor type with gas-insulated equipment
- Mixed technology substations
- Mobile substations

1.2 Need Determination

An active planning process is necessary to develop the business case for creating a substation or for making major modifications. Planners, operating and maintenance personnel, asset managers, and design engineers are among the various employees typically involved in considering such issues in substation design as load growth, system stability, system reliability, and system capacity; and their evaluations determine the need for new or improved substation facilities. Customer requirements, such as new factories, etc., should be considered, as well as customer relations and complaints. In some instances, political factors also influence this process, as is the case when reliability is a major issue. At this stage, the elements of the surrounding area are defined and assessed and a required in-service date is established.

It is usual for utilities to have long-term plans for the growth of their electric systems in order to meet the anticipated demand. Ten year forecasts are common and require significant input from the engineering staff. System planners determine the capacities of energy required and the requirements for shifting load around the system, but engineering personnel must provide cost info on how to achieve the planners' goals. Planners conduct studies that produce multiple options and all of these scenarios need to be priced in order to determine the most economical means of serving the customers.

A basic outline of what is required in what area can be summarized as follows: System requirements including

- Load growth
- System stability
- System reliability
- System capacity

Customer requirements including

- Additional load
- Power quality
- Reliability
- Customer relations
- Customer complaints
- Neighborhood impact

1.3 Budgeting

Part of the long-range plan involves what bulk power substations need to be created or expanded in order to move large blocks of energy around the system as necessary and where do they need to be located. Determinations have to be made as to the suitability of former designs for the area in question. To achieve this, most utilities rely on standardized designs and modular costs developed over time, but should these former designs be unsuitable for the area involved, that is, unlikely to achieve community acceptance, then alternative designs need to be pursued. In the case of bulk power substations, the equipment and land costs can differ greatly from standard designs. Distribution stations, however, are the most common on most systems and therefore have the best known installed costs. Since these are the substations closest to the customers, redesign is less likely to be required than screening or landscaping, so costs do not vary greatly.

Having established the broad requirements for the new station, such as voltages, capacity, number of feeders, etc., the issue of funding should then be addressed. This is typical when real estate investigations of available sites begin, since site size and location can significantly affect the cost of the facility. Preliminary equipment layouts and engineering evaluations are also undertaken at this stage to develop ballpark costs, which then have to be evaluated in the corporate budgetary justification system. Preliminary manpower forecasts of all disciplines involved in the engineering and construction of the substation should be undertaken, including identification of the nature and extent of any work that the utility may need to contract out. This budgeting process will involve evaluation of the project in light of corporate priorities and provide a general overview of cost and other resource requirements. Note that this process may be an annual occurrence. Any projects in which monies have yet to be spent are generally reevaluated every budget cycle.

Cost estimating also entails cash forecasting; for planning purposes, forecasts per year are sufficient. This means that every budget cycle, each proposed project must not only be reviewed for cost accuracy, but the cash forecast must also be updated. It is during these annual reviews that standardized or modularized costs also need to be reviewed and revised if necessary.

1.4 Financing

Once the time has arrived for work to proceed on the project, the process of obtaining funding for the project must be started. Preliminary detailed designs are required to develop firm pricing. Coordination between business units is necessary to develop accurate costs and to develop a realistic schedule. This may involve detailed manpower forecasting in many areas. The resource information has to be compiled in the format necessary to be submitted to the corporate capital estimate system and internal presentations must be conducted to sell the project to all levels of management.

Sometimes it may be necessary to obtain funding to develop the capital estimate. This may be the case when the cost to develop the preliminary designs is beyond normal departmental budgets, or if unfamiliar technology is expected to be implemented. This can also occur on large, complex projects or when a major portion of the work will be contracted. It may also be necessary to obtain early partial funding in cases where expensive, long lead-time equipment may need to be purchased such as large power transformers.

1.5 Traditional and Innovative Substation Design [1]

Substation engineering is a complex multidiscipline engineering function. It could include the following engineering disciplines:

- Environmental
- Civil
- Mechanical
- Structural
- Electrical—high voltage
- Protection and controls
- Communications

Traditionally, high-voltage substations are engineered based on preestablished layouts and concepts and usually conservative requirements. This approach may restrict the degree of freedom of introducing new solutions. The most that can be achieved with this approach is the incorporation of new primary and secondary technology in pre-engineered standards.

A more innovative approach is one that takes into account functional requirements such as system and customer requirements and develops alternative design solutions. System requirements include elements of rated voltage, rated frequency, existing system configuration (present and future), connected loads, lines, generation, voltage tolerances (over and under), thermal limits, short-circuit levels, frequency tolerances (over and under), stability limits, critical fault clearing time, system expansion, and interconnection. Customer requirements include environmental consideration (climatic, noise, aesthetic, spills, and right-of-way), space consideration, power quality, reliability, availability, national and international applicable standards, network security, expandability, and maintainability.

Carefully selected design criteria could be developed to reflect the company philosophy. This would enable, when desired, consideration and incorporation of elements such as life cycle cost, environmental impact, initial capital investment, etc., into the design process. Design solutions could then be evaluated based on preestablished evaluation criteria that satisfy the company interests and policies.

1.6 Site Selection and Acquisition

At this stage, a footprint of the station has been developed, including the layout of the major equipment. A decision on the final location of the facility can now be made and various options can be evaluated. Final grades, roadways, storm water retention, and environmental issues are addressed at this stage, and required permits are identified and obtained. Community and political acceptance must be achieved and details of station design are negotiated in order to achieve consensus. Depending on local zoning ordinances, it may be prudent to make settlement on the property contingent upon successfully obtaining zoning approval since the site is of little value to the utility without such approval. It is not unusual for engineering, real estate, public affairs, legal, planning, operations, and customer service personnel along with various levels of management to be involved in the decisions during this phase.

The first round of permit applications can now begin. Although the zoning application is usually a local government issue, permits for grading, storm water management, roadway access, and other environmental or safety concerns are typically handled at the state or provincial level and may be federal issues in the case of wetlands or other sensitive areas. Other federal permits may also be necessary, such as those for aircraft warning lights for any tall towers or masts in the station. Permit applications are subject to unlimited bureaucratic manipulation and typically require multiple submissions and could take many months to reach conclusion. Depending on the local ordinances, zoning approval may be automatic or may require hearings that could stretch across many months. Zoning applications with significant opposition could take years to resolve.

As a rule of thumb, the following site evaluation criteria could be used:

- Economical evaluation
- Technical evaluation
- Community acceptance

Economical evaluation should address the level of affordability, return on investment, initial capital cost, and life cycle cost.

Technical aspects that can influence the site selection process could include the following:

- *Land*: choose areas that minimize the need for earth movement and soil disposal.
- *Water*: avoid interference with the natural drainage network.
- *Vegetation*: choose low-productivity farming areas or uncultivated land.
- *Protected areas*: avoid any areas or spots listed as protected areas.
- *Community planning*: avoid urban areas, development land, or land held in reserve for future development.

- *Community involvement*: engage community in the approval process.
- *Topography*: flat but not prone to flood or water stagnation.
- *Soil*: suitable for construction of roads and foundations; low soil resistivity is desirable.
- *Access*: easy access to and from the site for transportation of large equipment, operators, and maintenance teams.
- *Line entries*: establishment of line corridors (alternatives: multi-circuit pylons, UG lines).
- *Pollution*: risk of equipment failure and maintenance costs increase with pollution level.

To address community acceptance issues it is recommended to

- Adopt a low profile layout with rigid buses supported on insulators over solid shape steel structures
- Locate substations in visually screened areas (hills, forest), other buildings, and trees
- Use gas-insulated switchgear (GIS)
- Use colors, lighting
- Use underground egresses as opposed to overhead

Other elements that may influence community acceptance are noise and oil leakages or spills.

To mitigate noise that may be emitted by station equipment, attention should be paid at station orientation with respect to the location of noise-sensitive properties and the use of mitigation measures such as noise barriers, sound enclosures, landscaping, and active noise cancellation.

Guidelines to address oil leakages or spills could be found in Chapter 8 as well as in Refs. [2,3].

1.7 Design, Construction, and Commissioning Process [4]

Having selected the site location, the design construction and commissioning process would broadly follow the steps shown in Figure 1.1. Recent trends in utilities have been toward sourcing design and construction of substations through competitive bidding process to ensure capital efficiency and labor productivity.

1.7.1 Station Design

Now the final detailed designs can be developed along with all the drawings necessary for construction. The electrical equipment and all the other materials can now be ordered and detailed schedules for all disciplines negotiated. Final manpower forecasts must be developed and coordinated with other business units. It is imperative that all stakeholders be aware of the design details and understand what needs to be built and by when to meet the in-service date. Once the designs are completed and the drawings published, the remaining permits can be obtained.

The following can be used as a guide for various design elements:

1. Basic layout
    a. Stage development diagram
    b. Bus configuration to meet single line requirements
    c. Location of major equipment and steel structures based on single line diagram
    d. General concept of station
    e. Electrical and safety clearances
    f. Ultimate stage
2. Design
    a. Site preparation
        i. Drainage and erosion, earth work, roads and access, and fencing
    b. Foundations
        i. Soils, concrete design, and pile design

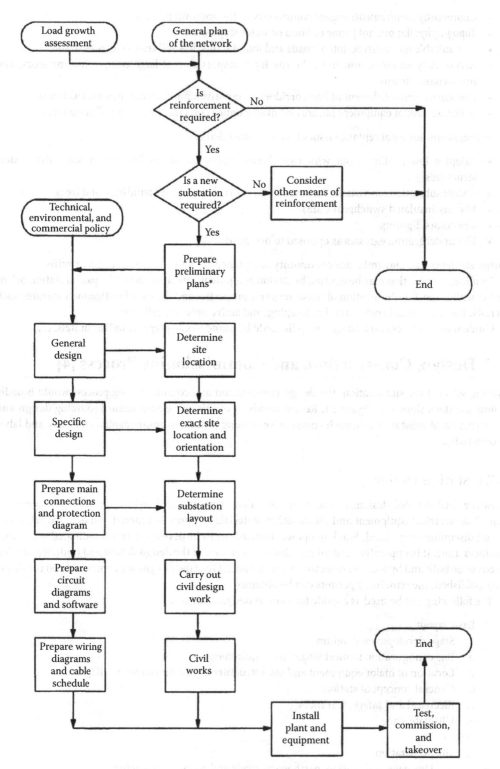

FIGURE 1.1 Establishment of a new substation. *General location, line directions, soil investigations, and transport routes.

c. Structures
   i. Materials, finishes, and corrosion control
d. Buildings
   i. Control, metering, relaying, and annunciation buildings—types such as masonry, prefabricated, etc.
   ii. Metalclad switchgear buildings
   iii. GIS buildings
e. Mechanical systems
   i. HVAC
   ii. Sound enclosure ventilation
   iii. Metalclad switchgear or GIS building ventilation
   iv. Fire detection and protection
   v. Oil sensing and spill prevention
f. Buswork
   i. Rigid buses
   ii. Strain conductors—swing, bundle collapse
   iii. Ampacity
   iv. Connections
   v. Phase spacing
   vi. Short-circuit forces
g. Insulation
   i. Basic impulse level and switching impulse level
h. Station insulators
   i. Porcelain post type insulators
   ii. Resistance graded insulators
   iii. Polymeric post insulators
   iv. Station insulator hardware
   v. Selection of station insulator—TR—ANSI and CSA standard
   vi. Pollution of insulators—pollution levels and selection of leakage distance
i. Suspension insulators
   i. Characteristics
   ii. Porcelain suspension insulators
   iii. Polymeric suspension insulators
   iv. Suspension insulators hardware
   v. Selection of suspension insulators
   vi. Pollution of insulators—pollution levels and selection of leakage distance
j. Clearances
   i. Electrical clearances
   ii. Safety clearances
k. Overvoltages
   i. Atmospheric and switching overvoltages
   ii. Overvoltage protection—pipe and rod gaps, surge arresters
   iii. Atmospheric overvoltage protection—lightning protection (skywires, lightning rods)
l. Grounding
   i. Function of grounding system
   ii. Step, touch, mesh, and transferred voltages
   iii. Allowable limits of body current
   iv. Allowable limits of step and touch voltages
   v. Soil resistivity
   vi. General design guidelines

m. Neutral systems
      i. Background of power system grounding
      ii. Three- and four-wire systems
      iii. HV and LV neutral systems
      iv. Design of neutral systems
   n. Station security
      i. Physical security [5]
      ii. Electronic security

1.7.2 Station Construction

With permits in hand and drawings published, the construction of the station can begin. Site logistics and housekeeping can have a significant impact on the acceptance of the facility. Parking for construction personnel, traffic routing, truck activity, trailers, fencing, and mud and dirt control along with trash and noise can be major irritations for neighbors, so attention to these details is essential for achieving community acceptance. All the civil, electrical, and electronic systems are installed at this time. Proper attention should also be paid to site security during the construction phase not only to safeguard the material and equipment but also to protect the public.

1.7.3 Station Commissioning

Once construction is complete, testing of various systems can commence and all punch list items addressed. To avoid duplication of testing, it is recommended to develop an inspection, testing, and acceptance plan (ITAP). Elements of ITAP include

- Factory acceptance tests (FATs)
- Product verification plan (PVP)
- Site delivery acceptance test (SDAT)
- Site acceptance tests (SATs)

Final tests of the completed substation in a partially energized environment to determine acceptability and conformance to customer requirements under conditions as close as possible to normal operation conditions will finalize the in-service tests and turnover to operations.

Environmental cleanup must be undertaken and final landscaping can be installed. Note that, depending upon the species of plants involved, it may be prudent to delay final landscaping until a more favorable season in order to ensure optimal survival of the foliage. Public relations personnel can make the residents and community leaders aware that the project is complete and the station can be made functional and turned over to the operating staff.

References

1. A. Carvalho et al. CIGRE SC 23. Functional substation as key element for optimal substation concept in a de-regulated market, *CIGRE SC 23 Colloquium*, Zurich, Switzerland, 1999.
2. H. Boehme (on behalf of WG 23-03). *General Guidelines for the Design of Outdoor AC Substations*, CIGRE WG 23-03 Brochure 161, 2000.
3. R. Cottrell (on behalf of WG G2). IEEE Std. 980. *Guide for Containment and Control of Spills in Substations*, 1994.
4. A.M. Sahazizian (on behalf of WG 23-03). CIGRE WG 23-03. The substation design process—An overview, *CIGRE SC 23 Colloquium*, Venezuela, 2001.
5. CIGRE WG B3.03. Substations physical security trends, 2004.

2
Gas-Insulated Substations

2.1 Sulfur Hexafluoride 2-1
2.2 Construction and Service Life 2-3
    Circuit Breaker • Current Transformers • Voltage
    Transformers • Disconnect Switches • Ground Switches •
    Interconnecting Bus • Air Connection • Power Cable
    Connections • Direct Transformer Connections • Surge
    Arrester • Control System • Gas Monitor System •
    Gas Compartments and Zones • Electrical and Physical
    Arrangement • Grounding • Testing • Installation •
    Operation and Interlocks • Maintenance
2.3 Economics of GIS 2-18
References 2-18

Phil Bolin
*Mitsubishi Electric
Power Products, Inc.*

A gas-insulated substation (GIS) uses a superior dielectric gas, sulfur hexafluoride (SF_6), at a moderate pressure for phase-to-phase and phase-to-ground insulation. The high-voltage conductors, circuit breaker interrupters, switches, current transformers (CTs), and voltage transformers (VTs) are encapsulated in SF_6 gas inside grounded metal enclosures. The atmospheric air insulation used in a conventional, air-insulated substation (AIS) requires meters of air insulation to do what SF_6 can do in centimeters. GIS can therefore be smaller than AIS by up to a factor of 10. A GIS is mostly used where space is expensive or not available. In a GIS, the active parts are protected from deterioration from exposure to atmospheric air, moisture, contamination, etc. As a result, GIS is more reliable, requires less maintenance, and will have a longer service life (more than 50 years) than AIS.

GIS was first developed in various countries between 1968 and 1972. After about 5 years of experience, the user rate increased to about 20% of new substations in countries where space was limited. In other countries with space easily available, the higher cost of GIS relative to AIS has limited its use to special cases. For example, in the United States, only about 2% of new substations are GIS. International experience with GIS is described in a series of CIGRE papers [1–3]. The IEEE [4,5] and the IEC [6] have standards covering all aspects of the design, testing, and use of GIS. For the new user, there is a CIGRE application guide [7]. IEEE has a guide for specifications for GIS [8].

2.1 Sulfur Hexafluoride

SF_6 is an inert, nontoxic, colorless, odorless, tasteless, and nonflammable gas consisting of a sulfur atom surrounded by and tightly bonded to six fluorine atoms. It is about five times as dense as air. SF_6 is used in GIS at pressures from 400 to 600 kPa absolute. The pressure is chosen so that the SF_6 will not condense into a liquid at the lowest temperatures the equipment experiences. SF_6 has two to three times the insulating ability of air at the same pressure. SF_6 is about 100 times better than air for interrupting arcs. It is the universally used interrupting medium for high-voltage circuit breakers, replacing the older mediums of oil and air. SF_6 decomposes in the high temperature of an electric arc or spark, but the decomposed gas recombines back into SF_6 so well that it is not necessary to replenish the SF_6 in GIS. There are some reactive decomposition by-products formed because of the interaction of sulfur and fluorine ions with trace amounts of moisture, air, and other contaminants. The quantities formed are very small. Molecular sieve absorbents inside the GIS enclosure eliminate these reactive by-products over time. SF_6 is supplied in 50 kg gas cylinders in a liquid state at a pressure of about 6000 kPa for convenient storage and transport.

Gas handling systems with filters, compressors, and vacuum pumps are commercially available. Best practices and the personnel safety aspects of SF_6 gas handling are covered in international standards [9].

The SF_6 in the equipment must be dry enough to avoid condensation of moisture as a liquid on the surfaces of the solid epoxy support insulators because liquid water on the surface can cause a dielectric breakdown. However, if the moisture condenses as ice, the breakdown voltage is not affected. So, dew points in the gas in the equipment need to be below about −10°C. For additional margin, levels of less than 1000 ppmv of moisture are usually specified and easy to obtain with careful gas handling. Absorbents inside the GIS enclosure help keep the moisture level in the gas low even though over time moisture will evolve from the internal surfaces and out of the solid dielectric materials [10].

Small conducting particles of millimeter size significantly reduce the dielectric strength of SF_6 gas. This effect becomes greater as the pressure is raised past about 600 kPa absolute [11]. The particles are moved by the electric field, possibly to the higher field regions inside the equipment or deposited along the surface of the solid epoxy support insulators—leading to dielectric breakdown at operating voltage levels. Cleanliness in assembly is therefore very important for GIS. Fortunately, during the factory and field power frequency high-voltage tests, contaminating particles can be detected as they move and cause small electric discharges (partial discharge) and acoustic signals—they can then be removed by opening the equipment. Some GIS equipment is provided with internal "particle traps" that capture the particles before they move to a location where they might cause breakdown. Most GIS assemblies are of a shape that provides some "natural" low electric-field regions where particles can rest without causing problems.

SF_6 is a strong greenhouse gas that could contribute to global warming. At an international treaty conference in Kyoto in 1997, SF_6 was listed as one of the six greenhouse gases whose emissions should be reduced. SF_6 is a very minor contributor to the total amount of greenhouse gases due to human activity, but it has a very long life in the atmosphere (half-life is estimated at 3200 years), so the effect of SF_6 released to the atmosphere is effectively cumulative and permanent. The major use of SF_6 is in electrical power equipment. Fortunately, in GIS the SF_6 is contained and can be recycled. By following the present international guidelines for the use of SF_6 in electrical equipment [12], the contribution of SF_6 to global warming can be kept to less than 0.1% over a 100 years horizon. The emission rate from use in electrical equipment has been reduced over the last decade. Most of this effect has been due to simply adopting better handling and recycling practices. Standards now require GIS to leak less than 0.5% per year. The leakage rate is normally much lower. Field checks of GIS in service after many years of service indicate that a leak rate objective lower than 0.1% per year is obtainable and is now offered by most manufacturers. Reactive, liquid (oil), and solid contaminants in used SF_6 are easily removed by filters, but inert gaseous contaminants such as oxygen and nitrogen are not easily removed. Oxygen and nitrogen are introduced during normal gas handling or by mistakes such as not evacuating all the air from the equipment before filling with SF_6. Fortunately, the purity of the SF_6 needs only be above 98% as established by international technical committees [12], so a simple field check of purity using commercially available percentage SF_6 meters will qualify the used SF_6 for reuse. For severe cases of contamination, the SF_6 manufacturers will take back the contaminated SF_6 and by putting it back into the production process in effect turn it back into "new" SF_6. Although not yet necessary, an end of life scenario for the eventual retirement of SF_6 is to incinerate the SF_6 with materials that will enable it to become part of environmentally acceptable gypsum.

The U.S. Environmental Protection Agency has a voluntary SF_6 emission reduction program for the electric utility industry that keeps track of emission rates, provides information on techniques to reduce emissions, and rewards utilities that have effective SF_6 emission reduction programs by high-level recognition of progress. Other counties have addressed the concern similarly or even considered banning or taxing the use of SF_6 in electrical equipment. Alternatives to SF_6 exist for medium-voltage electric power equipment (vacuum interrupters, clean air for insulation), but no viable alternative mediums have been identified for high-voltage electric power equipment in spite of decades of investigation. So far alternatives have had disadvantages that outweigh any advantage they may have in respect to a lower greenhouse gas effect. So for the foreseeable future, SF_6 will continue to be used for GIS where interruption of power system faults and switching is needed. For longer bus runs without any arcing gas-insulated line (GIL), a mixture of SF_6 with nitrogen is being used to reduce the total amount of SF_6 (see Chapter 18).

2.2 Construction and Service Life

GIS is assembled from standard equipment modules (circuit breaker, CTs, VTs, disconnect and ground switches, interconnecting bus, surge arresters, and connections to the rest of the electric power system) to match the electrical one-line diagram of the substation. A cross-section view of a 242 kV GIS shows the construction and typical dimensions (Figure 2.1).

The modules are joined using bolted flanges with an "O"-ring seal system for the enclosure and a sliding plug-in contact for the conductor. Internal parts of the GIS are supported by cast epoxy insulators. These support insulators provide a gas barrier between parts of the GIS or are cast with holes in the epoxy to allow gas to pass from one side to the other.

Up to about 170 kV system voltage, all three phases are often in one enclosure (Figure 2.2). Above 170 kV, the size of the enclosure for "three-phase enclosure" GIS becomes too large to be practical.

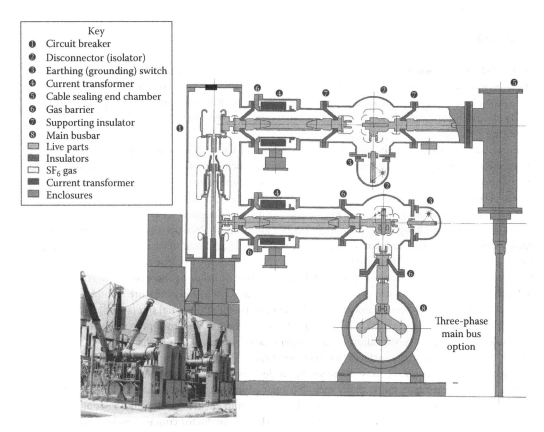

FIGURE 2.1 Single-phase enclosure GIS.

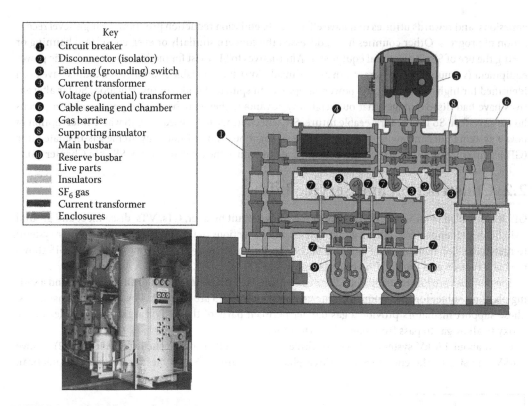

FIGURE 2.2 Three-phase enclosure GIS.

So a "single-phase enclosure" design (Figure 2.1) is used. There are no established performance differences between the three-phase enclosure and the single-phase enclosure GIS. Some manufacturers use the single-phase enclosure type for all voltage levels. Some users do not want the three phase-to-ground faults at certain locations (such as the substation at a large power plant) and will specify single-phase enclosure GIS.

Enclosures are today mostly cast or welded aluminum, but steel is also used. Steel enclosures are painted inside and outside to prevent rusting. Aluminum enclosures do not need to be painted but may be painted for ease of cleaning, a better appearance, or to optimize heat transfer to the ambient. The choice between aluminum and steel is made on the basis of cost (steel is less expensive) and the continuous current (above about 2000 A, steel enclosures require nonmagnetic inserts of stainless steel or the enclosure material is changed to all stainless steel or aluminum). Pressure vessel requirements for GIS enclosures are set by GIS standards [4,6], with the actual design, manufacture, and test following an established pressure vessel standard of the country of manufacture. Because of the moderate pressures involved, and the classification of GIS as electrical equipment, third party inspection and code stamping of the GIS enclosures are not required. The use of rupture disks as a safety measure is common although the pressure rise due to internal fault arcs in a GIS compartment of the usual size is predictable and slow enough that the protective system will interrupt the fault before a dangerous pressure is reached.

Conductors today are mostly aluminum. Copper is sometimes used for high continuous current ratings. It is usual to silver plate surfaces that transfer current. Bolted joints and sliding electrical contacts are used to join conductor sections. There are many designs for the sliding contact element. In general, sliding contacts have many individually sprung copper contact fingers working in parallel. Usually, the contact fingers are silver plated. A contact lubricant is used to ensure that the sliding contact surfaces do not generate particles or wear out over time. The sliding conductor contacts make assembly of the modules easy and also allow for conductor movement to accommodate differential thermal expansion

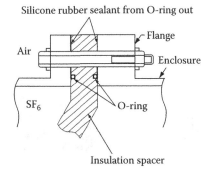

FIGURE 2.3 Gas seal for GIS enclosure. O-ring is primary seal; silicone rubber sealant is backup seal and protects O-ring and flange surfaces.

of the conductor relative to the enclosure. Sliding contact assemblies are also used in circuit breakers and switches to transfer current from the moving contact to the stationary contacts.

Support insulators are made of a highly filled epoxy resin cast very carefully to prevent formation of voids or cracks during curing. Each GIS manufacturer's material formulation and insulator shape has been developed to optimize the support insulator in terms of electric-field distribution, mechanical strength, resistance to surface electric discharges, and convenience of manufacture and assembly. Post, disk, and cone-type support insulators are used. Quality assurance programs for support insulators include a high-voltage power frequency withstand test with sensitive partial discharge monitoring. Experience has shown that the electric-field stress inside the cast epoxy insulator should be below a certain level to avoid aging of the solid dielectric material. The electrical stress limit for the cast epoxy support insulator is not a severe design constraint because the dimensions of the GIS are mainly set by the lightning impulse withstand level of the gas gap and the need for the conductor to have a fairly large diameter to carry to load currents of several thousand amperes. The result is enough space between the conductor and enclosure to accommodate support insulators having low electrical stress.

Service life of GIS using the construction described earlier, based on more than 30 years of experience to now, can be expected to be more than 50 years. The condition of GIS examined after many years in service does not indicate any approaching limit in service life. Experience also shows no need for periodic internal inspection or maintenance. Inside the enclosure is a dry, inert gas that is itself not subject to aging. There is no exposure of any of the internal materials to sunlight. Even the O-ring seals are found to be in excellent condition because there is almost always a "double-seal" system with the outer seal protecting the inner—Figure 2.3 shows one approach. This lack of aging has been found for GIS whether installed indoors or outdoors. For outdoor GIS special measures have to be taken to ensure adequate corrosion protection and tolerance of low and high ambient temperatures and solar radiation.

2.2.1 Circuit Breaker

GIS uses essentially the same dead tank SF_6 puffer circuit breakers as are used for AIS. Instead of SF_6-to-air bushings mounted on the circuit breaker enclosure, the GIS circuit breaker is directly connected to the adjacent GIS module.

2.2.2 Current Transformers

CTs are inductive ring type installed either inside the GIS enclosure or outside the GIS enclosure (Figure 2.4). The GIS conductor is the single turn primary for the CT. CTs inside the enclosure must be shielded from the electric field produced by the high-voltage conductor or high transient voltages can appear on the secondary through capacitive coupling. For CTs outside the enclosure, the enclosure itself

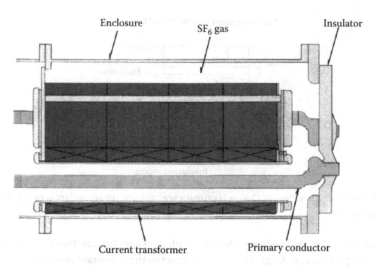

FIGURE 2.4 CTs for GIS.

must be provided with an insulating joint, and enclosure currents shunted around the CT. Both types of construction are in wide use.

Advanced CTs without a magnetic core (Rogowski coil) have been developed to save space and reduce the cost of GIS. The output signal is at a low level, so it is immediately converted by an enclosure-mounted device to a digital signal. It can be transmitted over long distances using wire or fiber optics to the control and protective relays. However, most protective relays being used by utilities are not ready to accept a digital input even though the relay may be converting the conventional analog signal to digital before processing. The Rogowski coil type of CT is linear regardless of current due to the absence of magnetic core material that would saturate at high currents.

2.2.3 Voltage Transformers

VTs are inductive type with an iron core. The primary winding is supported on an insulating plastic film immersed in SF_6. The VT should have an electric-field shield between the primary and secondary windings to prevent capacitive coupling of transient voltages. The VT is usually a sealed unit with a gas barrier insulator. The VT is either easily removable, so the GIS can be high voltage tested without damaging the VT, or the VT is provided with a disconnect switch or removable conductor link (Figure 2.5).

Advanced voltage sensors using a simple capacitive coupling cylinder between the conductor and enclosure have been developed. In addition to size and cost advantages, these capacitive sensors do not have to be disconnected for the Routine high-voltage withstand test. However, the signal level is low so it is immediately converted to a digital signal, encountering the same barrier to use as the advanced CT discussed in Section 2.2.2.

2.2.4 Disconnect Switches

Disconnect switches (Figure 2.6) have a moving contact that opens or closes a gap between stationary contacts when activated by an insulating operating rod that is itself moved by a sealed shaft coming through the enclosure wall. The stationary contacts have shields that provide the appropriate electric-field distribution to avoid too high a surface electrical stress. The moving contact velocity is relatively low (compared to a circuit breaker moving contact) and the disconnect switch can interrupt only low levels of capacitive current (e.g., disconnecting a section of GIS bus) or small inductive currents (e.g., transformer magnetizing current). For transformer magnetizing current interruption duty, the disconnect

Gas-Insulated Substations

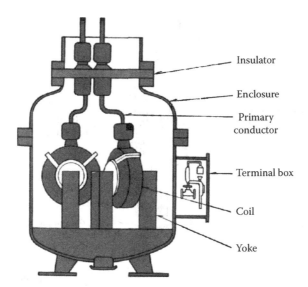

FIGURE 2.5 VTs for GIS.

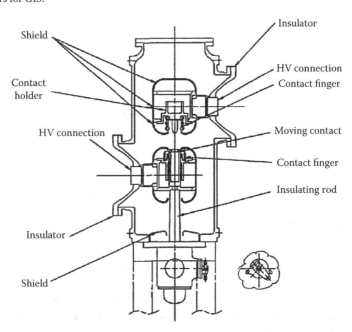

FIGURE 2.6 Disconnect switches for GIS.

switch is provided with a fast acting spring operating mechanism. Load break disconnect switches have been furnished in the past, but with improvements and cost reductions of circuit breakers, it is not practical to continue to furnish load break disconnect switches—a circuit breaker should be used instead.

2.2.5 Ground Switches

Ground switches (Figure 2.7) have a moving contact that opens or closes a gap between the high-voltage conductor and the enclosure. Sliding contacts with appropriate electric-field shields are provided at the enclosure and the conductor. A "maintenance" ground switch is operated either manually or by motor drive to close or open in several seconds. When fully closed, it can carry the rated short-circuit current

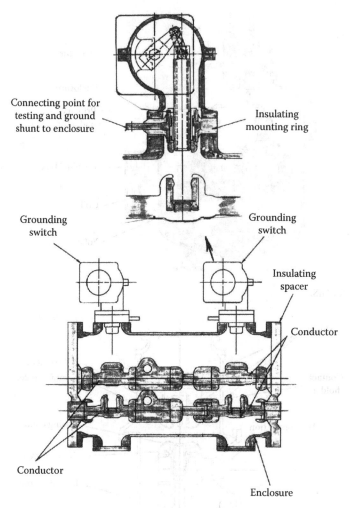

FIGURE 2.7 Ground switches for GIS.

for the specified time period (1 or 3 s) without damage. A "fast acting" ground switch has a high-speed drive, usually a spring, and contact materials that withstand arcing, so it can be closed twice onto an energized conductor without significant damage to itself or adjacent parts. Fast acting ground switches are frequently used at the connection point of the GIS to the rest of the electric power network, not only in case the connected line is energized, but also because the fast acting ground switch is better able to handle discharge of trapped charge.

Ground switches are almost always provided with an insulating mount or an insulating bushing for the ground connection. In normal operation, the insulating element is bypassed with a bolted shunt to the GIS enclosure. During installation or maintenance, with the ground switch closed, the shunt can be removed and the ground switch used as a connection from test equipment to the GIS conductor. Voltage and current testing of the internal parts of the GIS can then be done without removing SF_6 gas or opening the enclosure. A typical test is measurement of contact resistance using two ground switches (Figure 2.8).

2.2.6 Interconnecting Bus

To connect GIS modules that are not directly connected to each other, SF_6 bus consisting of an inner conductor and outer enclosure is used. Support insulators, sliding electrical contacts, and flanged enclosure

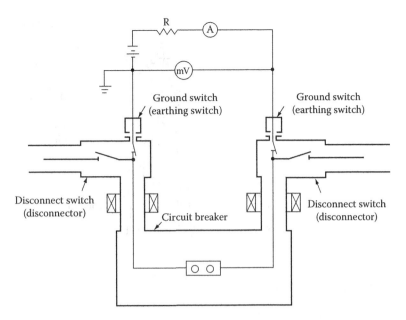

FIGURE 2.8 Contact resistance measured using ground switch. mV, voltmeter; A, ammeter; and R, resistor.

joints are usually the same as for the GIS modules, and the length of a bus section is normally limited by the allowable span between conductor contacts and support insulators to about 6 m. Specialized bus designs with section lengths of 20 m have been developed and are applied both with GIS and as separate transmission links (see Chapter 18).

2.2.7 Air Connection

SF_6-to-air bushings (Figure 2.9) are made by attaching a hollow insulating cylinder to a flange on the end of a GIS enclosure. The insulating cylinder contains pressurized SF_6 on the inside and is suitable for exposure to atmospheric air on the outside. The conductor continues up through the center of the insulating cylinder to a metal end plate. The outside of the end plate has provisions for bolting on an air-insulated conductor. The insulating cylinder has a smooth interior. Sheds on the outside improve the performance in air under wet or contaminated conditions. Electric-field distribution is controlled by internal metal shields. Higher voltage SF_6-to-air bushings also use external shields. The SF_6 gas inside the bushing has usually the same pressure as the rest of the GIS. The insulating cylinder has most often been porcelain in the past, but today many are a composite consisting of fiberglass epoxy inner cylinder with an external weathershed of silicone rubber. The composite bushing has better contamination resistance and is inherently safer because it will not fracture as will porcelain.

2.2.8 Power Cable Connections

Power cables connecting to a GIS are provided with a cable termination kit that is installed on the cable to provide a physical barrier between the cable dielectric and the SF_6 gas in the GIS (Figure 2.10). The cable termination kit also provides a suitable electric-field distribution at the end of the cable. Because the cable termination will be in SF_6 gas, the length is short and sheds are not needed. The cable conductor is connected with bolted or compression connectors to the end plate or cylinder of the cable termination kit. On the GIS side, a removable link or plug-in contact transfers current from the cable to the GIS conductor. For high-voltage testing of the GIS or the cable, the cable is disconnected from the GIS by removing the conductor link or plug-in contact. The GIS enclosure around the cable termination usually has an access port. This port can also be used for attaching a test bushing.

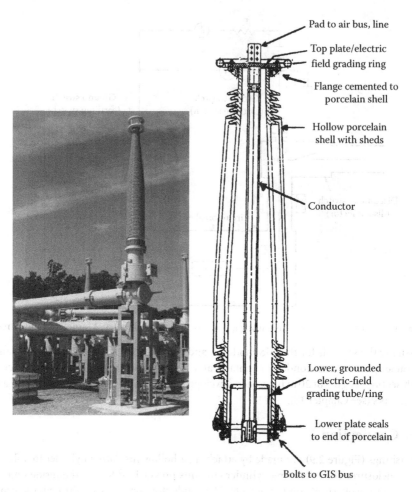

FIGURE 2.9 SF$_6$-to-air bushing.

For solid dielectric power cables up to system voltage of 170 kV "plug-in" termination kits are available. These have the advantage of allowing the GIS cable termination to have one part of the plug-in termination factory installed, so the GIS cable termination compartment can be sealed and tested at the factory. In the field, the power cable with the mating termination part can be installed on the cable as convenient and then plugged into the termination part on the GIS. For the test, the cable can be unplugged—however, power cables are difficult to bend and may be directly buried. In these cases, a disconnect link is still required in the GIS termination enclosure.

2.2.9 Direct Transformer Connections

To connect a GIS directly to a transformer, a special SF$_6$-to-oil bushing that mounts on the transformer is used (Figure 2.11). The bushing is connected under oil on one end to the transformer's high-voltage leads. The other end is SF$_6$ and has a removable link or sliding contact for connection to the GIS conductor. The bushing may be an oil-paper condenser type or, more commonly today, a solid insulation type. Because leakage of SF$_6$ into the transformer oil must be prevented, most SF$_6$-to-oil bushings have a center section that allows any SF$_6$ leakage to go to the atmosphere rather than into the transformer. For testing, the SF$_6$ end of the bushing is disconnected from the GIS conductor after

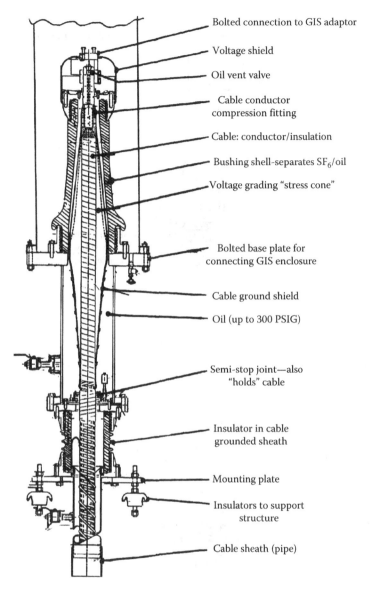

FIGURE 2.10 Power cable connection.

gaining access through an opening in the GIS enclosure. The GIS enclosure of the transformer can also be used for attaching a test bushing.

2.2.10 Surge Arrester

Zinc oxide surge-arrester elements suitable for immersion in SF_6 are supported by an insulating cylinder inside a GIS enclosure section to make a surge arrester for overvoltage control (Figure 2.12). Because the GIS conductors are inside in a grounded metal enclosure, the only way for lightning impulse voltages to enter is through the connection of the GIS to the rest of the electrical system. Cable and direct transformer connections are not subject to lightning strikes, so only at SF_6-to-air bushing connections is lightning a concern. Air-insulated surge arresters in parallel with the SF_6-to-air bushings usually provide adequate protection of the GIS from lightning impulse voltages at a much lower cost than

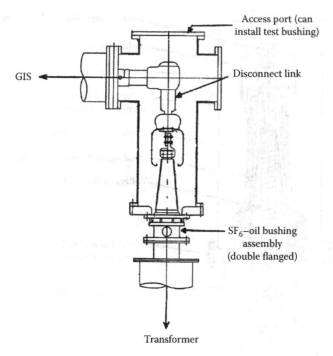

FIGURE 2.11 Direct SF_6 bus connection to transformer.

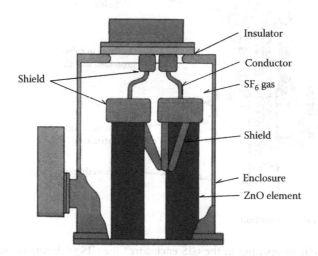

FIGURE 2.12 Surge arrester for GIS.

SF_6-insulated arresters. Switching surges are seldom a concern in GIS because with SF_6 insulation the withstand voltages for switching surges are not much less than the lightning impulse voltage withstand. In AIS, there is a significant decrease in withstand voltage for switching surges compared to lightning impulse because the longer time-span of the switching surge allows time for the discharge to completely bridge the long insulating distances in air. In the GIS, the short insulation distances can be bridged in the short time-span of a lightning impulse; so the longer time-span of a switching surge does not significantly decrease the breakdown voltage. Insulation coordination studies usually show there is not a need for surge arresters in a GIS; however, many users specify surge arresters at transformers and cable connections as the most conservative approach.

2.2.11 Control System

For ease of operation and convenience in wiring the GIS back to the substation control room, a local control cabinet (LCC) is usually provided for each circuit breaker position (Figure 2.13). The control and power wires for all the operating mechanisms, auxiliary switches, alarms, heaters, CTs, and VTs are brought from the GIS equipment modules to the LCC using shielded multiconductor control cables. In addition to providing terminals for all the GIS wiring, the LCC has a mimic diagram of the part of the GIS being controlled. Associated with the mimic diagram are control switches and position indicators for the circuit breaker and switches. Annunciation of alarms is also usually provided in the LCC. Electrical interlocking and some other control functions can be conveniently implemented in the LCC. Although the LCC is an extra expense, with no equivalent in the typical AIS, it is so well established and popular that elimination to reduce costs has been rare. The LCC does have the advantage of providing a very clear division of responsibility between the GIS manufacturer and user in terms of scope of equipment supply.

Switching and circuit breaker operation in a GIS produces internal surge voltages with a very fast rise time of the order of nanoseconds and peak voltage level of about 2 per unit. These "very fast transient" voltages are not a problem inside the GIS because the duration of this type of surge voltage is very short—much shorter than the lightning impulse voltage. However, a portion of the very fast transient voltages will emerge from the inside of the GIS at any places where there is a discontinuity of the metal enclosure—for example, at insulating enclosure joints for external CTs or at the SF_6-to-air bushings. The resulting "transient ground rise voltage" on the outside of the enclosure may cause some small sparks across the insulating enclosure joint or to adjacent grounded parts—these may alarm nearby personnel but are not harmful to a person because the energy content is very low. However, if these very fast

FIGURE 2.13 LCC for GIS.

transient voltages enter the control wires, they could cause misoperation of control devices. Solid-state controls can be particularly affected. The solution is thorough shielding and grounding of the control wires. For this reason, in a GIS the control cable shield should be grounded at both the equipment and the LCC ends using either coaxial ground bushings or short connections to the cabinet walls at the location where the control cable first enters the cabinet.

2.2.12 Gas Monitor System

The insulating and interrupting capability of the SF_6 gas depends on the density of the SF_6 gas being at a minimum level established by design tests. The pressure of the SF_6 gas varies with temperature, so a mechanical or electronic temperature compensated pressure switch is used to monitor the equivalent of gas density (Figure 2.14). GIS is filled with SF_6 to a density far enough above the minimum density for full dielectric and interrupting capability so that from 5% to 20% of the SF_6 gas can be lost before the performance of the GIS deteriorates. The density alarms provide a warning of gas being lost and can be used to operate the circuit breakers and switches to put a GIS that is losing gas into a condition selected by the user. Because it is much easier to measure pressure than density, the gas monitor system may be a pressure gage. A chart is provided to convert pressure and temperature measurements into density. Microprocessor-based measurement systems are available that provide pressure, temperature, density, and even percentage of proper SF_6 content. These can also calculate the rate at which SF_6 is being lost. However, they are significantly more expensive than the mechanical temperature compensated pressure switches, so they are supplied only when requested by the user.

2.2.13 Gas Compartments and Zones

A GIS is divided by gas barrier insulators into gas compartments for gas handling purposes. Due to the arcing that takes place in the circuit breaker, it is usually its own gas compartment. Gas handling systems are available to easily process and store about 1000 kg of SF_6 at one time, but the length of time needed to do this is longer than most GIS users will accept. GIS is therefore divided into relatively small gas compartments of less than several hundred kilograms. These small compartments may be

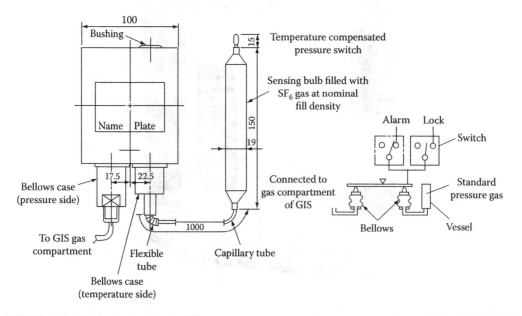

FIGURE 2.14 SF_6 density monitor for GIS.

connected with external bypass piping to create a larger gas zone for density monitoring. The electrical functions of the GIS are all on a three-phase basis, so there is no electrical reason to not connect the parallel phases of a single-phase enclosure type of GIS into one gas zone for monitoring. Reasons for not connecting together many gas compartments into large gas zones include a concern with a fault in one gas compartment causing contamination in adjacent compartments and the greater amount of SF_6 lost before a gas-loss alarm. It is also easier to locate a leak if the alarms correspond to small gas zones—on the other hand, a larger gas zone will, for the same size leak, give more time to add SF_6 between the first alarm and second alarm. Each GIS manufacturer has a standard approach to gas compartments and gas zones but, of course, will modify the approach to satisfy the concerns of individual GIS users.

2.2.14 Electrical and Physical Arrangement

For any electrical one-line diagram, there are usually several possible physical arrangements. The shape of the site for the GIS and the nature of connecting lines and cables should be considered. Figure 2.15 compares a "natural" physical arrangement for a breaker and a half GIS with a "linear" arrangement.

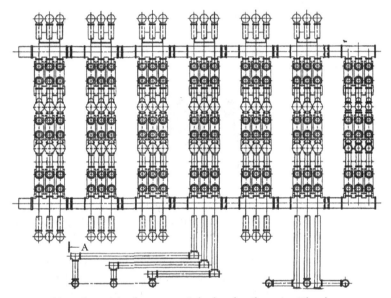

Natural—each bay between main busbars has three circuit breakers

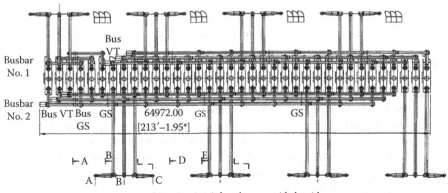

Linear—circuit breakers are side by side

FIGURE 2.15 One-and-one-half circuit breaker layouts.

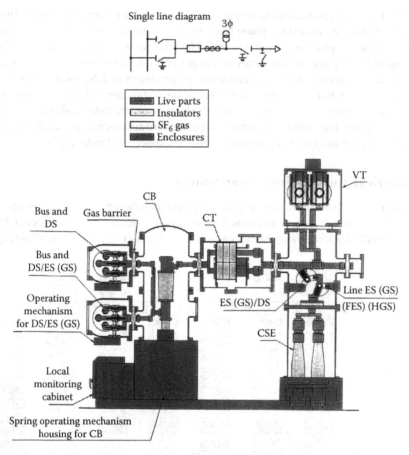

FIGURE 2.16 Integrated (combined function) GIS. Key: CB, circuit breaker; DS, disconnector; ES, earthing switch; GS, grounding switch; FES, fault making earthing switch; HGS, high-speed grounding switch; CT, current transformer; VT, voltage transformer; CSE, cable sealing end; and BUS, busbar.

Most GIS designs were developed initially for a double bus, single break arrangement (Figure 2.2). This widely used approach provides good reliability, simple operation, easy protective relaying, excellent economy, and a small footprint. By integrating several functions into each GIS module, the cost of the double bus, single breaker arrangement can be significantly reduced. An example is shown in Figure 2.16. Disconnect and ground switches are combined into a "three-position switch" and made a part of each bus module connecting adjacent circuit breaker positions. The cable connection module includes the cable termination, disconnect switches, ground switches, a VT, and surge arresters.

2.2.15 Grounding

The individual metal enclosure sections of the GIS modules are made electrically continuous either by the flanged enclosure joint being a good electrical contact in itself or with external shunts bolted to the flanges or to grounding pads on the enclosure. Although some early single-phase enclosure GIS were "single point grounded" to prevent circulating currents from flowing in the enclosures, today the universal practice is to use "multipoint grounding" even though this leads to some electrical losses in the enclosures due to circulating currents. The three enclosures of a single-phase GIS should be bonded to each other at the ends of the GIS to encourage circulating currents to flow—these circulating enclosure currents act to cancel the magnetic field that would otherwise exist outside the enclosure due to the conductor current. Three-phase enclosure GIS does not have circulating currents, does have eddy currents in the enclosure, and should also be multipoint grounded. With multipoint grounding and the many resulting parallel paths for the current from an internal fault to flow to the substation ground grid, it is easy to keep the touch and step voltages for a GIS to the safe levels prescribed in IEEE 80.

2.2.16 Testing

Test requirements for circuit breakers, CTs, VTs, and surge arresters are not specific for GIS and will not be covered in detail here. Representative GIS assemblies having all of the parts of the GIS except for the circuit breaker are design tested to show the GIS can withstand the rated lightning impulse voltage, switching impulse voltage, power frequency overvoltage, continuous current, and short-circuit current. Standards specify the test levels and how the tests must be done. Production tests of the factory-assembled GIS (including the circuit breaker) cover power frequency withstand voltage, conductor circuit resistance, leak checks, operational checks, and CT polarity checks. Components such as support insulators, VTs, and CTs are tested in accord with the specific requirements for these items before assembly into the GIS. Field tests repeat the factory tests. The power frequency withstand voltage test is most important as a check of the cleanliness of the inside of the GIS in regard to contaminating conducting particles, as explained in Section 2.1. Checking of interlocks is also very important. Other field tests may be done if the GIS is a very critical part of the electric power system—for example, a surge voltage test may be requested.

2.2.17 Installation

GIS is usually installed on a monolithic concrete pad or the floor of a building. The GIS is most often rigidly attached by bolting or welding the GIS support frames to embedded steel plates of beams. Chemical drill anchors can also be used. Expansion drill anchors are not recommended because dynamic loads when the circuit breaker operates may loosen expansion anchors. Large GIS installations may need bus expansion joints between various sections of the GIS to adjust to the fitup in the field and, in some cases, provide for thermal expansion of the GIS. The GIS modules are shipped in the largest practical assemblies; at the lower voltage level, two or more circuit breaker positions can be delivered fully assembled. The physical assembly of the GIS modules to each other using the bolted flanged enclosure joints and conductor contacts goes very quickly. More time is used for evacuation of air from gas compartments that have been opened, filling with SF_6 gas and control system wiring. The field tests are then done. For high-voltage GIS shipped as many separate modules, installation and test take about 2 weeks per circuit breaker position. Lower voltage systems shipped as complete bays, and mostly factory wired, can be installed more quickly.

2.2.18 Operation and Interlocks

Operation of a GIS in terms of providing monitoring, control, and protection of the power system as a whole is the same as that for an AIS except that internal faults are not self-clearing, so reclosing should not be used for faults internal to the GIS. Special care should be taken for disconnect and ground switch operation because if these are opened with load current flowing, or closed into load or fault current, the arcing between the switch moving and stationary contacts will usually cause a phase-to-phase fault in three-phase enclosure GIS or to a phase-to-ground fault in single-phase enclosure GIS. The internal fault will cause severe damage inside the GIS. A GIS switch cannot be as easily or quickly replaced as an AIS switch. There will also be a pressure rise in the GIS gas compartment as the arc heats the gas. In extreme cases, the internal arc will cause a rupture disk to operate or may even cause a burn-through of the enclosure. The resulting release of hot decomposed SF_6 gas may cause serious injury to nearby personnel. For the sake of both the GIS and the safety of personnel, secure interlocks are provided so that the circuit breaker must be open before an associated disconnect switch can be opened or closed, and the disconnect switch must be open before the associated ground switch can be closed or opened.

2.2.19 Maintenance

Experience has shown that the internal parts of the GIS are so well protected inside the metal enclosure that they do not age, and as a result of proper material selection and lubricants, there is negligible wear on the switch contacts. Only the circuit breaker arcing contacts and the Teflon nozzle of the interrupter experience wear proportional to the number of operations and the level of the load or fault currents being interrupted. The contacts and nozzle materials combined with the short interrupting time of modern circuit breakers provide typically for thousands of load current interruption operations and tens of full rated fault current interruptions before there is any need for inspection or replacement.

Except for circuit breakers in special use such as a pumped storage plant, most circuit breakers will not be operated enough to ever require internal inspection. So most GIS will not need to be opened for maintenance. The external operating mechanisms and gas monitor systems should be visually inspected, with the frequency of inspection determined by experience.

Replacement of certain early models of GIS has been necessary in isolated cases due to either inherent failure modes or persistent corrosion causing SF_6 leakage problems. These early models may no longer be in production, and in extreme cases the manufacturer is no longer in business. If space is available, a new GIS (or even AIS) may be built adjacent to the GIS being replaced and connections to the power system shifted over into the new GIS. If space is not available, the GIS can be replaced one breaker position at a time using custom designed temporary interface bus sections between the old GIS and the new.

2.3 Economics of GIS

The equipment cost of GIS is naturally higher than that of AIS due to the grounded metal enclosure, the provision of an LCC, and the high degree of factory assembly. A GIS is less expensive to install than an AIS. The site development costs for a GIS will be much lower than for an AIS because of the much smaller area required for the GIS. The site development advantage of GIS increases as the system voltage increases because high-voltage AIS takes very large areas because of the long insulating distances in atmospheric air. Cost comparisons in the early days of GIS projected that, on a total installed cost basis, GIS costs would equal AIS costs at 345 kV. For higher voltages, GIS was expected to cost less than AIS. However, the cost of AIS has been reduced significantly by technical and manufacturing advances (especially for circuit breakers) over the last 30 years, but GIS equipment has not shown significant cost reductions. So although GIS has been a well-established technology for a long time, with a proven high reliability and almost no need for maintenance, it is presently perceived as costing too much and only applicable in special cases where space is the most important factor. Currently, GIS costs are being reduced by integrating functions as described in Section 2.2.14. As digital control systems become common in substations, the costly electromagnetic CTs and VTs of a GIS will be replaced by less expensive sensors such as optical or capacitive VTs and Rogowski coil CTs. These less expensive sensors are also much smaller, reducing the size of the GIS, allowing more bays of GIS to be shipped fully assembled. Installation and site development costs are correspondingly lower. The GIS space advantage over AIS increases. An approach termed "mixed technology switchgear" (or hybrid GIS) that uses GIS breakers, switches, CTs, and VTs with interconnections between the breaker positions and connections to other equipment using air-insulated conductors is a recent development that promises to reduce the cost of the GIS at some sacrifice in space savings. This approach is especially suitable for the expansion of an existing substation without enlarging the area for the substation.

References

1. Cookson, A.H. and Farish, O., Particle-initiated breakdown between coaxial electrodes in compressed SF_6, *IEEE Transactions on Power Apparatus and Systems*, PAS-92(3), 871–876, May/June 1973.
2. IEC 1634: 1995. IEC technical report: High-voltage switchgear and controlgear—Use and handling of sulphur hexafluoride (SF_6) in high-voltage switchgear and controlgear.

3. IEEE Std. 1125-1993. IEEE Guide for Moisture Measurement and Control in SF_6 Gas-Insulated Equipment.
4. IEEE Std. C37.122.1-1993. IEEE Guide for Gas-Insulated Substations.
5. IEEE Std. C37.122-1993. IEEE Standard for Gas-Insulated Substations.
6. IEEE Std. C37.123-1996. IEEE Guide to Specifications for Gas-Insulated, Electric Power Substation Equipment.
7. IEC 62271-203: 1990. Gas-Insulated Metal-Enclosed Switchgear for Rated Voltages of 72.5 kV and above, 3rd edn.
8. Jones, D.J., Kopejtkova, D., Kobayashi, S., Molony, T., O'Connell, P., and Welch, I.M., GIS in service—Experience and recommendations, Paper 23–104 of *CIGRE General Meeting*, Paris, France, 1994.
9. Katchinski, U., Boeck, W., Bolin, P.C., DeHeus, A., Hiesinger, H., Holt, P.-A., Murayama, Y. et al. User guide for the application of gas-insulated switchgear (GIS) for rated voltages of 72.5 kV and above, CIGRE report 125, Paris, France, April 1998.
10. Kawamura, T., Ishi, T., Satoh, K., Hashimoto, Y., Tokoro, K., and Harumoto, Y., Operating experience of gas-insulated switchgear (GIS) and its influence on the future substation design, Paper 23–04 of *CIGRE General Meeting*, Paris, France, 1982.
11. Kopejtkova, D., Malony, T., Kobayashi, S., and Welch, I.M., A twenty-five year review of experience with SF_6 gas-insulated substations (GIS), Paper 23–101 of *CIGRE General Meeting*, Paris, France, 1992.
12. Mauthe, G., Pryor, B.M., Neimeyer, L., Probst, R., Poblotzki, J., Bolin, P., O'Connell, P., and Henriot, J., SF_6 recycling guide: Re-use of SF_6 gas in electrical power equipment and final disposal, CIGRE report 117, Paris, France, August 1997.

3
Air-Insulated Substations: Bus/Switching Configurations

3.1	Introduction	3-1
3.2	Single Bus Arrangement	3-1
3.3	Double Bus–Double Breaker Arrangement	3-2
3.4	Main and Transfer Bus Arrangement	3-3
3.5	Double Bus–Single Breaker Arrangement	3-4
3.6	Ring Bus Arrangement	3-5
3.7	Breaker-and-a-Half Arrangement	3-5
3.8	Comparison of Configurations	3-7

Michael J. Bio
Alstom Grid

3.1 Introduction

Various factors affect the reliability of an electrical substation or switchyard facility, one of which is the arrangement of switching devices and buses. The following are the six types of arrangements commonly used:

1. Single bus
2. Double bus–double breaker
3. Main and transfer (inspection) bus
4. Double bus–single breaker
5. Ring bus
6. Breaker-and-a-half

Additional parameters to be considered when evaluating the configuration of a substation or a switchyard are maintenance, operational flexibility, relay protection, cost, and also line connections to the facility. This chapter will review each of the six basic configurations and compare how the arrangement of switching devices and buses of each impacts reliability and these parameters.

3.2 Single Bus Arrangement

This is the simplest bus arrangement, a single bus and all connections directly to one bus (Figure 3.1).

Reliability of the single bus configuration is low: even with proper relay protection, a single bus failure on the main bus or between the main bus and circuit breakers will cause an outage of the entire facility.

With respect to maintenance of switching devices, an outage of the line they are connected to is required. Furthermore, for a bus outage the entire facility must be de-energized. This requires standby

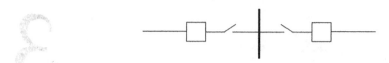

FIGURE 3.1 Single bus arrangement.

generation or switching loads to adjacent substations, if available, to minimize outages of loads supplied from this type of facility.

Cost of a single bus arrangement is relatively low, but also is the operational flexibility; for example, transfer of loads from one circuit to another would require additional switching devices outside the substation.

Line connections to a single bus arrangement are normally straight forward, since all lines are connected to the same main bus. Therefore, lines can be connected on the main bus in areas closest to the direction of the departing line, thus mitigating lines crossing outside the substation.

Due to the low reliability, significant efforts when performing maintenance, and low operational flexibility, application of the single bus configuration should be limited to facilities with low load levels and low availability requirements.

Since single bus arrangement is normally just the initial stage of a substation development, when laying out the substation a designer should consider the ultimate configuration of the substation, such as where future supply lines, transformers, and bus sections will be added. As loads increase, substation reliability and operational abilities can be improved with step additions to the facility, for example, a bus tie breaker to minimize load dropped due to bus outages.

3.3 Double Bus–Double Breaker Arrangement

The double bus–double breaker arrangement involves two breakers and two buses for each circuit (Figure 3.2).

With two breakers and two buses per circuit, a single bus failure can be isolated without interrupting any circuits or loads. Furthermore, a circuit failure of one circuit will not interrupt other circuits or buses. Therefore, reliability of this arrangement is extremely high.

Maintenance of switching devices in this arrangement is very easy, since switching devices can be taken out-of-service as needed and circuits can continue to operate with partial line relay protection and some line switching devices in-service, i.e., one of the two circuit breakers.

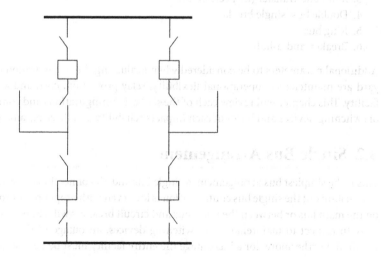

FIGURE 3.2 Double breaker–double bus arrangement.

Obviously, with double the amount of switching devices and buses, cost will be substantially increased relative to other more simple bus configurations. In addition, relaying is more complicated and more land is required, especially for low-profile substation configurations.

External line connections to a double breaker–double bus substation normally do not cause conflicts with each other, but may require substantial land area adjacent to the facility as this type of station expands.

This arrangement allows for operational flexibility; certain lines could be fed from one bus section by switching existing devices.

This bus configuration is applicable for loads requiring a high degree of reliability and minimum interruption time. The double breaker–double bus configuration is expandable to various configurations, for example, a ring bus or breaker-and-a-half configurations, which will be discussed later.

3.4 Main and Transfer Bus Arrangement

The main and transfer bus configuration connects all circuits between the main bus and a transfer bus (sometimes referred to as an inspection bus). Some arrangements include a bus tie breaker and others simply utilize switches for the tie between the two buses (Figure 3.3).

This configuration is similar to the single bus arrangement; in that during normal operations, all circuits are connected to the main bus. So the operating reliability is low; a main bus fault will de-energize all circuits.

However, the transfer bus is used to improve the maintenance process by moving the line of the circuit breaker to be maintained to the transfer bus. Some systems are operated with the transfer bus normally de-energized. When a circuit breaker needs to be maintained, the transfer bus is energized through the tie breaker. Then the switch, nearest the transfer bus, on the circuit to be maintained is closed and its breaker and associated isolation switches are opened. Thus transferring the line of the circuit breaker to be maintained to the bus tie breaker and avoiding interruption to the circuit load. Without a bus tie breaker and only bus tie switches, there are two options. The first option is by transferring the circuit to be maintained to one of the remaining circuits by closing that circuit's switch (nearest to the transfer bus) and carrying both circuit loads on the one breaker. This arrangement most likely will require special relay settings for the circuit breaker to carry the transferred load. The second option is by transferring the circuit to be maintained directly to the main bus with no relay protection from the substation.

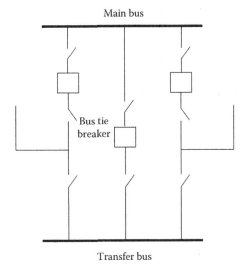

FIGURE 3.3 Main and transfer bus arrangement.

Obviously in the latter arrangement, relay protection (recloser or fuse) immediately outside the substation should be considered to minimize faults on the maintained line circuit from causing extensive station outages.

The cost of the main and transfer bus arrangement is more than the single bus arrangement because of the added transfer bus and switching devices. In addition, if a low-profile configuration is used, land requirements are substantially more.

Connections of lines to the station should not be very complicated. If a bus tie breaker is not installed, consideration as to normal line loading is important for transfers during maintenance. If lines are normally operated at or close to their capability, loads will need to be transferred or temporary generators provided similar to the single bus arrangement maintenance scenario.

The main and transfer bus arrangement is an initial stage configuration, since a single main bus failure can cause an outage of the entire station. As load levels at the station rise, consideration of a main bus tie breaker should be made to minimize the amount of load dropped for a single contingency.

Another operational capability of this configuration is that the main bus can be taken out-of-service without an outage to the circuits by supplying from the transfer bus, but obviously, relay protection (recloser or fuse) immediately outside the substation should be considered to minimize faults on any of the line circuit from causing station outages.

Application of this type of configuration should be limited to low reliability requirement situations.

3.5 Double Bus–Single Breaker Arrangement

The double bus–single breaker arrangement connects each circuit to two buses, and there is a tie breaker between the buses. With the tie breaker operated normally closed, it allows each circuit to be supplied from either bus via its switches. Thus providing increased operating flexibility and improved reliability. For example, a fault on one bus will not impact the other bus. Operating the bus tie breaker normally open eliminates the advantages of the system and changes the configuration to a two single bus arrangement (Figure 3.4).

Relay protection for this arrangement will be complex with the flexibility of transferring each circuit to either bus. Operating procedures would need to be detailed to allow for various operating arrangements, with checks to ensure the in-service arrangements are correct. A bus tie breaker failure will cause an outage of the entire station.

The double bus–single breaker arrangement with two buses and a tie breaker provides for some ease in maintenance, especially for bus maintenance, but maintenance of the line circuit breakers would still require switching and outages as described above for the single bus arrangement circuits.

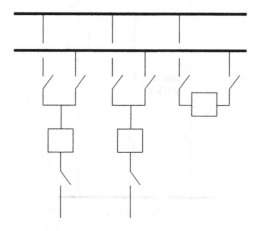

FIGURE 3.4 Double bus–single breaker arrangement.

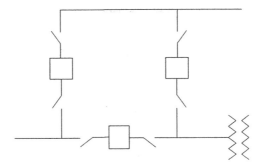

FIGURE 3.5 Ring bus arrangement.

The cost of this arrangement would be more than the single bus arrangement with the added bus and switching devices. Once again, low-profile configuration of this arrangement would require more area. In addition, bus and circuit crossings within the substation are more likely.

Application of this arrangement is best suited where load transfer and improved operating reliability are important. Though adding a transfer bus to improve maintenance could be considered, it would involve additional area and switching devices, which could increase the cost of the station.

3.6 Ring Bus Arrangement

As the name implies, all breakers are arranged in a ring with circuits connected between two breakers. From a reliability standpoint, this arrangement affords increased reliability to the circuits, since with properly operating relay protection, a fault on one bus section will only interrupt the circuit on that bus section and a fault on a circuit will not affect any other device (Figure 3.5).

Protective relaying for a ring bus will involve more complicated design and, potentially, more relays to protect a single circuit. Keep in mind that bus and switching devices in a ring bus must all have the same ampacity, since current flow will change depending on the switching device's operating position.

From a maintenance point of view, the ring bus provides good flexibility. A breaker can be maintained without transferring or dropping load, since one of the two breakers can remain in-service and provide line protection while the other is being maintained.

Similarly, operating a ring bus facility gives the operator good flexibility since one circuit or bus section can be isolated without impacting the loads on another circuit.

Cost of the ring bus arrangement can be more expensive than a single bus, main bus and transfer, and the double bus–single breaker schemes since two breakers are required for each circuit, even though one is shared.

The ring bus arrangement is applicable to loads where reliability and availability of the circuit is a high priority. There are some disadvantages of this arrangement: (a) a "stuck breaker" event could cause an outage of the entire substation depending on the number of breakers in the ring, (b) expansion of the ring bus configuration can be limited due to the number of circuits that are physically feasible in this arrangement, and (c) circuits into a ring bus to maintain a reliable configuration can cause extensive bus and line work. For example, to ensure service reliability, a source circuit and a load circuit should always be next to one another. Two source circuits adjacent to each other in a stuck breaker event could eliminate all sources to the station. Therefore, a low-profile ring bus can command a lot of area.

3.7 Breaker-and-a-Half Arrangement

The breaker-and-a-half scheme is configured with a circuit between two breakers in a three-breaker line-up with two buses; thus, one-and-a-half breakers per circuit. In many cases, this is the next development stage of a ring bus arrangement (Figure 3.6).

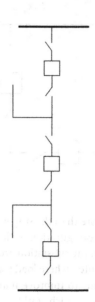

FIGURE 3.6 Breaker-and-a-half arrangement.

Similar to the ring bus, this configuration provides good reliability; with proper operating relay protection, a single circuit failure will not interrupt any other circuits. Furthermore, a bus section fault, unlike the ring bus, will not interrupt any circuit loads.

Maintenance as well is facilitated by this arrangement, since an entire bus and adjacent breakers can be maintained without transferring or dropping loads.

Relay protection is similar to the ring bus, and due to the additional devices, is more complex and costly than most of the previously reviewed arrangements.

TABLE 3.1 Bus/Switching Configuration Comparison Table

Configuration	Reliability/Operation	Cost	Available Area
Single bus	Least reliable—single failure can cause a complete outage. Limited operating flexibility	Least cost (1.0)—fewer components	Least area—fewer components
Double bus–double breaker	Highly reliable—duplicated devices; single circuit or bus fault isolates only that component. Greater operating and maintenance flexibility	High cost (2.17)—duplicated devices and more material	Greater area—more devices and more material
Main and transfer bus	Least reliable—reliability is similar to the single bus arrangement, but operating and maintenance flexibility improved with the transfer bus	Moderate cost (2.06)—more devices and material required than the single bus	Low area—high-profile configuration is preferred to minimize land use
Double bus–single breaker	Moderately reliable—with bus tie breaker, bus sections and line circuits are isolated. Good operating flexibility	High cost (2.15)—more devices and material	Greater area—more devices and more material
Ring bus	High reliability—single circuit or bus section fault isolated. Operation and maintenance flexibility good	Moderate cost (1.62)—additional components and materials	Moderate area—dependent on the extent of the substation development
Breaker-and-a-half	Highly reliable—bus faults will not impact any circuits, and circuit faults isolate only that circuit. Operation and maintenance flexibility best with this arrangement	Moderate cost (1.69)—cost is reasonable based on improved reliability and operational flexibility	Greater area—more components. Area increases substantially with higher voltage levels

The breaker-and-a-half arrangement can be expanded as needed. By detailed planning of the ultimate substation expansion with this configuration, line conflicts outside the substation can be minimized.

Cost of this configuration is commensurate with the number of circuits, but based on the good reliability, operating flexibility, and ease of maintenance, the price can be justified.

Obviously, the area required for this type of arrangement is significant, and the higher the voltage, the more clearances required and area needed.

3.8 Comparison of Configurations

As a summary to the discussion above, Table 3.1 provides a quick reference to the key features of each configuration discussed with a relative cost comparison. The single bus arrangement is considered the base, or 1 per unit cost with all others expressed as a factor of the single bus arrangement cost. Parameters considered in preparing the estimated cost were: (a) each configuration was estimated with only two circuits, (b) 138 kV was the voltage level for all arrangements, (c) estimates were based on only the bus, switches, and breakers, with no dead end structures, fences, land, or other equipment and materials, and (d) all were designed as low-profile stations.

Obviously, the approach used here is only a starting point for evaluating the type of substation or switching station to build. Once a type station is determined based on reliability, operational flexibility, land availability, and relative cost, a complete and thorough evaluation should take place. In this evaluation additional factors need be considered, such as, site development cost, ultimate number of feeders, land required, soil conditions, environmental impact, high profile versus low profile, ease of egress from substation with line circuits, etc. As the number of circuits increases, the relative difference in cost shown in the table may no longer be valid. These types of studies can require a significant amount of time and cost, but the end result will provide a good understanding of exactly what to expect of the ultimate station cost and configuration.

4
High-Voltage Switching Equipment

David L. Harris
SPX Transformer Solutions (Waukesha Electric Systems)

David Childress
David Childress Enterprises

4.1	Introduction	4-1
4.2	Ambient Conditions	4-1
4.3	Disconnect Switches	4-2
4.4	Load Break Switches	4-13
4.5	High-Speed Grounding Switches	4-17
4.6	Power Fuses	4-17
4.7	Circuit Switchers	4-19
4.8	Circuit Breakers	4-22

4.1 Introduction

The design of the high-voltage substation must include consideration for the safe operation and maintenance of the equipment. Switching equipment is used to provide isolation, no-load switching (including line charging current interrupting, loop splitting, and magnetizing current interrupting), load switching, and interruption of fault currents. The magnitude and duration of the load currents and the fault currents will be significant in the specification of the switching equipment used.

System and maintenance operations must also be considered when equipment is specified. One significant choice is the decision of single-phase or three-phase operation. High-voltage power systems are generally operated as a three-phase system, and the imbalance that will occur when operating equipment in a single-phase mode must be considered.

4.2 Ambient Conditions

Air-insulated high-voltage electrical equipment is usually covered by standards based on assumed ambient temperatures and altitude. Ambient temperatures are generally rated over a range from −40°C to +40°C for equipment that is air insulated and dependent on ambient cooling. Altitudes above 1000 m (3300 ft) may require derating.

At higher altitudes air density decreases; hence, the dielectric strength of air is also reduced and derating of the equipment is recommended. Operating clearances (strike distances) must be increased to compensate for the reduction in the dielectric strength of the ambient air. Current ratings generally decrease at higher elevations due to the decreased density of the ambient air, which is the cooling medium used for dissipating the heat generated by the losses associated with load current levels. However, the continuous current derating is slight in relation to the dielectric derating and in most cases it is negligible. In many cases, current derating is offset by the cooler temperature of the ambient air typically found at these higher elevations.

4.3 Disconnect Switches

A disconnect switch is a mechanical device that conducts electrical current and provides an open point in a circuit for isolation of one of the following devices:

- Circuit breakers
- Circuit switchers
- Power transformers
- Capacitor banks
- Reactors
- Other substation equipment

The three most important functions disconnect switches must perform are to open and close reliably when called to operate, to carry load currents continuously without overheating, and to remain in the closed operation position under fault current conditions. Disconnect switches are normally used to provide a point of visual isolation of the substation equipment for maintenance. Typically a disconnect switch would be installed on each side of a piece of substation equipment to provide a visible confirmation that the power conductors have been opened for personnel safety. Once the switches are operated to the open position, portable safety grounds can be attached to the de-energized equipment for worker protection. In place of portable grounds or in addition to portable grounds (as a means of redundant safety), switches can be equipped with grounding blades to perform the safety grounding function. The principal drawback of the use of grounding blades is that they provide a safety ground in a specific, unchangeable location, where portable safety grounds can be located at whatever location desired to achieve a ground point for personnel safety. A very common application of fixed position grounding blades is for use with a capacitor bank with the grounding blades performing the function of bleeding off the capacitor bank's trapped charge.

Disconnect switches are designed to continuously carry load currents and momentarily carry short-circuit currents for a specified duration (typically defined in seconds or cycles depending upon the magnitude of the short-circuit current). They are designed for no-load switching, opening or closing circuits where negligible currents are made or interrupted (including capacitive current [line charging current] and resistive or inductive current [magnetizing current]), or when there is no significant voltage across the open terminals of the switch (loop splitting [parallel switching]). They are relatively slow-speed operating devices and therefore are not designed for interruption of any significant magnitude current arcs. Disconnect switches are also installed to bypass breakers or other equipment for maintenance and can be used for bus sectionalizing. Interlocking equipment is available to prevent operating sequence errors, which could cause substation equipment damage, by inhibiting operation of the disconnect switch until the load current has been interrupted by the appropriate equipment. This interlocking equipment takes three basic forms:

- *Mechanical cam-action type* (see Figure 4.1)—used to interlock a disconnect switch and its integral grounding blades to prevent the disconnect switch from being closed when the grounding blades are closed and to prevent the grounding blades from being closed when the disconnect switch is closed.
- *Key type* (see Figure 4.2)—a mechanical plunger extension and retraction only or electromechanical equipment consisting of a mechanical plunger and either electrical auxiliary switch contacts or an electrical solenoid. It is used in a variety of applications including, but not limited to, interlocking a disconnect switch and its integral grounding blades, interlocking the grounding blades on one disconnect switch with a physically separate disconnect switch in the same circuit, or interlocking a disconnect switch with a circuit breaker (to ensure that the circuit breaker is open before the disconnect switch is allowed to open).
- *Solenoid type* (see Figure 4.3)—most commonly used to ensure that the circuit breaker is open before the disconnect switch is allowed to open.

High-Voltage Switching Equipment

FIGURE 4.1 Mechanical cam-action type interlock.

FIGURE 4.2 Key type interlock.

FIGURE 4.3 Solenoid type interlock.

FIGURE 4.4 Swing handle operator for disconnect switch.

Single-phase or three-phase operation is possible for some disconnect switches. Operating mechanisms are usually included to permit opening and closing of the three-phase disconnect switch by an operator standing at ground level. Common manual operating mechanisms include a swing handle (see Figure 4.4) or a gear crank (see Figure 4.5). Other manual operating mechanisms, which are less common, include a reciprocating or pump type handle or a hand wheel. The choice of which manual operating mechanism to use is made based upon the required amount of applied force necessary to permit operation of the disconnect switch. A general guideline is that disconnect switches rated 69 kV and below or 1200 A continuous current and below are typically furnished with a swing handle operating mechanism, whereas disconnect switches rated 115 kV and above or 1600 A continuous current and above are typically furnished with a

FIGURE 4.5 Gear crank operator for disconnect switch.

FIGURE 4.6 Motor operator for disconnect switch.

gear crank operating mechanism. This convention can vary based upon the type of disconnect switch used, as different types of disconnect switches have varying operating effort requirements. Motor-operating mechanisms (see Figure 4.6) are also available and are applied when remote switching is necessary or desired and when the disconnect switch's function is integrated into a comprehensive system monitoring and performance scheme such as a supervisory control and data acquisition (SCADA) system. These motor-operating mechanisms can be powered either via a substation battery source or via the input from an auxiliary AC source. Some motor-operating mechanisms have their own internal batteries that can be fed from an auxiliary AC source via an AC to DC trickle charger, thus providing multiple stored operations in the event of loss of auxiliary AC source supply. These stored energy motor operators (see Figure 4.7) are ideally suited for substations that do not have a control building to house substation batteries and for line

FIGURE 4.7 Stored energy motor operator for disconnect switch.

FIGURE 4.8 Horizontally upright mounted disconnect switch.

FIGURE 4.9 Vertically mounted disconnect switch.

installations where it is undesirable or economically infeasible to supply a DC battery source external to the motor operator. Remote terminal units are commonly used to communicate with the stored energy motor operator providing the remote electrical input signal that actuates the motor operator.

Disconnect switches can be mounted in a variety of positions, with the most common positions being horizontal upright (see Figure 4.8), vertical (see Figure 4.9), and under hung (see Figure 4.10). A disconnect switch's operation can be designed for vertical or horizontal operating of its switch blades. Several configurations are available, including

- Vertical break (see Figure 4.11)
- Double end break (also sometimes called double side break) (see Figure 4.12)
- Double end break "Vee" (also sometimes called double side break "Vee") (see Figure 4.13)
- Center break (see Figure 4.14)
- Center break "Vee" (see Figure 4.15)
- Single side break (see Figure 4.16)
- Vertical reach (also sometimes called pantograph, semipantograph, or knee-type switches) (see Figure 4.17)
- Grounding (see Figure 4.18)
- Hook stick (see Figure 4.19)

Each of these switch types has specific features that lend themselves to certain types of applications.

High-Voltage Switching Equipment 4-7

FIGURE 4.10 Under hung mounted disconnect switch.

FIGURE 4.11 Vertical break disconnect switch.

Vertical break switches are the most widely used disconnect switch design, are the most versatile disconnect switch design, can be installed on minimum phase spacing, are excellent for applications in ice environments due to their rotating blade design, and are excellent for installations in high fault current locations due to their contact design (see Figure 4.20).

Double end break switches can be installed on minimum phase spacing (the same phase spacing as for vertical break switches due to the disconnect switch blades being disconnected from both the source and the load when in the open position [see Figure 4.21]), can be installed in minimum overhead clearance locations (something that vertical break switch designs cannot do), do not require a counterbalance for the blades as the blades do not have to be lifted during operation (many vertical break switches

FIGURE 4.12 Double end break (double side break) disconnect switch.

FIGURE 4.13 Double end break "Vee" (double side break "Vee") disconnect switch.

utilize a counterbalance spring to control the blade movement during opening and closing operations and to reduce the operating effort required), are excellent for applications in ice environments due to their rotating blade design (even better, in fact, than vertical break switches are for this application due to the contact configuration of the double end break switch versus the vertical break switch), are excellent for installations in high fault current locations due to their contact design, and have the advantage

FIGURE 4.14 Center break disconnect switch.

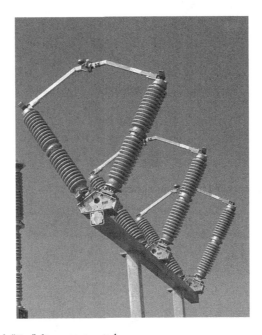

FIGURE 4.15 Center break "Vee" disconnect switch.

of being able to interrupt significantly more line charging current or magnetizing current than any single break type switch can due to their two break per phase design.

Double end break "Vee" switches share all of the same characteristics as the conventional double end break switches but with the additional feature advantage of consuming the smallest amount of substation space of any three-phase switch type as they can be installed on a single horizontal beam structure with one, two (see Figure 4.22), or three vertical columns (the quantity of which is determined by the kilovolt rating of the switch and other site-specific conditions such as seismic considerations).

Center break switches can be installed in minimum overhead clearance locations but require greater phase spacing than vertical break, double end break, or double end break "Vee" switches do (as center break switches have one of the two blades per phase energized when in the open position); require only six insulators per three-phase switch (versus the nine insulators per three-phase switch required for

FIGURE 4.16 Single side break disconnect switch.

FIGURE 4.17 Vertical reach (pantograph) disconnect switch.

FIGURE 4.18 Grounding switch.

High-Voltage Switching Equipment

FIGURE 4.19 Hook stick operated disconnect switch.

FIGURE 4.20 Contact design of vertical break disconnect switch.

FIGURE 4.21 Double end break disconnect switch in open position.

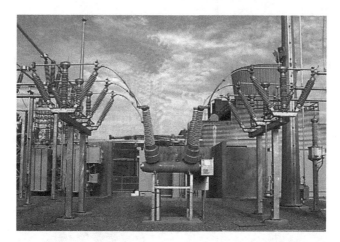

FIGURE 4.22 Double end break "Vee" disconnect switch on single horizontal member, two column structure.

FIGURE 4.23 Vertically mounted center break disconnect switch.

vertical break, double end break, and double end break "Vee" switches); do not require a counterbalance for the blades as the blades do not have to be lifted during operation; and are the best available three-phase switch design for vertical mounting (see Figure 4.23) as the two blades per phase self-counterbalance each other during opening and closing operations via the synchronizing pipe linkage.

Center break "Vee" switches share all of the same characteristics as the conventional center break switches but with the additional feature advantage of consuming a smaller amount of substation space as they can be installed on a single horizontal beam structure with one, two, or three vertical columns (the quantity of which is determined by the kV rating of the switch and other site-specific conditions such as seismic considerations).

Single side break switches can be installed in minimum overhead clearance locations but may require greater phase spacing than vertical break, double end break, or double end break "Vee" switches do; require only six insulators per three-phase switch (versus the nine insulators per three-phase switch required for vertical break, double end break, and double end break "Vee" switches); and do not require a counterbalance for the blades as the blades do not have to be lifted during operation.

Vertical reach switches are used most commonly in extra high-voltage (EHV) applications, typically for 345, 500, and 765 kV installations. The U.S. utility industry uses few of the vertical reach switches, but this switch design is fairly common in Europe and in other parts of the world.

High-Voltage Switching Equipment

FIGURE 4.24 Grounding switch integrally attached to vertical break disconnect switch.

Grounding switches can be furnished as an integral attachment to any of the previously mentioned disconnect switch types (see Figure 4.24) or can be furnished as a stand-alone device (i.e., not attached as an integral component of a disconnect switch) (see Figure 4.18). Grounding switches are commonly applied to perform safety grounding of disconnect switches, buses, and capacitor banks. As previously mentioned, when grounding switches are used, there is an interlocking scheme of some type normally employed to assure proper sequence of operations.

Hook stick switches are single-phase devices that provide isolation, bypassing (typically for a regulator, recloser, or current transformers), transferring (i.e., feeding a load from an alternate source), or grounding.

For all types of disconnect switches previously mentioned, phase spacing is usually adjusted to satisfy the spacing of the bus system installed in the substation. In order to attain proper electrical performance, the standards establish minimum metal-to-metal clearances to be maintained for a given switch type and kV rating.

Prior to about 1970, almost all switches had copper live part construction and met a standard that allows a 30°C temperature rise when the switch is energized and carrying its full nameplate current value. Subsequent to 1970, many switch designs of aluminum live part construction were created and a new governing standard that allows a 53°C temperature rise when the switch is energized and carrying its full nameplate current value came into existence. International standards allow a 65°C temperature rise when switches are energized and carrying their full nameplate current value. When it comes to the temperature rise capability of a switch, cooler is better as it means the switch has more inherent built-in current carrying capability; so a 30°C rise switch is more capable than a 53°C rise switch or a 65°C rise switch, and a 53°C rise switch is more capable than a 65°C rise switch.

4.4 Load Break Switches

A load break switch is a disconnect switch that has been equipped to provide breaking and making of specified currents. This is accomplished by the addition of equipment that changes what the last points of metal-to-metal contact upon opening and the first points of metal-to-metal contact upon closing are, that increases the switching speed at which the last points of metal-to-metal contact part in air, or that confines the arcing to a chamber which contains a dielectric medium capable of interrupting the arc safely and reliably.

FIGURE 4.25 Arcing horns on a vertical break switch.

Arcing horns (see Figure 4.25) are the equipment added to disconnect switches to allow them to interrupt very small amounts of charging or magnetizing current. The capability of arcing horns to perform current interruption is a function of arcing horn material (typically copper or stainless steel), switch break type (vertical break, double end break, double end break Vee, center break, center break Vee, or single side break), phase spacing, switch mounting position (horizontal upright, vertical, or under hung), and other factors. These "standard" arcing horns can be used on any kV-rated switch. Standard arcing horns do not have load breaking capability and should not be used to perform a load breaking function as damage to the disconnect switch will result. Also, standard arcing horns have no loop splitting rating.

High-speed arcing horns (see Figure 4.26) (sometimes called whip horns, quick breaks, buggy whips, or quick break whips) are the equipment added to disconnect switches to allow them to interrupt small amounts of charging or magnetizing current. The capability of these quick break whip horns is a function of arcing horn material (typically stainless steel or beryllium copper) and tip speed of the whip horn at the point when it separates from the fixed catcher on the jaw contact assembly of the switch. Quick break whip type arcing horns are suitable for use on disconnect switches rated 161 kV and below; above 161 kV quick break whip type arcing horns can produce visible or audible corona. Whip type arcing horns do not have load breaking capability and should not be used to perform a load breaking function as damage to the disconnect switch will result. Whip type arcing horns have no loop splitting rating.

FIGURE 4.26 High-speed arcing horns on a vertical break switch.

FIGURE 4.27 Load break switch with SF$_6$ interrupters.

If the need for interrupting loop currents, load currents, or large amounts of line charging current exists, then a disconnect switch can be outfitted with an interrupter (using either sulfur hexafluoride [SF$_6$] gas or vacuum as the interrupting medium) capable of performing these interrupting duties. Most commonly, the type of disconnect switch outfitted with these load/line/loop interrupters is a vertical break switch, although single side break switches are sometimes used. At 30 kV and below, center break switches and center break Vee switches can also be equipped with load/line/loop interrupters to perform these functions. While SF$_6$ gas load/line/loop interrupters (see Figure 4.27) are single gap type for all kV ratings (requiring no voltage division across multiple gaps per phase to achieve successful interruption), vacuum load/line/loop interrupters are multiple gap type for system voltages above 30 kV. At 34.5 and 46 kV, two vacuum bottles per phase are required; at 69 kV, three vacuum bottles per phase (see Figure 4.28); at 115 kV, five vacuum bottles per phase; at 138 kV, six vacuum bottles per phase; at 161 kV, seven vacuum bottles per phase; and at 230 kV, eight vacuum bottles per phase are necessary. An additional difference between vacuum interrupters and SF$_6$ interrupters is that SF$_6$ interrupters provide visual indication of the presence of adequate dielectric for successful interruption (see Figure 4.29), a feature not available on vacuum interrupters. This visual indication is a significant feature in the area of personnel safety, particularly on load break switches that may be manually operated.

In order to decide which of these attachments (arcing horns, quick break whips, or load/line/loop interrupters) is required for a given installation, it is necessary to be able to determine the amount of charging current and/or magnetizing current that exists. IEEE C37.32-2002 *American National Standard for High Voltage Switches, Bus Supports, and Accessories–Schedules of Preferred Ratings, Construction Guidelines, and Specifications*, Annex A provides a conservative rule of thumb regarding the calculation of the amount of available line charging current at a given kV rating as a function of miles of line as indicated in the following:

- 15 kV 0.06 A/mile of line
- 23 kV 0.10 A/mile of line
- 34.5 kV 0.14 A/mile of line
- 46 kV 0.17 A/mile of line
- 69 kV 0.28 A/mile of line
- 115 kV 0.44 A/mile of line
- 138 kV 0.52 A/mile of line

FIGURE 4.28 Load break switch with multibottle vacuum interrupters.

FIGURE 4.29 SF_6 interrupter's pressure indicator.

- 161 kV 0.61 A/mile of line
- 230 kV 0.87 A/mile of line
- 345 kV 1.31 A/mile of line

Many factors influence the amount of available line charging current, including the following:

- Phase spacing
- Phase-to-ground distance
- Atmospheric conditions (humidity, airborne contaminants, etc.)

- Adjacent lines on the same right-of-way (especially if of a different kV)
- Distance to adjacent lines
- Overbuild or underbuild on the same transmission towers (especially if of a different kV)
- Distance to overbuild or underbuild lines
- Conductor configuration (phase over phase, phase opposite phase, phase by phase [side by side], delta upright, delta inverted, etc.)

If it is desired to be more precise in the determination of the amount of available line charging current, exact values for a given installation can be calculated by analyzing all of the applicable system components and parameters of influence in lieu of using the rules of thumb shown previously.

When determining the amount of available magnetizing current at a given site, a conservative estimate is 1% of the full-load rating of the power transformer. For almost all power transformers, the actual value of magnetizing current is only a fraction of this amount; so if a more precise value is desired, the power transformer manufacturer can be consulted to obtain the specific value of magnetizing current for a given transformer. Just as there are a variety of factors that influence the amount of line charging current present in a given installation, so too are there various factors that affect the amount of available magnetizing current. These factors include, but are not limited to, transformer core design, transformer core material, transformer coil design, and transformer coil material.

4.5 High-Speed Grounding Switches

Automatic high-speed grounding switches are applied for protection of power transformers when the cost of supplying other protective equipment is deemed unjustifiable and the amount of system disturbance that the high-speed grounding switch creates is judged acceptable. The switches are generally actuated by discharging a spring mechanism to provide the "high-speed" operation. The grounding switch operates to provide a deliberate ground fault on one phase of the high-voltage bus supplying the power transformer, disrupting the normally balanced 120° phase shifted three-phase system by effectively removing one phase and causing the other two phases to become 180° phase shifted relative to each other. This system imbalance is remotely detected by protective relaying equipment that operates the transmission line breakers at the remote end of the line supplying the power transformer, tripping the circuit open to clear the fault. This scheme also imposes a voltage interruption to all other loads connected between the remote circuit breakers and the power transformer as well as a transient spike to the protected power transformer, effectively shortening the transformer's useful life. Frequently, a system utilizing a high-speed ground switch also includes the use of a motor-operated disconnect switch and a relay system to sense bus voltage. The relay system's logic allows operation of the motor-operated disconnect switch when there is no voltage on the transmission line to provide automatic isolation of the faulted power transformer and to allow reclosing operations of the remote breakers to restore service to the transmission line and to all other loads fed by this line.

The grounding switch scheme is dependent on the ability of the source transmission line relay protection scheme to recognize and clear the fault by opening the remote circuit breaker. Clearing times are necessarily longer since the fault levels are not normally within the levels appropriate for an instantaneous trip response. The lengthening of the trip time also imposes additional stress on the equipment being protected and should be considered when selecting this method for power transformer protection. High-speed grounding switches are usually considered when relative fault levels are low so that the risk of significant damage to the power transformer due to the extended trip times is mitigated.

4.6 Power Fuses

Power fuses are a generally accepted means of protecting small power transformers (i.e., power transformers of 15 MVA and smaller) (see Figure 4.30), capacitor banks, potential transformers, and/or station service transformers. The primary purpose of a power fuse is to provide interruption of permanent faults.

FIGURE 4.30 Power fuses protecting a power transformer.

Power fuses are an economical alternative to circuit switcher or circuit breaker protection. Fuse protection is generally limited to voltages from 15 to 69 kV but has been applied for the protection of equipment as large as 161 kV.

To provide the greatest protective margin, it is necessary to use the smallest fuse rating possible. The advantage of close fusing is the ability of the fuse unit to provide backup protection for some secondary faults. For the common delta-wye-connected transformer, a fusing ratio of 1.0 would provide backup protection for a phase-to-ground fault as low as 230% of the secondary full-load rating. Fusing ratio is defined as the ratio of the fuse rating to the transformer full-load current rating. With low fusing ratios, the fuse may also provide backup protection for line-to-ground faults remote to the substation on the distribution, subtransmission, or transmission network.

Fuse ratings also must consider other parameters than the full-load current of the transformer being protected. Coordination with other overcurrent devices, accommodation of peak overloading, and severe duty may require increased ratings of the fuse unit. The general purpose of the power transformer fuse is to accommodate, not interrupt, peak loads. Fuse ratings must consider the possibility of nuisance trips if the rating is selected too low for all possible operating conditions.

The concern of unbalanced voltages in a three-phase system must be considered when selecting fusing. The possibility of one or two fuses blowing must be reviewed. Unbalanced voltages can cause tank heating in three-phase power transformers and overheating and damage to three-phase motor loads. The potential for ferroresonance must be considered for some transformer configurations when using fusing.

Fuses are available in a number of time-to-melt and time-to-clear curves (standard, fast, medium, slow, and very slow) to provide coordination with other system protective equipment. Fuses are not voltage critical; they may be applied at any voltage equal to or less than their rated voltage. Fuses may not require additional structures as they are generally mounted on the incoming line structure (see Figure 4.31) and result in space savings in the substation layout. Power fuses are available in four mounting configurations—vertical, under hung, 45° under hung, and horizontal upright—with the vast majority of all power fuse installations being vertically mounted units (see Figure 4.32).

FIGURE 4.31 Incoming line structure mounted power fuses.

FIGURE 4.32 Vertically mounted power fuses.

4.7 Circuit Switchers

Circuit switchers have been developed to overcome some of the limitations of fusing for substation power transformers. Circuit switchers using SF_6 gas interrupters are designed to provide three-phase interruption (solving the unbalanced voltage considerations) and to provide protection for transient overvoltage and load current overloads at a competitive cost between the costs of power fuses and circuit breakers. Additionally, they can provide protection from power transformer faults based on differential, sudden pressure, and overcurrent relay schemes as well as critical operating constraints such as for low oil level, high oil or winding temperature, pressure relief device operation, etc.

The earliest circuit switchers were designed and supplied as a combination of a circuit breaking interrupter and an in-series isolating disconnect switch. These earliest models (see Figure 4.33) had multiple interrupter gaps per phase on above 69 kV interrupters and grading resistors, thus the necessity for the in-series disconnect switch. Later models have been designed with improved interrupters that have reduced the number of gaps required for successful performance to a single gap per phase, thus

FIGURE 4.33 Multiple interrupter gap per phase circuit switcher.

FIGURE 4.34 Vertical interrupter circuit switcher without integral disconnect switch.

eliminating the necessity of the disconnect switch blade in series with the interrupter. Circuit switchers are now available in vertical interrupter design (see Figure 4.34) or horizontal interrupter design configurations with (see Figure 4.35) or without (see Figure 4.36) an integral disconnect switch. The earliest circuit switchers had a 4 kA symmetrical primary fault current interrupting capability, but subsequent design improvements over the years have produced circuit switchers capable of 8, 10, 12, 16, 20, 25, 31.5, and 40 kA symmetrical primary fault current interrupting, with the highest of these interrupting values

FIGURE 4.35 Horizontal interrupter circuit switcher with integral vertical break disconnect switch.

FIGURE 4.36 Horizontal interrupter circuit switcher without integral disconnect switch.

being on par with circuit breaker capabilities. The interrupting times of circuit switchers have also been improved from their initial eight-cycle interrupting time to six to five to three cycles, with three cycles offering the same speed as the most commonly available circuit breaker interrupting time. Different model types, configurations, and vintages have different interrupting ratings and interrupting speeds. Circuit switchers have been developed and furnished for applications involving protection of power transformers, lines, cables, capacitor banks, and line connected or tertiary connected shunt reactors. Circuit switchers can also be employed in specialty applications such as series capacitor bypassing and for load/line/loop interrupting applications where fault-closing capability is required (as fault-closing capability is not a feature inherent in disconnect switch mounted load/line/loop interrupters or in the disconnect switches these interrupters are mounted on).

4.8 Circuit Breakers

A circuit breaker is defined as "a mechanical switching device capable of making, carrying, and breaking currents under normal circuit conditions and also making, carrying, and breaking for a specified time, and breaking currents under specified abnormal conditions such as a short circuit" (IEEE Standard C.37.100-1992).

Circuit breakers are generally classified according to the interrupting medium used to cool and elongate the electrical arc permitting interruption. The types of circuit breakers are

- Air magnetic
- Vacuum (see Figure 4.37)
- Air blast
- Oil (bulk oil [see Figures 4.38 and 4.39] and minimum oil)
- SF_6 gas (see Figures 4.40 and 4.41)

Air magnetic circuit breakers are limited to older switchgear and have generally been replaced by vacuum or SF_6 gas for switchgear applications. Vacuum is used for switchgear applications and for some outdoor breakers, generally 38 kV class and below.

FIGURE 4.37 Vacuum circuit breaker.

FIGURE 4.38 Single-tank bulk oil circuit breaker.

FIGURE 4.39 Three-tank bulk oil circuit breaker.

FIGURE 4.40 SF$_6$ gas dead tank circuit breaker.

FIGURE 4.41 SF$_6$ gas dead tank circuit breaker.

Air blast breakers, used for EHVs (≥345 kV), are no longer manufactured and have been replaced by breakers using SF_6 technology.

Oil circuit breakers have been widely used in the utility industry in the past but have been replaced by other breaker technologies for newer installations. Two designs exist: bulk oil (dead tank) designs dominant in the United States and minimum oil (live tank) designs prevalent in some other parts of the world. Bulk oil circuit breakers were designed as single-tank (see Figure 4.38) or three-tank (see Figure 4.39) devices, 69 kV and below ratings were available in either single-tank or three-tank configurations and 115 kV and above ratings in three-tank designs. Bulk oil circuit breakers were large and required significant foundations to support the weight and impact loads occurring during operation. Environmental concerns and regulations forced the necessity of oil containment and routine maintenance costs of the bulk oil circuit breakers coupled with the development and widespread use of the SF_6 gas circuit breakers have led to the selection of the SF_6 gas circuit breaker in lieu of the oil circuit breaker for new installations and the replacement of existing oil circuit breakers in favor of SF_6 gas circuit breakers in many installations.

Oil circuit breaker development had been relatively static for many years. The design of the interrupter employs the arc caused when the contacts are parted and the breaker starts to operate. The electrical arc generates hydrogen gas due to the decomposition of the insulating mineral oil. The interrupter is designed to use the gas as a cooling mechanism to cool the arc and also to use the pressure to elongate the arc through a grid (arc chutes) allowing extinguishing of the arc when the current passes through zero.

Vacuum circuit breakers use an interrupter that is a small cylinder enclosing the moving contacts under a hard vacuum. When the breaker operates, the contacts part and an arc is formed resulting in contact erosion. The arc products are immediately forced to be deposited on a metallic shield surrounding the contacts. Without a restrike voltage present to sustain the arc, it is quickly extinguished.

Vacuum circuit breakers are widely employed for metal clad switchgear up to 38 kV class. The small size of the vacuum breaker allows vertically stacked installations of vacuum breakers in a two-high configuration within one vertical section of switchgear, permitting significant savings in space and material compared to earlier designs employing air magnetic technology. When used in outdoor circuit breaker designs, the vacuum cylinder is housed in a metal cabinet or oil-filled tank for dead tank construction popular in the U.S. market.

Gas circuit breakers employ SF_6 as an interrupting and insulating medium. In "single puffer" mechanisms, the interrupter is designed to compress the gas during the opening stroke and use the compressed gas as a transfer mechanism to cool the arc and also use the pressure to elongate the arc through a grid (arc chutes), allowing extinguishing of the arc when the current passes through zero. In other designs, the arc heats the SF_6 gas and the resulting pressure is used for elongating and interrupting the arc. Some older dual pressure SF_6 breakers employed a pump to provide the high-pressure SF_6 gas for arc interruption.

Gas circuit breakers typically operate at pressures between 6 and 7 atm. The dielectric strength and interrupting performance of SF_6 gas reduce significantly at lower pressures, normally as a result of lower ambient temperatures. For cold temperature applications (ambient temperatures as cold as −40°C), dead tank gas circuit breakers are commonly supplied with tank heaters to keep the gas in vapor form rather than allowing it to liquefy; liquefied SF_6 significantly decreases the breaker's interrupting capability. For extreme cold temperature applications (ambient temperatures between −40°C and −50°C), the SF_6 gas is typically mixed with another gas, either nitrogen (N_2) or carbon tetra fluoride (CF_4), to prevent liquefaction of the SF_6 gas. The selection of which gas to mix with the SF_6 is based upon a given site's defining critical criteria, either dielectric strength or interrupting rating. An SF_6–N_2 mixture decreases the interrupting capability of the breaker but maintains most of the dielectric strength of the device, whereas an SF_6–CF_4 mixture decreases the dielectric strength of the breaker but maintains most of the interrupting rating of the device. Unfortunately, for extreme cold temperature applications of gas circuit breakers, there is no gas or gas mixture that maintains both full dielectric strength and full interrupting rating performance. For any temperature application, monitoring the density of the SF_6 gas is critical to the proper and reliable performance of gas circuit breakers. Most dead tank SF_6 gas circuit breakers have a density switch and a two-stage alarm system. Stage one (commonly known as the alarm stage) sends a signal to a remote monitoring location that the gas circuit breaker is experiencing a gas leak, while stage two sends a signal that the gas leak has caused the breaker to reach a gas level that can no longer assure proper operation of the breaker in the event of a fault current condition that must be cleared. Once the breaker reaches stage two (commonly known as the lockout stage), the breaker either will trip open and block any reclosing signal until the low-pressure condition is resolved or will block trip in the closed position and remain closed, ignoring any signal to trip, until the low-pressure condition is resolved. The selection of which of these two options, trip and block close or block trip, is desired is specified by the user and is preset by the breaker manufacturer.

Circuit breakers are available as live tank, dead tank, or grounded tank designs. Dead tank means interruption takes place in a grounded enclosure and current transformers are located on both sides of the break (interrupter contacts). Interrupter maintenance is at ground level and seismic withstand is improved versus live tank designs. Bushings (more accurately described as gas-filled weather sheds, because, unlike the condenser bushings found on bulk oil circuit breakers, gas breakers do not have true bushings) are used for line and load connections that permit installation of bushing current transformers for relaying and metering at a nominal cost. The dead tank breaker does require additional insulating oil or gas (i.e., more insulating oil or gas than just the amount required to perform successful interruption and to maintain adequate dielectric strength) to provide the insulation between the interrupter and the grounded tank enclosure.

Live tank means interruption takes place in an enclosure that is at line potential. Live tank circuit breakers consist of an interrupter chamber that is mounted on insulators and is at line potential. This approach allows a modular design as interrupters can be connected in series to operate at higher voltage levels. Operation of the contacts is usually through an insulated operating rod or rotation of a porcelain insulator assembly via an operating mechanism at ground level. This design minimizes the quantity of oil or gas required as no additional quantity is required for insulation of a grounded tank enclosure. The live tank design also readily adapts to the addition of pre-insertion resistors or grading capacitors when they are required. Seismic capability requires special consideration due to the high center of gravity of the live tank breaker design, and live tank circuit breakers require separate, structure mounted, free standing current transformers.

Grounded tank means interruption takes place in an enclosure that is partially at line potential and partially at ground potential. Although the grounded tank breaker's current transformers are on the same side of the break (interrupter contacts), the grounded tank breaker relays just like a dead tank breaker. The grounded tank breaker design came about as a result of the installation of a live tank breaker interrupter into a dead tank breaker configuration.

Interrupting times are usually quoted in cycles and are defined as the maximum possible delay between energizing the trip circuit at rated control voltage and the interruption of the circuit by the main contacts of all three poles. This applies to all currents from 25% to 100% of the rated short-circuit current.

Circuit breaker ratings must be examined closely. Voltage and interrupting ratings are stated at a maximum operating voltage rating, i.e., 38 kV rating for a breaker applied on a nominal 34.5 kV circuit. The breakers have an operating range designated as K factor per IEEE C37.06, Table 3 in the appendix. For a 72.5 kV breaker, the voltage range is 1.21, meaning that the breaker is capable of its full interrupting rating down to a voltage of 60 kV.

Breaker ratings need to be checked for some specific applications. Applications requiring reclosing operation should be reviewed to be sure that the duty cycle of the circuit breaker is not being exceeded. Some applications for out-of-phase switching or back-to-back switching of capacitor banks also require review and may require specific duty/special purpose/definite purpose circuit breakers to ensure proper operation during fault interruption.

5
High-Voltage Power Electronic Substations

5.1	Introduction	5-1
5.2	HVDC Converters	5-2
5.3	FACTS Controllers	5-18
5.4	Converter Technologies: For Smart Power and Grid Access	5-23
5.5	Control and Protection System	5-28
5.6	Losses and Cooling	5-31
5.7	Civil Works	5-32
5.8	Reliability and Availability	5-32
5.9	Outlook and Future Trends	5-33
	Acknowledgments	5-37
	References	5-37

Dietmar Retzmann
Siemens AG

Asok Mukherjee
Siemens AG

5.1 Introduction

The preceding sections on gas-insulated substations (GIS), air-insulated substations (AIS), and high-voltage switching equipment apply in principle also to the AC circuits in high-voltage power electronic substations. This section focuses on the specifics of power electronic controllers as applied in substations for power transmission purposes.

The dramatic development of power electronics with line and self-commutated converters in the past decades has led to significant progress in electric power transmission technology, resulting in advanced types of transmission systems, which require special kinds of substations. The most important high-voltage power electronic substations are converter stations, above all for high-voltage direct current (HVDC) transmission systems [1], and controllers for flexible AC transmission systems (FACTS) [2]. By means of power electronics, they provide features that are necessary to avoid technical problems in the power systems and they can increase the transmission capacity and system stability very efficiently [3,4]. The power grid of the future must be flexible, secure, as well as cost-effective, and environmentally compatible [5]. The combination of these three tasks can be tackled with the help of intelligent solutions as well as innovative technologies. HVDC and FACTS applications will consequently play an increasingly important role in the future development of power systems. This will result in efficient, low-loss AC/DC hybrid grids that will ensure better controllability of the power flow and, in doing so, do their part in preventing "domino effects" in case of disturbances and blackouts [6].

High-voltage power electronic substations consist essentially of the main power electronic equipment, that is, converter valves and FACTS controllers with their dedicated control and protection systems, including auxiliaries. Furthermore, in addition to the familiar components of conventional substations covered in the preceding sections, there are also converter transformers and reactive power compensation equipment, including harmonic filters (depending on technology) and buildings, or containerized solutions or platforms, including auxiliaries.

Most high-voltage power electronic substations are air insulated, although some use combinations of air and gas insulation. Typically, passive harmonic filters and reactive power compensation equipment as well as DC smoothing reactors are air insulated and usually outdoors, in case of specific environmental conditions or requirements indoors is an option for all passive components—whereas power electronic equipment (converter valves, FACTS controllers), control and protection electronics, active filters, and most communication and auxiliary systems are air insulated, but indoors or as containerized solutions.

Basic community considerations, grounding, lightning protection, seismic protection, and general fire protection requirements apply as with other substations. In addition, high-voltage power electronic substations may emit electric and acoustic noise and therefore require special shielding. Extra fire-protection is applied as a special precaution because of the high power density in the electronic circuits, although the individual components of today are mostly nonflammable and the materials used for insulation or barriers within the power electronic equipment are flame retardant [7,8].

International technical societies like IEEE, IEC, and CIGRE continue to develop technical standards, disseminate information, maintain statistics, and facilitate the exchange of know-how in this high-tech power engineering field. Within the IEEE, the group that deals with high-voltage power electronic substations is the IEEE Power Engineering Society (PES) Substations Committee, High Voltage Power Electronics Stations Subcommittee. On the Internet, it can be reached through the IEEE site (www.ieee.org).

5.2 HVDC Converters

Power converters make the exchange of power between systems with different phase angles and constant or variable frequencies possible. The most common converter stations are AC–DC converters for HVDC transmission. HVDC offers frequency- and phase-independent short- or long-distance overhead or underground power transmission with fast controllability [9]. Two basic types of HVDC converter stations exist: back-to-back AC–DC–AC converter stations and long-distance DC transmission terminal stations. They can be used to interconnect asynchronous AC systems or also be integrated into synchronous AC grids. Figure 5.1 depicts the configuration possibilities and the technologies of HVDC (examples from Siemens—other manufacturers have similar portfolio).

HVDC converters were originally established to transmit power between asynchronous AC systems. Such connections exist, for example, between the western and eastern grids of North America, with the

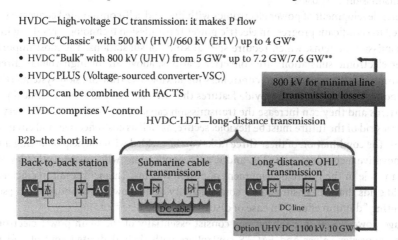

FIGURE 5.1 Configuration possibilities of HVDC. *with LTT, light-triggered thyristor—up to 4 kA; **with ETT, electrically triggered thyristor–up to 4.5/4.75 kA, depending on basic design. (Examples from Siemens AG, Erlangen, Germany.)

ERCOT system of Texas, with the grid of Quebec, and between the 50 and 60 Hz grids in South America and Japan. With these back-to-back HVDC converters, the DC voltage and current ratings are chosen to yield optimum converter costs. This aspect results in relatively low DC voltages, up to about several hundred kV, at power ratings up to 1000 MW and above. Figure 5.2 shows the schematic diagram of an HVDC back-to-back and Figure 5.3 a long-distance transmission converter station. Both applications are

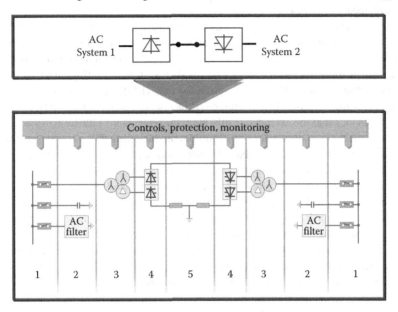

FIGURE 5.2 Schematic diagram of an HVDC back-to-back station. 1, AC switchyard; 2, AC filters, C-banks; 3, converter transform; 4, thyristor valves; and 5, smoothing reactors.

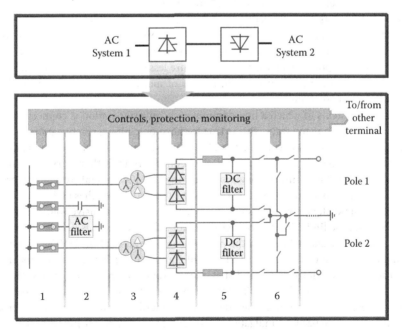

FIGURE 5.3 Schematic diagram of an HVDC long-distance transmission station. 1, AC switchyard; 2, AC filters, capacitor banks; 3, converter transformers; 4, thyristor valves; 5, smoothing reactor and DC filters; and 6, DC switchyard.

with line-commutated thyristor technology, with a DC smoothing reactor and reactive power compensation elements (including AC harmonic filters) on both AC buses. The term back-to-back indicates that rectifier (AC to DC) and inverter (DC to AC) are located in the same station. Typically, in HVDC long-distance transmission terminals, the two poles of a bipolar system can be operated independently, so that in case of component or equipment failures on one pole, power transmission with a part of the total rating can still be maintained ($n - 1$ redundancy).

Long-distance DC transmission terminal stations terminate DC overhead lines or cables and link them to the AC systems. Their converter voltages are governed by transmission efficiency considerations. A voltage level of ±500 kV has been a standard for long. Since 2009, the line-commutated converter technology has reached the ultra-high voltage (UHV) of ±800 kV with power ratings of 6400 MW and above, for a single bipolar transmission. There is an option for new 1100 kV UHV DC applications, which is currently under planning in China. This option offers the lowest losses and highest transmission capacity; however, it is obvious that the extended insulation requirements for 1100 kV will lead to an increase in the already huge mechanical dimensions of all equipment, including PTs, current transformers (CTs), DC bushings, breakers, disconnectors, busbars, transformers, and reactive power equipment.

Most HVDC converters of today are line-commutated 12-pulse converters. In Figures 5.2 and 5.3, typical 12-pulse bridge circuit with delta and wye transformer windings are used, which eliminate some of the harmonics typical for a 6-pulse Graetz bridge converter. The harmonic currents remaining are absorbed by adequately designed AC harmonic filters that prevent these currents from entering the power systems. At the same time, these AC filters meet most or all of the reactive power demand of the converters, which is typically 50% of the nominal active power. This high demand of reactive power compensation is the main drawback of the line-commutated, "classic" HVDC technology (relatively high space requirements). This means, a UHV DC transmission with 5000 MW needs approximately 2500 MVAr of reactive power compensation.

Converter stations connected to DC overhead lines often need DC harmonic filters as well. Traditionally, passive filters have been used, consisting of passive components like capacitors, reactors, and resistors. More recently, because of their superior performance, active (electronic) AC and DC harmonic filters [10–14]—as a supplement to passive filters—using insulated gate bipolar transistors (IGBTs) have been successfully implemented in some HVDC projects. IGBTs have also led to the recent development of self-commutated converters, also called voltage-sourced converters (VSCs) [15–25]. They do not need reactive power from the grid and require less or—in case of new multilevel technologies [26–37]—even no harmonic filtering at all. Details of multilevel DC and FACTS technology are discussed in the following sections.

The AC system or systems to which a line-commutated converter station is connected have significant impact on its design in many ways. This is true for harmonic filters, reactive power compensation devices, fault duties, and insulation coordination. Weak AC systems (i.e., with low short-circuit ratios) represent special challenges for the design of HVDC converters with thyristors [38]. Some stations include temporary overvoltage limiting devices consisting of MOV (metal oxide varistors) arresters with forced cooling for permanent connection or using fast insertion switches [39].

HVDC systems, long-distance transmissions in particular, require extensive voltage insulation coordination, which cannot be limited to the converter stations themselves. It is necessary to consider the configuration, parameters, and behavior of the AC grids on both sides of the HVDC, as well as the DC line connecting the two stations. Internal insulation of equipment such as transformers and bushings must take into account the voltage-gradient distribution in solid and mixed dielectrics. The main insulation of a converter transformer has to withstand combined AC and DC voltage stresses. Substation clearances and creepage distances must be adequate. Standards for indoor and outdoor clearances and creepage distances have been promulgated [40,41] and are constantly being further developed, for example [42]. DC electric fields are static in nature, thus enhancing the pollution of exposed surfaces. This pollution, particularly in combination with water, can adversely influence the voltage-withstand capability and voltage distribution of the insulating surfaces. In converter stations, therefore, it is often necessary to engage in adequate cleaning practices of the insulators and bushings, to apply protective greases, and to protect them with booster sheds. Initial insulation problems with extra-high voltage (EHV) at former ±600 kV DC bushings have been a matter of concern and studies [43–46].

However, latest developments in ±660 kV DC EHV technology in China, project Ningdong–Shandong, have shown that this EHV DC level is fully feasible now by using advanced insulation technologies, already proven in UHV DC applications [47–49]. The Ningdong–Shandong DC system is equipped with converters from Alstom Grid, EHV DC switchyards from Siemens (sending station indoor, receiving station outdoor), and converter controls from Xuji, China (license agreement with Siemens).

A specific issue with long-distance DC transmission is the use of ground return. Used during contingencies, ground (and sea) return can increase the economy and availability of HVDC transmission. The necessary electrodes are usually located at some distance from the station, with a neutral line leading to them, refer to Figure 5.3. The related neutral bus, switching devices, and protection systems form part of the station. Electrode design depends on the soil or water conditions [50,51]. The National Electric Safety Code (NESC) in the United States does not allow the use of earth as a permanent return conductor. Monopolar HVDC operation in ground-return mode is permitted only under emergencies and for a limited time. Also environmental issues are often raised in connection with HVDC submarine cables using sea water as a return path. This has led to the concept of metallic return path provided by a separate low-voltage line. The IEEE–PES has given support to introduce changes to the NESC to better meet the needs of HVDC transmission while addressing potential side effects to other systems.

Mechanical switching devices on the DC side of a typical bipolar long-distance converter station comprise metallic return transfer breakers (MRTB) and ground return transfer switches (GRTS). No true DC breakers exist till date, but developments and prototype tests have been carried out since long, refer to [52]. Basically, DC breakers can be realized by means of

- Pyrotechnics type of breakers
- Traditional AC breakers modified for DC
- Electronic current control by semiconductors
- Combination(s) of these solutions

At present, DC fault currents are still best and most swiftly interrupted by the converters themselves. MRTBs with limited DC current interrupting capability have been developed since long [53] and are applied successfully in many DC schemes worldwide. They include commutation circuits, that is, parallel reactor/capacitor (L/C) resonance circuits that create artificial current zeroes across the breaker contacts. For UHV DC applications, fast bypass switches are used to bypass one or more of the series converters, for system redundancy or for voltage reduction of the DC line, whenever it is necessary.

The conventional grid-connecting equipment in the AC switchyard of a converter station is covered in the preceding sections. In addition, reactive power compensation and harmonic filter equipment are connected to the AC buses of the line-commutated converter station. Circuit breakers used for switching these shunt capacitors and filters must be specially designed for capacitive switching (high-voltage stresses). A back-to-back converter station does not need any mechanical DC switching device (see Figure 5.1).

Figures 5.4 through 5.7 show photos of different converter stations. The back-to-back station in Figure 5.4 is one of several earlier asynchronous links between the former Eastern and Western European power grids, which were located in Austria and Germany. Due to the later synchronization of parts of the Eastern Grids with the Western Grid (former UCTE, now ENTSO-E), these B2Bs have been taken out of operation.

Figure 5.4 shows the converter station of one such B2B, interconnecting the German and Czech Power Grids. The converter transformers are arranged on both sides "back-to-back" of the converter hall. The control building is directly connected to its right side; two outdoor DC smoothing reactors are on the left side of the converter hall; the AC filter circuits are on the lower part, left side, and upper part, right side; and the AC buses are at the outer lower and upper part of the photo. An additional AC line compensation reactor can be seen in the upper middle part of the picture.

FIGURE 5.4 View of a typical HVDC back-to-back converter station (600 MW Etzenricht, Germany—now out of operation).

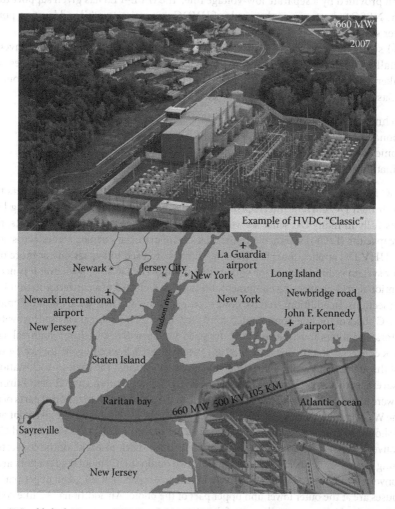

FIGURE 5.5 DC cable link Neptune RTS, Fairfield, CT. (Photo courtesy of the station Sayreville, NJ.)

High-Voltage Power Electronic Substations

FIGURE 5.6 Neptune HVDC—a view of Station Duffy Avenue inside: the cable, indoor smoothing reactor, and thyristor converter.

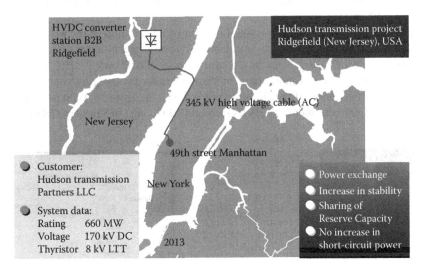

FIGURE 5.7 Hudson Transmission Project—second HVDC (B2B) to strengthen the power supply system of New York.

After the 2003 blackout in the United States and parts of Canada and (shortly after) a number of blackouts in Europe, new HVDC and FACTS projects are gradually coming up to enhance the system security and to "squeeze" more power out of the grids [4,6,9,32,54,55]. One example is the Neptune HVDC project in the United States. The task given by Neptune Regional Transmission System LLC (RTS) in Fairfield, Connecticut, was to construct an HVDC transmission link between Sayreville, New Jersey, and Duffy Avenue, Long Island/New York. As new overhead lines could not be built in this densely populated area, power had to be brought directly to Long Island by an HVDC cable transmission, bypassing the AC sub-transmission network. For various reasons, environmental protection in particular, it was decided not to build a new power plant on Long Island near the city in order to cover the power demand of Long Island with its districts Queens and Brooklyn, which is particularly high in summer. The Neptune HVDC interconnection is an environmentally compatible, cost-effective solution that helps meet these needs.

The low-loss power transmission provides access to various energy resources, including renewable ones. The interconnection is carried out via a combination of submarine and subterranean cable directly to the network of Nassau County, which borders on the city area of New York. Figure 5.5 shows in the upper part a photo of the Sayreville station, which is connected via world's first 500 kV DC MI cable (mass impregnated, lower part of the figure) to the station Duffy Avenue, Long Island. A view of the station Duffy Avenue inside the valve hall with the thyristor converters suspended from the ceiling, the indoor DC smoothing reactor (indoor for reasons of noise reduction) and the monopolar cable is given in Figure 5.6.

During trial operation, 2 weeks ahead of schedule, Neptune HVDC proved its security functions for system support and blackout prevention in megacities in a very impressive way. On June 27, 2007, a blackout occurred in New York City. Over 380,000 people were without electricity in Manhattan and Bronx for up to 1 h, subway came to a standstill and traffic lights were out of operation. In this situation, Neptune HVDC successfully supported the power supply of Long Island and thus of 700,000 households.

In Figure 5.7, a second DC project in the area of New York, in this case with B2B and a short AC cable through the Hudson River, is shown. The benefits of this additional DC "energy bridge" in the Megacity New York are depicted in the figure.

The grid developments in Europe progress in a similar way. A number of new DC projects are already in operation and more are coming up, for the same reasons as in the United states, refer to Figure 5.8. The DCs provide power trading opportunities by means of excellent controllability of HVDC, they build power highways across the ocean (Figure 5.8, lower part) and connect Northern and Southern parts of Europe, and they enable interconnection between the different asynchronous parts of ENTSO-E (e.g., in Denmark, Figure 5.8, upper part).

China is presently the country with the largest number of HVDC links in the world. Due to rapidly growing industries in this emerging country, the demand on power generation as well as on power transmission is continuously increasing. Nowadays, a large number of bulk power UHV AC and DC transmission schemes over distances of more than 2000 km are under planning for connection of various large hydro power stations. Some of the DC projects are shown in Figure 5.9, right part. In the left part of the figure, an example of the 3000 MW HVDC project Gui-Guang I in Southern China is depicted.

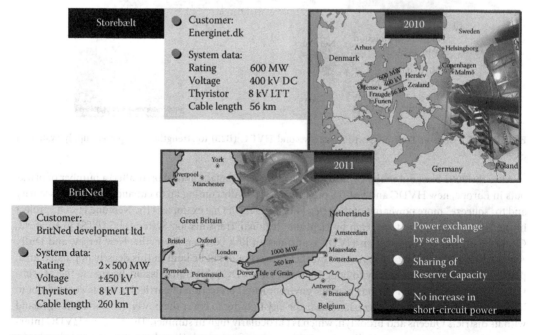

FIGURE 5.8 Europe—the HVDC portfolio is growing too.

FIGURE 5.9 HDVC Projects in China enable low-loss West-to-East transmission of hydro-power-based electrical energy produced in the country's interior to coastal load centers (right side). Sending station of long-distance overhead line transmission Guizhou-Guangdong I (left side).

FIGURE 5.10 UHV DC Yunnan–Guangdong: from "3D" to reality—view of the sending station Chuxiong.

The next generation HVDC is the Yunnan–Guangdong ±800 kV UHV DC Project, world's first HVDC project in the UHV range and one of the most important power links in Southern China. It connects Chuxiong in the Yunnan province to Suidong in the Guangdong province over a distance of 1418 km. This project and its new technology present a major step forward in the HVDC technology and provide a new level of efficiency in power transmission [85]. Figure 5.10 shows a view of the UHV DC sending station Chuxiong with valve halls and the relatively small DC yard on the right side and the huge AC yard equipment on the left and middle part of the picture. This gigantic project was, in fact, the kickoff for the DC Super Grid development, worldwide. The utility China Southern Power Grid and Siemens succeeded to put pole 1 of this first UHV DC into operation in December 2009 and pole 2 in June 2010.

A simplified single line diagram for the basic configuration is shown in Figure 5.11. It can be seen that the solution for the Yunnan–Guangdong project consists of a series connection of two 12-pulse bridges with 400 kV rated voltage each. In order to enable uninterruptible power transfer during connection and disconnection of individual groups, DC bypass switches, DC bypass disconnect switches, and group disconnect switches are included in the arrangement. It should be noted that even though the transportation limitation is the most important reason for selecting such an arrangement with smaller transformer units, also system security aspects are a crucial issue: increased power availability is possible compared to single 12-pulse bridge designs since any outage of a single group does only affect 25% of the installed power capability. Especially for a link with such large power rating of 5000 MW or more, this is an important aspect for the system redundancy. This high redundancy will be even more important when the next generation of a UHV DC voltage level of 1100 kV with an increased power output of 10 GW comes into application.

A view of the 400 and 800 kV converter buildings is depicted in Figure 5.12 and the 800 kV converter hall inside is shown in Figure 5.13—the left side shows the large transformer bushings entering the valve hall, where they are connected to the converter on the right side. The 800 kV DC wall bushing (right side

High-Voltage Power Electronic Substations 5-11

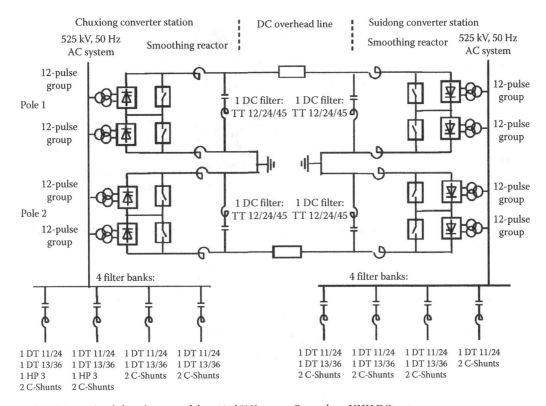

FIGURE 5.11 Single line diagram of the ±800 kV Yunnan–Guangdong UHV DC system.

FIGURE 5.12 800 kV UHV DC Yunnan–Guangdong—view of the bipolar valve halls with two 400 kV systems in series to build 800 kV.

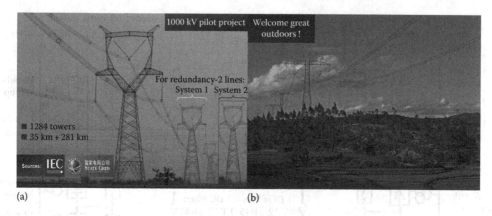

FIGURE 5.13 View of transformer bushings (a) and thyristor valve towers with 800 kV DC wall bushing (b) in the 800 kV valve hall.

FIGURE 5.14 China—comparison of UHV AC (1000 kV pilot project, left side) (Photo courtesy of State Grid, Beijing, China.) and UHV DC lines (800 kV, right side). (Photo courtesy of Siemens AG, Erlangen, Germany.)

of Figure 5.13) finally connects the valves from inside to outside in the DC yard and feeds the DC line, after passing through the DC smoothing reactor and DC filters (Figure 5.11).

In Figure 5.14, a comparison of UHV AC at 1000 kV (left part, protype project with three substations) and UHV DC transmission (right part) is depicted. While the UHV AC system would finally need two three-phase lines for $n-1$ redundancy, the DC achieves redundancy with just two poles, which means two bundles of conductors on one tower only, a strong reduction in the right-of-way requirements, refer to Figure 5.14. A third conductor for metallic return at low voltage level would be an option. In China and India, however, due to the very long transmission distances of more than 1000 up to over 3000 km, decisions have been made to do without a metallic return conductor—for economic reasons. In India, there are similar prospects for UHV DC as in China due to the large extension of the grid. Regarding AC system, however, India has plans to realize UHV levels up to 1200 kV.

The Yunnan–Guangdong project helps save around 33 million tons CO_2 in comparison with a local power generation, which, in view of the current energy mix in China, would have involved a relatively high carbon amount. In the upper part of Figure 5.15, a view of the control room during the inauguration ceremony of pole 1 is given. Operating and commissioning area are split, to allow testing of pole 2, while pole 1 is already in operation and transmits power. The lower part of the figure shows the control desk with both poles at full power of 5000 MW, in April 2011. A very important security feature of the UHV DC is its high continuous overload capability of 20%, 25% for 2 h, and 50% for valuable 3 s, see Figure 5.16. This means,

High-Voltage Power Electronic Substations 5-13

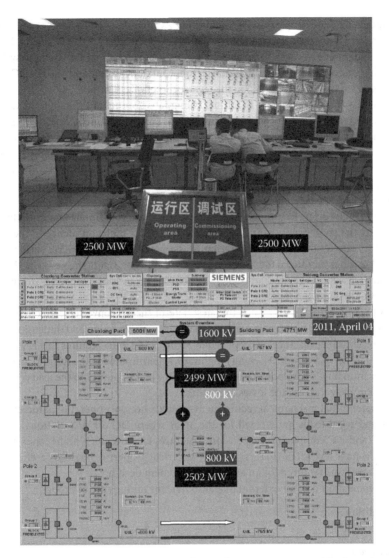

FIGURE 5.15 Yunnan–Guangdong. Upper part: control room during inauguration of the first pole on December 29, 2009; lower part: control desk at full power transfer 5000 MW with both poles in operation.

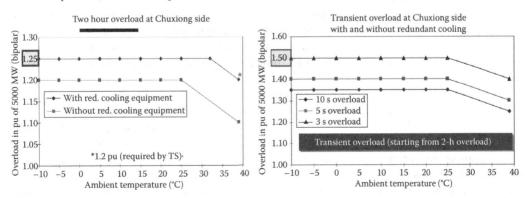

FIGURE 5.16 Example of UHV DC "Bulk" overload characteristics. *Note*: max. ambient temperature is 34.2°C in Chuxiong.

in emergency situations, for example, loss of neighboring AC lines or power plants, the HVDC is able to transmit an overload of 6000 MW permanently, 6250 MW for 2 h, and 7500 MW for 3 s. High overload capacity and redundant system design for emergency situations are a crucial issue for system security [56,57]. Although the normal power flow direction is from Chuxiong to Suidong, the HVDC system is also designed to transmit up to 4500 MW (90%) in the reverse direction.

The second 800 kV UHV DC project in China, Xiangjiaba–Shanghai of State Grid Corporation of China, which also involves Siemens as well as ABB and Chinese partners, boasts significantly high yearly CO_2 savings of over 40 million tons, thanks to the very high hydro power transmission capacity of 6400 MW. This currently world's biggest UHV DC in operation started full commercial operation in July 2010. Siemens and its Chinese partners delivered all HVDC transformers and thyristor valves with new 6 in. thyristors for the sending station Fulong, 1 year ahead of schedule. These are the biggest HVDC transformers and power converters in operation, worldwide.

The third UHV DC project, Jinping of State Grid Corporation of China, was initially planned for 6400 MW, but after the good results of the first 800 kV testings, a high confidence in the new technology came up, and State Grid increased the rating to the new level of 7200 MW, with a transmission distance of 2095 km, a big step ahead toward the Super Grid concept. Using the new 6 in. thyristors, even 7500 MW would be feasible (refer to Figure 5.1), without losing overload capacity—for more power, a voltage increase would be reasonable.

The UHV HVDC systems at 800 kV require the latest state-of-the-art converter technology [83,84,86,87]. The different components of this kind of installations boast impressive design and dimensions owing to the required insulation clearance distances, as depicted in Figure 5.17. China requires this HVDC technology to construct a number of high-power DC energy highways, superimposed to the AC grid, in order to transmit electric power from huge hydro power plants in the center of the country to the load centers located as far as 2000–3000 km away with as little losses as possible.

It is well understood that mechanical requirements include not only operational forces but also seismic conditions and wind loads anticipated for the areas where DC (or AC) stations are located. By nature, UHV AC and DC equipment, suspension structures etc. are much higher than today's equipment for existing voltage levels. Due to this, not only electrical properties but also careful consideration of mechanical stresses was required for an adequate equipment design.

Designing equipment for correct external insulation means taking care of proper

- Flashover distances
- Creepage distances

of the equipment housings.

Required flash distances determine the axial length of the equipment. Flash distances can be calculated fairly well based on the specified insulation levels for the equipment. For UHV equipment, the switching impulse level became the dimensioning factor. DC voltages are not decisive with respect to the flash distance. Corrections were included for equipment to be installed at higher altitudes above sea level. Flash distances increase more than linearly with increasing switching impulse voltages. Finally, the correct design (for the DC projects) was verified by corresponding type tests of the equipment [58].

As far as DC converter valves and associated equipment are concerned, the design of creepage distance is not a problem as such equipment is installed indoors in the converter valve hall like in many 500 kV projects. The valve hall provides a controlled environment. For UHV DC equipment inside the valve hall, the same specific creepage distance can be selected as for the existing 500 kV DC equipment. In case of outdoor installation of the DC yard, the most appropriate solution is using composite insulators with silicone housing and sheds. This provides the major advantage of hydrophobic insulator surfaces that significantly improves operation of the DC system in polluted areas. However, also indoor configuration of the DC yard is feasible. In this case, the specific creepage distance can be reduced to some extent compared to the outdoor installation. Relative to existing 500 kV equipment, it can be

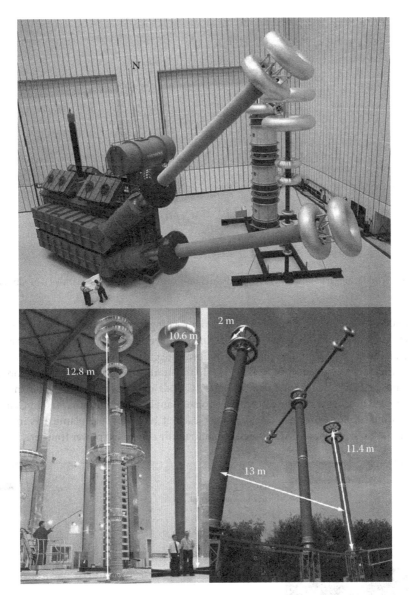

FIGURE 5.17 UHV DC station equipment during testing. Upper part: converter transformer; lower part, from left to right: 800 kV DC arrester, DC voltage divider, and DC disconnector.

stated that the specific creepage distances both for outdoor and for indoor installation need not to be increased for UHV DC equipment in order to ensure safe performance against pollution flashovers.

Converter transformers are one of the very important components for UHV DC application. It is quite understood that the existing technology and know-how of converter transformers can manage higher DC voltages. Yet, there are critical areas that did need careful consideration and further development in order to keep the electrical stresses at a safe level. Above all the windings and the transformer internal part of bushings on the valve side of the converter transformers with the barrier systems and cleats and leads required very careful attention.

In the project Xiluodo–Guangdong (China Southern Power Grid), two bipolar 500 kV DC systems, each at 3200 MW are connected in parallel (Figure 5.18). In total, the DC power output of the two 500 kV systems is the same as in the aforementioned Xiangjiaba–Shanghai project. On a global scale, till date,

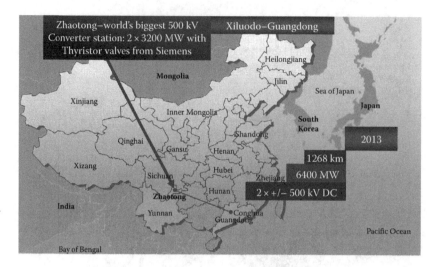

FIGURE 5.18 World's biggest 500 kV HVDC Transmission Project: 2 Systems in parallel—in China Southern Power Grid.

UHV DC at 800 kV is applied in China only, but India is going to follow, for the same reasons of high power transmission over long distances. In the future, however, in other regions, for example, in the Americas, Asia, Russia, Africa, and possibly also in Europe (e.g., for "Green Energy" from Africa—with DESERTEC) similar developments can be expected. In the meantime, at a smaller scale, more and more interconnections and DC systems integrated into the AC grids are being built worldwide. In Figure 5.19, upper part, a new B2B scheme enables interconnection of Bangladesh with India, and in Canada (lower part of the figure), two long-distance transmissions, each at 1000 MW, are planned as integrated DC energy bridges in the synchronous AC systems, enabling fully controlled power transfer.

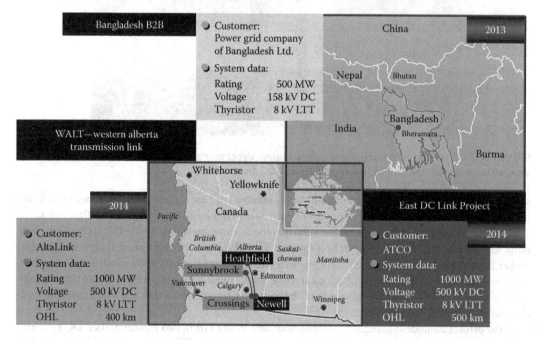

FIGURE 5.19 From DC interconnection in Asia back-to-DC transmission in the Americas—new projects with HVDC offering many benefits.

In addition to the well-proven "Classic" DC systems with line-commutated thyristors, new schemes with VSCs have progressed and are realized. Main focus of this "Smart Grid" [5] technology is Grid Access of Renewable Energy Source (RES); however, it is suitable in the same way for transmission and system interconnection. Preferences of VSC DC and AC applications are, especially when the new modular multilevel converter (MMC—HVDC PLUS and STATCOM, e.g., SVC PLUS [34,35]) technology is applied as follows:

- Space reduction of 50%, typically—essential for Offshore DC
- Enhanced control features, including independent P-Q control for DC and fast V control for AC application
- Grid access to weak or passive networks and Blackstart capability

Furthermore, with VSC, multiterminal DC applications and meshed DC networks are much easier with the current reversion capability of the IGBTs. More details of the converters are depicted in Section 5.4.

An overview of the first MMC HVDC project with a ±200 kV XLPE DC sea cable transmission is given in Figure 5.20. The goal of this project was to eliminate bottlenecks in the overloaded Californian grid (upper part of the figure): new power plants cannot be constructed in this densely populated area

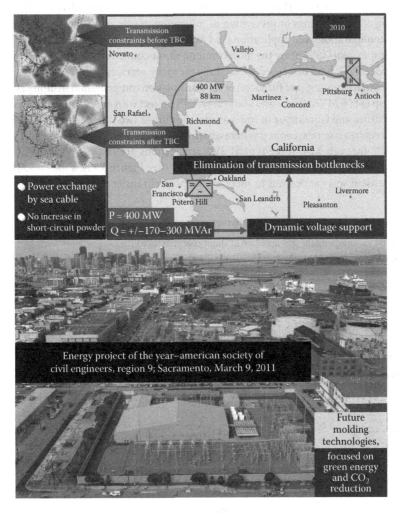

FIGURE 5.20 The "Trans Bay Cable" Project in the United States—world's first VSC HVDC with MMC technology and ±200 kV XLPE cable.

and there is no right-of-way for new lines or land cables. This is the reason why a DC cable is laid through the bay, and the power flows through it by means of the HVDC PLUS (VSC) technology in an environmentally compatible way. This project received an important Energy Award of the American Society of Civil Engineers (see Figure 5.20), and it is an outstanding "showcase" for excellent VSC performance using independent control for active and reactive power.

An important development in the power supply of megacities is the outsourcing of power generation to close or more distant surrounding regions. That is, transmission networks and distribution systems are forced to interconnect increasingly longer distances. Furthermore, efficiency and reliability of supply play an important role in every planning, particularly in the face of increasing energy prices and almost incalculable safety risks during power blackouts.

5.3 FACTS Controllers

The acronym FACTS stands for Flexible AC Transmission Systems. These systems add some of the virtues of DC, that is, system stability improvement and fast controllability to AC transmission by means of electronic controllers. Such controllers can be shunt or series connected or both. They represent variable reactances or AC voltage sources. They can provide load-flow control and, by virtue of their fast controllability, damping of power swings or prevention or mitigation of subsynchronous resonance (SSR). The most common configurations and applications of FACTS are depicted in Figure 5.21.

Static VAr compensators (SVCs) are the most common shunt-connected controllers. They are, in effect, variable reactances. SVCs have been used successfully for many years, either for load (flicker) compensation of large industrial loads (e.g., arc furnaces) or for transmission compensation in utility systems. Rating of SVCs can go up to 800 MVAr; the world's largest FACTS project with series compensation (TCSC/FSC) as of date is at Purnea and Gorakhpur in India at a total rating of two times 1.7 GVAr. Like HVDC converters, FACTS requires controls, cooling systems, harmonic filters, transformers, and related civil works.

In Figures 5.22 through 5.24, three typical SVC applications are shown. With the Mead-Adelanto and the Mead-Phoenix Transmission Project (MAP/MPP—see Figure 5.22), a major 500 kV transmission system extension was carried out to increase the power transfer opportunities between Arizona

FACTS—flexible AC transmission systems: support of power flow

- SVC, static var compensator (the standard of shunt compensation)
- SVC PLUS (=STATCOM—static synchronous compensator, with VSC)
- FSC, fixed series compensation
- TCSC, thyristor controlled series compensation
- TPSC, thyristor protected series compensation*
- GPFC, grid power flow controller (FACTS-B2B)
- UPFC, unified power flow controller (with VSC)
- CSC, convertible synchronous compensator (with VSC)

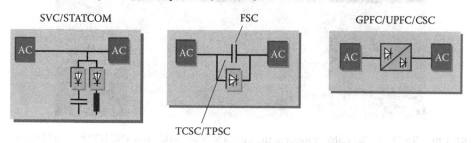

FIGURE 5.21 FACTS—configurations and applications. * denotes special high power LTT hyristors.

High-Voltage Power Electronic Substations

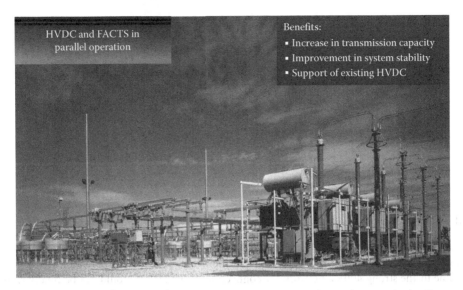

FIGURE 5.22 FACTS shunt compensation for system upgrade: 2 SVCs Mead-Adelanto, 500 kV, each at 388 MVAr cap.—United States, 1995.

FIGURE 5.23 SVC Siems: the first HV SVC in Germany (400 kV) with 100 MVAr ind./200 MVAr cap. for AC grid support of Baltic Cable HVDC.

and California [59]. The extension includes two main series compensated 500 kV line segments and two equally rated SVCs supplied by Siemens, in close cooperation with GE, for system studies and delivery of equipment at the Adelanto and the Marketplace SVC substations and the series compensation. The SVCs (photo of Marketplace) enabled the integrated operation of the already existing highly compensated EHV AC system and the large HVDC transmission (IPP). The SVC installation was an essential prerequisite for the overall system stability at an increased power transfer rate.

In Figure 5.23, an innovative FACTS application with SVC, also in combination with HVDC, for transmission enhancement in Germany is shown [3]. Nearby the SVC where offices and homes located, some special measures for noise abatement had been taken for SVC transformer and thyristor-controlled reactor (TCR), see Figure 5.23. This was the first high-voltage FACTS controller in the German network.

FIGURE 5.24 SVC São Luis, ELETRONORTE, Brazil: TCR, TSC/FC, 230 kV, 150 MVAr cap./100 MVAr ind.

The reasons for the SVC installation at Siems substation near the landing point of the Baltic Cable HVDC were unforeseen right-of-way restrictions in the neighboring area, where an initially planned new tie-line to the strong 400 kV network for connection of the HVDC could not be finally realized. Therefore, with the existing reduced network voltage of 110 kV, only a reduced amount of power transfer of the DC link was possible since its commissioning in 1994, in order to avoid repetitive HVDC commutation failures and voltage problems in the grid. In an initial step toward grid access improvement, an additional transformer for connecting the 400 kV HVDC AC bus to the 110 kV bus was installed. Finally, in 2004, with the new SVC equipped with a fast coordinated control, the HVDC could fully utilize its transmission capacity up to the initial design rating of 600 MW. In addition to this measure, a new cable to the 220 kV grid was installed to increase the system stability with regard to performance improvement of the HVDC controls. Prior to commissioning, intensive studies were carried out: first with the computer program PSS™NETOMAC and then with the RTDS real-time simulator by using the physical SVC controls and simplified models for the HVDC.

In Figure 5.24, a containerized SVC solution in Brazil is shown. This SVC also contributes significantly to the grid stability, as indicated in the figure. In Figure 5.25, an SVC indoor solution in Denmark is depicted. Reasons for the challenging indoor application were extremely high requirements for noise abatement in a touristic area. Task of the SVC is voltage stabilization for a large nearby offshore wind farm. In the upper part of the figure, the single-line diagram of the SVC is given, with two TCRs and two harmonic filters, which also serve as fixed capacitors. The thyristor controller provides fast control of the overall SVC reactance between its capacitive and inductive design limits. As a consequence, the SVC improves the voltage quality at the wind farm's grid connection point.

Like the classical fixed series compensation (FSC), thyristor-controlled series compensation (TCSC) [60,61] is normally located on insulated platforms, one per phase, at phase potential. Whereas the FSC compensates a fixed portion of the line inductance, TCSC's effective capacitance and compensation level can be varied statically and dynamically. The variability is accomplished by a TCR connected in parallel with the main capacitor. This circuit and the related main protection and switching components of the TCSC are shown in the photo of Figure 5.26 of a 500 kV project in Brazil. The thyristors are located in weatherproof housings on the platforms. Communication links exist between the platforms and ground. Liquid cooling is provided through ground-to-platform pipes made of insulating material. Auxiliary platform power, where needed, is extracted from the line current via CTs. Like most conventional FSCs, TCSCs are typically integrated into existing substations. In the figure, the benefits and features of both fixed and controlled series compensation are depicted.

Figure 5.27 shows a photo of a 500 kV TCSC installation in the United States. In the picture, the platform-mounted valve housings are clearly visible. Slatt (United States) has six equal TCSC modules per phase, with two valves combined in each of the three housings per bank. At the Serra da Mesa installation (refer to Figure 5.26), each platform has one single valve housing.

High-Voltage Power Electronic Substations 5-21

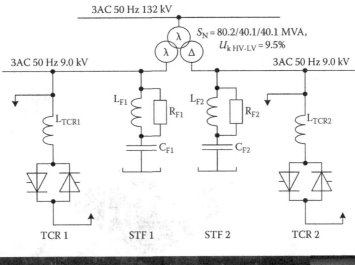

FIGURE 5.25 SVC Radsted: an innovative indoor solution for wind farm support—80 MVAr cap./65 MVAr ind., with extremely high requirements for noise abatement (upper part: single-line diagram; lower part: view of the substation).

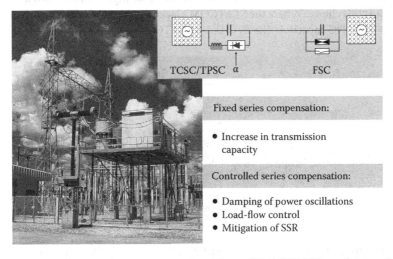

FIGURE 5.26 FACTS application of series compensation. (Photo of 500 kV TCSC Serra da Mesa, Furnas/Brazil.)

FIGURE 5.27 Aerial view of BPA's Slatt, Oregon, 500 kV TCSC. (Photo courtesy of GE, Fairfield, CT.)

A new development in series compensation is the thyristor protected series compensation (TPSC). The circuit is basically the same as for TCSC, but without any controllable reactor and with self-cooled thyristors (no water cooling). The thyristors of a TPSC are used only as a bypass switch to protect the capacitors against overvoltage, thereby avoiding large MOV arrester banks with relatively long cooldown intervals. The thyristor is specially designed for high valve currents; they can withstand up to 110 kA peak with a fast cooldown time by means of a special heat sink. A number of these TPSCs are installed in the Californian grid. Figure 5.28 shows one such installation at 500 kV (right side) in comparison with an FSC (left side, also at 500 kV, in China).

A new type of controlled shunt compensator, a static synchronous compensator called STATCOM, uses VSCs, initially with high-power Gate-Turn-Off thyristors (GTO), now mostly with IGBTs [18,19]. Figure 5.29 shows the related one-line diagram of a new type of STATCOM, the multilevel SVC PLUS [30,31,34,35], in comparison with line-commutated SVC "Classic." STATCOM is the electronic equivalent of the well-known (rotating) synchronous condenser, and one application of STATCOM is the replacement of old synchronous condensers. The need for high control speed and low maintenance can support this choice. Where the STATCOM's lack of inertia is a problem, it can be overcome by a sufficiently large DC capacitor. The SVC PLUS STATCOM requires only small (high-frequency

FIGURE 5.28 FSC: fixed series compensation, Fengjie, China, 500 kV, 2 × 610 MVAr and TPSCs Vincent, Midway, and El Dorado/USA, seven systems at 500 kV.

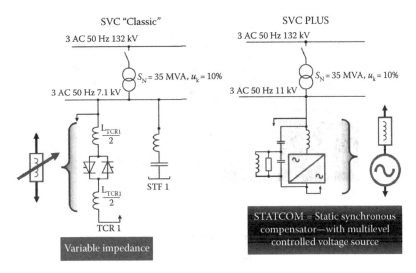

FIGURE 5.29 STATCOM with MMC (SVC PLUS) versus SVC "Classic."

harmonic filters) and uses MMC technology with distributed capacitors, similar to HVDC PLUS. This makes the footprint of an SVC PLUS station significantly more compact than that of the conventional SVC. The configuration possibilities of the SVC PLUS STATCOM are depicted in Figure 5.30. Part (a) shows the containerized solution, part (b) the project Kikiwa in New Zealand for voltage quality improvement in the 220 kV AC system (voltage dip compensation), and part (c) the various combinations from containerized solution to open rack solution (for buildings) and hybrid solution with MSR and/or MSC combinations. These possibilities offer a high degree of flexibility for system applications, from ±25 to ±200 MVAr, and above.

The ease with which FACTS stations, in particular with the MMC technology, can be reconfigured or even relocated is an important factor and can influence the substation design [22,62]. Changes in generation and load patterns can make such flexibility desirable.

The principle of the MMC technology is explained in the next section and additional projects are depicted there.

5.4 Converter Technologies: For Smart Power and Grid Access

The entire MMC PLUS system for HVDC as well as for SVC has a modular structure and can be flexibly configured, what simplifies its standardization, see Figure 5.31. The converter modules are connected on the secondary side of a high-voltage coupling transformer (for simplification not shown in the figure) to build the HVDC or the SVC. Due to the MMC configuration, there is almost no—or, in the worst case, very small—need for AC voltage filtering to achieve a clean voltage. The system configuration is very compact and normally occupies 50% less space than a "classic" HVDC or SVC system.

The VSC application for AC substations (STATCOM with SVC PLUS) is shown on the left part of Figure 5.31 and for DC substations on the right side. Due to its compact and modular design, the MMC PLUS technology is ideally suited for offshore grid access applications, where space is always a crucial issue [36,37,63].

The grid-compatible connection of wind power is a crucial criterion when it comes to the security of power supply. This grid compatibility already starts at an early planning stage—a wide range of test criteria must be determined by means of design studies. When providing grid coupling by means of AC cables, dynamic stabilization of grid voltage in order to comply with the Grid Code cannot be done without. This is where SVCs or STATCOMS come into play. Switching compensators with mechanical

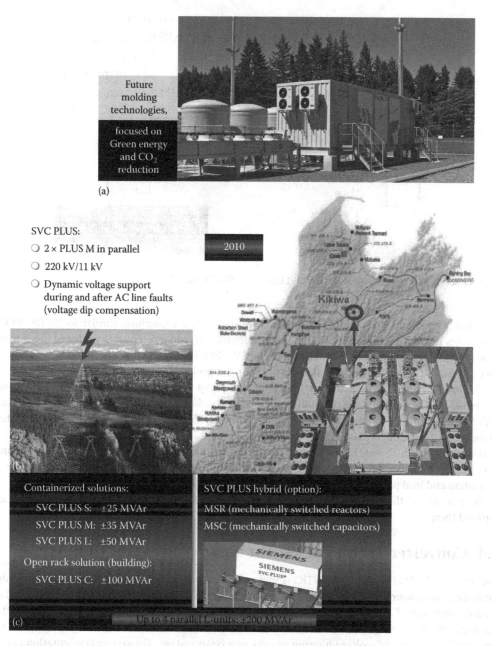

FIGURE 5.30 STATCOM with SVC PLUS: configurations and applications. (a) Containerized solution; (b) power quality in HV AC systems—Kikiwa Project, South Island, New Zealand; (c) from containerized to open rack and hybrid solutions.

elements alone are usually too slow; however, when used in combination with the SVC or SVC PLUS, they can be of advantage.

When using long cables, the HVDC is the preferred solution for grid access, preferably with VSC and MMC technology (space savings). The application of line-commutated power converters particularly for offshore installations is basically possible both with the "Classic" HVDC and "Classic" FACTS; it is, however, not the best solution. Particularly due to the fact that the commutation processes in the power converter are determined by the grid voltage, the grid conditions at the point of grid coupling must be

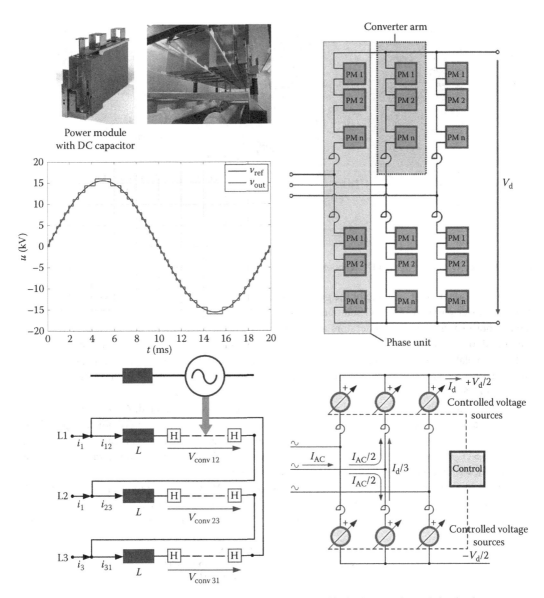

FIGURE 5.31 From FACTS to HVDC—power electronic building blocks (PEBBs) for multilevel voltage generation with MMC: SVC PLUS and HVDC PLUS. multilevel voltage-sourced converters for AC substations (left side) and for DC substations (right side).

ideal, for example, an adequate high short-circuit power of the grid. Self-commutated VSCs require no "driving" grid voltage—they develop it themselves by means of DC voltage (Black-Start Capability)—therefore, they are more suitable to provide grid access from and to the offshore platforms.

Up till now the VSC for HVDC and FACTS applications have mainly been implemented with two- or three-level power converters, which, however, required comparatively many filters for reactive power compensation and to provide sufficient voltage quality. The multilevel VSCs boast of significant advantages with regard to dynamics and harmonics. This is the main reason why the new MMC technology has been developed, with significant advantages in high-voltage applications including the fact that minimal filtering or even no filtering is required, making the installations of this kind extremely compact.

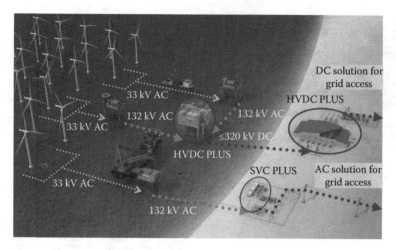

FIGURE 5.32 Overview: AC and DC technologies for grid access of wind farms.

Figure 5.32 depicts an example of the grid access of an offshore wind farm installation with the AC solution (lower part) and the DC solution (middle part) by means of the compact MMC technology [36]. In both cases, the wind generators are interconnected via medium-voltage AC cables and then connected to an intermediate offshore station (AC platform) in order to be led to the coast or to a DC station on a platform with one or several parallel high-voltage cables, which helps increase the efficiency. At the coast the grid access takes place, including voltage stabilization by means of the HVDC PLUS or the SVC PLUS.

In Figure 5.33, two examples of grid access with the AC solution are depicted. A 500 MW wind farm is installed at Greater Gabbard, and the London Array project of 630 MW, with an option for upgrade to 1000 MW. Both projects are off the south coast of England. The grid stabilization is carried out on land with three SVC PLUS systems for Greater Gabbard and four SVC PLUS for London Array, each system at 50 MVAr, with mechanically switched elements in addition.

Figure 5.34 shows the first three offshore projects for the HVDC PLUS, among them the SylWin 1 with the world's first MMC VSC HVDC at 864 MW. The world's biggest MMC VSC is going to be the France–Spain interconnector INELFE at 2 × 1000 MW. All three offshore projects in Figure 5.34 have

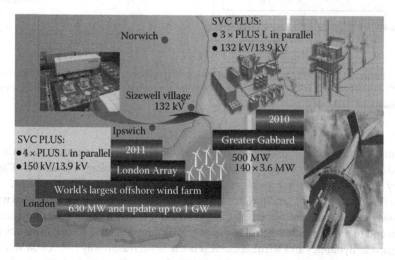

FIGURE 5.33 United Kingdom: many large wind farm projects with AC grid connection—short distances make this feasible.

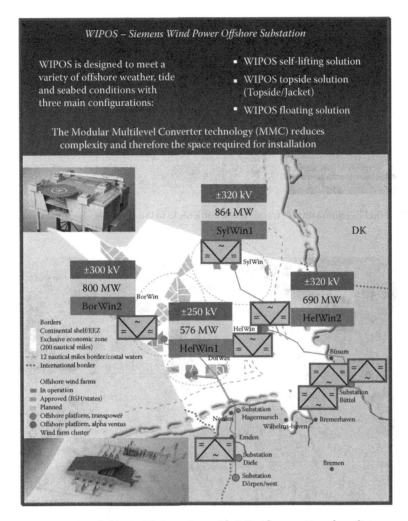

FIGURE 5.34 Germany: many large wind farm projects with DC grid connection—long distances need DC cable transmission.

been submitted to the Consortium of Prysmian (Cable) and Siemens (Grid Access) by Transpower, a subsidiary of the Dutch grid operator TenneT. The DC voltage of the SylWin 1 project is at ±320 kV, till date it is the highest level for an XLPE cable. The commercial operation of the projects is scheduled for 2013–2015. For the offshore DC stations, a new system, referred to as WIPOS (Wind Power Offshore Station), has been developed for the needs of the HVDC PLUS, which constitutes an innovative, floating, and self-lifting platform solution; see left and upper part of the figure.

The INELFE project, Figure 5.35, is a dedicated solution of Siemens to enhance the power transfer and stability of the France–Spain grid interconnection—a transmission bottleneck since long. This makes the HVDC project an important link in the expansion of the Trans-European network. Therefore, the project is partly funded by the EU and is scheduled for commissioning in late 2013. The XLPE cable (from Prysmian) will be designed for ±320 kV DC.

While SVC and STATCOM controllers are shunt devices, and TCSCs are series devices, the so-called unified power flow controller (UPFC) is a combination of both [20]. The UPFC uses a shunt-connected transformer and a transformer with series-connected line windings, both interconnected to a DC capacitor via related voltage-source-converter circuitry within the control building. A further development

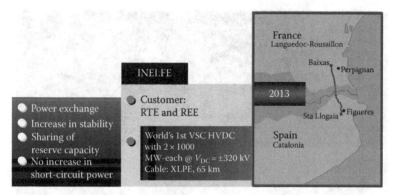

FIGURE 5.35 INELFE—elimination of transmission bottlenecks in the 400 kV AC grid by means of HVDC cable transmission.

FIGURE 5.36 CSC at NYPA's 345 kV Marcy, New York.

[21,54,64] involves similar shunt and series elements as the UPFC, and this can be reconfigured to meet changing system requirements. This configuration is called a convertible static compensator (CSC). Figure 5.36 depicts this CSC system at the 345 kV Marcy substation in New York state.

5.5 Control and Protection System

Today's state-of-the-art HVDC and FACTS controls—fully digitized and processor-based—allow steady-state, quasi steady-state, dynamic, and transient control actions and provide important equipment and system protection functions. Fault monitoring and sequence-of-event recording devices are used in most power electronics stations. Typically, these stations are remotely controlled and offer full local controllability as well. HMI interfaces are highly computerized, with extensive supervision and control via monitor and keyboard. All of these functions exist in addition to the basic substation secondary systems described in Chapters 6 and 7.

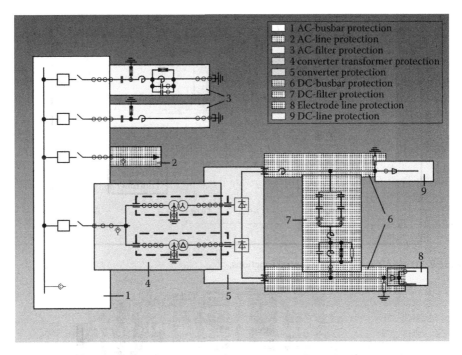

FIGURE 5.37 AC and DC yard: the protection zones—example of HVDC "Classic."

HVDC control and protection algorithms are usually rather complex. Real power, reactive power, AC bus frequency and voltage, startup and shutdown sequences, contingency and fault-recovery sequences, remedial action schemes, modulation schemes for power oscillation and (optional) SSR damping, and loss of communication are some of the significant control parameters and conditions. Fast dynamic performance is standard. Special voltage vs. current (v/i) control characteristics are used for converters in multiterminal HVDC systems to allow safe operation even under loss of interstation communication. Furthermore, HVDC controls provide equipment and system protection, including thyristor overcurrent, thyristor overheating, and DC line and cable fault protection. Control and protection reliability are enhanced through redundant and fault-tolerant design. HVDC stations can often be operated from different control centers. An example of the typical HVDC "Classic" protection zones of one station and one pole of a long-distance transmission is given in Figure 5.37. Each protection zone is covered by at least two independent protective units—the primary protective unit and the secondary (backup) protective unit. Protection systems are separated from the control software and hardware. Some control actions are initiated by the protection scheme via fast signals to the control system.

Figure 5.38 illustrates the basic control functions of a bipolar HVDC long-distance transmission scheme. Valve control at process level is based on phase-angle synchronization for the firing control (Trigger-set) for gating of the thyristors (or other semiconductors) precisely timed with respect to the related AC phase voltages. The firing control determines the converter DC voltages and, per Ohm's law, DC currents and load flow. A stable operating point of the HVDC can be achieved by putting one of the two stations into current control and the other into voltage control. The current and voltage set points ($I_{d\,ref}$, $V_{d\,ref}$) are generated by the power controller. Additional control functions, for example, for AC voltage or frequency control (and others) can be added by software in flexible way. Usually, the sending station (rectifier) is in current control mode, and the receiving station (inverter) is in voltage control. Adaptions with modified settings are made during transient system conditions, for example, with VDCL (voltage-dependent current limits) when one of the two AC system voltages is temporarily reduced. In HVDC "Classic," the tap changer is an important function to match the optimal operation points for both rectifier and inverter station, when the AC system voltages are changing.

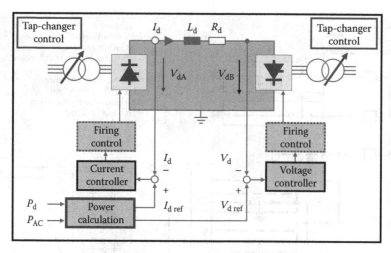

FIGURE 5.38 HVDC "Classic"—basic control functions.

FIGURE 5.39 HVDC "Classic" control hardware—SIMATIC WinCC and SIMATIC TDC (Technology and Drive Control). (Photo courtesy of Siemens AG, Erlangen, Germany.)

Figure 5.39 shows an example of HVDC "Classic" control hardware, with digital processor racks and system interfaces in cubicles. Control monitors are displaying the main parameters of voltages, currents, and active and reactive power as well—for the whole bipolar HVDC system, each station.

The control and protection schemes of FACTS stations are tailored to the related circuits and tasks. Industrial SVCs have open-loop, direct, load-compensation control. In transmission systems, FACTS controllers are designed to provide closed-loop steady-state and dynamic control of reactive power and bus voltage, as well as some degree of load-flow control, with modulation loops for stability and optional SSR mitigation. In addition, the controls include equipment and system protection functions. An example of the basic SVC control functions is given in Figure 5.40. Like in HVDC, control options can be added in a flexible way, using software functions.

With SVC and TCSC, the main control determines the effective shunt and series reactance, respectively. This fast reactance control, in turn, has the steady-state and dynamic effects listed earlier. STATCOM control is a phase-angle-based inverter AC voltage injection, with output magnitude control. The AC output is essentially in phase with the system voltage. The amplitude determines whether the STATCOM acts in a capacitive or inductive mode.

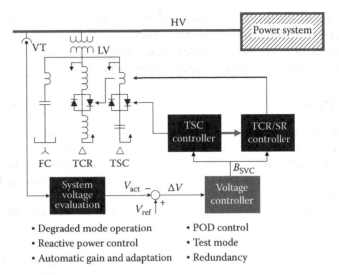

FIGURE 5.40 FACTS: SVC "Classic" controls. Upper part: main function is voltage control; lower part: control options.

Most controllers included here have the potential to provide power oscillation damping, that is, to improve system stability. By the same token, if not properly designed, they may add to or even create system undamping, especially SSR. It is imperative to include proper attention to SSR in the control design and functional testing of power electronic stations, especially in the vicinity of existing or planned large turbogenerators.

Principally, the control and protection systems described earlier comprise the following distinctive hardware and software subsystems:

- Valve firing and monitoring circuits
- Main (closed-loop) control
- Open-loop control (sequences, interlocks, etc.)
- Protective functions
- Monitoring and alarms
- Diagnostic functions
- Operator interface and communications
- Data handling

5.6 Losses and Cooling

Valve losses in high-voltage power electronic substations are comparable in magnitude to those of the associated transformers. Typical HVDC converter efficiency, including auxiliaries, smoothing reactors, AC and DC filters, transformers, and converters, exceeds 99% in each station. The total converter station losses at full load (i.e., those of both converter stations, with all components except DC line) amount to 1.3%–1.5% of the rated power only (depending on design). This means that the losses in each terminal of a 1000 MW long-distance transmission system can approach less than 10 MW. Those of a 200 MW back-to-back station (both conversions AC–DC–AC in the same station) can be less than 2 MW.

Deionized water circulated in a closed loop is generally used as primary valve coolant. Various types of dry or evaporative secondary coolers dissipate the heat, usually into the surrounding air.

Regarding transmission line losses, there are many benefits when using UHV DC at a voltage of ±800 kV: the line losses drop by approximately 60% compared to ±500 kV DC at the same power;

for 660 kV DC, the loss reduction is 43%. When comparing transmission losses of AC and DC systems, it becomes apparent that the latter typically has 30%–50% less losses.

Standard procedures to determine and evaluate high-voltage power electronic substation losses, HVDC converter station losses in particular, have been developed [65].

5.7 Civil Works

High-voltage power electronic substations are special because of the valve rooms and buildings required for converters and controls, respectively. Insulation clearance requirements can lead to very large valve rooms (halls). The valves are connected to the yard through wall bushings. Converter transformers are often placed adjacent to the valve building, with the valve-side bushings penetrating through the walls in order to save space.

The valves require controlled air temperature, humidity, and cleanness inside the valve room. Although the major part of the valve losses is handled by the valve cooling system, a fraction of the same is dissipated into the valve room and adds to its air-conditioning or ventilation load. The periodic fast switching of electronic converter and controller valves causes a wide spectrum of harmonic currents and electromagnetic fields, as well as significant audible noise. Therefore, valve rooms are usually shielded electrically with wire meshes in the walls and windows. Electric interference with radio, television, and communication systems can usually be controlled with power-line carrier filters and harmonic filters.

Sources of audible noise in a converter station include the transformers, capacitors, reactors, and coolers. To comply with the contractually specified audible noise limits within the building (e.g., in the control room) and outdoors (in the yard, at the substation fence), low-noise equipment, noise-damping walls, barriers, and special arrangement of equipment in the yard may be necessary. Examples of noise reduction have been shown in Figures 5.23 and 5.25. The theory of audible noise propagation is well understood [66], and analytical tools for audible noise design are available [67]. Specified noise limits can thus be met, but doing so may have an impact on total station layout and cost. Of course, national and local building codes also apply. In addition to the actual valve room and control building, power electronic substations typically include rooms for coolant pumps and water treatment, for auxiliary power distribution systems, air-conditioning systems, battery rooms, and communication rooms.

Extreme electric power flow densities in the valves create a certain risk of fire. Valve fires with more or less severe consequences have occurred in the past [7]. Improved designs as well as the exclusive use of flame-retardant materials in the valve, coordinated with special fire detection and protection devices, reduce this risk to a minimum [8]. The converter transformers have fire walls in between and dedicated sprinkler systems around them as effective fire-fighting equipment.

Many high-voltage power electronic stations have spare transformers to minimize interruption times following a transformer failure. This leads to specific arrangements and bus configurations or extended concrete foundations and rail systems in some HVDC converter stations.

Some HVDC schemes use outdoor valves with individual housings. They avoid the cost of large valve buildings at the expense of more complicated valve maintenance. TCSC stations also have similar valve housings on insulated platforms together with the capacitor banks and other equipment.

5.8 Reliability and Availability

Power electronic systems in substations have reached levels of reliability and availability comparable with all the other substation components. System availability is influenced by forced outages due to component failures and by scheduled outages for preventive maintenance or other purposes. By means of built-in redundancy, detailed monitoring, self-supervision of the systems, segmentation and automatic switch-over strategies, together with consistent quality control and a prudent operation and maintenance philosophy, almost any level of availability is achievable. The stations are usually designed for unmanned operation. The different subsystems are subjected to an automatic internal control routine, which logs and evaluates any deviations or abnormalities and relays them to remote control centers for eventual actions if necessary. Any guaranteed level of availability is based on built-in redundancies in key subsystem components. With redundant thyristors and IGBT modules in the valves, spare converter transformers at each station, a completely redundant control and protection system, available spare parts for other important subsystems, maintenance equipment, and trained maintenance personnel at hand, an overall availability level as high as 99% can be attained, and the average number of annual forced outages can be kept below five.

The outage time for preventive maintenance of the substation depends mainly on a utility's practices and philosophy. Most of the substation equipment, including control and protection, can be overhauled in coordination with the valve maintenance, so that no additional interruption of service is necessary. Merely a week annually is needed per converter station of an HVDC link.

Because of their enormous significance in the high-voltage power transmission field, HVDC converters enjoy the highest level of scrutiny, systematic monitoring, and standardized international reporting of reliability design and performance. CIGRE has developed a reporting system [68] and publishes biannual HVDC station reliability reports [69]. At least one publication discusses the importance of substation operation and maintenance practices on actual reliability [70]. The IEEE has issued a guide for HVDC converter reliability [71]. Other high-voltage power electronic technologies have benefited from these efforts as well. Reliability, availability, and maintainability (RAM) have become frequent terms used in major high-voltage power electronic substation specifications [72] and contracts.

High-voltage power electronic systems warrant detailed specifications to assure successful implementation. In addition to applicable industry and owner standards for conventional substations and equipment, many specific conditions and requirements need to be defined for high-voltage power electronic substations. To facilitate the introduction of advanced power electronic technologies in substations, the IEEE and IEC have developed and continue to develop applicable standard specifications [73,74].

Operation and maintenance training are important for the success of high-voltage power electronic substation projects. A substantial part of this training is best performed on site during commissioning. The IEEE and other organizations have, to a large degree, standardized high-voltage power electronic component and substation testing and commissioning procedures [75–77].

An essential part of HVDC and FACTS implementation is the design verification of the main control and protection functions under real system conditions, with physical control and protection hardware and detailed AC and DC system models, without any risk of damage. Real-time digital system simulators have therefore become a major tool for the off-site function tests of all controls, thus reducing the amount of actual on-site testing. An example of such a simulator set-up is depicted in Figure 5.41. An example of an HVDC off-site test program (termed FPT: functional performance test) is given in Table 5.1.

Nonetheless, staged fault tests are still performed with power electronic substations including, for example, with the Kayenta TCSC [78].

5.9 Outlook and Future Trends

For interconnecting asynchronous AC networks and for transmission of bulk power over long distances, HVDC systems remain economically, technically, and environmentally the preferred solution at least in the near future. One can expect continued growth of power electronics applications in transmission systems. Innovations such as the VSCs [23] or the capacitor-commutated converter [79], active filters, outdoor valves [80], or the transformerless converter [81] may reduce the complexity and size of HVDC converter stations [82]. VSCs technology combined with innovative DC cables can make converter stations economically viable also at lower power levels (up to 300 MW).

New and more economical FACTS and HVDC technologies, such as the multilevel technology, have already been introduced. Self-commutated converters and active filters changed the footprint of

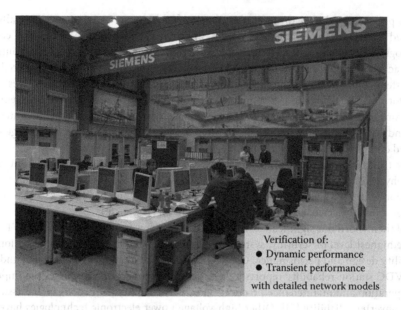

FIGURE 5.41 Example of HVDC/FACTS off-site testing in a real-time digital simulator.

TABLE 5.1 HVDC Off-Site Testing—Example of a Master Test Plan

HVDC protection
Energization of reactive power elements
Open cable test
Converter block/de-block performance
Steady-state performance
DC power ramp
Power step response
Power reversal
DC voltage step response
Current step response
Extinction angle step response
Control mode transfer (ΔV/Id/Gamma)
AC and DC fault performance
Commutation failure/misfiring
Stability functions

high-voltage power electronic substations. STATCOMs may replace rotating synchronous condensers. TCSCs, UPFC, and VSC HVDCs may replace phase-shifting transformers to some degree. New developments such as electronic transformer tap changers, semiconductor breakers, electronic fault-current limiters, and arresters may even affect the "conventional" parts of the substation. As a result, the high-voltage power electronic substations of the future will be more common, more effective, more compact, easier to relocate, and found in a wider variety of settings.

A summary of the existing AC and DC transmission technologies with overhead lines, cables, and GIL (gas-insulated lines), in comparison with HVDC, is given in Figure 5.42. For long-distance bulk power transmission, HVDC is still the best solution today, offering minimal losses. The figure includes

```
┌─────────────────────────────────────────────┐
│ Solutions with overhead lines               │
└─────────────────────────────────────────────┘
  • High-voltage DC transmission:
      ○ HVDC "Classic" with 500 kV (HV)/660 kV (EHV)—3 to 4 GW
      ○ HVDC "Bulk" with 800 kV (UHV)—5 to 7.6 GW     Option UHV DC 1100 kV: 10 GW
    For comparison: HVDC PLUS (VSC) ≤ 1100 MVA        The winner is HVDC!
  • AC transmission:
      ○ 400 kV (HV)/500 kV AC (EHV)—1.5/2 GVA
      ○ 800 kV AC (EHV)—3 GVA
      ○ 1000 kV AC (UHV)—6 to 8 GVA
```

```
┌─────────────────────────────────────────────┐
│ Solutions with DC Cables*                   │
└─────────────────────────────────────────────┘
  • 500/600 kV DC—per Cable, mass impregnated: 1 GW to 2 GW (actual-prospective)
```

```
┌─────────────────────────────────────────────┐
│ Solutions with GIL–gas insulated lines      │
└─────────────────────────────────────────────┘
  • 400 kV AC (HV)—1.8 GVA/2.3 GVA (directly buried/tunnel or outdoor)
  • 500 kV AC (EHV)—2.3 GVA/2.9 GVA (directly buried/tunnel or outdoor)
  • 550 kV AC (EHV)—Substation: standard 3.8 GVA/special 7.6 GVA**
  • 800 kV AC (EHV)—Tunnel: 5.6 GVA***
```

FIGURE 5.42 Power capacities of existing solutions for DC and AC transmission. *Note:* Power AC @1 System 3~, Power DC @ Bipole ±. * denotes distances over 80 km: AC Cables too complex. ** denotes reference: Bowmanville, Canada, 1985 - Siemens. *** denotes reference: Huanghe Laxiwa Hydropower Station, China, 2009-CGIT (USA).

the previously mentioned option for a 1100 kV UHV DC application, which is currently under discussion in China. This option offers the lowest losses and highest transmission capacity.

With high penetration of strongly fluctuating RESs, AC grids will need additional enhancements. AC and DC overlay grids are in discussion in some countries: China, India, for example, and, to some degree, also Brazil. Figure 5.43 shows an example of Germany, where the planned offshore wind farms might not reach the load centers of the country due to existing AC bottlenecks. In studies, the idea of a DC overlay network has already been developed, as depicted in the figure.

In conclusion, the features and benefits of HVDC can be summarized as follows:

- Three HVDC options are available: VSC, "Classic," and "Bulk."
- With DC, overhead line losses are typically 30%–50% less than with AC.
- For cable transmission (over about 80 km), HVDC is the only solution.
- HVDC can be integrated into the existing AC systems.
- HVDC supports AC in terms of system stability.
- System interconnection with HVDC and integration of HVDC:
  - DC is a Firewall against cascading disturbances.
  - Bidirectional control of power flow.
  - Frequency, voltage, and power oscillation damping control available.
  - Staging of the links—quite easy.
  - No increase in short-circuit power.
  - DC is a stability booster.

A combination of FACTS and classic line-commutated HVDC technology is feasible as well. In the present state-of-the-art VSC-based HVDC technologies, the FACTS function of reactive power control is already integrated, that means, additional FACTS controllers are superfluous. However, bulk power transmission up to the GW range remains still reserved for classic, line-commutated thyristor-based HVDC systems.

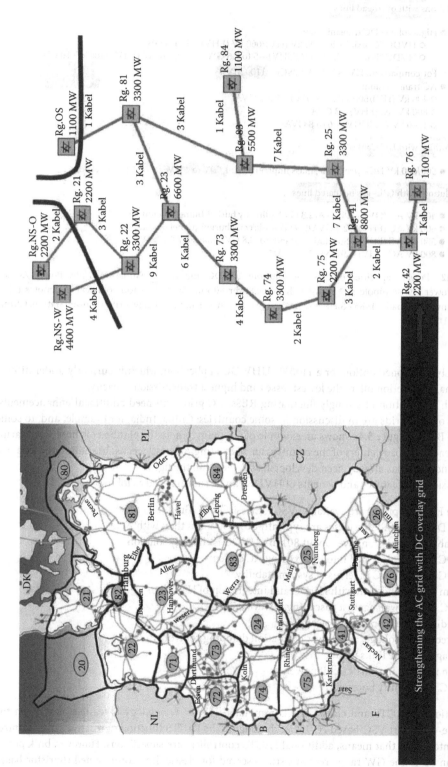

FIGURE 5.43 Studies for multiterminal—example of Germany: impact of high amounts of future wind power transfer from Northern sea regions to Southern load centers. (From DENA Grid study II, Berlin, Germany, November 2010.)

Acknowledgments

The authors would like to thank their former colleague Gerhard Juette (retired) for his valuable contributions to the former editions of this chapter. Photos included in this third edition are mostly from Siemens, in all other cases, by courtesy of the related contributors mentioned.

References

1. Economic Assessment of HVDC Links, CIGRE Brochure Nr.186 (Final Report of WG 14–20), June 2001.
2. Hingorani, N.G., Flexible AC transmission, *IEEE Spectrum*, 30, 40–45, April 1993.
3. Kirschner, L., Retzmann, D., and Thumm, G., Benefits of FACTS for power system enhancement, *IEEE/PES T & D Conference*, Dalian, China, August 14–18, 2005, p. 7.
4. Breuer, W., Povh, D., Retzmann, D., and Teltsch, E., Trends for future HVDC applications, *16th CEPSI*, Mumbai, India, November 6–10, 2006.
5. European Technology Platform SmartGrids, *Vision and Strategy for Europe's Electricity Networks of the Future*, Luxembourg, Belgium, 2006.
6. Beck, G., Povh, D., Retzmann, D., and Teltsch, E., Global blackouts: Lessons learned, *Power-Gen Europe*, Milan, Italy, June 28–30, 2005.
7. Krishnayya, P.C.S. et al., Fire aspects of HVDC Thyristor valves and valve halls, Paper presented at *CIGRE 35th International Conference on Large High Voltage Electric Systems*, Paris, France, 1994, p. 13.
8. Holweg, J., Lips, H.P., and Sachs, G., Flame resistance of HVDC thyristor valves: Design aspects and flammability tests, Paper 410-04 presented at *CIGRE Symposium*, Tokyo, Japan, 1995.
9. Liu, W.J., Breuer, W., and Retzmann, D., Prospects of HVDC and FACTS for sustainability and security of power systems, *APSCOM 2009—The 8th IET International Conference on Advances in Power System Control, Operation and Management*, Challenges and Solutions in Electricity Supply Industry for Tomorrow, Hong Kong, Japan, November 8–11, 2009, p. 13.
10. Pereira, M. and Sadek, K., Application of power active filters for damping harmonics, Paper presented at *CIGRE Study Committee 14 International Colloquium on HVDC and FACTS*, Montreal, Quebec, Canada, 1995.
11. Andersen, N., Gunnarsson, S., Pereira, M., Fritz, P., Damstra, G.C., Enslin, J.H.R., and O'Lunelli, D., Active filters in HVDC applications, Paper presented at *CIGRE International Colloquium on HVDC and FACTS*, Johannesburg, South Africa, 1997.
12. CIGRE Working Group 14.28, Active dc filters in HVDC applications, *Electra*, 187, 1999.
13. Westerweller, T., Pereira, M., Huang, H., and Wild, G., Performance calculations and operating results of active harmonic filters for HVDC transmission systems, Paper presented at *IEEE-PES Summer Meeting*, Vancouver, British Columbia, Canada, 2001.
14. Sørensen, P.L., Pereira, M., Sadek, K., and Wild, G., Active AC filters for HVDC applications, Paper presented at *CIGRE International Conference on Power Systems (ICPS)*, Wuhan, China, 2001, p. 5.
15. Torgerson, D.R., Rietman, T.R., Edris, A., Tang, L., Wong, W., Mathews, H., and Imece, A.F., A transmission application of back-to-back connected voltage source converters, Paper presented at *EPRI Conference on the Future of Power Delivery in the 21st Century*, La Jolla, CA, 1997.
16. Asplund, G., Eriksson, K., Svensson, K., Jiang, H., Lindberg, J., and Palsson, R., DC transmission based on voltage source converters, Paper 14-302, CIGRE, Paris, France, 1998.
17. Electric Power Research Institute and Western Area Power Administration, Modeling development of converter topologies and control for BTB voltage source converters, TR-111182, EPRI, Palo Alto, CA, and Western Area Power Administration, Golden, CO, 1998.
18. Schauder, C.D. et al., TVA STATCON installation, Paper presented at *CIGRE Study Committee 14 International Colloquium on HVDC and FACTS*, Montreal, Quebec, Canada, 1995.

19. Schauder, C.D. et al., Operation of ±100 MVAr TVA STATCON, *IEEE Transactions on Power Delivery*, 12, 1805–1811, 1997.
20. Mehraban, A.S. et al., Application of the world's first UPFC on the AEP system, Paper presented at *EPRI Conference on the Future of power Delivery*, Washington, DC, 1996.
21. Arabi, S., Hamadanizadeh, H., and Fardanesh, B., Convertible static compensator performance studies on the NY state transmission system, *IEEE Transactions on Power Systems*, 17, July 2002.
22. Knight, R.C., Young, D.J., and Trainer, D.R., Relocatable GTO-based static VAr compensator for NGC substations, Paper 14–106 presented at *CIGRE Session*, Paris, 1998.
23. Zhao, Z. and Iravani, M.R., Application of GTO voltage source inverter in a hybrid HVDC link, *IEEE Transactions on Power Delivery*, 9, 369–377, 1994.
24. Schettler, F., Huang, H., and Christl, N., HVDC transmission systems using voltage-sourced converters—Design and applications, *IEEE Power Engineering Society Summer Meeting*, Vancouver, British Columbia, Canada, July 2000.
25. Hingorani, N., *IEEE PES Working Group I8 on Power Electronics Building Block Concepts—High Voltage Power Electronics Stations Subcommittee (I0) of the Substations Committee.*
26. Marquardt, R. and Lesnicar, A., New concept for high voltage-modular multilevel converter, *PESC Conference*, Aachen, Germany, 2004, p. 5.
27. Bernet, S., Meynard, T., Jakob, R., Brückner, T., and McGrath, B., Tutorial multi-level converters, *Proceedings of the IEEE-PESC Tutorials*, Aachen, Germany, 2004, p. 16.
28. VSC Transmission, CIGRE Brochure No. 269, CIGRE Working Group B4.37, April 2005.
29. Pérez de Andrés, J.M., Dorn, J., Retzmann, D., Soerangr, D., and Zenkner, A., Prospects of VSC converters for transmission system enhancement, *PowerGrid Europe*, Madrid, Spain, June 26–28, 2007.
30. Dorn, J., Huang, H., and Retzmann, D. Novel voltage-sourced converters for HVDC and FACTS applications, *CIGRÉ Symposium*, Osaka, Japan, November 1–4, 2007.
31. Gemmell, B., Dorn, J., Retzmann, D., and Soerangr, D., Prospects of multilevel VSC technologies for power transmission, *IEEE PES Transmission and Distribution Conference & Exposition*, Chicago, IL, April 21–24, 2008, p. 16.
32. Povh, D. and Retzmann, D., *Part 1 HVDC 'Classic & Bulk'*; Kreusel, J., *Part 2 VSC HVDC: Integrated AC/DC Transmission Systems—Benefits of Power Electronics for Security and Sustainability of Power Supply*; PSCC, Survey Session 2, Glasgow, U.K., July 14–18, 2008.
33. Dorn, J., Huang, H., and Retzmann, D., *A New Multilevel Voltage-Sourced Converter Topology for HVDC Applications*, CIGRÉ, Paris, France, August 24–29, 2008.
34. Retzmann, D., *Part 1 Modular Multilevel Converter—Technology & Principles—Part 2 HVDC/FACTS Using VSC—Applications & Prospects*, Cigré-Brazil B4 "Tutorial on VSC in Transmission Systems—HVDC & FACTS, Rio de Janeiro, Brazil," October 6–7, 2009.
35. Gambach, H. and Retzmann, D., PEBB concepts—From medium voltage drives to high voltage applications, *IEEE - ISIE*, Bari, Italy, July 2010.
36. Lemes, M., Retzmann, D., Schultze, A., and Uecker, K., Challenges and solutions for grid access of wind power—The Smart Grid, *International CIGRE Symposium on Assessing and Improving Power System Security, Reliability and Performance in Light of Changing Energy Sources*, Recife, Pernambuco, Brazil, April 3–6, 2011.
37. Claus, M., Mcdonald, S., Cahill, P., Retzmann, D., Pereira, M., and Uecker, K., Innovative VSC technology for integration of "green energy"—Without impact on system protection and power quality, *CIRED 21st International Conference on Electricity Distribution*, Frankfurt, Germany, June 6–9, 2011, p. 4.
38. Institute of Electrical and Electronics Engineers, *Guide for Planning DC Links Terminating at AC System Locations Having Low Short-Circuit Capacities, Parts I and II*, IEEE Std. 1204–1997, IEEE, Piscataway, NJ, 1997.
39. deLaneuville, H., Haidle, L., McKenna, S., Sanders, S., Torgerson, D., Klenk, E., Povh, D., Flairty, C.W., and Piwko, R., Miles City Converter Station and Virginia Smith Converter Station Operating Experiences, Paper presented at *IEEE Summer Power Meeting*, San Diego, CA, 1991.

40. CIGRE Working Group 33–05, Application guide for insulation coordination and arrester protection of HVDC converter stations, *Electra*, 96, 101–156, 1984.
41. *Impacts of HVDC Lines on The Economics of HVDC Projects*, CIGRE Brochure No. 366, Final Report of Joint Working Group B2/B4/C1.17, August 2009.
42. Povh, D., High voltage direct current (HVDC) transmission for DC voltages above 100 kV, IEC TC 115.
43. Schneider, H.M. and Lux, A.E., Mechanism of HVDC wall bushing flashover in nonuniform rain, *IEEE Transactions on Power Delivery*, 6, 448–455, 1991.
44. Porrino, A. et al., Flashover in HVDC bushings under nonuniform rain, Paper presented at *9th International Symposium on High Voltage Engineering*, Graz, Austria, 1995, Vol. 3, p. 3204.
45. HVDC Converter Stations for Voltages above ±600 kV, CIGRÉ Working Group 14.32, December 2002.
46. Åström, U., Weimers, L., Lescale, V., and Asplund, G., Power transmission with HVDC at voltages above 600 kV, *IEEE/PES Transmission and Distribution Conference & Exhibition: Asia and Pacific*, Dalian, China, August 14–18, 2005, p. 6.
47. Åström, U., Lescale, V., and Gao, L., Status and special aspects of Xiangjiaba-Shanghai 800 kV UHVDC project, *International Conference on UHV Power Transmission*, Beijing, China, May 2009, p. 6.
48. Technological assessment of 800 kV HVDC applications, CIGRÉ Brochure No. 417, CIGRE Working Group B4.45, June 2010.
49. Lemes, M., Retzmann, D., Soerangr, D., and Uecker, K., Security and sustainability with UHV DC super grid solutions—World's first project in operation, *International CIGRE Symposium on Assessing and Improving Power System Security, Reliability and Performance in Light of Changing Energy Sources*, Recife, Pernambuco, Brazil, April 3–6, 2011.
50. Tykeson, K., Nyman, A., and Carlsson, H., Environmental and geographical aspects in HVDC electrode design, *IEEE Transactions on Power Delivery*, 11, 1948–1954, 1996.
51. Holt, R.J., Dabkowski, J., and Hauth, R.L., HVDC power transmission electrode siting and design, ORNL/Sub/95-SR893/3, Oak Ridge National Laboratory, Oak Ridge, TN, 1997.
52. Bachmann, B., Mauthe, G., Ruoss, E., Lips, H.P., Porter, J., and Vithayathil, J., Development of a 500 kV airblast HVDC circuit breaker, *IEEE Transactions on Power Apparatus and Systems*, Vol. PAS-104, No. 9, September 1985.
53. Vithayathil, J.J., Courts, A.L., Peterson, W.G., Hingorani, N.G., Nilson, S., and Porter, J.W., HVDC circuit breaker development and field tests, *IEEE Transactions on Power Apparatus and Systems*, 104, 2693–2705, 1985.
54. Edris, A., Zelingher, S., Gyugyi, L., and Kovalsky, L., Squeezing more power from the grid, *IEEE Power Engineering Review*, 22, June 2002.
55. Breuer, W., Retzmann, D., and Uecker, K., Highly efficient solutions for smart and bulk power transmission of green energy, *21st World Energy Congress*, Montreal, Quebec, Canada, September 12–16, 2010.
56. Aggarwal, R.K., Mukherjee, A., and Retzmann, D., Improved power system performance by co-ordinated GMPC control for parallel AC/DC operation, *International Conference on Power Generation, System Planning and Operation*, Indian Institute of Technology, New Delhi, India, December 12–13, 1997, p. 11.
57. Mukherjee, A., Nietsch, C., Povh, D., Retzmann, D., and Weinhold, M., Advanced technologies for power transmission and distribution in South Asia—Present state and future trends, *South Asia Power Conference & Exhibition*, Calcutta, India, November 20–22, 1996.
58. Haeusler, M., Huang, H., and Papp, K., Design and testing of 800 kV HVDC equipment, *Proceedings of CIGRE Conference*, Paris, France, 2008, p. 8.
59. Braun, K., Karlecik-Maier, F., Retzmann, D.W., Hormozi, F.J., and Piwko, R.J., Advanced AC/DC real-time and computer simulation for the Mead-Adelanto Static VAR compensators, *11th Conference on Electric Power Supply Industry CEPSI*, Kuala Lumpur, Malaysia, October 21–25, 1996, p. 10.

60. Piwko, R.J. et al., The slatt thyristor-controlled series capacitor, Paper 14–104 presented at *CIGRE 35th International Conference on Large High Voltage Electric Systems*, Paris, France, 1994, p. 5.
61. Montoya, A.H., Torgerson, D.A., Vossler, B.A., Feldmann, W., Juette, G., Sadek, K., and Schultz, A., 230 kV advanced series compensation Kayenta substation (Arizona), project overview, Paper presented at *EPRI FACTS Workshop*, Cincinnati, OH, 1990, p. 5.
62. Renz, K.W. and Tyll, H.K., Design aspects of relocatable SVCs, Paper presented at *VIII National Power Systems Conference*, New Delhi, India, 1994, p. 5.
63. Koch, H. and Retzmann, D., Connecting large offshore wind farms to the transmission network, *IEEE PES T&D Conference*, New Orleans, LA, April 19–22, 2010, p. 5.
64. Fardanesh, B., Henderson, M., Shperling, B., Zelingher, S., Gyugi, L., Schauder, C., Lam, B., Mountford, J., Adapa, R., and Edris, A., *Feasibility Studies for Application of a FACTS Device on the New York State Transmission System*, CIGRE, Paris, France, 1998.
65. International Electrotechnical Commission, *Determination of Power Losses in HVDC Converter Stations*, IEC Std. 61803.
66. Beranek, L.L., *Noise and Vibration Control*, rev. edn., McGraw Hill, New York, 1971, Institute of Noise Control Engineering, 1988.
67. Smede, J., Johansson, C.G., Winroth, O., and Schutt, H.P., Design of HVDC converter stations with respect to audible noise requirements, *IEEE Transactions on Power Delivery*, 10, 747–758, 1995.
68. CIGRE Working Group 04, Protocol for reporting the operational performance of HVDC transmission systems, CIGRE Report WG 04, presented in Paris, France, 2004.
69. Christofersen, D.J., Vancers, I., Elahi, H., and Bennett, M.G., A survey of the reliability of HVDC systems throughout the world, presented at *CIGRE Session*, Paris, France, 2004.
70. Cochrane, J.J., Emerson, M.P., Donahue, J.A., and Wolf, G., A survey of HVDC operating and maintenance practices and their impact on reliability and performance, *IEEE Transactions on Power Delivery*, 11(1), 1995.
71. Institute of Electrical and Electronics Engineers, *Guide for the Evaluation of the Reliability of HVDC Converter Stations*, IEEE Std. 1240–2000, IEEE, Piscataway, NJ, 2000.
72. Vancers, I. et al., A summary of North American HVDC converter station reliability specifications, *IEEE Transactions on Power Delivery*, 8, 1114–1122, 1993.
73. Canelhas, A., Peixoto, C.A.O., and Porangaba, H.D., Converter station specification considering state of the art technology, Paper presented at *IEEE/Royal Institute of Technology Stockholm Power Tech: Power Electronics Conference*, Stockholm, Sweden, 1995, p. 6.
74. Institute of Electrical and Electronics Engineers, *Guide for a Detailed Functional Specification and Application of Static VAr Compensators*, IEEE Std. 1031–2000, IEEE, Piscataway, NJ, 2000.
75. Institute of Electrical and Electronics Engineers, *Recommended Practice for Test Procedures for High-Voltage Direct Current Thyristor Valves*, IEEE Std. 857–1996, IEEE, Piscataway, NJ, 2000.
76. Institute of Electrical and Electronics Engineers, *Guide for Commissioning High-Voltage Direct Current Converter Stations and Associated Transmission Systems*, IEEE Std. 1378–1997, IEEE, Piscataway, NJ, 1997.
77. Institute of Electrical and Electronics Engineers, *Guide for Static VAr Compensator Field Tests*, IEEE Std. 1303–2000, IEEE, Piscataway, NJ, 2000.
78. Weiss, S. et al., Kayenta staged fault tests, Paper presented at *EPRI Conference on the Future of Power Delivery*, Washington, DC, 1996, p. 6.
79. Bjorklund, P.E. and Jonsson, T., Capacitor commutated converters for HVDC, Paper presented at *IEEE/Royal Institute of Technology Stockholm Power Tech: Power Electronics Conference*, Stockholm, Sweden, 1995, p. 6.
80. Asplund, G. et al., Outdoor thyristor valve for HVDC, *Proceedings of the IEEE/Royal Institute of Technology Stockholm Power Tech; Power Electronics*, Stockholm, Sweden, June 18–22, 1995, p. 6.
81. Vithayathil, J.J. et al., DC systems with transformerless converters, *IEEE Transactions on Power Delivery*, 10, 1497–1504, 1995.

82. Carlsson, L. et al., Present trends in HVDC converter station design, Paper presented at *Fourth Symposium of Specialists in Electrical Operation and Expansion Planning (IV SEPOPE)*, Curitiba, Brazil, 1994.
83. Balbierer, S., Haeusler, M., Ramaswami, V., Hong, R., Chun, Sh., and Tao, Sh., Basic design aspects of the 800 kV UHVDC project Yunnan-Guangdong, *Sixth International Conference of Power Transmission and Distribution Technology*, Guangzhou, China, 2007, p. 6.
84. Kumar, A., Wu, D., and Hartings, R., Experience from first 800 kV HVDC test installation, *International Conference on Power Systems (ICPS-2007)*, Bangalore, India, December 12–14, 2007, p. 5.
85. Claus, M., Retzmann, D., Sörangr, D., and Uecker, K., Solutions for smart and super grids with HVDC and FACTS, *17th CEPSI*, Macau, SAR of China, October 27–31, 2008, p. 9.
86. Zhang, D., Haeusler, M., Rao, H., Shang, Ch., and Shang, T., Converter station design of the ±800 kV UHVDC Project Yunnan-Guangdong, *17th CEPSI*, Macau, China, October 27–31, 2008, p. 5.
87. Nayak, R.N., Sehgal, Y.K., and Sen, S., Planning and design studies for +800 kV, 6000 MW HVDC system, *CIGRÉ*, Article B4-117, 2008, p. 11.

6
Interface between Automation and the Substation

	6.1	Physical Challenges **6**-1
		Components of a Substation Automation System • Locating Interfaces • Environment • Electrical Environment
	6.2	Measurements **6**-5
		What Measurements Are Needed • Performance Requirements • Characteristics of Digitized Measurements • Instrument Transformers • New Measuring Technology • Substation Wiring Practices • Measuring Devices • Scaling Measured Values • Integrated Energy Measurements: Pulse Accumulators
	6.3	State (Status) Monitoring.. **6**-17
		Contact Performance • Ambiguity • Wetting Sources • Wiring Practices
	6.4	Control Functions.. **6**-19
		Interposing Relays • Control Circuit Designs • Latching Devices • Intelligent Electronic Devices for Control
	6.5	Communication Networks inside the Substation **6**-22
		Point-to-Point Networks • Point-to-Multipoint Networks • Peer-to-Peer Networks • Optical Fiber Systems • Communications between Facilities • Communication Network Reliability • Assessing Channel Capacity
	6.6	Testing Automation Systems..**6**-26
		Test Facilities • Commissioning Test Plan • In-Service Testing
James W. Evans | 6.7 | Summary... **6**-29 |
The St. Claire Group, LLC | | References..**6**-29 |

An electric utility substation automation (SA) system depends on the interface between the substation and its associated equipment to provide and maintain the high level of confidence demanded for power system operation and control. It must also serve the needs of other corporate users to a level that justifies its existence. This chapter describes typical functions provided in utility SA systems and some important aspects of the interface between substation equipment and the automation system components.

6.1 Physical Challenges

6.1.1 Components of a Substation Automation System

The electric utility SA system uses any number of devices integrated into a functional array by a communication technology for the purpose of monitoring, controlling, and configuring the substation.

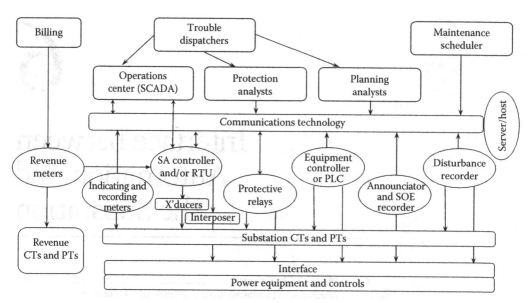

FIGURE 6.1 Power station SA system functional diagram.

SA systems incorporate microprocessor-based intelligent electronic devices (IEDs), which provide inputs and outputs to the system while performing some primary control or processing service. Common IEDs are protective relays, load survey and/or operator indicating meters, revenue meters, programmable logic controllers (PLCs), and power equipment controllers of various descriptions. Other devices may also be present, dedicated to specific functions for the SA system. These may include transducers, position sensors, and clusters of interposing relays. Dedicated devices often use a controller (SA controller) or interface equipment such as a conventional remote terminal unit (RTU) as a means to connect into the SA system. The SA system typically has one or more communication connections to the outside world.

Common communication connections include utility operations centers, maintenance offices, and/or engineering centers. A substation display or users station, connected to or part of a substation host computer, may also be present. Most SA systems connect to a traditional supervisory control and data acquisition (SCADA) system master station serving the real-time needs for operating the utility network from one or more operations center. SA systems may also incorporate a variation of SCADA remote terminal unit (RTU) for this purpose or the RTU function may appear in an SA controller or substation host computer. Other utility users usually connect to the system through a bridge, gateway, or network processor. The components described here are illustrated in Figure 6.1.

6.1.2 Locating Interfaces

The SA system interfaces to control station equipment through interposing relays and to measuring circuits through meters, protective relays, transducers, and other measuring devices as indicated in Figure 6.1. These interfaces may be associated with, or integral to, an IED or dedicated interface devices for a specific automation purpose. The interfaces may be distributed throughout the station or centralized within one or two cabinets. Finding space to locate the interfaces can be a challenge depending on available panel space and layout of station control centers. Small substations can be more challenging than large ones. Choices for locating interfaces also depend on how much of the substation will be modified for automation and the budget allocated for modification. Individual utilities rely on engineering and economic judgment for guidance in selecting a design.

The centralized interface simplifies installing an SA system in an existing substation since the placement of the interface equipment affects only one or two panels housing the new SA controller, substation host, human machine interface (HMI), discrete interface, and new IED equipment. However, cabling will be required from each controlled and monitored equipment panel, which meets station panel wiring standards for insulation, separations, conductor sizing, and interconnection termination. Centralizing the SA system–station equipment interface has the potential to adversely affect the security of the station as many control and instrument transformer circuits become concentrated in a single panel or cabinet and can be seriously compromised by fire and invite mishaps from human error. This practice has been widely used for installing earlier SCADA systems where all the interfaces are centered around the SCADA RTU and often drives the configuration of an upgrade from SCADA to automation.

Placing the interface equipment on each monitored or controlled panel is much less compromising but may be more costly and difficult to design. Each interface placement must be individually located, and more panels are affected. If a low-energy interface (less than 50 V) is used, a substantial savings in cable cost may be realized since interconnections between the SA controller and the interface devices may be made with less expensive cable and hardware. Low-energy interconnections can lessen the impact on the cabling system of the substation, reducing the likelihood that additional cable trays, wireways, and ducts will be needed.

The distributed approach is more logical when the SA system incorporates protective relay IEDs, panel-mounted indicating meters, or control function PLCs. Protection engineers usually insist on separating protection devices into logical groups based on substation configuration for security. Similar concerns often dictate the placement of indicating meters and PLCs. Many utilities have abandoned the "time-honored" operator "bench board" in substations in favor of distributing the operator control and indication hardware throughout the substation. The interface to the SA system becomes that of the IED on the substation side and a communication channel on the SA side. Depending on the communications capability of the IEDs, the SA interface can be as simple as a shielded, twisted pair cable routed between IEDs and the SA controllers. The communication interface can also be complex where multiple short haul RS-232 connections to a communication controller are required. "High-end" IEDs often have Ethernet network capability that will integrate with complex networks. These pathways may also utilize optical fiber systems and unshielded twisted pair (UTP) Ethernet cabling or even coaxial cable or some combination thereof.

As the cabling distances within the substation increase, system installation costs increase, particularly if additional cable trays, conduit, or ducts are required. Using SA communication technology and IEDs can often reduce interconnection cost. Distributing multiple, small, SA "hubs" throughout the substation can reduce cabling to that needed for a communication link to the SA controller. Likewise, these hubs can be electrically isolated using fiber–optic (F/O) technology for improved security and reliability. More complex SA systems use multiple communication systems to maintain availability should a channel be compromised.

External influences may dictate design and construction choices. These could be operating practices or even accounting rules. For example, in some regulatory jurisdictions in order for an addition or upgrade to substation to be considered "capital improvement" such that it can be added to the utility rate base, the project must include wholesale replacement of a unit of property such as a control panel. The accounting practice makes upgrading a component of a panel or an addition to a panel an "Expense" item rather than a "Capital" item; hence, the improvement cannot be added to the rate base and is paid for out of earnings. With this practice in place the only reasonable design incorporates wholesale replacements and precludes partial upgrades.

6.1.3 Environment

The environment of a substation is another challenge for SA equipment. Substation control buildings are seldom heated or air-conditioned. Ambient temperatures may range from well below freezing to above 100°F (40°C). Metal clad switchgear substations can reach ambient temperatures in excess of 140°F (50°C) even in temperate climates. Temperature changes stress the stability of measuring components in IEDs, RTUs, and transducers. Good temperature stability is important in SA system equipment and needs to be defined in the equipment purchase specifications. IEEE Standard 1613 defines environmental requirements for SA system communication components. Designers of SA systems for substations need to pay careful attention to the temperature specifications of the equipment selected for SA. In many environments, self-contained heating or air-conditioning is advisable.

When equipment is installed in outdoor enclosures, not only is the temperature cycling problem aggravated, but also moisture from precipitation and condensation becomes troublesome. Outdoor enclosures usually need heaters to control their temperature to prevent condensation. The placement of heaters should be reviewed carefully when designing an enclosure, as they can aggravate temperature stability and even create hot spots within the cabinet that can damage components and shorten life span. Heaters near the power batteries help improve low-temperature performance but adversely affect battery life span at high ambient temperatures. Obviously, keeping incident precipitation out of the enclosure is very important. Drip shields and gutters around the door seals will reduce moisture penetration. Venting the cabinet helps limit the possible buildup of explosive gases from battery charging but may pose a problem with the admittance of moisture. Incident solar radiation shields may also be required to keep enclosure temperature manageable. Specifications that identify the need for wide temperature range components, coated circuit boards, and corrosion-resistant hardware are part of specifying and selecting SA equipment for outdoor installation.

Environmental factors also include airborne contamination from dust, dirt, and corrosive atmospheres found at some substation sites. Special noncorrosive cabinets and air filters may be required for protection against the elements. Insects and wildlife also need to be kept out of equipment cabinets. In some regions, seismic requirements are important enough to be given special consideration.

6.1.4 Electrical Environment

The electrical environment of a substation is severe. High levels of electrical noise and transients are generated by the operation of power equipment and their controls. Operating high-voltage disconnect switches can generate transients that couple onto station current, potential, and control wiring entering or leaving the switchyard and get distributed throughout the facility. Operating station controls for circuit breakers, capacitors, and tap changers can also generate transients that can be found throughout the station on battery power and station service wiring. Extra high voltage (EHV) stations also have high electrostatic field intensities that couple to station wiring. Finally, ground rise during faults or switching can damage electronic equipment in stations. IEEE Standard 1613-2003 defines testing for SA system components for electrical environment. IEEE Standard C37.90-2005, IEEE Standard C37.90.1-2002, and IEEE Standard C37.90.2-2004 and IEEE Standard C37.90.3-2001 define electrical environmental testing requirements for protective devices.

Effective grounding is critical to controlling the effects of substation electrical noise on electronic devices. IEDs need a solid ground system to make their internal suppression effective. Ground systems should be radial to a single point with signal and protective grounds separated. Signal grounds require large conductors for "surge" grounds. They must be as short as possible and establish a single ground point for logical groupings of equipment. These measures help to suppress the introduction of noise and transients into measuring circuits. A discussion of this topic is usually found in the IED manufacturer's installation instructions and their advice should be heeded.

The effects of electrical noise can be controlled with surge suppression, shielded and twisted pair cabling, as well as careful cable separation practices. Surges can be suppressed with capacitors, metal oxide varistors (MOVs), and semiconducting over voltage "Transorbs" applied to substation instrument transformer and control wiring. IEDs qualified under IEEE C37.90-2005 include surge suppression within the device to maintain a "surge fence." However, surge suppression can create reliability problems as well. Surge suppressors must have sufficient energy-absorbing capacity and be coordinated so that all suppressors clamp around the same voltage. Otherwise, the lowest dissipation, lowest voltage suppressor will become sacrificial. Multiple failures of transient suppressors can short circuit important station signals to ground, leading to blown voltage transformer (VT) fuses, shorted current transformers (CTs), and shorted control wiring; even false tripping. At a minimum, as the lowest energy, lowest clamp voltage devices fail; the effectiveness of the suppression plan degrades, making the devices suspect damage and misoperation.

While every installation has a unique noise environment, some testing can help prevent noise problems from becoming unmanageable. IEEE surge withstand capability test IEEE Standard C37.90-2005 for protective devices and IEEE Standard 1613-2003 for automation devices address the transients generated by operating high-voltage disconnect switches and the operation of electromechanical control devices. These tests can be applied to devices in a laboratory or on the factory floor. They should be included when specifying station interface equipment. Insulation resistance and high potential test are also sometimes useful and are standard requirements for substation devices for many utilities.

6.2 Measurements

Electric utility SA systems gather power system performance parameters (i.e., volts, amperes, watts, and vars) for system generators, transmission lines, transformer banks, station buses, and distribution feeders. Energy output and usage measurements (i.e., kilowatt-hours and kilovar-hours) are also important for the exchange of financial transactions. Other measurements such as transformer temperatures, insulating gas pressures, fuel tank levels for on-site generation, or head level for hydro generators might also be included in the system's suite of measurements. Often, transformer tap positions, regulator positions, or other multiple position measurements are also handled as if they were measurements. These values enter the SA system through IEDs, transducers, and sensors of many descriptions. The requirements for measurements are best defined by the users of the measurements. Consider reviewing IEEE Standard C37.1-2007 Clause 5 for insight on defining measurement requirements.

IEDs, meters, and transducers measure electrical parameters (watts, vars, volts, amps) with instrument transformers shown in Figure 6.2. They convert instrument transformer outputs to digitized values for a communication method or DC voltages or currents that can be readily digitized by a traditional SCADA RTU or SA controller.

Whether a measurement is derived from the direct digital conversion of AC input signals by an IED or from a transducer analog with an external digitizing process, the results are functionally the same. However, IEDs that perform signal processing and digital conversion directly as part of their primary function use supplementary algorithms to process measurements. Transducers use analog signal processing technology to reach their results. IEDs use a communication channel for passing digitized data to the SA controller instead of conventional analog signals and an external device to digitize them.

6.2.1 What Measurements Are Needed

The suite of measurements included in the SA system serve many users with differing requirements. It is important to assess those requirements when specifying measuring devices and designing their placement, as all IED measurements are not functionally equal and may not serve specific users. For example, it is improbable that measurements made by a protective relay will serve the needs for measuring energy interchange accounting unless the device has been qualified under the requisite revenue measuring standards. Other examples where measurement performance differences are important might not be as obvious but can have significant impact on the usability of the data collected and the results from including that data in a process.

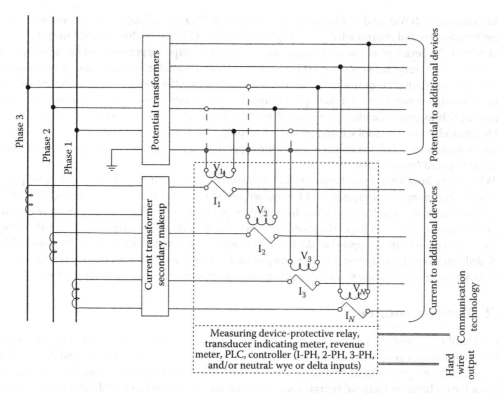

FIGURE 6.2 SA system electrical measuring interface.

Another case in point, planners prefer measurements that are averaged with an algorithm that mimics the heating of conductors, not instantaneous "snapshot" or averaged instantaneous values.

Likewise, the placement of sensors for the IED's primary function may not be the correct location for the measurement required. For example, the measurements made by a recloser control made on the secondary side of a power transformer or at a feeder will not suffice when the required measurement should be made at the primary of the transformer. Notably, the recloser measurement includes both real and reactive power losses of the transformer that would not be present in a measurement made on the primary. Voltage sensors can be on the adjacent bus separated from the measurement current sensors by a section breaker or reactor and will give erroneous measurement. There are many subtleties to sensor placement for measurements as well as their connected relationship.

The task of defining the requirements for measurements belongs to the user of the measurement. System users need to specify the specific set of measurements they require. Along with those measurements they should supply the details of where those measurements must be made within the electrical network. They need to specify the accuracy requirements and the applicable standards that must be applied to those measurements. Users must also define the performance parameters such as latency and refresh rates that are discussed in more detail in the following sections. The system designer needs to have these requirements in hand before rendering the system design. Without the specifics the designer must guess at what will be sufficient. Unfortunately, many systems are constructed without this step taking place and the results bring dissatisfaction to the user, discredit to the designer, and added cost to correct problems.

6.2.2 Performance Requirements

In the planning stages of an SA system, the economic value of the data to be acquired needs to be weighed against the cost to measure it. A balance must be struck to achieve the data quality required to suit the users and functions of the system. This affects the conceptual design of the measuring interface and provides input to the performance specifications for IEDs and transducers as well as the measuring practices applied. This step is important. Specifying a higher performance measuring system than required raises the overall system cost. Conversely, constructing a low-performance system adds costs when the measuring system must be upgraded. The tendency to select specific IEDs for the measuring system without accessing the actual measuring technology can lead to disappointing performance.

The electrical relationship between measurements and the placement of available instrument transformer sources deserves careful attention to insure satisfactory performance. Many design compromises can be made when installing SA monitoring in an existing power station because of the availability of measuring sensors. This is especially true when using protective relays as load-monitoring data sources (IEDs). Protection engineers often ignore current omissions or contributions at a measuring point, as they may not materially affect fault measurements. These variances are often intolerable for power flow measurements. Measuring source placement may also result in measurements that include or exclude reactive contributions of a series or shunt reactor or capacitor. Measurements could also include unwanted reactive component contributions of a transformer bank. Measurements might also become erroneous when a section breaker is open if the potential source is on an adjacent bus. Power system charging current and unbalances also influence measurement accuracy, especially at low load levels. Some placement issues are illustrated in Figure 6.3. The compromises are endless and each produces an unusual operating condition in some state. When deficiencies are recognized, the changes to correct them can be very costly, especially, if instrument transformers must be installed, moved, or replaced to correct the problem.

The overall accuracy of measured measurements is affected by a number of factors. These include instrument transformer errors, IED or transducer performance, and analog-to-digital (A/D) conversion. Accuracy is not predictable based solely on the IED, transducer, or A/D converter specifications.

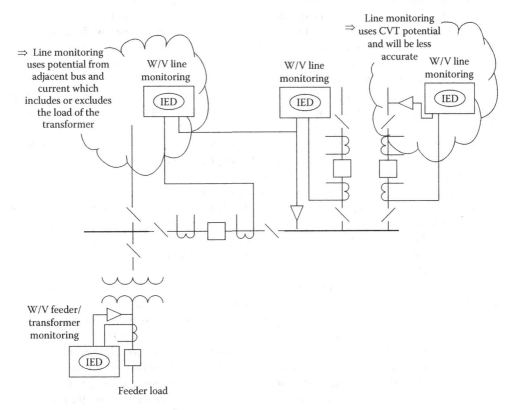

FIGURE 6.3 Measurement sensor placement.

Significant measuring errors often result from instrument transformer performance and errors induced in the scaling and digitizing process. IEEE Standard C57.13 describes instrument transformer specifications.

Revenue metering accuracy is usually required for monitoring power interchange at interconnection points and where the measurements feed economic area interchange and dispatch systems. High accuracy, revenue metering grade, instrument transformers, and 0.25% accuracy class IEDs or transducers can produce consistent real power measurements with accuracy of 1% or better at 0.5–1.0 power factor and reactive power measurements with accuracy of 1% or better at 0–0.5 power factors. Note that real power measurements at low power factors and reactive power measurements at high power factors are difficult to make accurately.

When an SA system provides information for internal power flow telemetering, revenue grade instrument transformers are not usually available. SA IEDs and transducers must often share lesser accuracy instrument transformers provided for protective relaying or load monitoring. Overall accuracy under these conditions can easily decrease to 2%–3% for real power, voltage, and current measurements and 5% or greater for reactive power.

6.2.3 Characteristics of Digitized Measurements

The processing of analog AC voltages and currents into digitized measurements for an SA system adds some significant characteristics to the result. Processing analog DC signals into digitized measurements adds many of the same characteristics. As measurements pass through the SA system, more characteristics are added that can have a significant impact on their end use.

In the analog environment, a signal may have any value within its range. In the digital environment, signals may have only discrete values within their range. The set of values is imparted by the A/D conversion process. The increments within the digital value set are determined by the minimum resolution of the A/D converter and number of states into which it can resolve. For discussion, consider an A/D converter whose minimum resolution is 1.0 mV and whose range is 4095 increments (4.095 V). The increment figure is usually expressed in the converter's basic binary format, in this case, 12 bits. In order to perform its conversion, each input it converts must be scaled so that the overall range of the input falls within the range of 0–4.095 V. If the input can assume values that are both positive and negative, then the converter range is split by offsetting the converter range by one-half (2.047 V) giving the effective range of positive and negative 2.047 V. The minimum resolution of a measurement processed by this converter is then the full scale of the input divided by the number of states, 4095 for unipolar and 2047 for bipolar. For example, if a bus voltage of 13,200 V were to be converted, the full-scale range of 15,000 would be a reasonable choice. The minimum resolution of this measurement is then 15,000/4,095 or 3.66 V. A display that showed values to the nearest volt, or less, is thus misleading and inappropriate since the value displayed cannot be resolved to 1.0 V or less but only to 3.66 V. If the measurements in this example are assumed to be bipolar, as would be direct input AC signals, then the minimum resolution for the voltage measurement is 7.32 V. A better choice for displaying this value would be to truncate the last two digits on the display making the value appear as XY.ZW kV.

The minimum resolution of any digitized measurement significantly impacts usability of that measurement in any process or calculation in which it appears. The minimum resolution can be improved by adding resolution to the A/D converter, such as using a 16 bit converter that has 65,535 states. In the aforementioned example, the minimum resolution becomes 15,000/32,768 or 0.46 V. The higher resolution converter is more expensive and may or may not be economically required. Minimum resolution also affects the dynamic range of a measurement. For the 12 bit converter that range is approximately 2,000/1 and 32,000/1 for the 16 bit converter. More realistically, the dynamic range for a 12 bit converter is 200/1 and 3200/1 for a 16 bit converter. However, more characteristics control the usable range of the measurement.

The problem with representing large numbers is more complex with power measurements. Once power levels reach into the megawatt region, the numbers are so large that they cannot be transported easily in a 16 bit format with the unit of watts. Scaling these larger numbers becomes imperative and as a result the resolution of smaller numbers suffers. Scaling is discussed later in this chapter.

A/D converters have their accuracy specifications stated at full scale but generally do not state their expected performance at midrange or at the lower end of their range where it is common for electrical measurements to reside during much of their life. In addition, converters that are offset to midrange to allow conversion of inputs, which are bipolar or are AC suffer difficulty measuring inputs that are near zero (converter midrange). These measurements often have an offset of several times their minimum resolution increment and declining accuracy in this portion of their range. Figure 6.4 illustrates an accuracy band for converted measurements as a function of range. Converters may also introduce fluctuations around their measured value on the order of several times their minimum increment. This "bounce," as it is called, gives annoying changes to the observed measurement that makes the lower significant digits of a measurement unusable.

IEDs and automation controllers frequently have software and self-calibration components to help minimize the effects of converter performance on measurements. This software may correct for offsets and drift as well as filter out or average some of the bounce. It is important to define the performance requirements across the range of measurements to be encountered by the system before selecting a measurement technology. Note that some performance limitations can be mitigated by carefully selecting the scaling applied to measurements so as to avoid measurements made at the low end of the converter range.

There are two other important characteristics of digitized measurements to be considered in the design of an automation system that are not directly related to the conversion process. These are more aligned with how the system handles measurements than how they are made. These are latency and time skew.

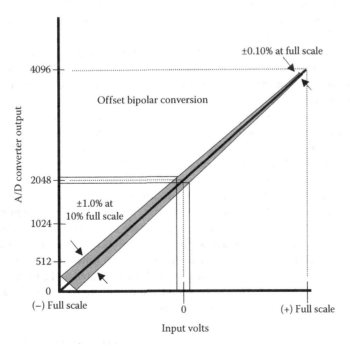

FIGURE 6.4 A/D conversion accuracy.

Latency is the time from which inputs presented at a measuring device are measured and available to the user, whether the user is a human or a software program that requires it. Contributors to latency include

- The time required to process the inputs into measurements and make them available to the measuring device's communication process
- The time to move the measurement from the measuring device's communication process across the network to the user's communication process
- The time for the user's communication process to make the measurement available to the user

While we would expect the inputs at a measuring device to be measured and available at its communications port on demand that may not be the case. Many IEDs and automation controllers have functions, which take precedence to processing measurements and moving them out to the communication process. This is especially true of protection devices where the protective functions take precedence over any other function when it senses possible fault processing requirements. In some devices, there can be a delay in the order of seconds incurred in refreshing the measurements stored in the communication process memory with new values. Until the communication process has new values, it will send old values on request or when its scheduled report occurs. Where latency is an unknown it can be measured by continuously requesting a measurement and observing the time required to see a step change on a measurement input in the returned value. For some devices, this time delay is variable and not easily predicted.

There may be multiple users of measurements made in an automation system. Each user has a pathway to retrieve their measurements based on one or more communication technologies and links. Users can expect latency in their measurements to result from these components that often differ between users based on the differences in the technology, pathways, and links. That suggests that each user needs to define the acceptable latency as part of their performance definitions.

Communication technology can have a profound impact on latency. Different communication technologies and channels have different base speeds at which they transport data. Contributing characteristics include basic communication bit rate, channel speed, message handling procedures, and protocol characteristics. Simple systems poll for measurements on a predictable schedule that suggests that user measurements are transported in one or two poll cycles. During the poll cycle, the delays are directly predicted from the channel and procedural characteristics.

More complex systems may have less predictable message handling characteristics or intermediary devices that are temporary residences of measurements while they move along the pathway. Intermediaries may reformat and rescale measurements for users and divergent pathways and technologies. For example, a user pathway may include a moderately high-speed technology from the measuring device to the first intermediary and a store-and-forward technology for the link to user via a remote server or host. Thus, the first part of the journey may take one-tenth of seconds to complete and the additional hops might take minutes. This emphasizes the importance for users to define their expectations and requirements. This latency may be variable depending on the technology and loading factors and could be difficult to predict or measure.

The last contributor to latency is the time for the measurement to be available to the user at its final destinations. Moving the values to the user host does not fully define the time required for the value to be available to the user. The destination host may introduce delays while it processes the values and places them in the resting place for user access. This time may also be a variable based on the activity level of the host. The host may simply buffer new data until it has time to process it and drop it in its database. Delays can also occur as the values are retrieved and presented to the user.

Latency can have substantial impacts on the users of automation measurements. Users need to specify what latency is tolerable and specify a means to detect when that requirement is not being met to the extent that the added delays affect their process.

Time skew can be important to many data users that are looking at a broad set of values. Time skew is the time difference between a measurement in a data set and any other measurement in that set.

The data set might include measurements taken within a single substation or a subset thereof. However, a data set may include measurements made across a wide geographic area, which might include multiple substations, generating stations, and even multiple utilities taken to perform generation dispatch and system security. Clearly, the user (or applications program) cannot simply assume all values in the data set are taken precisely at the same time instant unless some special provisions have been made to assure that happens. Part of the performance requirement definition is to determine what an acceptable time skew is for each user data set. As with other characteristics, different users will have different requirements. The data transport scheme used to move the values to the user plays a very significant role in determining and controlling time skew.

Some systems have provisions to assure time skew is minimized. A simple method to minimize time skew samples measurements at a specific time by "freezing and holding" them and saves them until they can be retrieved without taxing the communication link. While it is more common to apply this method to sets of energy interchange measurements (kilowatt-hours or kilovar-hours) across interconnections and generating sources, the same principle can be used for other data sets. "Freeze and hold" schemes rely on a system-wide broadcast command or a high-accuracy clock to synchronize sampling the measurements. An updated version of this concept uses a high-accuracy time source such as a GPS to synchronize sampling. The sample data set then has a time tag attached so that the user knows when the sample set was taken. This concept is being applied to the measurement of voltage phase angles across large areas to measure and predict stability.

6.2.4 Instrument Transformers

Electrical measurements on a high-voltage transmission or distribution network cannot be made practically or safely with direct contact to the power carrying conductors. Instead, the voltages and currents must be brought down to a safe and usable level that can be input into measuring instruments. This is the task of an instrument transformer. They provide replica voltages and currents scaled to more manageable levels. They also bring their replicas to a safe ground potential reference. The most common output range is 0–150 V for voltages and 0–5.0 A for currents based on their nominal inputs. Other ranges are used as well. The majority of these devices are iron core transformers. However, other sensor technologies can perform this function that are discussed further in this section.

6.2.4.1 Current Transformers

CTs of all sizes and types find their way into substations to provide the current replicas for metering, controls, and protective relaying. Some will perform well for SA applications and some may be marginal. CT performance is characterized by turns ratio, turns ratio error (ratio correction factor), saturation voltage, phase angle error, and rated secondary circuit load (burden). CTs are often installed around power equipment bushings, as shown in Figure 6.5. They are the most common types found in medium- and high-voltage equipment. Bushing CTs are toroidal, having a single primary turn (the power conductor), which passes through their center. The current transformation ratio results from the number of turns wound on the core to make up the primary and secondary. Lower voltage CTs are often a "wound" construction with both a multiturn primary and secondary winding around their "E-form" or "shell-form" core. Their ratio is the number of secondary turns divided by the number of primary turns. CT secondary windings are often tapped to provide multiple turns ratios. The core cross-sectional area, diameter, and magnetic properties determine the CT's performance. As the CT is operated over its nominal current ranges, its deviations from specified turns ratio are characterized by its ratio correction curve sometimes provided by the manufacturer. At low currents, the exciting current of the iron core causes ratio errors that are predominant until sufficient primary magnetic flux overcomes the effects of core magnetizing. Thus, watt or var measurements made at very low load may be substantially in error both from ratio error and phase shift. Exciting current errors are a function of individual CT construction. They are generally higher for protection CTs than revenue metering CTs by design.

FIGURE 6.5 Bushing current transformer installation.

Revenue metering CTs are designed with core cross sections chosen to minimize exciting current effects and their cores are allowed to saturate at fault currents. Protection CTs use larger cores as high current saturation must be avoided for the CT to faithfully reproduce high currents for fault sensing. The exciting current of the larger core at low primary current is not considered important for protection but can be a problem for measuring low currents. Core size and magnetic properties determine the ability of CTs to develop voltage to drive secondary current through the circuit load impedance (burden). This is an important consideration when adding SA IEDs or transducers to existing metering CT circuits, as added burden can affect accuracy. The added burden of SA devices is less likely to create metering problems with protection CTs at load levels but could have undesirable effects on protective relaying at fault levels. In either case, CT burdens are an important consideration in the design. Experience with both protection and metering CTs wound on modern high-silicon steel cores has shown, however, that both perform comparably once the operating current sufficiently exceeds the exciting current if secondary burden is kept low.

CT secondary windings are generally uncommitted. They can be connected in any number of configurations so long as they have a safety ground connection to prevent the windings from drifting toward the primary voltage. It is common practice to connect CTs in parallel so that their current contribution can be summed to produce a new current such as one representing a line current where the line has two circuit breaker connections such as in a "breaker-and-a-half" configuration.

CTs are an expensive piece of equipment and replacing them to meet new measuring performance requirements is usually cost prohibitive. However, new technology has developed, which makes it possible for an IED to compensate for CT performance limitations. This technology allows the IED to "learn" the properties of the CT and correct for ratio and phase angle errors over the CT's operating range. Thus, a CT designed to feed protection devices can be used to feed revenue measuring IEDs and meet the requirements of IEEE Standard C57.13.

Occasions arise where it is necessary to obtain current from more than one source by summing currents with auxiliary CTs. There are also occasions where auxiliary CTs are needed to change the overall ratio or shift phase relationships from a source from a wye to a delta or vice versa to suit a particular measuring scheme. These requirements can be met satisfactorily only if the auxiliaries used are adequate. If the core size is too small to drive the added circuit burden, the auxiliaries will introduce excessive ratio and phase angle errors that will degrade measurement accuracy. Using auxiliary transformer must be approached with caution.

6.2.4.2 Voltage Sources

The most common voltage sources for power system measurements are either wound transformer (VTs) or capacitive divider devices (capacitor voltage transformers [CVTs] or bushing voltage devices). Some new applications of resistor dividers and magneto-optic technologies are also becoming available. All provide scaled replicas of their primary high voltage. They are characterized by their ratio, load capability (burden), and phase angle response. Wound VTs provide the best performance with ratio and phase angle errors suitable for revenue measurements. Even protection-type VTs can provide revenue metering performance if the burden is carefully controlled. VTs are usually capable of supplying large secondary circuit loads without degradation, provided their secondary wiring is of adequate size. For SA purposes, VTs are unaffected by changes in temperature and only marginally affected by changes in load. They are the preferred source for measuring voltages. VTs operating at 69 kV and above are almost always connected with their primary windings connected phase to ground. At lower voltages, VTs can be purchased with primary windings that can be connected either phase to ground or phase to phase. VT secondary windings are generally uncommitted and can be connected with wye, delta, or in a number of different configurations so long as they have a suitable ground reference.

CVTs use a series stack of capacitors, connected as a voltage divider to ground, along with a low VT to obtain a secondary voltage replica. They have internal reactive components that are adjusted to compensate for the phase angle and ratio errors. CVTs are less expensive than wound transformers and can approximate wound transformer performance under controlled conditions. While revenue grade CVTs are available, CVTs are less stable and less accurate than wound VTs. Older CVTs may be too unstable to perform satisfactorily. Secondary load and ambient temperature can affect CVTs. CVTs must be individually calibrated in the field to bring their ratio and phase angle errors within specifications and must be recalibrated whenever the load is changed. Older CVTs can change ratio up to ±5% with significant phase angle changes as well resulting from ambient temperature variation. In all, CVTs are a reluctant choice for SA system measuring. When CVTs are the only choice, consideration should be given to using modern devices for better performance and a periodic calibration program to maintain their performance at satisfactory levels.

Bushing capacitor voltage devices (BCVDs) use a tap made in the capacitive grading of a high-voltage bushing to provide the voltages replica. They can supply only very limited secondary load and are very load sensitive. They can also be very temperature sensitive. As with CVTs, if BCVDs are the only choice, they should be individually calibrated and periodically checked.

6.2.5 New Measuring Technology

There are new technologies appearing in the market that are not based on the iron core transformer. Each of these technologies has its particular application. Table 6.1 lists some of these technologies and their particular characteristic of interest.

TABLE 6.1 Characteristics of New Technology Measurement Sensors

Air core current transformers	Wide operating range without saturation
Rogowsky coils	Wide operating range without saturation, can be embedded in insulators
Hall effect sensor	Can be embedded in line post insulators
Magneto-optic	Very high accuracy, dynamic range, reduced size and weight
Resistor divider	Low cost

6.2.6 Substation Wiring Practices

VTs and CTs are the interfaces between the power system and the substation. Their primary connections must meet all the applicable standards for loadability, safety, and reliability. Utilities have generally adopted a set of practices for secondary wiring that meets their individual needs. Generally, VT secondaries are wired with #12 AWG conductors, or larger, depending on the distance they must run. CTs are generally wired with #10 AWG or larger conductors, also depending on the length of the wire run. Where secondary cable distance or instrument transformer burden is a problem, utilities often specify multiple parallel conductors. Once inside the substation control center, wiring practices differ between utilities but #12 AWG conductors are usually specified. Instrument transformer wiring is generally 600 V class insulation. Utilities generally have standards that dictate acceptable terminal blocks and wire terminals as well as how these devices are to be used.

6.2.7 Measuring Devices

The integrated substation has changed the way measurements for automation are made significantly. Early SCADA and monitoring systems relied on transducers to convert CT and VT signals to something that could be handled by a SCADA RTU or monitoring equipment. While this technology still has many valid applications, it is increasingly common practice to collect measurements from IEDs in the substation via a communication channel. Where IEDs and the communication channel can meet the performance requirements of the system, transducers and separate conversion devices become redundant; thus, savings can be accrued by deleting them from the measuring plan.

6.2.7.1 Transducers

Transducers measure power system parameters by sampling instrument transformer secondaries. They provide scaled, low-energy signals that represent power system measurements that the SA interface controller can easily accept. Transducers also isolate and buffer the SA interface controller from the power system and substation environments. Transducer outputs are DC voltages or currents in the range of a few tens of volts or milliamperes. A SCADA RTU or other such device processes and transmits digitized transducer signals. In older systems, a few transducer signals were sometimes transmitted to a central location using analog technology.

Transducers measuring power system electrical measurements are designed to be compatible with instrument transformer outputs. Voltage inputs are based around 120 or 115 VAC and current inputs accept 0–5 A. Many transducers can operate at levels above their normal ranges with little degradation in accuracy provided their output limits are not exceeded. Transducer input circuits share the same instrument transformers as the station metering and protection systems; thus, they must conform to the same wiring standards as any switchboard component. Special termination standards also apply in many utilities. Test switches for "in-service" testing are often provided to make it possible to test transducers without shutting down power equipment. Most transducers require an external power source to supply their power requirements. The reliability of these sources is crucial to maintaining data available.

Transducer outputs are voltage or current sources specified to supply a rated voltage or current into a specific load. For example, full output may correspond to 10 V at up to 10 mA output current or 1.0 mA into a maximum 10 kΩ load resistance, up to 10 V maximum. Some over-range capability is provided in transducers so long as the maximum current or voltage capability is not exceeded. The over-range may vary from 20% to 100%, depending on the transducer; however, accuracy is usually not specified for the over-range area.

Transducer outputs are usually wired with shielded, twisted pair cable to minimize stray signal pickup. In practice, #18 AWG conductors or smaller are satisfactory, but individual utility practices differ. It is common to allow transducer output circuits to remain isolated from ground to reduce their susceptibility to transient damage, although some SA controller suppliers require a common ground for all signals, to accommodate semiconductor multiplexers. Some transducers may also have a ground reference associated with their outputs. Double grounds, where transducer and controller both have ground references, can cause major reliability problems. Practices also differ somewhat on shield grounding with some shields utilities grounding at both ends but are more common to ground shields at the SA controller end only. When these signals must cross a switchyard, however, it is a good practice to not only provide the shielded, twisted pairs but also provide a heavy gauge overall cable shield. This shield should be grounded where it leaves a station control house to enter a switchyard and where it reenters another control house. These grounds are terminated to the station ground mass and not the SA analog grounds bus.

6.2.7.2 Intelligent Electronic Devices as Analog Data Sources

Technological advancements have made it practical to use electronic substation meters, protective relays, and even reclosers and regulators as sources for measurement. IED measurements are converted directly to digital form and passed to the SA system via a communication channel while the IED performs its primary function. In order to use IEDs effectively, it is necessary to assure that the performance characteristics of the IED fit the requirements of the system. Some IEDs designed for protection functions, where they must accurately measure fault currents, do not measure low load accurately. Others, where measuring is part of a control function, may lack overload capability or have insufficient resolution. Sampling rates and averaging techniques will affect the quality of data and should be evaluated as part of the system product selection process. With reclosers and regulators the measuring CTs and VTs are often contained within the equipment. They may not be accurate enough to meet the measuring standards set for the SA system. Regulators may only have a single-phase CT and VT, which limits their accuracy for measuring three-phase loads. These issues challenge the SA system integrator to deliver a quality system.

The IED communication channel becomes an important data highway and needs attention to security, reliability, and, most of all, throughput. A communication interface is needed in the SA system to retrieve and convert the data to meet the requirement of the data users.

6.2.8 Scaling Measured Values

In an SA system, the transition of power system measurements to database or display values is a process that entails several steps of scaling, each with its own dynamic range and scaling constants. CT and VT first scale power system parameters to replicas, then an IED or transducer scales them again. In the process an A/D conversion occurs as well. Each of these steps has its own proportionality constant that, when combined, relates the digital coding of the data value to the primary measurements. At the data receiver or master station, coded values are operated on by one or more constants to convert the data to user-acceptable values for processes, databases, and displays. In some system architectures, data values must be rescaled in the process of protocol conversion. Here, an additional scaling process manipulates the data value coding and may add or truncate bits to suit the conversion format.

SA system measuring performance can be severely affected by data value scaling. Optimally, under normal power system conditions, each IED or transducer should be operating in its most linear range and utilize as much A/D conversion range as possible. Scaling should take into account the minimum, normal, and maximum value for the measurement, even under abnormal or emergency conditions. Optimum scaling balances the expected value at maximum, the CT and VT ratios, the IED or transducer range, and the A/D range, to utilize as much of the IED or transducer output and A/D range as possible under normal power system conditions without driving the conversion over its full scale at maximum value. This practice minimizes the quantizing error of the A/D conversion process and provides the best measurement resolution. Some measurements with excessive dynamic ranges may even need to be duplicated at different scaling in order to meet the performance required.

Some IEDs perform scaling locally such as for user displays and present scaled measurements at their communication ports. Under some circumstance, the prescaled measurements can be difficult to use in an SA system, particularly, when a protocol conversion must take place that cannot handle large numbers. A solution to this problem is to set the IED scaling to unity and apply all the scale factors at the data receivers. The practical restraints imposed when applying SA to an existing substation using available instrument transformer ratios will compromise scaling within A/D or IED ranges.

6.2.9 Integrated Energy Measurements: Pulse Accumulators

Energy transfer measurements are derived from integrating instantaneous values over an arbitrary time period, usually 15 min values for 1 h. The most common of these is watt-hours, although var-hours and amp-squared-hours are not uncommon. They are usually associated with energy interchange over interconnecting tie lines, generator output, at the boundary between a transmission provider and distribution utility or the load of major customers. In most instances, they originate from a revenue measuring package, which includes revenue grade instrument transformers and one or more watt-hour and var-hour meters. Most utilities provide remote and/or automatic reading through the utility meter reading system. They also can be interfaced to an SA system.

Integrated energy transfer values are traditionally recorded by counting the revolutions of the disk on an electromechanical watt-hour meter. Newer technology makes this concept obsolete but the integrated interchange value continues as a mainstay of energy interchange between utilities and customers. In the old technology, a set of contacts opens and closes in direct relation to the disk rotation, either mechanically from a cam driven by the meter disk shaft or through the use of opto-electronics and a light beam interrupted by or reflected off the disk. These contacts may be standard form "A," form "B," form "C," or form "K," which is peculiar to watt-hour meters. Modern revenue meters often mimic this feature, as do some transducers. Each contact transfer (pulse) represents an increment of energy transfer as measured by a watt-hour meter. Pulses are accumulated over a period of time in a register and then the total is recorded on command from a clock.

When applied to SA systems, energy transfer measurements are processed by metering IEDs, pulse accumulators (PAs) in an RTU, or SA controllers. The PA receives contact closures from the metering package and accumulates them in a register. On command, the pulse count is frozen, then reported to an appropriate data user. The register is sometimes reset to zero to begin the cycle for the next period. This command is synchronized to a master clock, and all "frozen" accumulator measurements are reported some time later when time permits. Some RTUs can freeze and store their PAs from an internal or local external clock should the master "freeze-and-read command" be absent. These may be internally "time tagged" for transmission when commanded by the master station. High-end meter IEDs retain interval accumulator reads in memory that can be retrieved by the utility automatic meter reading system. They may share multiple ports and supply data to the SA system. Other options may include the capability to arithmetically process several demand measurements to derive a resultant. Software to "de-bounce" the demand contacts is also sometimes available.

Integrated energy transfer telemetering is almost always provided on tie lines between bordering utilities and at the transmission–distribution or generation–transmission boundary. The location of the measuring point is usually specified in the interconnection agreement contract, along with a procedure to insure metering accuracy. Some utilities agree to share a common metering point at one end of a tie and electronically transfer the interchange reading to the bordering utility. Others insist on having their own duplicate metering, sometimes specified to be a "backup" service. When a tie is metered at both ends, it is important to verify that the metering installations are within expected agreement. Even with high-accuracy metering, however, some disagreement can be expected, and this is often a source of friction between utilities.

6.3 State (Status) Monitoring

State indications are an important function of SA systems. Any system indication that can be resolved into a small number of discrete states can be handled as state (status or binary) indication. These are items where the monitored device can assume states like "on or off," "open or closed," "in or out," but states in between are unimportant or not probable. Examples include power circuit breakers, circuit switchers, reclosers, motor-operated disconnect switches, pumps, battery chargers, and a variety of other "on–off" functions in a substation. Multiple on–off states are sometimes grouped to describe stepping or sequential devices. In some cases, status points might be used to convey a digital value such as a register where each point is one bit of the register.

Status points may be provided with status change memory so that changes occurring between data reports are observable. State changes may also be "time tagging" to provide sequence of events. State changes can also be counted in a register and be reported in several different formats. Many status indications originate from auxiliary switch contacts that are mechanically actuated by the monitored device. Interposing relay contacts are also used for status points where the interposer is driven from auxiliary switches on the monitored equipment. This practice is common depending on the utility and the availability of spare contacts. Interposing relays are also used to limit the exposure of status point wiring to the switchyard environment. Many IEDs provide state indication from internal electronic switches that function as contacts.

6.3.1 Contact Performance

The mechanical behavior of either relay or auxiliary switch contacts can complicate state monitoring. Contacts may electrically open and close several times as the moving contact bounces against a stationary contact when making the transition (mechanically bounce). Many transitions can occur before the contacts finally settling into their final position. The system input point may interpret the bouncing contact as multiple operations of the primary device. A number of techniques are used to minimize the effects of bouncing contacts. Some systems rely on "C" form contacts for status indications so that status changes are recognized only when one contact closes proceeded by the opening of its companion. Contact changes occurring on one contact only are ignored. "C" contact arrangements are more immune to noise pulses. Another technique to deal with bouncing is to wait for a period of time after the first input change before resampling the input, giving the contact a chance to bounce into its final state.

Event recording with high-speed resolution is particularly sensitive to contact bounce as each transition is recorded in the log. When the primary device is subject to pumping or bouncing induced from its mechanical characteristics, it may be difficult to prevent excessive status change reporting. When interposing devices are used, event contacts can also experience unwanted delays that can confuse interpretation of event timing sequences. While this may not be avoidable, it is important to know the response time of all event devices so that event sequences can be correctly interpreted.

IEDs often have de-bounce algorithms in their software to filter contact bouncing. These algorithms allow the user to "tune" the de-bouncing to be tolerant of bouncing contacts. However tempting the "tuning" out the bouncing might be, tuning might cover a serious equipment problem that is the root cause.

6.3.2 Ambiguity

State monitoring can be subject to a certain degree of ambiguity. Where a monitored device is represented by a single input, a change of state is inferred when that input changes. However, the single input does not really indicate that the state has changed but that the previous state is no longer valid. Designers need to consider the consequences of this ambiguity. Devices that have significant impact if their state is misrepresented should have two inputs so that two changes must occur that are complimentary to better insure the state of the end device is known. Some designers flag the instances where the two input configurations assume a state where inputs are ambiguous as alarm points.

It is also possible to inject ambiguous state indications into a system by operating practices of the devices. For example, if a circuit breaker is being test operated, its indication points may be observed by the system as valid changes that result in log entries or alarm indications. Likewise, a device may show an incorrect state when it has been removed from service as when a circuit breaker or switcher has its control power disconnected or is "racked" into an inoperative or disconnected position in switchgear. Ambiguity may also result from the loss of power to the monitoring device. Loss of a communication link or a software restart on an intermediary device can also introduce ambiguity.

As much as it is possible, it is important to insure users of state data that the data are not misrepresenting reality. The consequences of ambiguous data should be evaluated as part of any system design.

6.3.3 Wetting Sources

Status points are usually monitored from isolated, "dry," contacts and the monitoring power (wetting) is supplied from the input point. Voltage signals from a station control circuits can also be monitored by SA controllers and interpreted as status signals. Equipment suppliers can provide a variety of status point input options. When selecting between options, the choice balances circuit isolation against design convenience. The availability of spare isolated contacts often becomes an issue when making design choices. Voltage signals may eliminate the need for spare contacts but can require circuits from various parts of the station and from different control circuits be brought to a common termination location. This compromises circuit isolation within the station and raises the possibility of test personnel causing circuit misoperation. Usually, switchboard wiring standards would be required for this type of installation, which could increase costs. Voltage inputs are often fused with small fuses at the source to minimize the risk that the exposed wiring will compromise the control circuits.

In installations using isolated dry contacts, the wetting voltage is sourced from the station battery, an SA controller, IED supply, or, on some occasions, station AC service. Each monitored control circuit must then provide an isolated contact for status monitoring. Isolating circuits at each control panel improves the overall security of the installation. It is common for many status points to share a common supply, either station battery or a low-voltage supply provided for this purpose. When status points are powered from the station battery, the monitored contacts have full control voltages appearing across their surfaces and thus can be expected to be more immune to open circuit failures from contact surface contamination. Switchboard wiring standards would be required for this type of installation. An alternative source for status points is a low-voltage wetting supply. Wiring for low-voltage sourced status points may not need to be switchboard standard in this application, which may realize some economies. Usually, shielded, twisted pairs are used with low-voltage status points to minimize noise effects. Concern over contact reliability due to the lower "wetting" voltage can be partially overcome by using contacts that are closed when the device is in its normal position, thereby maintaining a loop current through the contact. Some SA systems provide a means to detect wetting supply failure for improved reliability. Where multiple IEDs are status point sources, it can be difficult to detect a lost wetting supply. Likewise, where there are multiple points per IED and multiple IED sources, it can be challenging to maintain isolation.

In either approach, the status point loop current is determined by the monitoring device design. Generally, the loop current is 1.0–20 mA. Filter networks and/or software filtering is usually provided to reduce noise effects and false changes resulting from bouncing contacts.

Interface between Automation and the Substation 6-19

6.3.4 Wiring Practices

When wiring status points, it is important to ensure that the cable runs radially between the monitor and the monitored device. Circuits where status circuit loops are not parallel pairs are subject to induced currents that can cause false status changes. Circular loops most often occur when existing spare cable conductors in multiple cables are used or when using a common return connection for several status points. Designers should be wary of this practice. The resistance of status loops can also be an important consideration. Shielded, twisted pairs make the best interconnection for status points, but this type of cable is not always readily available in switchboard standard sizes and insulation for use in control battery-powered status circuits.

Finally, it is important to provide for testing status circuits. Test switches or jumper locations for simulating open or closed status circuits are needed as well as a means for isolating the circuit for testing.

6.4 Control Functions

The control functions of electric utility SA systems permit routine and emergency switching, local and remote operating capability for station equipment, and action by programmed logic. SA controls are most often provided for circuit breakers, reclosers, and switchers. Control for voltage regulators, tap-changing transformers, motor-operated disconnects, valves, or even peaking units through an SA system is also common.

A variety of different control outputs are available from IEDs and SA controllers, which can provide both momentary timed control outputs and latching-type interposing. Latching is commonly associated with blocking of automatic breaker reclosing or voltage controllers for capacitor switching. A typical interface application for controlling a circuit breaker is shown in Figure 6.6.

6.4.1 Interposing Relays

Electric power station controls often require high power levels and operate in circuits powered from 48, 125, or 250 VDC station batteries or from 120 or 240 VAC station service. Control circuits often must switch 10 or 20 A to affect their action that imposes constraints on the interposing devices.

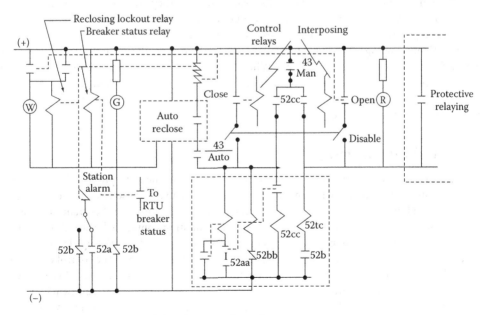

FIGURE 6.6 Schematic diagram of a breaker control interface.

The interposing between an SA controller or IED and station controls commonly use large electromechanical relays. Their coils are driven by the SA control system through static or pilot duty relay drivers and their contacts switch the station control circuits. Interposing relays are often specified with 25 A, 240 VAC contact rating to insure adequate interrupting duty. Smaller interposing relays are also used, however, often with only 10 or 3 A contacts, where control circuits allow. When controlling DC circuits, the large relays may be required, not because of the "close into and carry" current requirements, but to provide the long contact travel needed to interrupt the arc associated with interrupting an inductive DC circuit. Note that most relays, which would be considered for the interposing function, do not carry DC interrupting ratings. "Magnetic blowout" contacts, contacts fitted with small permanent magnets, which lengthen the interruption arc to aid in extinguishing it, may also be used to improve interrupting duty. They are polarity sensitive, however, and work only if correctly wired. Correct current flow direction must be observed.

Many SA controllers and IEDs require the use of surge suppression to protect their control output contacts. A fast switching diode may be used across the driven coil to commute the coil collapse transient when the coil is de-energized. As the magnetic field of the device coil collapses on de-energization, the coil voltage polarity reverses and the collapsing field generates a back electric and magnetic field (EMF) in an attempt to sustain the coil current. The diode conducts the back EMF and prevents the build up of high voltage across the coil. The technique is very effective. It requires the diode to handle the steady-state current of the coil and be able to withstand at least three times the steady-state voltage. However, control circuits that have a high capacitive impedance component often experience ringing transients that simple diodes do not commute. Rather, the simple diode causes the transient to be offset and allows it to continue ringing. These applications require clamping for both the discharge and oscillatory transients where zener diodes are better suited. Failure of a suppression diode in the shorted mode will disable the device and cause a short circuit to the control driver, although some utilities use a series resistor with the diode to reduce this reliability problem. Some designers select the diode to be "sacrificial" so that if it fails, it is vaporized to become an open circuit instead of a short.

Other transient suppressors that are effective for ringing transients include Transorbs and MOV. Transorbs behave like back-to-back zener diodes and clamp the voltage across their terminals to a specified voltage. They are reasonably fast. MOVs are very similar but the suppression takes place as the voltages across them cause the metal oxide layers to break down and they are not as fast. Both Transorbs and MOVs have energy ratings in Joules. They must be selected to withstand the energy of the device coil. In many applications, designers choose to clamp the transients to ground with Transorbs and MOVs rather than clamping them across the coil. In this configuration, the ground path must be of low impedance for the suppression to be effective. Where multiple suppressors are used, it is important to insure surge suppressors are coordinated so that they clamp around the same voltage and have similar switching and dissipation characteristics. Without coordination, the smallest, lowest voltage device will become sacrificial and become a common point of failure.

6.4.2 Control Circuit Designs

Many station control circuits can be designed so that the interrupting duty problem for interposing devices is minimized thereby allowing smaller interposers to be used. These circuits are designed so that once they are initiated, some other contact in the circuit interrupts control current in preference to the initiating device. The control logic is such that the initiating contact is bypassed once control action begins and remains bypassed until control action is completed. The initiating circuit current is then interrupted, or at least greatly reduced, by a device in another portion of the control circuit. This eliminates the need for the interposing relay to interrupt heavy control circuit current. This is typical of modern circuit breaker closing circuits, motor-operated disconnects, and many circuit switchers. Other controls that "self-complete" are breaker tripping circuits, where the tripping current is interrupted by the breaker auxiliary switch contacts long before the initiating contact opens. This is not true of circuit breaker closing circuits, however. Closing circuits usually must interrupt the coil current of the "anti-pump" elements in the circuit that can be highly inductive.

Redesigning control circuits often simplifies the application of control. The need for large interposing relay contacts can be eliminated in many cases by simple modifications to the controlled circuit to make them "self-completing." An example of this would be the addition of any auxiliary control relay to a breaker control circuit, which maintains the closing circuit until the breaker has fully closed and provides antipumping should it trip free. This type of revision is often desirable, anyway, if a partially completed control action could result in some equipment malfunction.

Control circuits may also be revised to limit control circuit response to prevent more than one action from taking place while under supervisory control. This includes preventing a circuit breaker from "pumping" if it were closed into a fault or failed to latch. Another example is to limit tap changer travel to move only one tap per initiation. Many designers try to insure that a device cannot give simultaneous complementary control signals such as giving a circuit breaker a close and trip signal at the same time. This can be important in controlling standby or peaking generator where the control circuits might not be designed with remote control in mind.

6.4.3 Latching Devices

It is often necessary to modify control circuit behavior when SA control is used to operate station equipment. Control mode changes that would ordinarily accompany a local operator performing manual operation must also occur when action occurs through SA control. Many of these require latched interposing relays that modify control behavior when supervisory control is exercised and can be reset through SA or local control. The disabling of automatic circuit breaker reclosing when a breaker is opened through supervisory control action is an example. Automatic reclosing must also be restored and/or reset when a breaker is closed through supervisory control. This concept also applies to automatic capacitor switcher controls that must be disabled when supervisory control is used and can be restored to automatic control through local or supervisory control.

These types of control modifications generally require a latching-type interposing design. Solenoid-operated control switches have become available, which can directly replace a manual switch on a switchboard and can closely mimic manual control action. These can be controlled through supervisory control and can frequently provide the proper control behavior.

6.4.4 Intelligent Electronic Devices for Control

IEDs often have control capability accessible through their communication ports. Protective relays, panel meters, recloser controls, and regulators are common devices with control capability. They offer the opportunity to control substation equipment without a traditional RTU and/or interposing relay cluster for the interface, sometimes without even any control circuit additions. Instead, the control interface is embedded in the IED. When using embedded control interfaces, the SA system designer needs to assess the security and capability of the interface provided. These functional requirements for a control interface should not change just because the interface devices are within an IED. External interposing may be required to meet circuit loads or interrupting duty.

When controlling equipment with IEDs over a communication channel, the integrity of the channel and the security of the messaging system become important factors. Not all IEDs have select-before-operate capability common to RTUs and SCADA systems. Their protocols may also not have efficient error detection that could lead to misoperation. In addition, the requirements to have supervisory control disabled for test and maintenance should not impact the IED's primary function.

Utilities are showing increasing interest in using PLCs in substations. PLCs have broad application in any number of industrial control applications and have a wide variety of input/output modules, processors, and communication options available. They are also well supported with development tools and have a programming language standards, IEC Standard 61131-1,2,3, which propose to offer significant portability to PLC user's software. In substation applications, designers need to be wary of the stresses that operating DC controls may impose on PLC I/O modules. PLCs also may require special power supply considerations to work reliably in the substation environment. Still, PLCs are a flexible platform for logic applications such as interlocking and process control applications such as voltage regulation and load shedding.

6.5 Communication Networks inside the Substation

SA systems are based on IEDs that share information and functionality by virtue of their communications capability. The communication interconnections may use hard copper, optical fiber, wireless, or a combination of these. The communication network is the glue that binds the system together. The communications pathways may vary in complexity depending on the end goals of the system. Ultimately, the internal network passes information and functionality around the substation and upward to the utility enterprise. Links to the enterprise may take a number of different forms and will not be discussed in this chapter.

6.5.1 Point-to-Point Networks

The communication link from an IED to the SA system may be a simple point to point connection where the IED connects directly to an SA controller. Many IEDs connect point to point to a multiported controller or data concentrator, which serves as the SA system communication hub. In early integrations, these connections were simple EIA-232 (RS-232) serial pathways similar to those between a computer and a modem. RS-232 does not support multiple devices on a pathway. Some IEDs will not communicate on a party line since they do not support addressing and have only primitive message control. RS-232 is typically used for short distance, only 50 ft. Most RS-232 connections are also solid device to device. Isolating RS-232 pathways requires special hardware. Often, utilities use point-to-point optical fiber links to connect RS-232 ports together to insure isolation.

6.5.2 Point-to-Multipoint Networks

Many automation systems rely on point-to-multipoint connections for IEDs. IEDs that share a common protocol often support a "party line" communication pathway where they share a channel. An SA controller may use this as a "master–slave" communication bus where the SA controller controls the traffic on the channel. All devices on a common bus must be addressable so that only one device communicates at a time. The SA controller communicates to each device one at a time so as to prevent communication collisions.

EIA-485 (RS-485) is the most common point-to-multipoint bus. It is a shielded, twisted copper pair terminated at each end of the bus with a termination resistor equal to the characteristic impedance of the bus cable, typically 120 Ω. RS-485 buses support 32 standard load devices on the channel. Channel length is typically 1500 ft maximum length. More devices can be connected to the bus if they are fractional standard load devices; up to 256 at 1/8 standard load. The longer the bus, the more likely communication error will occur because of reflections on the transmission line; therefore, the longer the bus, the slower it normally runs. RS-485 may run as fast as 10 MBPS on short buses although most operate closer to 19.2 kbps or slower on long buses. The RS-485 bus must be linear, end to end. Stubs or taps will cause reflections and are not permitted. RS-485 devices are wired in a "daisy chain" arrangement. RS-422 is similar to RS-485 except it is two pairs: one outbound and one inbound. This is in contrast to RS-485 where messages flow in both directions as the channel is turned around when each device takes control of the bus while transmitting.

6.5.3 Peer-to-Peer Networks

There is a growing trend in IEDs' communication to support peer-to-peer messaging. Here, each device has equal access to the communication bus and can message any other device. This is substantially different than a master–slave environment even where multiple masters are supported. A peer-to-peer network must provide a means to prevent message collisions or to detect them and mitigate the collision.

PLC communications and some other control systems use a token passing scheme to give control to devices along the bus. This is often called "token ring." A permissive message (the token) is passed from device to device along the communication bus that gives the device authority to transmit messages. Different schemes control the amount of access time each "pass" allows. While the device has the "token," it may transmit messages to any other device on the bus. These busses may be RS-485 or higher speed coaxial cable arrangements. When the token is lost or a device fails, the bus must restart. Therefore, token ring schemes must have a mechanism to recapture order.

Another way to share a common bus as peers is to use a carrier sense multiple access with collision detection (CSMA/CD) scheme. Ethernet, IEEE Standard 802.x is such a scheme. Ethernet is widely used in the information technology environment and is finding its way into substations. Ethernet can be coaxial cable or twisted pair cabling. UTP cable, Category V or VI (CAT V, CAT VI), is widely used for high-speed Ethernet local area networks (LANs). Some utilities are extending their wide area networks (WANs) to substations where it becomes both an enterprise pathway and a pathway for SCADA and automation. Some utilities are using LANs within the substation to connect IEDs together. A growing number of IEDs support Ethernet communication over LANs. Where IEDs cannot support Ethernet, some suppliers offer network interface modules (Terminal Server) to make the transition. A number of different communication protocols are appearing on substation LANs, embedded in a general purpose networking protocol such as TCP/IP (Internet Protocol).

While Ethernet can be a device to multiple device network like RS-485, it is more common to wire devices to a hub, switch, or router. Each device has a "home run" connection to the hub. In the hub, the outbound path of each device connects to inbound path of all other devices. All devices hear a message from one device. Hubs can also acquire intelligence and perform a switching service. A switched hub passes outbound messages only to the intended recipient. That allows more messages to pass through without busying all devices with the task of figuring out for whom the message is intended. Switched hubs also mitigate collisions such that individual devices can expect its channel to be collision free. Switched hubs can also add delays in message passing, as the hub must examine every message address and direct it to the addressee's port. Routers connect segments of LANs and WANs together to get messages in the right place and to provide security and access control. Hubs and routers require operating power and therefore must be provided with a high reliability power source in order to function during interruptions in the substation.

IEEE Standard 1613—Standards for Communication Networks in Substations covers environmental and testing for electric power substation networks.

6.5.4 Optical Fiber Systems

Optical fiber is an excellent medium for communicating within the substation. It isolates devices electrically because it is nonconducting. This is very important because high levels of radiated electromagnetic fields and transient voltages are present in the substation environment.

Optical fiber can be used in place of copper cable runs to make point-to-point connections. A fiber media converter is required to make the transition from the electrical media to the fiber. They are available in many different configurations. The most common are Ethernet and RS-232 to fiber but they are also available for RS-485 and RS-422. Fiber is ideal for connecting devices in different substation buildings or out in the switchyard. Figure 6.7 illustrates a SCADA system distributed throughout a substation connected together with a fiber network.

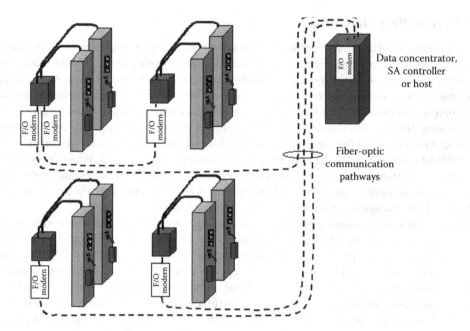

FIGURE 6.7 A F/O network for distributed SCADA and automation.

6.5.4.1 Fiber Loops

Low-speed fiber communication pathways are often provided to link multiple substation IEDs together on a common channel. The IEDs could be recloser controls, PLCs, or even protective relays distributed throughout the switchyard. While fiber is a point-to-point connection, fiber modems are available that provide a repeater function. Messages pass through the modem, in the RX port and out the TX port, to form a loop as illustrated in the Figure 6.8. When an IED responds to a message, it breaks the loop and sends its message on toward the head of the loop. The fiber cabling is routed around to all devices to make up the loop. However, a break in the loop will make all IEDs inaccessible. Another approach to

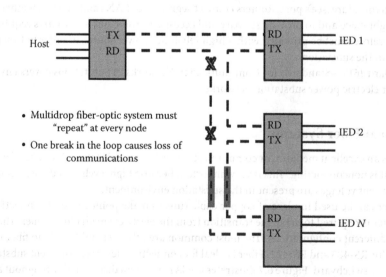

FIGURE 6.8 F/O loop.

Interface between Automation and the Substation 6-25

this architecture is to use bidirectional modems that have two paths around the loop. This technique is immune to single-fiber breaks. It is also easy to service.

Some utilities implement bidirectional loops to reach multiple small substations close to an access point to save building multiple access points. When the access point is a wire line that requires isolations, the savings can be substantial. Also, the devices may not be accessible except through a power cable duct system such as urban areas that are served by low-voltage networks. Here, the extra cost of the bidirectional fiber loop is often warranted.

6.5.4.2 Fiber Stars

Loop topology does not always fit substation applications. Some substation layouts better fit a "star" configuration where all the fiber runs are home runs to a single point. To deal with star topologies there are alternatives. The simplest is to use multistrand fiber cables and make a loop with butt splices at the central point. While the cable runs can all be home runs, the actual configuration is a loop. However, there are star configuration fiber modems available, which eliminate the need for creating loops. This modem supports multiple F/O ports and combines them to single port. Typically, the master port is an RS-232 connection where outgoing messages on the RS-232 port are sent to all outgoing optical ports and returning messages are funneled from the incoming optical ports to the receive side of the RS-232 port. Another solution is to make the modems at the central point all RS-485 where the messages can be distributed along a short RS-485 bus.

6.5.4.3 Message Limitations

In the earlier discussion, there are two limitations imposed by the media. First, there is no provision for message contention and collision detection. Therefore, the messaging protocol must be master slave or the modems must deal with the possibility of collisions. Unsolicited reporting will not work because of the lack of collision detection. In fiber loop topologies, outgoing message will be injected into the loop at the head device and travel the full circumference of the loop and reappear as a received message to the sender. This can be confusing to some communication devices at the head end. That device must be able to ignore its own messages.

6.5.4.4 Ethernet over Fiber

As IEDs become network ready and substation SCADA installations take a more network-oriented topology, F/O links for Ethernet will have increasing application in substations. Just as with slower speed fiber connections, Ethernet over fiber is great for isolating devices and regions in the substation. There are media converters and fiber-ready routers, hubs, and switches readily available for these applications. Because Ethernet has a collision detection system, the requirement to control messaging via a master–slave environment is unnecessary. The routers and switches take care of that problem. The star configuration is also easily supported with a multiport fiber router.

6.5.5 Communications between Facilities

Some utilities have leveraged their right of ways into optical fiber communication systems. F/O technology is very wide band and therefore capable of huge data throughputs of which SCADA and automation messaging might represent only a tiny fraction of the available capacity.

Utilities have taken different paths in dealing with F/O opportunities. Some have chosen to leverage the value of their existing right-of-way by building their own F/O communication networks and leasing services to others. Still other utilities have leased just the right-of-way to a telecommunications provider for income or F/O access for their own use. Using a piece of the F/O highway for SCADA or automation is an opportunity. But, if the highway needs to be extended to reach the substation, the cost can get high.

Typical F/O systems are based around high-capacity synchronous optical network (SONET) communication technology. Telecommunication people see these pathways as high-utilization assets and tend to try to add as many services to the network as possible. SCADA and automation communications can certainly be part of such a network. Some industry experts believe that the power system operations communications, SCADA, ought to reside on its own network for security and not share the close proximity to corporate traffic that is part of an F/O network.

F/O technology has another application that is very valuable for SCADA and automation. Because F/O is nonconductive, it is a perfect medium for connecting communicating devices that may not share a solid ground plane. This is typical of substation equipment. These applications do not need the high-bandwidth properties and use simple low-speed F/O modems. F/O cable is also low cost. This allows devices in outbuildings to be safely interconnected. It is also an excellent method to isolate radio equipment from substation devices to lessen the opportunity for lightening collected by radios to damage substation devices.

6.5.6 Communication Network Reliability

The more the functionality of the SA system is distributed to IEDs, the more critical the communication network becomes. The network design can easily acquire single points of failure sensitivities that can cripple the entire system and even affect substation functions. System designers need to make a risk assessment of their proposed communication architecture to assure users it can meet their expectations for reliability. Designers may need to duplicate critical components and pathways to meet their goals. They may also choose to segment IEDs into parallel networks to maintain high reliability. It may be appropriate to separate critical IEDs from those that are not as critical. Still, designers need to look after details such as power sources, cable separation, panel assignments, and pathway routes to maintain adequate performance.

6.5.7 Assessing Channel Capacity

A necessary task in designing a communication network for a substation is to assess the channel capacity required. This entails accounting for the message size for each device and message type as it passes data to other devices on the network along with whatever overhead is required by the messaging protocol. Along with the message size, the update rate must be factored in. The sum of the message sizes with overhead and channel control times multiplied by the update rate and divided by the channel bit rate will dictate how many devices can share a channel. The larger the sum of the message sizes and the faster the update rate, the fewer devices a channel can support. With Ethernet networks easily reaching speeds of 100 MBPS it is perhaps tempting to ignore assessing channel capacity. However, the fast channel is often plagued by messaging that transfers large blocks data that can temporarily, at least, clog the high-speed channel and impair the overall performance of the system. The network protocols used can substantially add to the overhead in the overall messaging system. Routers and switches can also become choke points in the network and should be evaluated for their potential to cause degradation.

6.6 Testing Automation Systems

Testing assures the quality and readiness of substation equipment. An SA system will require testing at several points along its life span. It is important to make allowances for testing within the standard practices of the utility. While testing practices are part of the utility "culture," designing the testing facilities for SA system with enough flexibility to allow for culture change in the future will be beneficial. Surely, testing can have a great impact on the availability of the automation system and under some circumstances, the availability of substation power equipment and substation reliability. Testing can be a big contributor to operation and maintenance (O & M) cost.

6.6.1 Test Facilities

SA systems integrate IEDs whose primary function may be protection, operator interface, equipment control, and even power interchange measurement for monetary exchange. A good test plan allows for the automation functions to be isolated from the substation while the primary functions of the IEDs remain in operation.

6.6.1.1 Control

It is necessary to test automation control to confirm control point mapping to operator interfaces and databases. This is also necessary for programmed control algorithms. Utilities want to be sure that the right equipment operates when called upon. Having the wrong equipment operate, or nothing at all operate, will severely hamper confidence in the system. Since any number of substation IEDs may be configured to control equipment, test methods must be devised to facilitate testing without detrimental impact on the operation of the substation. Disconnect points and operation indicators may be needed for this purpose. For example, if a breaker failure relay is also the control interface for local and remote control of its associated circuit breaker, then it should be possible to test the control functions without having to shut the breaker down because it would be without breaker failure protection. If the breaker control portion of the breaker failure relay can be disabled without disabling its protective function, then testing may be straightforward. However, some utilities solve this problem by disabling all the breaker failure outputs and allowing the circuit breaker to remain in service without protection for short periods of time while control is being tested. Other utilities rely on a redundant device to provide protection while one device is disabled for testing. These choices are made based on the utility's experience and comfort level. Work rules sometimes dictate testing practices.

While being able to disable control output is necessary, it is also important to be able to verify the control output has occurred when it is stimulated. With IEDs, it is often not possible to view the control output device since it is buried within the IED. It may be useful to install indicators to show the output device is active. Otherwise, at least a temporary indicating device is needed to verify that control has taken place. At least once during commissioning, every control interface should operate its connected power equipment to assure that interface actually works.

6.6.1.2 Status Points

Status point mapping must also be tested. Status points appear on operator interface displays, logs of various forms, and maybe data sources for programmed logic or user algorithms. They are important for knowing the state of substation equipment. Any number of IEDs may supply state information to an automation system. Initially, it is recommended that the source equipment for status points be exercised so that the potential for contact bounce to cause false indications is evaluated. Simulating contact state changes at the IED input by shorting or opening the input circuits is often used for succeeding tests. Disconnect points make that task easier and safer.

As with control points, some care must be exercised when simulating status points. Status changes will be shown on operator interfaces and entered into logs. Operators will have to know to disregard them and cleanse the logs after testing is completed. Since the IED monitors the status point for its own function, the IED may need to be disabled during status point testing. If the automation system has programmed logic processes running, it is possible that status changes will propagate into the algorithms and cause unwanted actions to take place. These processes need to be disabled or protected from the test data.

6.6.1.3 Measurements

Measurements may also come from many different substation IEDs. They feed operator interfaces, databases, and logs. They may also feed programmed logic processes. Initially, measuring IEDs need to have their measurements checked for reasonability. Reasonability tests include making sure the sign of the measurement is as expected in relation to the power system and that the data values accurately represent the measurements. Utilities rarely calibrate measuring IEDs as they once did transducers, but reasonability testing should target uncovering scaling errors and incorrectly set CT and PT ratios. Most utilities provide disconnect and shorting switches (test switches) so that measuring IEDs can have test sources connected to them. That allows known voltage and currents to be applied and the results checked against the expected value. Test switches can be useful in the future if the accuracy of the IEDs falls into question. They also simplify replacing the IED without shutting down equipment if it fails in service.

Some IEDs allow the user to substitute test values in place of "live" measurements. Setting test values can greatly simplify checking the mapping of values through the system. By choosing a "signature" value, it is easy to discern test from live values as they appear on screens and logs. This feature is also useful for checking alarm limits and for testing programmed logic.

During testing of measuring IEDs, some care must be exercised to prevent test data from causing operator concerns. Test data will appear on operator interface displays. It may trigger alarm messages and make log entries. These must be cleansed from logs after testing is completed. Since measuring IEDs may feed data to programmed logic processes, it is important to disable such processes during testing to prevent unwanted actions. Any substituted values need to be returned to live measurements at the end of testing as well.

6.6.1.4 Programmed Logic

Many SA systems include programmed logic as a component of the system. Programmed logic obtains data from substation IEDs and provides some output to the substation. Output often includes control of equipment such as voltage regulation, reactive control, or even switching. Programmed logic is also used to provide interlocks to prevent potentially harmful actions from taking place. These algorithms must be tested to insure they function as planned. This task can be formidable. It requires that data inputs are provided and the outputs checked against expected result. A simulation mode in the logic host can be helpful in this task. Some utilities use a simulator to monitor this input data as the source IEDs are tested. This verifies the point mapping and scaling. They may also use a simulator to monitor the result of the process based on the inputs. Simulators are valuable tools for testing programmed logic. Many programs are so complex that they cannot be fully tested with simulated data; therefore, their results may not be verifiable. Some utilities allow their programmed logic to run off of live data with a monitor watching the results for a test period following commissioning to be sure the program is acceptable.

6.6.2 Commissioning Test Plan

Commissioning an SA system requires a carefully thought-out test plan. There needs to be collaboration between users, integrators, suppliers, developers, and constructors. Many times, the commissioning test plan is an extension of the factory acceptance test (FAT), assuming a FAT was performed. Normally, the FAT does not have enough of the substation pieces to be comprehensive; therefore, the real "proof test" will be at commissioning. Once the test plan is in place it should be rigorously adhered to. Changes to the commissioning test plan should be documented and accepted by all parties. Just as in the FAT, a record of deviations from expected results should be documented and later remedied.

A key to a commissioning test plan is to make sure every input and output that is mapped in the system is tested and verified. Many times this cannot be repeated once the system is in service.

6.6.3 In-Service Testing

Once an automation system is in service, it will become more difficult to thoroughly test. Individual IEDs may be replaced or updated without a complete end-to-end check because of access restriction to portions of the system. Utilities often feel exchanging "like for like" is not particularly risky.

However, this assumes the new device has been thoroughly tested to insure it matches the device being replaced. Often the same configuration file for the old device is used to program the new device, hence further reducing the risk. Some utilities purchase an automation simulator to further test new additions and replacements.

However, new versions of IEDs, databases, and communication software should make the utility wary of potential problems. It is not unusual for new software to include bugs that had previously been corrected as well as new problems in what were previously stable features. Utilities must decide to what level they feel new software versions need to be tested. A thorough simulator and bench test is in order before beginning to deploy new software in the field. It is important to know what versions of software are resident in each IED and the system host. Keeping track of the version changes and resulting problems may lead to significant insights.

Utilities must expect to deal with in-service support issues that are common to integrated systems.

6.7 Summary

The addition of SA systems control impacts station security and deserves a great deal of consideration. It should be recognized that SA control can concentrate station controls in a small area and can increase the vulnerability of station control to human error and accident. This deserves careful attention to the control interface design for SA systems. The security of the equipment installed must insure freedom from false operation, and the design of operating and testing procedures must recognize these risks and minimize them.

References

IEC-61131-3 Programmable Languages PLC Software Structure, Languages and Program Execution.
IEEE Standard C57.13-1993 IEEE Standard Requirements for Instrument Transformers.
IEEE Standard C37.90-2005 IEEE Standard for Relays and Relay Systems Associated with Electric Power Apparatus.
IEEE Standard C37.90.1-2002 IEEE Standard for Surge Withstand Capability (SWC) Tests for Relays and Relay Systems Associated with Electric Power Apparatus.
IEEE Standard C37.90.2-2004 IEEE Standard for Withstand Capability of Relay Systems to Radiated Electromagnetic Interference from Transceivers.
IEEE Standard C37.90.3-2001 IEEE Standard Electrostatic Discharge Tests for Protective Relays.
IEEE Standard 1613-2003 IEEE Standard Environmental and Testing Requirements for Communications Networking Devices in Electric Power Substations.
IEEE Standard C37.1-2007 IEEE Standard for SCADA and Automation Systems.

7
Substation Integration and Automation

7.1	Introduction	7-1
7.2	Open Systems	7-2
7.3	Operational versus Nonoperational Data	7-2
	Operational Data • Nonoperational Data • Configuration Data	
7.4	Data Flow	7-3
	Level 1: Field Devices • Level 2: Data Concentrator • Level 3: SCADA and Data Warehouse • Communications with the Substation (Layer 2 to Layer 3)	
7.5	Asset Management	7-4
7.6	Redundancy	7-5
7.7	System Integration Technical Issues	7-5
	Protocol Considerations • Understanding System Architecture: Documentation • System Architecture Design Considerations • Serial Communications • Highly Available Networks • Factory Acceptance Test	
7.8	System Components	7-10
	Remote Terminal Unit • Data Concentrators • Substation Gateways • Protocol Convertors • Remote Input/Output Devices • Logic Processors • Bay Controllers • Human Machine Interface • Ethernet Switches • Routers and Layer 3 Switches	
7.9	Cyber Security	7-14
7.10	Automation Applications	7-14
7.11	OSI Communications Model	7-15
	Application (Layer 7) • Presentation (Layer 6) • Session (Layer 5) • Transmission (Layer 4) • Network (Layer 3) • Data Link (Layer 2) • Physical (Layer 1)	
7.12	Protocol Fundamentals	7-16
	DNP 3.0 • Proprietary Protocols • IEC 60870 • Modbus • IEC 61850	
7.13	Synchrophasors	7-19
	Wide Area Situational Awareness • Phasor Measurement Units • Phasor Data Concentrator	
7.14	Summary	7-21
	Bibliography	7-21

Eric MacDonald
GE Energy–Digital Energy

7.1 Introduction

The desire to provide highly available power with reduced staff drives utilities and system owners to automate substations. The desire for more rapid restoration is a constant in today's business environment. Reductions in staff and advances in technology result in the unmanned automated substation.

Electricity networks cover vast geographic territories. To effectively manage the substation hubs that connect the different sections of the system, communications networks are used to provide control over the switches, circuit breakers, and other primary equipment that control the flow of power in the system. This control can be automated based on predetermined criteria or can be executed based on the manual intervention of controllers. The controllers can be located within the fences of the substation or remotely positioned at a central control house sending control messages across the communications system.

7.2 Open Systems

In much of the world, there is a push toward open systems, those that use nonproprietary interfaces and protocols. The idea is to simplify the integration of devices through common communications standards allowing the system designer to choose the best solution for the situation based on the equipment available at the time. This is a very favorable approach for the system owner. They can plan to change devices and manufacturers and not be locked into a relationship with any supplier or manufacturer.

In practice, manufacturers continue to differentiate themselves through proprietary advancements. While the open system approach is a lofty goal for the system owner, it is at odds with the goals of manufacturers seeking long-term relationships and rapid developments of differentiating features. Propriety developments can often be done faster due to the reduced testing requirements and the ability to control all aspects of the design. In addition, proprietary systems often provide better performance as they can be developed without the additional overhead required to adhere to an open standard. Open systems pose a threat to manufactures that their highly specialized devices become commoditized. Standards continue to be developed by organizations such as the Institute of Electric and Electronics Engineers (IEEE) and the International Electrotechnical Committee (IEC) to provide more complete feature sets. There are often provisions put into the standards (private parts) to allow the manufacturers to add differentiating features.

Standards such as Ethernet, DNP 3.0, IEC 60870, and IEC 61850 are discussed later. They are examples of successful open standards that have been widely adopted in the substations.

7.3 Operational versus Nonoperational Data

Information and communications in the substation can be grouped into three distinct groups:

1. Operational: position of breakers and switches, as well as voltages, currents, and calculated power
2. Nonoperational: information related to the power system that is not analyzed or utilized in close to real time (such as fault records)
3. Configuration: used to alter settings or update configurations of equipment

7.3.1 Operational Data

The focus of the data communicated out of the substation has been for many years focused on operational data such as circuit breaker positions, volts, and amperes. It is usually time stamped/sequenced and delivered in close to real time. This information is considered critical to the operation of the power system and is used by system operators.

7.3.2 Nonoperational Data

There is a wealth of data that is not consumed by the system operators that is important to the long-term management of the power system. Digital fault records, circuit breaker contact wear indications,

FIGURE 7.1 Intelligent electronic device.

dissolved gas, and moisture in oil are examples of nonoperational data. This information is important to engineering and maintenance personnel to ensure the reliable delivery of power, but it is not critical that this information be delivered in close to real time.

7.3.3 Configuration Data

The configuration and settings for the intelligent electronic devices (IEDs) in the substation are at least as important as the operational and nonoperational data produced by fully functional IEDs. This may sound intuitive but may be an afterthought to the novice. System owners must have up-to-date archives of the configuration files for each and every IED in the substation if they want to be able to rapidly recover the effects of an IED failure. Further, future maintenance updates to the system can be greatly inhibited if backups are not revision controlled and up to date. Figure 7.1 shows an example of an IED.

7.4 Data Flow

Data are commonly obtained from the substation using one or more of the following methods:

1. Direct communications to IEDs by modem or through a network
2. Pass-through communication to IEDs through a data gateway
3. Communications to a data concentrator
4. Physically visiting a substation and connecting a computer to the IED

To ensure near-real-time operational data arrive to its destination without delaying a system, it is important to segregate it from nonoperational and configuration data. There are many ways to do this. It is undesirable for a large file transfer to impede the data flow of time-sensitive data. This can be accomplished by providing alternate data paths into the substation. Currently, much of the world still uses a Bell 202 modem and a leased copper line to connect remote substations to control centers. Using this line for operational data and accessing IEDs by physically going to the substation for configuration and nonoperational data is a simple way to accomplish data segregation.

There is a change taking place in the telecommunications world that suggests this type of arrangement (modems and leased lines) will not last much longer. As telecommunications advance further into digital networks, the analog leased lines are more difficult and expensive to maintain. Many utilities are migrating their systems to use fiber-optic networks rather than continue with leased lines. As a transmission or distribution system owner, it may be more economical to install a fiber-optic network than to continue leasing lines from telecom companies. These fiber-optic networks are usually transmission control protocol/Internet protocol (TCP/IP) or synchronous optical network (SONET) based. Data segregation should be accomplished in SONET through bandwidth allocation; in Ethernet IP networks, virtual local area networks (VLANs) can be used.

Sending maintenance staff to substations is generally undesirable for system owners. It is preferable to utilize communications instead, particularly where skilled workers are not only expensive but also hard to find. Also, where large numbers of substations are involved or they are separated by great distances, travel becomes extremely inefficient.

7.4.1 Level 1: Field Devices

Field devices are IEDs used to protect and control the power grid. They can be protective relays, capacitor bank controllers, meters, voltage regulators, or any other electronic device used for the management of the grid. These devices can be connected to the power grid through a variety of instrument transformers, digital contact inputs, digital outputs, or milliamp signals.

Field devices are directly connected to the primary equipment of the substation through these connections. They communicate information critical to the operation of the substation such as

1. Circuit breaker position
2. Switch position
3. Equipment health
4. Voltage measurements
5. Current measurements

7.4.2 Level 2: Data Concentrator

The number of IEDs that are required to manage the power grid is immense. In some substations, hundreds of protective relays and other devices are required. It is impractical for the grid control centers to actively manage and monitor each individual IED. In most cases, a tiered system with level 2 data concentration is more effective. The data-hub for the substation has many names: remote terminal unit (RTU), data concentrator, data gateway, or substation host processors are currently popular monikers. There are some distinctions between these devices, but often their functionality is combined into one device with a common purpose: to provide a data path for data to the system controllers and maintenance personnel. The functionality and distinctions between these devices are discussed later in this chapter.

7.4.3 Level 3: SCADA and Data Warehouse

The overall operation of the electrical grid takes place in regional control centers. The region can be as small as a local municipality or could span multiple countries. It depends very much on the laws and practices of the region. Information from the substation is required to manage the power system. Therefore, the data concentrator layer 2 devices must provide information to the level 3 control center. This control center may also include historical records of occurrences.

7.4.4 Communications with the Substation (Layer 2 to Layer 3)

To communicate with the substation, a variety of methods are currently in use globally. Currently, this communication is not considered time critical. This means that updated information may flow through the substation to the grid control center with a delay of seconds rather than milliseconds or microseconds. The most common physical connections to the substation are through the following communications interfaces:

1. Bell 202 modem (leased line)
2. IP networks
3. SONET

7.5 Asset Management

To improve the reliability and reduce the costly power outages, utilities are implementing methods for better managing the life cycle of their substations. This means that greater intelligence is required for monitoring the condition of equipment: primary (circuit breakers, switches, transformers, etc.) and secondary (protective relays, IEDs, etc.). The implementation of equipment condition monitoring (ECM) as part of a broad-based asset management policy and plan offers the opportunity to improve reliability. It also requires a number of tools to proactively identify problems before they cause outages.

Equipment monitoring can include the monitoring of such simple alarms as a power supply failure or as complex as identifying the characteristics of a breaker on the verge of failure. In some cases IEDs provide their own self-diagnostics and allow access to the alarms through communication protocols or hardwire contacts. The modern microprocessor-based relay often produces enormous amounts of information. A challenge for system owner is to effectively manage this information to optimize reliability and the cost of maintenance. Nonoperational data analysis is the key to this process.

7.6 Redundancy

One of the most important concepts in the provision of reliable power is redundancy. It is a simple fact that the devices designed to provide control over the electrical system fail. These failures can be explained by a myriad of reasons such as the following:

1. Harsh conditions
2. Design flaws
3. Aging components
4. Voltage transients
5. Rodent interference
6. Manufacturing errors
7. Old age

Although good manufacturers design products for reliability, there is no way all to ensure perfection. To ensure reliable operation of the system, the components can be duplicated, giving an alternate method of control in the event of a failure. The application of dual alternate controls is referred to as redundancy. In this situation, the dual control apparatuses are referred to as the primary and alternate for the remainder of this chapter. (Terminology varies widely in real-world applications, e.g., "A" and "B" systems are often found in North America.)

While it is true that there is a chance that both primary and alternate systems could fail at the same time, it is impractical to continually build backups. The substation engineer generally designs the station based on the provision of dual contingency from the power systems protection world. This suggests that the chances of a power system event (or fault) and the failure of a primary control system are likely enough to require a second contingency, the alternate system. The chances of the alternate failing at the same time are sufficiently low that another backup for the alternate is not a considered reasonable expense in most cases (there are always exceptions).

Redundancy is not always required. The substation engineer must decide on whether or not a backup control system is worth the added complexity and cost. This decision must be based on the following:

1. Criticality of the application
2. Regulatory pressures
3. Cost

Cost comes not only in the form of added equipment but also in greater engineering time required to design, test, commission, and maintain the system. Procedurally, a nonredundant system is much easier to maintain.

7.7 System Integration Technical Issues

The integration of IEDs in a substation protection and control system is a technical issue. Although great strides have been made in applying standards, each manufacturer or design engineer can interpret standards in different ways. Not all devices have implemented the same protocols, and not all versions of devices come equipped with the same standard protocols.

7.7.1 Protocol Considerations

Selecting the right protocol for an application is important to the success of a given substation project. This may be sound simple, or even intuitive; however, there are many factors that play into the selection. As described in Section 7.2, it may be best to select a protocol that is based on an international standard. There are other factors to consider, such as the following:

1. What was the protocol designed for?
2. Is the standard widely implemented and accepted?
3. Does the system operator/owner have the staff with the experience to design and maintain a system based on this standard or protocol?
4. Do the devices which I would like to utilize support the protocol?
5. Is their additional cost to add the protocol to the device?
6. Are there specific desired functions or features and how are they supported by the protocol?
7. What are the tools available to troubleshoot the communications should there be a problem?
8. Is there training required by design and maintenance staff and is the training available in the time period required?
9. What system architecture is required to support this protocol?

For example, the IEC 60870-5-103 protocol may be well suited to communication from a bay controller to a protective relay, but it may not be the best protocol for a control center to communicate with remote substation RTUs. IEC 61850 may be well suited to a fast moving and modern thinking utility but may not be the best for a cautious, "keep it simple" organization.

7.7.2 Understanding System Architecture: Documentation

Documentation of the substation control system is critical. While this is true in general for engineering activities, it is also true that many automation systems are not adequately documented. This leads to great challenges in maintenance and makes future alteration and expansion extremely difficult and expensive. A set of design documents should be produced to ensure the system is implemented and tested adequately. These documentations should provide insight into not only how the individual components are configured but also how the components work as a system.

The architecture of a substation system is more than just the physical connections between devices. It is also relationship between those devices. In an Ethernet-based system, the physical connections, in fact, tell very little about the nature of the destinations and sources of the messages traveling over the Ethernet. For troubleshooting and maintenance, it is important to document both the physical and logical relationships between devices. This can be done through drawings, spread sheets, or a combination of the two. Often drawings help provide a system understanding that is difficult to achieve through spread sheets and tables alone.

To describe the substation automation system, the following documents should be produced:

1. *System architecture (physical connection diagram)*: This drawing should show the details of the physical connections and often the network addresses of devices. Different physical connection media, such as fiber optics or RS-232 communications cables, should be distinguished by different line types. (Colors could be used; however, printed copies may not be distributed in color, so line types are often more practical.)
2. *System architecture (logical relationship diagram)*: This diagram should show the application layer connection between devices and the protocol addresses associated with them. It is useful to include a differentiation between protocols (serial and local area network [LAN]) through different line types.

3. *Points list*: A master list is usually produced that captures the information that is relevant to the system owner. The list should capture the details of the source of the information, the specifics of the protocols used, and where the information will be sent
4. *Logic drawings*: Any logic that is implemented within a substation automation system, including control logic.

7.7.3 System Architecture Design Considerations

The design of the substation system architecture is important to the sustained reliability of the system. The control and visibility of the electricity grid is dependent on accurate information from the substations. As in any engineering activity, decisions must be made based on the cost of implementation and criticality of the system.

7.7.3.1 Green Field versus Brown Field

Substation automation projects vary greatly when comparing a new substation (green field) and the upgrading of the equipment in an existing station (brown field). Green field projects allow the engineer greater freedom in design. New products, techniques, and standards are available and accessible in these installations. In addition, physical space is usually less of a concern as space can be properly allocated for devices and cables in advance of construction.

Brownfield projects require more time in researching and understanding the existing equipment. Many projects that appear simple end up over budget due to a lack of detailed analysis and planning.

Interoperability of legacy devices (existing devices) and new automation devices is a real concern. It is of particular importance to test these connections prior to installation and commissioning if possible. Sometimes this is not possible due to the challenges associated with working on live equipment (e.g., taking a substation out of service requires careful planning and expense to the owner). In these cases, special care must be taken to analyze communications parameters. Even protocols that are well documented and designed based on international standards often run into interoperability issues.

7.7.4 Serial Communications

Serial communications have played an important role in substations for many years. There are a number of different protocols used today. The technical aspects of these protocols are described in detail in many text and web references, and this chapter will not summarize the technical details of the international standards. This chapter will focus on the practical application of these standards within the substation. The most common forms of serial communications in the modern substations are as follows.

7.7.4.1 RS-232/EIA-232

RS-232 is expected to remain an important standard in substation communication for many years to come. The standard has been a popular option since the 1970s and remained so well into the 2000s. The current installed base of devices that use the RS-232 standard will ensure that substation engineers need tools to set up and troubleshoot RS-232 communications indefinitely.

RS-232 is a point-to-point protocol, or in simple terms a means for two electronic devices to communicate. The connection is done through a number of wires grouped together as a cable. Most commonly, each end of the cable is terminated in a 9 or 25 pin D-shaped connector (named DB-9 and DB-25, respectively). The connectors are classed as either male (with pins) or female (with receptor holes for the pins). While there are recommended pin configurations (commonly called "pinouts"), it should not be assumed that all RS-232 connections follow the recommendations. Many manufacturers use nonstandard pinouts and often technicians are required to make their own cables and adapters to facilitate troubleshooting and/or system integration. The allowable cable length is dependent on the data rate used.

The standard was developed for communication between a data circuit-terminating equipment (DCE) and data terminal equipment (DTE). Laptops and personal computers (PCs) were often equipped with DCE serial ports prior to 2005, but now it is more common to use a USB to serial port adapter to provide an RS-232 port to a laptop. The DCE was typically a computer; the DTE was a modem. This paradigm does not fit most applications in the substation where RS-232 is often used for communications between two IEDs, which are in reality computers themselves. This means that careful attention must be paid to pinouts. The following is a list of simple tools suggested for the use in troubleshooting an RS-232 connection in a substation:

1. *Breakout box*: a nonintelligent device that allows the user to connect two serial cables to the box and through either dip switches or jumper cables allow the user to alter the pinouts of each cable, often includes light emitting diodes (LEDs) to show voltage activity visually to the user
2. *Null modem*: a cable or connector used to cross the TX (transmit) and RX (receive) signals (usually pins 3 and 2, respectively, on both DB-9 and DB-25 connectors) and a common pin to facilitate DCE to DCE communications
3. *Pin extractors*: tools used to change pinouts in connectors
4. *"Straight through" cable*: a cable with a connector on each end of the cable where pins 1–9 are connected straight through to pins 1–9 on the other end of the cable. (pin 1 to pin 1, pin 2 to pin 2, etc.)
6. *Male–female adapters*: These are used to change connections from male to female, or female to male
7. *DB-25 to DB-9 adapters*: They allow the user flexibility to connect to different devices

7.7.4.2 RS-422

RS-422 is a serial communication standard using four wires very similar to RS-485 (described in the next section). It is a point to multipoint standard allowing one master to speak to up to 32 slave devices. RS-422 is not as popular as RS-232 or RS-485 in substation communications.

7.7.4.3 RS-485/EIA-485/TIA-485

This standard is very popular in substation applications. It can be implemented in either a two-wire or four-wire implementation. There are a number of reasons for the popularity of RS-485. It is simple to wire and can support up to 32 devices on a single network and can support a wide variety of protocols. There are a number of challenges that arise from the use of multiple different devices on the same network, particularly when the devices are different makes or models. To simplify integration, the data concentrator should be configurable to allow different numbers of stop bits, parity, and baud rates. Because slave devices are often not as configurable as data concentrators, it is sometimes impossible to connect some devices on the same network. A simple solution is to split RS-485 networks by device type connected. This practice costs marginally more in wiring and ports but is often economical compared to troubleshooting.

7.7.4.4 Fiber Optics

Fiber-optic cables are an excellent solution for transmitting data within an industrial or substation environment. Because the cables are made of glass or plastic, they are insulators. This means that the communications are less susceptible to noise from varying electric fields and do not pose a danger to connected equipment in the event of a high-voltage transient event.

The advantages of fiber optics come at the cost. Not only does the connected equipment increase in cost due to the addition of transmitters, but also the maintenance cost increases as it is more challenging to "break into" the cable to monitor communications. This means another fiber-optic convertor must be purchased in order to connect a troubleshooting computer and allow it to monitor the communication. Serial fiber optics are usually point-to-point communications. Although point to multipoint can be done, it is usually a proprietary implementation (each device acts as a repeater).

Substation Integration and Automation

7.7.5 Highly Available Networks

Communications in the substation can be used to manage information that requires extremely high levels of reliability. Messages essential to the protection of the power system are of particular importance. To ensure a high level of reliability in an Ethernet system, there are a number of technologies and topologies that are of particular interest. Some of the most important are described in the following.

7.7.5.1 Ring Topology

Ring topology Ethernet networks are often preferred in substation environments. The reasons for their success are related to the intuitive redundancy and fast calculable reroute times. The simplicity of the network is appealing for maintenance and design.

7.7.5.2 High-Availability Seamless Redundancy

High-availability seamless redundancy (HSR) is described in the IEC 62439-3 standard. Its application promises a networked communications system that can withstand the failure of one of its core communications devices without losing data. The technology is primarily designed for a ring topology. The network is designed without switches; instead, each of the devices in the network performs switching itself. When a message is sent, it sends out packets in both directions around the ring. The receiving device makes a decision on receipt, forwards packets destined for other devices, accepts the packet if it is the first to arrive and it has reached its intended destination or discards the packet if it is the second packet to arrive.

While there are many interesting benefits to the application of HSR, it is not clear that the benefits outweigh the drawbacks. HSR eliminates the need for switches and as such appears to simplify the network. However, in application when HSR is applied in a substation, it may prove to add complexity in maintenance. If one of the devices connected in the network must be removed from service for maintenance (e.g., feeder protection is taken out of service and thus the protective relay in the HSR network is powered down), the HSR ring is broken and the redundancy is no longer functional. While this may be acceptable for a maintenance window, it becomes more challenging if multiple bays or devices need maintenance at the same time. For this reason the technology may be better suited to hybrid architectures with an alternate highly available network such as its sister (described in the same IEC standards) parallel redundancy protocol (PRP).

7.7.5.3 Parallel Redundancy Protocol

Like HSR, PRP is described in the IEC 62439 standard. It also offers a very high level of redundancy. However, unlike HSR, PRP does not eliminate the need for switches. Instead, PRP solves the redundancy puzzle by creating a second parallel network of switches. When a device sends a message, it sends out a packet on both networks. The switched network transfers the message to the designated receiving device and the destination device then accepts the first packet to arrive and discards the duplicate on arrival.

While the initial cost of the hardware required for a PRP network may not be appealing, simplicity of maintenance should be considered in the design phase. The PRP network is not compromised when the IEDs (e.g., protective relays) are taken out of service for maintenance.

7.7.5.4 Star Topology

Star topologies are simple. They offer very fast transmit times. Unfortunately, they also result in a single point of failure, the center of the star. For this reason, the star topology is not popular in critical installations.

7.7.5.5 Hybrid Topologies

The use of multiple ring and star configurations is appealing when examining methods for applying redundancy to a network. While there are protocols that facilitate very complex physical architectures, one must consider the application carefully. In architectures that depart from a simple ring topology, it becomes increasingly difficult, if not impossible to accurately calculate the impact of the failure of one of the switches in the network. In a simple ring topology, the general rule of thumb for a rapid spanning tree protocol (RSTP) reroute is in the order of 5 ms per switch (at the time of this writing). In more complex architectures, this reroute can be in the order of minutes. A decision must be taken during the design stage if this type of delay can be accepted in the event of a failure.

7.7.6 Factory Acceptance Test

To increase the probability of success of substation automation project, the system should be exhaustively tested prior to its installation in the field. While this is not always possible or practical, it is extremely beneficial when executed properly.

Where possible, a mockup of the system should be created. All communications interfaces should be tested prior to the shipping of the devices. It is common practice for vendors to provide an opportunity for their customers to participate in a demonstration of the system prior to accepting shipment. The demonstration test or factory acceptance test (FAT) should be executed according to a detailed plan. The plan should be written by the vendor based on consultation with the client. Although in practice it is unlikely that the system will pass the FAT without any augmentations requested, there is an expectation that the FAT will be executed without the uncovering of major errors or omissions.

7.8 System Components

7.8.1 Remote Terminal Unit

The traditional heart of the automation system was the RTU. For many years, the RTU has been a mainstay of distributed automation and Supervisory and data acquisition (SCADA) systems. This section naturally focuses on substation applications, although much of the information presented is also true for other industries. Figure 7.2 shows RTUs suitable for deployment in substations.

The main requirements for an RTU are a communications interface and the ability to monitor digital status points and analog values (currents, voltages, etc.). The RTU is important in widely spaced geographic regions. Its main function is to provide information about the power system to a central control system through a communications interface and to provide remote control of switches and circuit breakers. Many RTUs are also used to monitor current and voltage and to calculate power.

In addition, the RTU can be used to monitor many other signal and status points in the substation such as a door alarm and a battery failure alarm. These types of signals are applicable to the entire substation and as such are difficult to group into any bay or protection group. For maintenance purposes, it is simpler to connect these signals to a substation RTU rather than another IED or microprocessor relay.

The overwhelming popularity of the microprocessor relay has led many to challenge the place of the RTU in the modern substation. The argument against the RTU is that the increased computational power, abundance of physical input/output (IO), and the required instrument transformer connections (for current and voltage readings) already exist in the protective relays and therefore should be reused in the SCADA system for monitoring. While the initial savings on physical hardware and decreased wiring may be appealing,

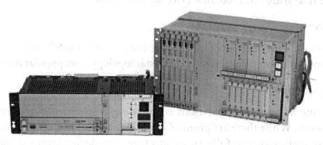

FIGURE 7.2 Remote terminal unit.

the engineer must consider the maintenance and operational environment that the system must perform within. The use of an RTU in a substation often simplifies design, commissioning, and maintenance.

7.8.2 Data Concentrators

The data concentrator aggregates information and provides a subset of that information to another device or devices. It is similar in function to the RTU and often can be the same device. The main difference in the terms is that a data concentrator does not necessarily have physical interfaces to monitor contact statuses and analog values. The data concentrator uses communication protocols to acquire data from other devices rather than through a direct connection.

7.8.3 Substation Gateways

The term "gateway" is unfortunately applied in a couple of similar, although distinctly different applications in the substation. The first and simplest is the router. In IP networking, the gateway is a device that allows communication between different subnets. This is called layer 3 switching or routing. This terminology is common when dealing with information technology (IT) departments who spend their days and nights fixing IP networks.

In the substation, the gateway is something different to the protection and control staff. In the modern smart grid, the gateway serves as the substations security access point. It manages and logs access to the information available in the substation. The substation gateway can be thought of as a superset of the data concentrator and the RTU (although the RTU can be a separate device, it does not have to be in most substation applications). Figure 7.3 shows a substation gateway.

7.8.4 Protocol Convertors

Although not ideal, protocol convertors can be used to solve the problem of two devices that do not speak the same "language." The protocol convertor can be a simple two-port device providing conversion between protocols such as a Modbus-TCP (networked) to Modbus-RTU (serial) convertor. They can also be as intricate as a large-scale RTU or data concentrator that converts many protocols simultaneously on different ports. The standalone protocol convertor is not preferred as it adds another possible point of failure to the system. However, not all devices are created with all protocols and therefore the protocol convertor can be the integrator's best friend. Protocol convertors are particularly useful when interfacing with older generation equipment to new more modern systems. Figure 7.4 shows a substation-hardened media convertor.

FIGURE 7.3 Gateway.

FIGURE 7.4 Media convertor.

7.8.5 Remote Input/Output Devices

Remote IO units are of particular interest in very large applications. The cost of the copper cabling and the pulling, and terminating of those cable can be very high. For this reason there is a lot of interest in distributing IO devices throughout a station and communicating to them through a network. The remote IO device presents a number of challenges to the integrator.

When IO is distributed, the time stamping of signals can be problematic. It may be necessary (depending on the region and local regulations) to have a very high resolution and accurate time stamp recorded with any status change. This means that the remote IO unit must have some way to synchronize with an external clock.

The remote IO unit may require more intelligence that its name suggests. Generic object-oriented system event (GOOSE) message IO devices, for example, require some sort of intelligence (timers) to ensure that a loss of communications does not permanently close a contact. For this reason, careful attention should be paid to the protocols selected.

7.8.6 Logic Processors

Most IEDs in the modern substation include some sort of logic processing capabilities. Where programmable logic controllers (PLCs) are popular in many automation systems, they are less popular in the substation. This may be in part due to the challenging physical environment that the substation presents. It may be in part due to devices that have been developed to provide functionality specific to the substation space. Substation logic processors are usually in the form of data concentrators and RTUs and provide advanced protocols and automation functions that need to be developed from scratch in a PLC.

7.8.7 Bay Controllers

The concept of "Bays" is not popular in North America, but has been very successfully adapted in countries that favor the IEC standards, particularly in Europe. Bays are logical groupings of IEDs and primary equipment in a system. For example, a substation could be grouped into a bay for each transmission line, transformer, or feeder to which it is connected. In its simplest form, the bay controller is much like an RTU. It monitors digital status points and offers control over the switches and breakers within its bay. More complex bay controllers monitor currents and voltages and offer some protective relaying functionality, synchronization checks, and even human machine interface (HMI) control functionality. Figure 7.5 shows a substation bay controller.

7.8.8 Human Machine Interface

The HMI is an important piece of many substations. It provides a window into the substation where operators and maintenance personnel can find a centralized view of the current state of the substation

FIGURE 7.5 Bay controller.

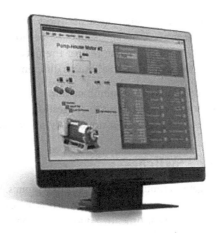

FIGURE 7.6 Typical HMI screen.

including breaker and switch position. Figure 7.6 shows an HMI screen developed for motor management. Like most technology, the capabilities of the HMI are dependent not only on the technology used but also on the capabilities and budget allocated to the implementation team.

The HMI usually starts with a representation of the single line diagram of the station. This diagram shows the position of circuit breakers and switches. Analog values such as currents, voltages, and power should also be displayed. The HMI can be used to provide control over the system or be simply used for monitoring. More powerful HMI products allow different levels of control to different users. The HMI can also be used to monitor the sequence of events that occur in a substation and to monitor and manage alarms.

7.8.9 Ethernet Switches

LANs are primarily built using Ethernet switches and used to allow small numbers of devices to communicate with each other. Ethernet communications are an important part of the modern substation. While Ethernet technology has been the standard in home and office computing for many years, it has not been so in the substation. Switches built for the harsh environment that exists in substation (extremes in temperature, electromagnet interference, and voltage transients) were not in production and the products designed for office use were prone to failure in the substation. In addition, most utilities require equipment that does not use a fan for cooling. (Moving mechanical parts such as fans are expected to fail in greater frequency than static electronics.) Figure 7.7 shows a 32-port substation grade communications switch.

When considering switches for use in the substation, a thorough examination of the devices desired for connection to the network is important. In general, a fiber-optic system is preferred as it does not conduct and is not subject to electromagnetic interference. It is important to check the ports available on the end devices and be sure the same interfaces are available in the switch. Although media convertors can be used, they introduce another point of failure and should be avoided. Dual power supplies for added redundancy should also be considered for substation communications switches.

FIGURE 7.7 Ethernet switch.

Ethernet communications are useful in small area networks. When large numbers of devices or bandwidth intense applications are used, Ethernet traffic can be slow. To ensure important data are transferred and not affected by congestion in the rest of the network managed switches are used. The following terminology is commonly used to distinguish different Ethernet technology:

Managed switches—allow prioritization of communications
Unmanaged switches—provide switching subject to congestion
Hubs—repeat all received messages on all ports, not usually suitable for advanced applications

7.8.10 Routers and Layer 3 Switches

In some cases, it is useful to segregate communicating devices into groups. When large numbers of devices are connected in a LAN, the performance of a network is degraded as there is a finite amount of data that can be transmitted by the LAN switches at any one time. To reduce congestion, LANs can be split up into separate networks. These segregated LANs are then less subject to delays associated with congestion. To allow communication between multiple LANS, a bridging device must be used that operates at the network layer of the open systems interconnection (OSI) model for communications (see Section 7.11). The router or layer 3 switch is the device used to perform this bridging function. Traffic from one LAN can be routed to other LANs connected though a router to create a metropolitan area network (MAN) or wide area network (WAN).

It is important to group devices appropriately to manage bandwidth. It is of little use to group devices that need to use great bandwidth to intercommunicate on separate LANs. This leads only to congestion in the router. Often each substation is set up with a single LAN and is connected to a wider corporate network MAN or WAN through a router. This segregates the data within a substation and can allow communications from an engineer's desk to multiple substations. Layer 3 switches operate in the same manner as routers from a system perspective but process the routing functions in hardware, making the process much faster.

While very appealing from an operational perspective, the use of WANs requires system owners to pay particular attention to cyber security.

7.9 Cyber Security

Cyber security is a natural extension of the substation control system. Because the protection and control system of a substation is in control of the electricity grid, it could be a target for miscreants and terrorists. To deal with increased security threats, utilities have a variety of choices. They can halt progress and rely on aging technology such as point-to-point serial communications and roll trucks to every substation every time there is a problem or they can adapt new techniques and devices for maintaining a secure operating environment. Security is not just an issue for utilities. It is an issue for banks, governments, armies, and virtually anyone with a computer. Other industries continue to push forward with new cyber security techniques and technologies, so should the power industry. Security will be examined in detail in Chapter 17.

7.10 Automation Applications

Automation of the substation provides fast resolution for power outages. Changes can be made to the position of breakers and switches using algorithms running either locally (in one of the substation logic processors or gateways) or centrally in the control center computers. In the substation, controls can be executed very rapidly to adjust to changing conditions (such as faults). Centralized controls take longer due to delays in transmission of information but can be executed based on a much wider area. Some examples of automation applications are as follows:

1. Fault detection isolation restoration (FDIR)
2. Autotransfer
3. Volt–Var control (VVC)

4. Alarm management
5. Redundant point monitoring
6. Reclosing
7. Autosectionalizing

7.11 OSI Communications Model

The OSI model for communication is the basis for most computer communications systems. It is described in the IEC 7498-1. The model is used to facilitate open communications between different computer systems. It separates communications into the following seven layers:

1. Physical layer
2. Data-link layer
3. Network layer
4. Transport layer
5. Transmission layer
6. Session layer
7. Application layer

In the OSI model, only adjacent layers may communicate. Then through each level of abstraction, a "virtual" connection is made between the layers of communicating devices (e.g., the transport layer processes on a sending device communicate with the transport layer processes on a receiving device.)

It is simpler to explain the functionality of the seven layers starting with the application layer and that is the way it is presented later. It is, however, more effective to troubleshoot systems starting with the physical layer. Although the OSI model is fundamental in most communications systems, it is worth noting that not all communications protocols utilize all seven layers. Often functionality can be combined or slightly realigned. Despite this fact, it is still useful to understand the model.

Because this model has been so widely accepted, knowledge of the model has become important to troubleshooting and system design. This model is not the focus of this chapter, and as such will not be covered in detail. Those that wish to specialize in system integration and communications should invest in further research in this area. It is particularly useful in troubleshooting faulty communications to be able to segregate and troubleshoot each incremental layer starting with the physical.

7.11.1 Application (Layer 7)

The highest layer of the model interacts directly with process or program and is the single access point into the open systems interconnection environment (OSIE) providing services to facilitate communications. The application layers is responsible not only for providing data transfer but also for identifying communications partners, authority of other devices to communicate, and establishing levels for quality of service. The application layer communicates directly with and has access to the services provided by the presentation layer and has access to the services provided by it.

7.11.2 Presentation (Layer 6)

The presentation layer is primarily concerned with syntax; that is, it is designed to provide a common representation of data across platforms. It provides services to the application layer such as compression of data and encryption. The presentation layer utilizes the services provided by the session layer.

7.11.3 Session (Layer 5)

The session layer helps manage connections between the presentation layers of communicating devices. It is responsible for making and breaking communications between systems. The session layer should

7.11.4 Transmission (Layer 4)

The transmission layer provides a simple interface to the session layer, abstracting the details of the lower three layers and packetizing data. It should provide the ability to implement flow control, allowing a receiving system to request a reduced data rate from a sender and ensure the receiving system is not overwhelmed. The transmission layer should provide optimal use of the available networks, balancing capacity against the demands of each session entity requiring transport services. The transmission level also deals with end-to-end quality of service and error recovery.

7.11.5 Network (Layer 3)

The network layer handles routing and relaying of messages. The network layer includes a system-wide unique network address that is used to identify end user devices. The network layer manages data paths through a network. It is responsible for negotiating networks and subdivision of networks to ensure data can find an appropriate path from one device to another. Networks can be large as simple as a point-to-point communication link or as grand as the Internet, which is a massive interconnection of multiple networks. (Note: The Internet is primarily a TCP/IP network. It is built on a model similar to the OSI model, although some of the OSI functions are grouped into different layers.)

7.11.6 Data Link (Layer 2)

The data-link layer detects and (sometimes) corrects errors introduced in the physical layer. It is not concerned with the same wide scale as the network layer. It is concerned with the rapid management of the data coming through the physical layer from one device to another. In the substation, it is worth noting that GOOSE messaging operates in the data-link layer and that Ethernet switches primarily operate in switching data-link layer information.

7.11.7 Physical (Layer 1)

The physical layer operates directly on the connection media (e.g., copper or fiber-optic cable). Data are transmitted in digital pulses. Physical connections can be between two or more end devices.

7.12 Protocol Fundamentals

When working in substation automation and systems integration, there is no way to avoid discussions on protocols. To ensure that the IEDs in a substation work as a system, there must be communications between the devices. This communication is usually through either copper wires of fiber-optic cables (wireless is not commonly used within the substation due to problems with interference and security). Signals are sent through these physical media using binary signals grouped according to an agreed protocol. A protocol is at its simplest an agreement on terms of engagement. In human terms, the shaking of hands when meeting is an example of a protocol.

This section describes some of the most important protocols to the substation engineer.

7.12.1 DNP 3.0

Distributed network protocol (DNP) is one of the most successful protocols in energy management globally. It is the preferred protocol for communications between substations and control centers through most of the Americas, Africa, and Asia. DNP 3.0 is an evolving protocol designed specifically for communications between master control stations RTUs and IEDs. It is managed by the DNP3 users group. The protocol is open. This means that any vendor has the information required to implement DNP 3.0.

The protocol is well suited to control and monitoring of electricity networks. It is loosely based on the OSI model but utilizes an enhanced performance architecture (EPA) to reduce bandwidth. Thus, it focuses on the physical, data-link, network, and application layers. Because protocol requirements are very clearly documented and the DNP3 users group details different levels of compliance, the DNP 3.0 protocol is well established and a natural choice when communications are required between devices from different manufacturers.

7.12.2 Proprietary Protocols

Proprietary protocols have been developed by many manufacturers. The reasons are varied. For the most part, propriety protocols are developed for one of the following reasons:

1. Speed of development
2. Market differentiation
3. Open protocols do not exist or insufficient for the desired performance
4. Open protocols do not exist or insufficient for the desired purpose

Much advancement has come through the implementation of proprietary protocols. Although open protocols are preferable for interoperability, it should be remembered that many current standards started as proprietary protocols.

7.12.3 IEC 60870

The IEC TC (Technical Committee) 57 is responsible for many of the advancements in open communications. IEC 60870 is no exception. This set of standards describes an open system for SCADA communications. The breadth of the IEC 60870 standards goes beyond the substation and includes teleportation and intercontrol center communications. In the substation, the following companion standards are of particular interest:

1. IEC 60870-5-101—serial communications from a control center to a data concentrator
2. IEC 60870-5-103—serial communications from a data concentrator or controller to a protection IED
3. IEC 60870-5-104—networked (LAN-based) communications from a control center to a data concentrator

The IEC 60870 protocols are of particular importance in Europe and North Africa and Asia and parts of South America, although not popular in the Americas.

7.12.4 Modbus

Industrial applications have made use of the modbus protocol for many years. Although it is used in some utility stations, modbus remains less popular than IEC 60870 and DNP 3.0. Modbus is a very versatile protocol. Information is stored in registers that are accessed through modbus for reading and writing. Functions such as time stamping are not natively built into the modbus protocol. If such functionality is required, it must be stored in registers like any other data.

Modbus is an excellent choice for its flexibility and simplicity. It can be used for anything from process control to file transfers. Unfortunately, the standard has been interpreted in different ways by many companies. As a general rule, modbus is not best selection when seeking interoperability.

7.12.5 IEC 61850

Of all the technological developments in protection and control engineering in the twenty-first century, none generated more excitement and interest across the industry than IEC 61850. The standard is the result of the efforts of the IEC Technical Committee 57 to produce an open standard for substation modeling and communications. The standard has gained a lot of popularity in Europe and continues to spread throughout the rest of the world (although it has been slow to catch on in North America).

System owners and engineers continue to desire systems that are simple to integrate and provide high performance and flexibility. They look to IEC 61850 to fill this desire. The IEC 61850-6 standard describes an XML (extensible mark-up language)-based syntax for modeling substations called substation configuration language (SCL).

The standard is extensive in its reach. It covers communications between IEDs for protection and control (Station Bus), replacing traditional copper wire connection between IEDs (GOOSE messages) and replacing copper wiring from instrument transformers to IEDs (process bus) in addition to providing a standard for an object-oriented modeling of the substation. IEC 61850 edition 2 expands the reach of the protocol to include routable GOOSE messages, synchrophasors, cyber security, and more. It remains to be seen to what extent the IEC 61850 standard will be implemented.

As with any engineering activity, the implementation of the IEC 61850 protocol should be carefully examined. While there may be a desire to use the latest and greatest technology, it should be noted that there is a new paradigm that needs to be learned by design and maintenance staff as well as a new set of tools and standards. The initial cost of implementing the IEC 61850 standard will almost certainly be higher for the system owner than simply repeating what has been done before. As the technology advances and staff learn new skills, the advantages should start to be found.

7.12.5.1 IEC 61850 Configuration Paradigm

The IEC 61850 station uses a system configuration paradigm. Each device in the system is described by an IED capability description (ICD) file written in the SCL. This file identifies the functionality that the IED can perform in the system. To configure the system, each ICD file is imported into a substation configuration description (SCD) file, which is used to link IED configurations together. The resulting configuration is exported out of the SCD file and becomes a series of configured IED description (CID) files to be loaded into the IED.

In practice, this procedure is not always practical. Often it is simpler to manually configure parameters or impossible to set parameters from a system configuration tool. Gateways may import SCL files directly or establish communications through IEC 61850's self-description functions.

Configuration of devices in IEC 61850 presents a challenge for utilities. Where standard practice has previously been to recommission protection and control devices when any change was made to a configuration, new standard practices may be required. A change to the configuration parameters of any devices in the system should result in a change to the SCD file. It is certainly impractical to recommission an entire station for a simple change; the peace of mind that the right change was made may be difficult to attain. There may continue to be resistance to the system configuration paradigm in favor of continuing the current modular approach to system maintenance.

7.12.5.2 GOOSE

One of the most clearly beneficial pieces of the IEC 61850 standard is the GOOSE. The GOOSE message is designed to utilize the high-speed networks to replace copper wires. There are many cost-saving advantages to changing from copper wire to communications signals such as

1. Reduced material cost
2. Reduced cost of changes post construction
3. Simplified physical construction

Communications-based data can be transferred in speeds equal to or faster than hardwired signals. To do this, GOOSE messages require managed switches as they are considered high priority signals. The GOOSE message is passed through a managed switch before noncritical data.

7.12.5.3 Station Bus and Process Bus

Part of the IEC 61850 standard is the separation of communications into two main groups, station bus for monitoring and control and process bus that is intended to replace the copper wiring used for connections to instrument transformers. Most of the initial IEC 61850 installations included only the station bus. Process bus has been more challenging to implement using Ethernet networks.

Relay manufacturers utilize different methods and algorithms for digitizing and sampling analog measurement from instrument transformers. The often-patented and proprietary algorithms used to protect the power system rely on these sampled values. The vision of the process bus is to provide a "Merging Unit," which samples the values at the physical location of the instrument transformer and publishes the sampled values to an Ethernet network for any protective relay needing the information. In further efforts to improve interoperability, the standard was further extended with the IEC 61850-9-2-LE technical recommendation that gives greater detail as to the nature of the sampled values.

For the Ethernet-based process bus to provide the same functionality as existing conventional systems, enormous bandwidth will be required. The current high-speed internal buses used in microprocessor-based relays must be replaced by high-speed networks. These networks introduce more overhead to the communication data. The data must be extremely reliable, thus PRP and HSR, (explained elsewhere in this chapter) and are being introduced to the IEC 61850 standard. Ethernet has not been sufficient for the challenge.

Non-Ethernet-based solutions have been implemented. The first only merging unit commercially available today does not use Ethernet. Instead the manufacturer applied the principal of removing the copper wires between instrument transformers and IEDs and replacing them with fiber optics based on a published interpretation of the IEC 61850 protocol. While this solves the problem, many IEC 61850 purists do not support this development as it does not utilize traditional switched networks and instead replaces them with point-to-point fiber optics. This solution utilizes the power of the OSI model for communications, which accounts for advancements in technology by abstracting the lower layers of the protocol communications. If designs are implemented with these principals, it should not matter what lower level technologies are utilized. The methods best suited to the application should be available if the technology exists.

7.13 Synchrophasors

Synchrophasors are phasor (rotating vector) measurements synchronized to Coordinated Universal Time (UTC). The synchrophasor is a modern innovation for describing alternating current (AC) networks. Synchrophasors were first established in the IEEE 1344 standard in 1995 but have not been widely deployed. Updates to the standards were issued in 2011 and have started to achieve widespread acceptance. Synchrophasors promise to add a new level of understanding of the operation of the power system through showing never before details. Where traditional protocols have been centered around SCADA data updates every few seconds, synchrophasors can be used to record measurements many times as second. The primary applications are related to wide area monitoring systems (WAMS) that allow utilities to capture and analyze the effects of power system events (such as faults—undesired and uncontrolled current flow) over great geographical areas.

The current standard for transmission of synchrophasor data is C37.118. This protocol can be used for transmitting synchrophasors but presents a number of challenges. It is very effective as a point-to-point protocol but is not well suited to network-based communications. For this reason, the IEC 61850-90-5 standard was designed to provide routable synchrophasors using UDP. IEC 61850 also adds security enhancements beyond the scope of C37.118.

7.13.1 Wide Area Situational Awareness

In AC networks, synchrophasors can be used to estimate the stress put on the electrical grid. Power flows from high voltage to low voltage. Because the voltage levels in an AC network are constantly changing, slight variations in frequency across great distances cause power to flow. Changes in the power system (such as loss of transmission lines or generators starting or stopping) lead to rapid changes in the phase angles of voltages at different places in the power system that puts stress on the system.

Although WAMS are not strictly substation based, they are discussed here as substations are the hubs that connect the electrical grid. Therefore, the measurement equipment and data concentrators required to provide synchrophasors reside in substations. Figure 7.8 shows an example of an energy management system for wide area monitoring.

7.13.2 Phasor Measurement Units

Phasor measurement units (PMUs) measure voltages and currents from the grid and publish them to a subscriber using a synchrophasor-based protocol such as C37.118 or IEC 61850-90-5. PMU data are data concentrated by phasor data concentrator (PDC) units. This is primarily done because of the large bandwidth that would be required by many PMUs on a network. The scale of a large system could easily grow to a level that would overwhelm a single centralized PDC (known as a super-PDC). PMUs can be integrated into protective relays. Figure 7.9 shows a relay-based PMU.

7.13.3 Phasor Data Concentrator

PDCs are used to data concentrate a specific set of data transmitted using the technology known as synchrophasors. Synchrophasors and the associated protocols are examined previously.

FIGURE 7.8 Wide area management.

FIGURE 7.9 N60 PMU.

The PDC is important in a synchrophasor system as synchrophasors require great band width. When synchrophasors are implemented over a large-scale grid, it is impractical to send all the information from the PMUs to the central controller. The PDC is designed to provide a localized archive and filter and to connect to a "super-PDC" at the control center.

7.14 Summary

There has been a lot of attention paid to smart grids and adding intelligence to power management. Substation automation systems engineers and techs have been quietly doing this for the past 25 years. The new focus has led to increased investment and expectations. The automation and communications systems in the electrical grid provide the "glue" that holds together modern power and energy management systems. Automation of substations affects industrial, commercial, and residential customers. The substation is the hub of the power system; control and automation of these connections require a diverse skill set.

Bibliography

Adamiak, M., Premerlani, W., and Kasztenny, B., Synchrophasors: Definition, measurement and application, *Protection and Control Journal*, GE Multilin Publications, pp. 1–13, 2006.

Blackburn, L., *Protective Relaying: Principals and Applications*, 2nd edn., CRC Press, Boca Raton, FL, 1998.

Dogger, G., Tennese, G., Kakoske, D., and MacDonald, E., Designing a new IEC 61850 architecture, Presented at Distributech, Tampa, FL, 2010.

Institute of Electrical and Electronics Engineers, *IEEE Standard Definition, Specification and Analysis of Systems Used for Supervisory Control, Data Acquisition, and Automatic Control*, IEEE Std. C37.1-1994, IEEE, Piscataway, NJ, 1994.

Institute of Electrical and Electronics Engineers, *IEEE Standard Electrical Power System Device Function Numbers and Contact Designations*, IEEE Std. C37.2-1996, IEEE, Piscataway, NJ, 1996.

International Electrotechnical Committee, *Industrial Communication Networks—High Availability Automation Networks. Part 3: Parallel Redundancy Protocol (PRP) and High-Availability Seamless Redundancy (HSR)*, IEC 62439-3:2010/Amd1, IEC, Geneva, Switzerland, 2010.

International Electrotechnical Committee, *Information Technology—Open Systems Interconnection—Basic Reference Model: The Basic Model*, ISO/IEC 7498-1, 2nd edn., IEC, Geneva, Switzerland, 1996.

Madani, V., Martin, K., and Novosel, D., Synchrophasor standards and guides, Presented at *NASPI General Meeting*, Ft. Worth, TX, 2011, pp. 1–21.

McDonald, J., North Carolina municipal power agency boosts revenue by replacing SCADA, *Electricity Today*, 15(7), 2003.

McDonald, J., Substation integration and automation—Fundamentals and best practices, *IEEE Power and Energy*, 1, March 2003.

McDonald, J., Caceres, D., Borlase, S., Janssen, M., and Olaya, J.C., ISA embraces open architecture, *Transmission and Distribution World*, 51(9), 68–75, October 1999.

McDonald, J., Doghman, M., and Dahl, B., Present and future integration of diagnostic equipment monitoring at OPPD, Paper presented at *EPRI Substation Equipment Diagnostics Conference IX*, Palo Alto, CA, 2001, pp. 1–5.

McDonald, J., Daugherty, R., and Ervin, S., On the road of intelligent distribution, *Transmission and Distribution World*, September 2006.

McDonald, J.D., Acquiring operational and non-operational data from substation IEDs, SCADA, Substation and Feeder Automation in Electric Utilities Short Course, 2008.

McDonald, J. et al., *Electric Power Engineering Handbook*, CRC Press, Boca Raton, FL, 2000.

Parashar, M. and Bilke, T., North American synchrophasor initiative, phasor technology overview, NASPI OITT Webcast, January 22, 2008.

Patel, M. et al., *Real-Time Application of Synchrophasors for Improving Reliability*, 2010 [Online]. Available: www.naspi.org/resources/papers/rapir_final_20101017.pdf [Accessed: April 12, 2011].

Sidhu, T., Kanabar, M., and Parikh, P., Implementation issues with IEC 61850 based substation automation systems, Presented at *Fifteenth National Power Systems Conference (NPSC)*, IIT Bombay, India, 2008, pp. 473–478.

Wester, C. and Adamiak, M., Practical applications of Ethernet in substations and industrial facilities, Presented at *Power Systems 2011 Conference*, Clemson University, Clemson, SC, 2011, pp. 1–12.

8
Oil Containment*

8.1	Oil-Filled Equipment in Substation 8-2
	Large Oil-Filled Equipment • Cables • Mobile Equipment • Oil-Handling Equipment • Oil Storage Tanks • Other Sources • Spill Risk Assessment
8.2	Containment Selection Consideration 8-4
8.3	Oil Spill Prevention Techniques 8-5
	Containment Systems • Discharge Control Systems
8.4	Warning Alarms and Monitoring 8-14
References 8-15

Thomas Meisner
Hydro One Networks, Inc.

Containment and control of oil leaks and spills at electric supply substations is a concern for electric utilities. The environmental impact of oil spills and their cleanup is governed by regulatory authorities necessitating increased attention in substations to the need for secondary oil containment and a Spill Prevention Control and Countermeasure (SPCC) plan. Beyond the threat to the environment, cleanup costs associated with oil spills could be significant, and the adverse community response to any spill is becoming increasingly unacceptable.

The probability of an oil spill occurring in a substation is very low. However, certain substations, due to their proximity to navigable waters or designated wetlands, the quantity of oil on site, surrounding topography, soil characteristics, etc., have or will have a higher potential for discharging harmful quantities of oil into the environment. At minimum, an SPCC plan will probably be required at these locations, and installation of secondary oil-containment facilities might be the right approach to mitigate the problem.

Before an adequate spill prevention plan is prepared and a containment system is devised, the engineer must first be thoroughly aware of the regulatory requirements.

The federal requirements of the United States for discharge, control, and countermeasure plans for oil spills are contained in the Code of Federal Regulations, Title 40 (40CFR), Parts 110 and 112. The aforementioned regulations only apply if the facility meets the following conditions:

1. Facilities with aboveground storage capacities greater than 2500 L (approximately 660 gal) in a single container or 5000 L (approximately 1320 gal) in aggregate storage
2. Facilities with a total storage capacity greater than 159,000 L (approximately 42,000 gal) of buried oil storage

---

* Sections of this chapter reprinted from IEEE Std. 980-1994 (R2001), *IEEE Guide for Containment and Control of Oil Spills in Substations*, 1995, Institute of Electrical and Electronics Engineers, Inc. (IEEE). The IEEE disclaims any responsibility or liability resulting from the placement and use in the described manner. Information is reprinted with permission of the IEEE.

3. *Any* facility that has spilled more than 3786 L (1000 gal) of oil in a single event or spilled oil in two events occurring within a 12 month period
4. Facilities that, due to their location, could reasonably be expected to discharge oil into or upon the navigable waters of the United States or its adjoining shorelines

In other countries, applicable governmental regulations will cover the aforementioned requirements.

8.1 Oil-Filled Equipment in Substation

A number of electrical apparatus installed in substations are filled with oil that provides the necessary insulation characteristics and assures their required performance (IEEE, 1994). Electrical faults in this power equipment can produce arcing and excessive temperatures that may vaporize insulating oil, creating excessive pressure that may rupture the electrical equipment tanks. In addition, operator errors, sabotage, or faulty equipment may also be responsible for oil releases.

The initial cause of an oil release or fire in electrical apparatus may not always be avoidable, but the extent of damage and the consequences of such an incident can be minimized or prevented by adequate planning in prevention and control.

Described in the following are various sources of oil spills within substations. Spills from any of these devices are possible. The user must evaluate the quantity of oil present, the potential impact of a spill, and the need for oil containment associated with each oil-filled device.

8.1.1 Large Oil-Filled Equipment

Power transformers, oil-filled reactors, large regulators, and circuit breakers are the greatest potential source of major oil spills in substations, since they typically contain the largest quantity of oil.

Power transformers, reactors, and regulators may contain anywhere from a few hundred to 100,000 L or more of oil (500 to approximately 30,000 gal), with 7,500–38,000 L (approximately 2,000–10,000 gal) being typical. Substations usually contain one to four power transformers, but may have more.

The higher voltage oil circuit breakers may have three independent tanks, each containing 400–15,000 L (approximately 100–4,000 gal) of oil, depending on their rating. However, most circuit breaker tanks contain less than 4500 L (approximately 1,200 gal) of oil. Substations may have 10–20 or more oil circuit breakers.

8.1.2 Cables

Substation pumping facilities and cable terminations (potheads) that maintain oil pressure in pipe-type cable installations are another source of oil spills. Depending on its length and rating, a pipe-type cable system may contain anywhere from 5,000 L (approximately 1,500 gal) up to 38,000 L (approximately 10,000 gal) or more of oil.

8.1.3 Mobile Equipment

Although mobile equipment and emergency facilities may be used infrequently, consideration should be given to the quantity of oil contained and associated risk of oil spill. Mobile equipment may contain up to 30,000 L (approximately 7,500 gal) of oil.

8.1.4 Oil-Handling Equipment

Oil filling of transformers, circuit breakers, cables, etc., occurs when the equipment is initially installed. In addition, periodic reprocessing or replacement of the oil may be necessary to ensure that proper insulation qualities are maintained. Oil pumps, temporary storage facilities, hoses, etc., are brought in to accomplish this task. Although oil-processing and oil-handling activities are less common, spills from these devices can still occur.

8.1.5 Oil Storage Tanks

Some consideration must be given to the presence of bulk oil storage tanks (either aboveground or belowground) in substations as these oil tanks could be responsible for an oil spill of significant magnitude. Also, the resulting applicability of the 40CFR, Part 112 rules for these storage tanks could require increased secondary oil containment for the entire substation facility. The user may want to reconsider storage of bulk oil at substation sites.

8.1.6 Other Sources

Station service, voltage, and current transformers, as well as smaller voltage regulators, oil circuit reclosers, capacitor banks, and other pieces of electrical equipment typically found in substations, contain small amounts of insulating oil, usually less than the 2500 L (approximately 660 gal) minimum for a single container.

8.1.7 Spill Risk Assessment

The risk of an oil spill caused by an electric equipment failure is dependent on many factors, including

- Engineering and operating practices (i.e., electrical fault protection, loading practices, switching operations, testing, and maintenance)
- Quantities of oil contained within apparatus
- Station layout (i.e., spatial arrangement, proximity to property lines, streams, and other bodies of water)
- Station topography and site preparation (i.e., slope, soil conditions, ground cover)
- Rate of flow of discharged oil

Each facility must be evaluated to select the safeguards commensurate with the risk of a potential oil spill.

The engineer must first consider whether the quantities of oil contained in the station exceed the quantities of oil specified in the regulations, and secondly, the likelihood of the oil reaching navigable waters if an oil spill or rupture occurs. If no likelihood exists, no SPCC plan is required.

SPCC plans must be prepared for each piece of portable equipment and mobile substations. These plans have to be general enough that the plan may be used at any and all substations or facility location.

Both the frequency and magnitude of oil spills in substations can be considered to be very low. The probability of an oil spill at any particular location depends on the number and volume of oil containers and other site-specific conditions.

Based on the applicability of the latest regulatory requirements, or when an unacceptable level of oil spills has been experienced, it is recommended that a program be put in place to mitigate the problems. Typical criteria for implementing oil spill containment and control programs incorporate regulatory requirements, corporate policy, frequency and duration of occurrences, cost of occurrences, safety hazards, severity of damage, equipment type, potential impact on nearby customers, substation location, and quality-of-service requirements (IEEE, 1994).

The decision to install secondary containment at new substations (or to retrofit existing substations) is usually based on predetermined criteria. A 1992 IEEE survey addressed the factors used to determine where oil spill containment and control programs are needed. Based on the survey, the criteria in Table 8.1 are considered when evaluating the need for secondary oil containment.

TABLE 8.1 Secondary Oil-Containment Evaluation Criteria

Criteria	Utilities Responding That Apply These Criteria (%)
Volume of oil in individual device	88
Proximity to navigable waters	86
Total volume of oil in substation	62
Potential contamination of groundwater	61
Soil characteristics of the station	42
Location of substation (urban, rural, remote)	39
Emergency response time if a spill occurs	30
Failure probability of the equipment	21
Age of station or equipment	10

Source: IEEE, *IEEE Guide for Containment and Control of Oil Spills in Substations*, IEEE Std. 980-1994 (R2001), 1994.

TABLE 8.2 Secondary Oil-Containment Equipment Criteria

Equipment	Utilities Responding That Provide Secondary Containment (%)
Power transformers	86
Aboveground oil storage tanks	77
Station service transformers	44
Oil circuit breakers	43
Three-phase regulators	34
Belowground oil storage tanks	28
Shunt reactors	26
Oil-filling equipment	22
Oil-filled cables and terminal stations	22
Single-phase regulators	19
Oil circuit reclosers	15

Source: IEEE, *IEEE Guide for Containment and Control of Oil Spills in Substations*, IEEE Std. 980-1994 (R2001), 1994.

The same 1992 IEEE survey provided no clear-cut limit for the proximity to navigable waters. Relatively, equal support was reported for several choices over the range of 45–450 m (150–1500 ft).

Rarely is all of the equipment within a given substation provided with secondary containment. Table 8.2 lists the 1992 IEEE survey results identifying the equipment for which secondary oil containment is provided.

Whatever the criteria, each substation has to be evaluated by considering the criteria to determine candidate substations for oil-containment systems (both new and retrofit). Substations with planned equipment change-outs and located in environmentally sensitive areas have to be considered for retrofits at the time of the change-out.

8.2 Containment Selection Consideration

Containment selection criteria have to be applied in the process of deciding the containment option to install in a given substation (IEEE, 1994). Criteria to be considered include operating history of the equipment, environmental sensitivity of the area, the solution's cost–benefit ratio, applicable governmental regulations, and community acceptance.

The anticipated cost of implementing the containment measures must be compared to the anticipated benefit. However, cost alone can no longer be considered a valid reason for not implementing containment and/or control measures because any contamination of navigable waters may be prohibited by government regulations.

Economic aspects can be considered when determining which containment system or control method to employ. Factors such as proximity to waterways, volume of oil, response time following a spill, etc., can allow for the use of less effective methods at some locations.

Due to the dynamic nature of environmental regulations, some methods described in this section could come in conflict with governmental regulations or overlapping jurisdictions. Therefore, determination of which containment system or control method to use must include research into applicable laws and regulations.

Community acceptance of the oil spill containment and control methods is also to be considered. Company policies, community acceptance, customer relations, etc., may dictate certain considerations. The impact on adjacent property owners must be addressed and, if needed, a demonstration of performance experiences could be made available.

8.3 Oil Spill Prevention Techniques

Upon an engineering determination that an oil spill prevention system is needed, the engineer must weigh the advantages and disadvantages that each oil retention system may have at the facility in question. The oil retention system chosen must balance the cost and sophistication of the system to the risk of the damage to the surrounding environment. The risks, and thus the safeguards, will depend on items such as soil, terrain, relative closeness to waterways, and potential size of discharge. Each of the systems that are described in the following may be considered based on their relative merits to the facility under consideration. Thus, one system will not always be the best choice for all situations and circumstances.

8.3.1 Containment Systems

The utility has to weigh the advantages and disadvantages that each oil retention system may have at the facility in question. Some of the systems that could be considered based on their relative merits to the facility under consideration are presented in the following.

8.3.1.1 Yard Surfacing and Underlying Soil

100–150 mm (4–6 in.) of rock gravel surfacing are normally required in all electrical facility yards. This design feature benefits the operation and maintenance of the facility by providing proper site drainage, reducing step and touch potentials during short-circuit faults, eliminating weed growth, improving yard working conditions, and enhancing station aesthetics. In addition to these advantages, this gravel will aid in fire control and in reducing potential oil spill cleanup costs and penalties that may arise from federal and state environmental laws and regulations.

Yard surfacing is not to be designed to be the primary or only method of oil containment within the substation, but rather has to be considered as a backup or bonus in limiting the flow of oil in the event that the primary system does not function as anticipated.

Soil underlying power facilities usually consists of a non-homogeneous mass that varies in composition, porosity, and physical properties with depth.

Soils and their permeability characteristics have been adapted from typical references and can be generalized as in Table 8.3.

8.3.1.2 Substation Ditching

One of the simplest methods of providing total substation oil spill control is the construction of a ditch entirely around the outside periphery of the station. The ditch has to be of adequate size as to contain all surface runoffs due to rain and insulating oil. These ditches may be periodically drained by the use of valves.

TABLE 8.3 Soil Permeability Characteristics

Permeability (cm/s)	Degree of Permeability	Type of Soil
Over 10^{-1}	High	Stone, gravel, and coarse- to medium-grained sand
10^{-1} to 10^{-3}	Medium	Medium-grained sand to uniform, fine-grained sand
10^{-3} to 10^{-6}	Low	Uniform, fine-grained sand to silty sand or sandy clay
Less than 10^{-6}	Practically impermeable	Silty sand or sandy clay to clay

Source: IEEE, *IEEE Guide for Containment and Control of Oil Spills in Substations,* IEEE Std. 980-1994 (R2001), 1994.

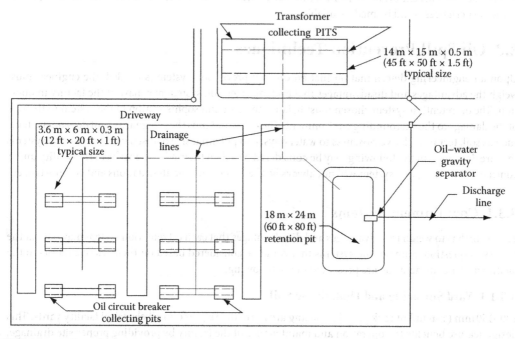

FIGURE 8.1 Typical containment system with retention and collection pits.

8.3.1.3 Collecting Ponds with Traps

In this system, the complete design consists of a *collection pit* surrounding the protected equipment, drains connecting the collection pits to an open *containment pit*, and an oil trap that is sometimes referred to as a skimming unit and the discharge drain. Figure 8.1 (IEEE, 1994) presents the general concept of such a containment solution. The collection pit surrounding the equipment is filled with rocks and designed only deep enough to extinguish burning oil. The bottom of this pit is sloped for good drainage to the drainpipe leading to an open containment pit. This latter pit is sized to handle all the oil of the largest piece of equipment in the station. To maintain a dry system in the collecting units, the invert of the intake pipe to the containment pit must be at least the maximum elevation of the oil level. In areas of the country subject to freezing temperatures, it is recommended that the trap (skimmer) be encased in concrete, or other similar means available, to eliminate heaving due to ice action.

8.3.1.4 Oil-Containment Equipment Pits

Probably one of the most reliable but most expensive methods of preventing oil spills and insuring that oil will be contained on the substation property is by placing all major substation equipment on

Oil Containment

or in containment pits. This method of oil retention provides a permanent means of oil containment. These containment pits will confine the spilled oil to relatively small areas that in most cases will greatly reduce the cleanup costs.

One of the most important issues related to an equipment pit is to prevent escape of spilled oil into underlying soil layers. Pits with liners or sealers may be used as part of an oil containment system capable of retaining any discharged oil for an extended period of time. Any containment pit must be constructed with materials having medium to high impermeability (above 10^{-3} cm/s) and be sealed in order to prevent migration of spilled oil into underlying soil layers and groundwater. These surfaces may be sealed and/or lined with any of the following materials:

1. *Plastic or rubber*—Plastic or rubber liners may be purchased in various thickness and sizes. It is recommended that a liner be selected that is resistant to mechanical injury, which may occur due to construction and installation, equipment, chemical attacks on surrounding media, and oil products.
2. *Bentonite (clay)*—Clay and bentonite may also be used to seal electrical facility yards and containment pits. These materials can be placed directly in 100–150 mm (4–6 in.) layers or may be mixed with the existing subsoil to obtain an overall soil permeability of less than 10^{-3} cm/s.
3. *Spray-on fiberglass*—Spray-on fiberglass is one of the most expensive pit liners available, but in some cases, the costs may be justifiable in areas, which are environmentally sensitive. This material offers very good mechanical strength properties and provides excellent oil retention.
4. *Reinforced concrete*—200 mm (7 7/8 in.) of reinforced concrete may also be used as a pit liner. This material has an advantage over other types of liners in that it is readily available at the site at the time of initial construction of the facility. Concrete has some disadvantages in that initial preparation is more expensive and materials are not as easily workable as some of the other materials.

If materials other than those listed earlier are used for an oil-containment liner, careful consideration must be given to selecting materials, which will not dissolve or become soft with prolonged contact with oil, such as asphalt.

8.3.1.5 Fire-Quenching Considerations

In places where the oil-filled device is installed in an open pit (not filled with stone), an eventual oil spill associated with fire will result in a pool fire around the affected piece of equipment (IEEE, 1994). If a major fire occurs, the equipment will likely be destroyed. Most utilities address this concern by employing active or passive quenching systems or drain the oil to a remote pit. Active systems include foam or water spray deluge systems.

Of the passive fire-quenching measures, pits filled with 19–37.5 mm (3/4 to 1½ in.) clear stone with a maximum of 5% passing at the 19 mm sieve are the most effective. The stone-filled pit provides a fire-quenching capability designed to extinguish flames in the event that a piece of oil-filled equipment catches on fire. An important point is that in sizing a stone-filled collecting or retention pit, the final oil level elevation (assuming a total discharge) has to be situated approximately 300 mm (12 in.) below the top elevation of the stone.

All the materials used in construction of a containment pit have to be capable of withstanding the higher temperatures associated with an oil fire without melting. If any part of the containment (i.e., discharge pipes from containment to a sump) melts, the oil will be unable to drain away from the burning equipment, and the melted materials may pose an environmental hazard.

8.3.1.6 Volume Requirements

Before a substation oil-containment system can be designed, the volume of oil to be contained must be known. Since the probability of an oil spill occurring at a substation is very low, the probability of simultaneous spills is extremely low. Therefore, it would be unreasonable and expensive to design a containment system to hold the sum total of all of the oil contained in the numerous oil-filled pieces of equipment normally installed in a substation. In general, it is recommended that an oil-containment system be sized to contain the volume of oil in the single largest oil-filled piece of equipment, plus any accumulated water from sources such as rainwater, melted snow, and water spray discharge from fire protection systems. Interconnection of two or more pits to share the discharged oil volume may provide an opportunity to reduce the size requirements for each individual pit.

Typically, equipment containment pits are designed to extend 1.5–3 m (5–10 ft) beyond the edge of the tank in order to capture a majority of the leaking oil. A larger pit size is required to capture all of the oil contained in an arcing stream from a small puncture at the bottom of the tank (such as from a bullet hole). However, the low probability of the event and economic considerations govern the 1.5–3 m (5–10 ft) design criteria. For all of the oil to be contained, the pit or berm has to extend 7.5 m (25 ft) or more beyond the tank and radiators.

The volume of the pit surrounding each piece of equipment has to be sufficient to contain the spilled oil in the air voids between the aggregate of gravel fill or stone. A gravel gradation with a nominal size of 19–50 mm (3/4 to 2 in.) that results in a void volume between 30% and 40% of the pit volume is generally being used. The theoretical maximum amount of oil that can be contained in 1 ft^3 or 1 m^3 of stone is given by the following formulae:

$$\text{Oil volume (gal)} = \frac{\text{Void volume of stone}(\%)}{100 \times 0.1337\,\text{ft}^3} \quad (8.1)$$

$$\text{Oil volume (L)} = \frac{\text{Void volume of stone}(\%)}{100 \times 0.001\,\text{m}^3} \quad (8.2)$$

where
1 gal = 0.1337 ft^3
1 L = 0.001 m^3 = 1 dm^3

If the pits are not to be automatically drained of rainwater, then an additional allowance must be made for precipitation. The additional space required would depend on the precipitation for that area and the frequency at which the facility is periodically inspected. It is generally recommended that the pits have sufficient space to contain the amount of rainfall for this period plus a 20% safety margin.

Expected rain and snow accumulations can be determined from local weather records. A severe rainstorm is often considered to be the worst-case event when determining the maximum volume of short-term water accumulation (for design purposes). From data reported in a 1992 IEEE survey, the storm water event design criteria employed ranged from 50 to 200 mm (2 to 8 in.) of rainfall within a short period of time (1–24 h). Generally, accepted design criteria is assuming a one in a 25 year storm event.

The area directly surrounding the pit must be graded to slope away from the pit to avoid filling the pit with water in times of rain.

8.3.1.7 Typical Equipment Containment Solutions

Figure 8.2 illustrates one method of pit construction that allows the equipment to be installed partially belowground. The sump pump can be manually operated during periods of heavy rain or automatically operated. If automatic operation is preferred, special precautions must be included to insure that oil is not pumped from the pits. This can be accomplished with either an oil-sensing probe or by having all major equipment provided with oil-limit switches (an option available from equipment suppliers). These limit switches are located just below the *minimum* top oil line in the equipment and will open when the oil level drops below this point.

A typical above-grade pit and/or berm, as shown in Figure 8.3, has maintenance disadvantages but can be constructed relatively easily after the equipment is in place at new and existing electrical facilities.

Oil Containment

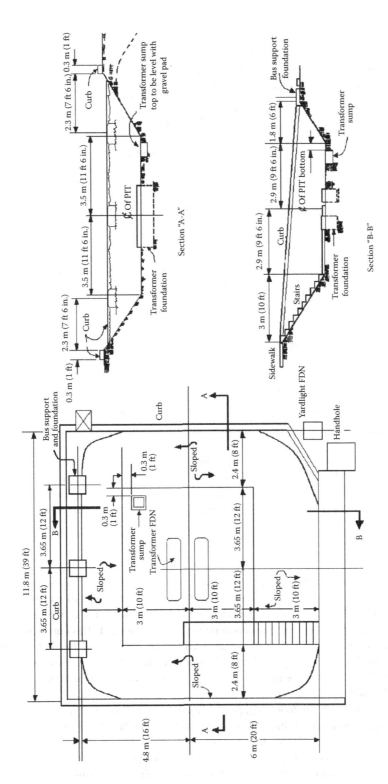

FIGURE 8.2 Typical below-grade containment pit.

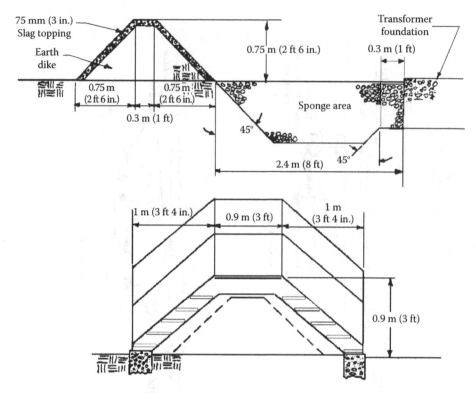

FIGURE 8.3 Typical above-grade berm/pit.

These pits may be emptied manually by gate valves or pumps depending on the facility terrain and layout or automatically implemented by the use of equipment oil limit switches and dc-operated valves or sump pumps.

Another method of pit construction is shown in Figure 8.4. The figure shows all-concrete containment pits installed around transformers. The sump and the control panel for the oil pump (located inside the sump) are visible and are located outside the containments. Underground piping provides the connection between the two adjacent containments and the sump. The containments are filled with fire-quenching stones.

8.3.2 Discharge Control Systems

An adequate and effective station drainage system is an essential part of any oil-containment design (IEEE, 1994). Drains, swales, culverts, catch basins, etc., provide measures to ensure that water is diverted away from the substation. In addition, the liquid accumulated in the collecting pits or sumps of various electrical equipment or in the retention pit has to be discharged. This liquid consists mainly of water (rainwater, melted snow or ice, water spray system discharges, etc.). Oil will be present only in case of an equipment discharge. It is general practice to provide containment systems that discharge the accumulated water into the drainage system of the substation or outside the station perimeter with a discharge control system.

These systems, described in the following, provide methods to release the accumulated water from the containment system while blocking the flow of discharged oil for later cleanup. Any collected water has to be released as soon as possible so that the entire capacity of the containment system is available for oil containment in the event of a spill. Where the ambient temperatures are high enough, evaporation may eliminate much of the accumulated water. However, the system still should be designed to handle the worst-case event.

Oil Containment

FIGURE 8.4 All-concrete containment pits.

8.3.2.1 Oil–Water Separator Systems

Oil–water separator systems rely on the difference in specific gravity between oil and water (IEEE, 1994). Because of that difference, the oil will naturally float on top of the water, allowing the water to act as a barrier and block the discharge of the oil.

Oil–water separator systems require the presence of water to operate effectively and will allow water to continue flowing even when oil is present. The presence of emulsified oil in the water may, under some turbulent conditions, allow small quantities of oil to pass through an oil–water separator system.

Figure 8.5 (IEEE, 1994) illustrates the detail of an oil–water gravity separator that is designed to allow water to discharge from a collecting or retention pit, while at the same time retaining the discharged oil.

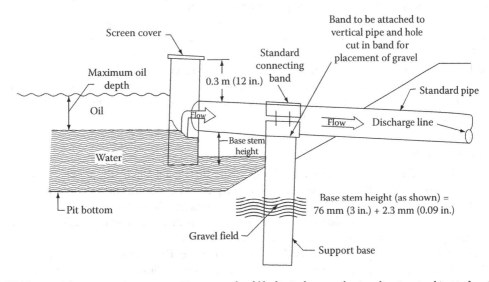

FIGURE 8.5 Oil–water gravity separator. *Note*: usage should be limited to areas having climate not subject to freezing.

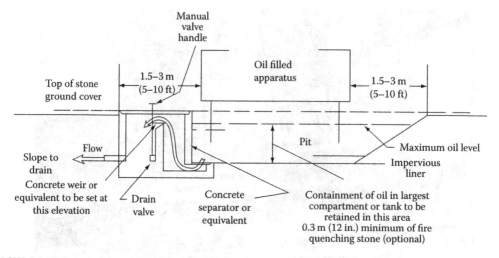

FIGURE 8.6 Equipment containment with oil–water separator.

Figure 8.6 (IEEE, 1994) illustrates another type of oil–water separator. This separator consists of a concrete enclosure, located inside a collecting or retention pit and connected to it through an opening located at the bottom of the pit. The enclosure is also connected to the drainage system of the substation. The elevation of the top of the concrete weir in the enclosure is selected to be slightly above the maximum elevation of discharged oil in the pit. In this way, the level of liquid in the pit will be under a layer of fire-quenching stones where a stone-filled pit is used. During heavy accumulation of water, the liquid will flow over the top of the weir into the drainage system of the station. A valve is incorporated in the weir. This normally closed, manually operated valve allows for a controlled discharge of water from the pit when the level of liquid in the pit and enclosure is below the top of the weir.

Figure 8.7 (IEEE, 1994) provides typical detail of an oil trap type oil–water separator. In this system, the oil will remain on top of the water and not develop the head pressure necessary to reach the bottom of the inner vertical pipe. In order for this system to function properly, the water level in the manhole portion of the oil trap must be maintained at an elevation no lower than 0.6 m (2 ft) below the inlet elevation. This will ensure that an adequate amount of water is available to develop the necessary hydraulic head within the inner (smaller) vertical pipe, thereby preventing any discharged oil from leaving the site. It is important to note that the inner vertical pipe should be extended downward past the calculated water–oil interface elevation sufficiently to ensure that oil cannot discharge upward through the inner pipe. Likewise, the inner pipe must extend higher than the calculated oil level elevation in the manhole to ensure that oil does not drain downward into the inner pipe through the vented plug. The reason for venting the top plug is to maintain atmospheric pressure within the vertical pipe, thereby preventing any possible siphon effect.

8.3.2.2 Flow Blocking Systems

Described in the following are two oil flow blocking systems that do not require the presence of water to operate effectively (IEEE, 1994). These systems detect the presence of oil and block all flow (both water and oil) through the discharge system. The best of these systems have been shown to be the most sensitive in detecting and blocking the flow of oil. However, they are generally of a more complex design and may require greater maintenance to ensure continued effectiveness.

Figure 8.8 illustrates an oil stop valve installed inside a manhole. The valve has only one moving part: a ballasted float set at a specific gravity between that of oil and water. When oil reaches the manhole, the

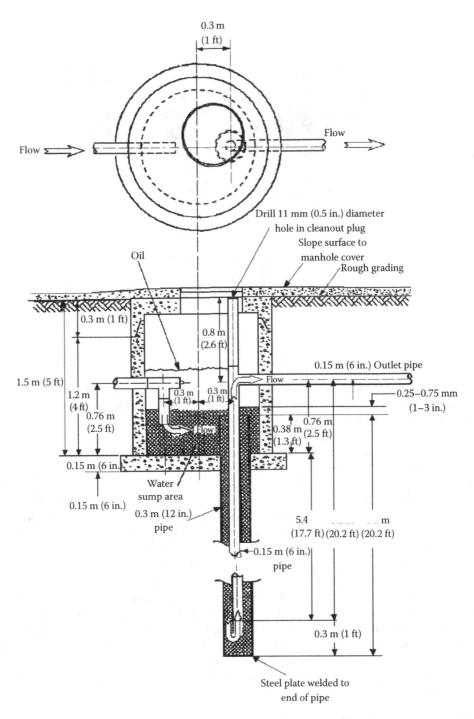

FIGURE 8.7 Oil trap type oil–water separator. To ensure proper functioning of the oil trap structure, a minimum water level shall be maintained to a depth not less than mid depth of the 0.3 m (12 in.) diameter steel pipe from a practical point of view. The minimum water level should be approximately 25–75 mm (1–3 in.) above the top of the steel pipe which projects through the bottom of the manhole.

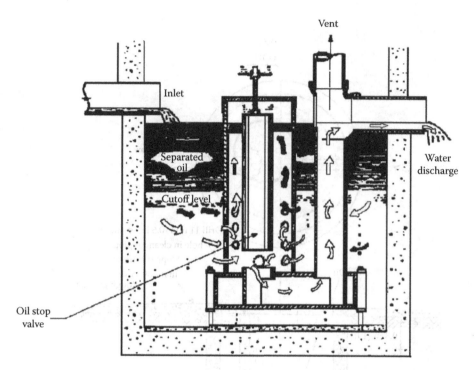

FIGURE 8.8 Oil stop valve installed in manhole.

float in the valve loses buoyancy and sinks as the oil level increases until it sits on the discharge opening of the valve and blocks any further discharge. When the oil level in the manhole decreases, the float will rise automatically and allow discharge of water from the manhole. Some of the oil stop valves have a weep hole in the bottom of the valve that allows the ballasted float to be released after the oil is removed. This can cause oil to discharge if the level of the oil is above the invert of the discharge pipe.

Figure 8.9 illustrates a discharge control system consisting of an oil-detecting device and a pump installed in a sump connected to the collecting or retention pits of the oil-containment system. The oil-detecting device may use different methods of oil sensing (e.g., capacitance probes, turbidimeters, and fluorescence meters). The capacitance probe shown detects the presence of oil on the surface of the water, based on the significant capacitance difference of these two liquids and, in combination with a logic of liquid level switches, stops the sump pump when the water–oil separation layer reaches a preset height in the sump. Transformer low oil-level or gas protection can be added into the control diagram of the pump in order to increase the reliability of the system during major spills.

Some containment systems consist of collecting pits connected to a retention pit or tank that have no link to the drainage system of the substation. Discharge of the liquid accumulated in these systems requires the use of permanently installed or portable pumps. However, should these probes become contaminated, they may cease to function properly. Operating personnel manually activate these pumps. This system requires periodic inspection to determine the level of water accumulation. Before pumping any accumulated liquid, an inspection is required to assess whether the liquid to be pumped out is contaminated.

8.4 Warning Alarms and Monitoring

In the event of an oil spill, it is imperative that cleanup operations and procedures be initiated as soon as possible to prevent the discharge of any oil or to reduce the amount of oil reaching navigable waters (IEEE, 1994). Hence, it may be desirable to install an early detection system for alerting responsible

Oil Containment

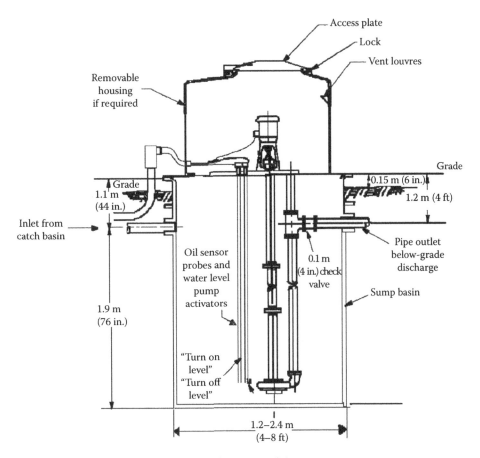

FIGURE 8.9 Sump pump water discharge (with oil sensing probe).

personnel of an oil spill. Some governmental regulations may require that the point of discharge (for accumulated water) from a substation be monitored and/or licensed.

The most effective alarms are the ones activated by the presence of oil in the containment system. A low oil-level indicator within the oil-filled equipment can be used; however, it may not activate until 3%–6% of the transformer oil has already discharged. In cases where time is critical, it may be worthwhile to also consider a faster operating alarm such as one linked to the transformer sudden gas pressure relay. Interlocks have to be considered as a backup to automatic pump or valve controls.

Alarms are transmitted via supervisory equipment or a remote alarm system to identify the specific problem. The appropriate personnel are then informed so that they can determine if a spill has occurred and implement the SPCC contingency plan.

References

Design Guide for Oil Spill Prevention and Control at Substations, U.S. Department of Agriculture, Rural Electrification Administration Bulletin 65-3, January, 1981.
IEEE Guide for Containment and Control of Oil Spills in Substations, IEEE Std. 980-1994 (R2001).

9
Community Considerations*

James H. Sosinski
(retired)
Consumers Energy

9.1	Community Acceptance	9-1
9.2	Planning Strategies and Design	9-2
	Site Location and Selection, and Preparation • Aesthetics • Electric and Magnetic Fields • Safety and Security • Permitting Process	
9.3	Construction	9-11
	Site Preparation • Noise • Safety and Security • Site Housekeeping • Hazardous Material	
9.4	Operations	9-12
	Site Housekeeping • Fire Protection • Hazardous Material	
9.5	Defining Terms	9-14
	References	9-14

9.1 Community Acceptance

Community acceptance generally encompasses the planning, design, and construction phases of a substation as well as the in-service operation of the substation. It takes into account those issues that could influence a community's willingness to accept building a substation at a specific site. New substations or expansions of existing facilities often require extensive review for community acceptance. Government bodies typically require a variety of permits before construction may begin.

For community acceptance, several considerations should be satisfactorily addressed, including the following:

- Noise
- Site preparations
- Aesthetics
- Fire protection
- Potable water and sewage
- Hazardous materials
- Electric and magnetic fields
- Safety and security

This chapter on Community Considerations is essentially a condensed version of IEEE Standard 1127-1998.

---

* Chapters 4, 5, 6, 7, and 8 (excluding Sections 5.3.2.2, 5.3.5, 5.4.2.1, 5.4.2.2, 5.4.2.3, 5.4.3.1, 5.4.3.2, 5.4.3.3, 5.4.3.4, 5.4.3.5, 6.1, 6.2, 7.1.4, 7.4, 8.2.1., 8.2.2, Tables 8.1 and 8.2, and Figs. 8.1 and 8.2) reprinted from IEEE Std. 1127–1998, "IEEE Guide for the Design, Construction, and Operation of Electric Power Substations for Community Acceptance and Environmental Compatibility" Copyright © 1998, by the Institute of Electrical and Electronics Engineers, Inc. (IEEE). The IEEE disclaims any responsibility or liability resulting from the placement and use in the described manner. Information is reprinted with the permission of the IEEE.

9.2 Planning Strategies and Design

Planning is essential for the successful design, construction, and operation of a substation. The substation's location and proximity to wetlands, other sensitive areas, and contaminated soils; its aesthetic impact; and the concerns of nearby residents over noise and electric and magnetic fields (EMF) can significantly impact the ability to achieve community acceptance. Public perceptions and attitudes toward both real and perceived issues can affect the ability to obtain all necessary approvals and permits.

These issues can be addressed through presentations to governmental officials and the public. Failure to obtain community acceptance can delay the schedule or, in the extreme, stop a project completely.

9.2.1 Site Location and Selection, and Preparation

The station location (especially for new substations) is the key factor in determining the success of any substation project. Although the site location is based on electric system load growth studies, the final site location may ultimately depend upon satisfying the public and resolving potential community acceptance concerns. If necessary, a proactive public involvement program should be developed and implemented. The best substation site selection is influenced by several factors including, but not limited to, the following:

1. Community attitudes and perceptions
2. Location of nearby wetlands, bodies of water, or environmentally sensitive areas
3. Site contamination (obvious or hidden)
4. Commercial, industrial, and residential neighbors, including airports
5. Permit requirements and ordinances
6. Substation layout (including future expansions) and placement of noise sources
7. Levels of electric and magnetic fields
8. Availability and site clearing requirements for construction staging
9. Access to water and sewer
10. Drainage patterns and storm water management
11. Potential interference with radio, television, and other communication installations
12. Disturbance of archaeological, historical, or culturally significant sites
13. Underground services and geology
14. Accessibility
15. Aesthetic and screening considerations

9.2.1.1 Wetlands

A site-development plan is necessary for a substation project that borders wetlands. Such a plan for the site and its immediate surroundings should include the following:

1. Land-use description
2. Grades and contours
3. Locations of any wetland boundaries and stream-channel encroachment lines
4. Indication of flood-prone areas and vertical distance or access to ground water
5. Indication of existing wildlife habitats and migratory patterns

The plan should describe how site preparation will modify or otherwise impact these areas and what permanent control measures will be employed, including ground water protection.

9.2.1.2 Site Contamination

Soil borings should be taken on any proposed substation site to determine the potential presence of soil contaminants.

There are many substances that, if found on or under a substation site, would make the site unusable or require excessive funds to remediate the site before it would be usable. Some of the substances are as follows:

1. Polychlorinated biphenyls (PCBs)
2. Asbestos
3. Lead and other heavy metals
4. Pesticides and herbicides
5. Radioactive materials
6. Petrochemicals
7. Dioxin
8. Oil

Governmental guidelines for the levels of these substances should be used to determine if the substance is present in large enough quantities to be of concern.

The cost of removal and disposal of any contaminants should be considered before acquiring or developing the site. If a cleanup is needed, the acquisition of another site should be considered as governmental regulations can hold the current owner or user of a site responsible for cleanup of any contamination present, even if substances were deposited prior to acquisition. If a cleanup is initiated, all applicable governmental guidelines and procedures should be followed.

9.2.1.3 Potable Water and Sewage

The substation site may need potable water and sewage disposal facilities. Water may be obtained from municipal or cooperative water utilities or from private wells. Sewage may be disposed of by municipal services or septic systems, or the site could be routinely serviced by portable toilet facilities, which are often used during construction. Where municipal services are used for either water or sewer service, the requirements of that municipality must be met. Septic systems, when used, should meet all applicable local, state, and federal regulations.

9.2.2 Aesthetics

Aesthetics play a major role where community acceptance of a substation is an issue. Sites should be selected that fit into the context of present and future community patterns.

Community acceptability of a site can be influenced by

1. Concerns about compatibility with present and future land uses
2. Building styles in the surrounding environment
3. Landscape of the site terrain
4. Allowance for buffer zones for effective blending, landscaping, and safety
5. Site access that harmonizes with the community

In addition, the site may need to be large enough to accommodate mobile emergency units and future expansions without becoming congested.

9.2.2.1 Visual Simulation

Traditionally, a site rendering was an artist's sketch, drawing, painting, or photomontage with airbrush retouching, preferably in color, and as accurate and realistic as possible. In recent years, these traditional techniques, although still employed, have given way to two- and three-dimensional computer-generated images, photorealism, modeling, and animation to simulate and predict the impact of proposed developments.

This has led to increased accuracy and speed of image generation in the portrayal of new facilities for multiple-viewing (observer) positions, allowing changes to be made early in the decision-making process while avoiding costly alterations that sometimes occur later during construction.

A slide library of several hundred slides of aesthetic design choices is available from the IEEE. It is a compilation of landscaping, decorative walls and enclosures, plantings, and site location choices that have been used by various utilities worldwide to ensure community acceptance and environmental compatibility.

9.2.2.2 Landscaping and Topography

Landscaping: Where buffer space exists, landscaping can be a very effective aesthetic treatment. On a site with little natural screening, plantings can be used in concert with architectural features to complement and soften the visual effect.

All plantings should be locally available and compatible types, and should require minimum maintenance. Their location near walls and fences should not compromise either substation grounding or the security against trespass by people or animals.

Topography: Topography or land form, whether shaped by nature or by man, can be one of the most useful elements of the site to solve aesthetic and functional site development problems.

Use of topography as a visual screen is often overlooked. Functionally, earth forms can be permanent, visual screens constructed from normal on-site excavating operations. When combined with plantings of grass, bushes, or evergreens and a planned setback of the substation, berms can effectively shield the substation from nearby roads and residents.

Fences and walls: The National Electrical Safety Code® ([NESC®] [Accredited Standards Committee C2-2007]) requires that fences, screens, partitions, or walls be employed to keep unauthorized persons away from substation equipment.

Chain-link fences: This type of fence is the least vulnerable to graffiti and is generally the lowest-cost option. Chain-link fences can be galvanized or painted in dark colors to minimize their visibility, or they can be obtained with vinyl cladding. They can also be installed with wooden slats or colored plastic strips woven into the fence fabric. Grounding and maintenance considerations should be reviewed before selecting such options.

Wood fences: This type of fence should be constructed using naturally rot-resistant or pressure-treated wood, in natural color or stained for durability and appearance. A wood fence can be visually overpowering in some settings. Wood fences should be applied with caution because wood is more susceptible to deterioration than masonry or metal.

Walls: Although metal panel and concrete block masonry walls cost considerably more than chain-link and wood fences, they deserve consideration where natural or landscaped screening does not provide a sufficient aesthetic treatment. Brick and precast concrete can also be used in solid walls, but these materials are typically more costly. These materials should be considered where necessary for architectural compatibility with neighboring facilities. Walls can offer noise reduction (discussed later) but can be subject to graffiti. All issues should be considered before selecting a particular wall or fence type.

9.2.2.3 Color

When substations are not well screened from the community, color can have an impact on the visual effect.

Above the skyline, the function of color is usually confined to eliminating reflective glare from bright metal surfaces. Because the sun's direction and the brightness of the background sky vary, no one color can soften the appearance of substation structures in the course of changing daylight. Below the skyline, color can be used in three aesthetic capacities. Drab coloring, using earth tones and achromatic hues, is a technique that masks the metallic sheen of such objects as chain-link fences and steel structures, and reduces visual contrast with the surrounding landscape. Such coloring should have very limited variation in hues, but contrast by varying paint saturation is often more effective than a monotone coating. Colors and screening can often be used synergistically. A second technique is to use color to direct visual attention to more aesthetically pleasing items such as decorative walls and enclosures. In this use, some brightness is warranted, but highly saturated or contrasting hues should be avoided. A third technique is to brightly color equipment and structures for intense visual impact.

9.2.2.4 Lighting

When attractive landscaping, decorative fences, enclosures, and colors have been used to enhance the appearance of a highly visible substation, it may also be appropriate to use lighting to highlight some of these features at night. Although all-night lighting can enhance substation security and access at night, it should be applied with due concern for nearby residences.

9.2.2.5 Structures

The importance of aesthetic structure design increases when structures extend into the skyline. The skyline profile typically ranges from 6 to 10 m (20 to 35 ft) above ground. Transmission line termination structures are usually the tallest and most obvious. Use of underground line exits will have the greatest impact on the substation's skyline profile. Where underground exits are not feasible, low-profile station designs should be considered. Often, low-profile structures will result in the substation being below the nearby tree line profile.

For additional cost, the most efficient structure design can be modified to improve its appearance. The following design ideas may be used to improve the appearance of structures:

1. Tubular construction
2. Climbing devices not visible in profile
3. No splices in the skyline zone
4. Limiting member aspect ratio for slimmer appearance
5. Use splices other than pipe-flange type
6. Use of gusset plates with right-angle corners not visible in profile
7. Tapering ends of cantilevers
8. Equal length of truss panel
9. Making truss diagonals with an approximate 60° angle to chords
10. Use of short knee braces or moment-resistant connections instead of full-height diagonal braces
11. Use of lap splice plates only on the insides of H-section flanges

9.2.2.6 Enclosures

Total enclosure of a substation within a building is an option in urban settings where underground cables are used as supply and feeder lines. Enclosure by high walls, however, may be preferred if enclosure concealment is necessary for community acceptance.

A less costly design alternative in nonurban locales that are served by overhead power lines is to take advantage of equipment enclosures to modify visual impact. Relay and control equipment, station batteries, and indoor power switchgear all require enclosures. These enclosures can be aesthetically designed and strategically located to supplement landscape concealment of other substation equipment. The exterior appearance of these enclosures can also be designed (size, color, materials, shape) to match neighboring homes or buildings.

Industrial-type, pre-engineered metal enclosures are a versatile and economic choice for substation equipment enclosures. Concrete block construction is also a common choice for which special shaped and colored blocks may be selected to achieve a desired architectural effect. Brick, architectural metal panels, and precast concrete can also be used.

Substation equipment enclosures usually are not exempt from local building codes. Community acceptance, therefore, requires enclosure design, approval, and inspection in accordance with local regulation.

9.2.2.7 Bus Design

Substations can be constructed partly or entirely within aboveground or belowground enclosures. However, cost is high and complexity is increased by fire-protection and heat-removal needs. The following discussion deals with exposed aboveground substations.

Air-insulated substations: The bus and associated substation equipment are exposed and directly visible. An outdoor bus may be multitiered or spread out at one level. Metal or wood structures and insulators support such bus and power line terminations. Space permitting, a low-profile bus layout is generally best for aesthetics and is the easiest to conceal with landscaping, walls, and enclosures. Overhead transmission line terminating structures are taller and more difficult to conceal in such a layout. In dry climates, a low-profile bus can be achieved by excavating the earth area, within which outdoor bus facilities are then located for an even lower profile.

Switchgear: Metal-enclosed or metal-clad switchgear designs that house the bus and associated equipment in a metal enclosure are an alternative design for distribution voltages. These designs provide a compact low-profile installation that may be aesthetically acceptable.

Gas-insulated substation (GIS): Bus and associated equipment can be housed within pipe-type enclosures using sulfur hexafluoride or another similar gas for insulation. Not only can this achieve considerable compactness and reduced site preparation for higher voltages, but it can also be installed lower to the ground. A GIS can be an economically attractive design where space is at a premium, especially if a building-type enclosure will be used to house substation equipment (see IEEE Std. C37.123-1996).

Cable bus: Short sections of overhead or underground cables can be used at substations, although this use is normally limited to distribution voltages (e.g., for feeder getaways or transformer-to-switchgear connections). At higher voltages, underground cable can be used for line-entries or to resolve a specific connection problem.

Noise: Audible noise, particularly continuously radiated discrete tones (e.g., from power transformers), is the type of noise that the community may find unacceptable. Community guidelines to ensure that acceptable noise levels are maintained can take the form of governmental regulations or individual/community reaction (permit denial, threat of complaint to utility regulators, etc.). Where noise is a potential concern, field measurements of the area background noise levels and computer simulations predicting the impact of the substation may be required. The cost of implementing noise reduction solutions (low-noise equipment, barriers or walls, noise cancellation techniques, etc.) may become a significant factor when a site is selected.

Noise can be transmitted as a pressure wave either through the air or through solids. The majority of cases involving the observation and measurement of noise have dealt with noise being propagated through the air. However, there are reported, rare cases of audible transformer noise appearing at distant observation points by propagating through the transformer foundation and underground solid rock formations. It is best to avoid the situation by isolating the foundation from bedrock where the conditions are thought to favor transmission of vibrations.

9.2.2.8 Noise Sources

Continuous audible sources: The most noticeable audible noise generated by normal substation operation consists of continuously radiated audible discrete tones. Noise of this type is primarily generated by power transformers. Regulating transformers, reactors, and emergency generators, however, could also be sources. This type of noise is most likely to be subject to government regulations. Another source of audible noise in substations, particularly in extra high voltage (EHV) substations, is corona from the bus and conductors.

Continuous radio frequency (RF) sources: Another type of continuously radiated noise that can be generated during normal operation is RF noise. These emissions can be broadband and can cause interference to radio and television signal reception on properties adjacent to the substation site. Objectionable RF noise is generally a product of unintended sparking, but can also be produced by corona.

Impulse sources: While continuously radiated noise is generally the most noticeable to substation neighbors, significant values of impulse noise can also accompany normal operation. Switching operations will cause both impulse audible and RF noise with the magnitude varying with voltage, load, and operation speed. Circuit-breaker operations will cause audible noise, particularly operation of air-blast breakers.

9.2.2.9 Typical Noise Levels

Equipment noise levels: Equipment noise levels may be obtained from manufacturers, equipment tendering documents, or test results. The noise level of a substation power transformer is a function of the MVA and BIL rating of the high voltage winding. These transformers typically generate a noise level ranging from 60 to 80 dBA.

Transformer noise will "transmit" and attenuate at different rates depending on the transformer size, voltage rating, and design. Few complaints from nearby residents are typically received concerning substations with transformers of less than 10 MVA capacity, except in urban areas with little or no buffers. Complaints are more common at substations with transformer sizes of 20–150 MVA, especially within the first 170–200 m (500–600 ft). However, in very quiet rural areas where the nighttime ambient can reach 20–25 dBA, the noise from the transformers of this size can be audible at distances of 305 m (1000 ft) or more. In urban areas, substations at 345 kV and above rarely have many complaints because of the large parcels of land on which they are usually constructed.

Attenuation of noise with distance: The rate of attenuation of noise varies with distance for different types of sound sources depending on their characteristics. Point sound sources that radiate equally in all directions will decrease at a rate of 6 dB for each doubling of distance. Cylindrical sources vibrating uniformly in a radial direction will act like long source lines and the sound pressure will drop 3 dB for each doubling of distance. Flat planar surfaces will produce a sound wave with all parts of the wave tracking in the same direction (zero divergence). Hence, there will be no decay of the pressure level due to distance only. The designer must first identify the characteristics of the source before proceeding with a design that will take into account the effect of distance.

A transformer will exhibit combinations of all of the above sound sources, depending on the distance and location of the observation point. Because of its height and width, which can be one or more wavelengths, and its nonuniform configuration, the sound pressure waves will have directional characteristics with very complex patterns. Close to the transformer (near field), these vibrations will result in lobes with variable pressure levels. Hence, the attenuation of the noise level will be very small. If the width (W) and height (H) of the transformer are known, then the near field is defined, from observation, as any distance less than $2\sqrt{WH}$ from the transformer.

Further from the transformer (far field), the noise will attenuate in a manner similar to the noise emitted from a point source. The attenuation is approximately equal to 6 dB for every doubling of the distance. In addition, if a second adjacent transformer produces an identical noise level to the existing transformer (e.g., 75 dBA), the total sound will be 78 dBA for a net increase of only 3 dB. This is due to the logarithmic effect associated with a combination of noise sources.

9.2.2.10 Governmental Regulations

Governmental regulations may impose absolute limits on emissions, usually varying the limits with the zoning of the adjacent properties. Such limits are often enacted by cities, villages, and other incorporated urban areas where limited buffer zones exist between property owners. Typical noise limits at the substation property line used within the industry are as follows:

- Industrial zone <75 dBA
- Commercial zone <65 dBA
- Residential zone <55 dBA

Additional governmental noise regulations address noise levels by limiting the increase above the existing ambient to less than 10 dB. Other regulations could limit prominent discrete tones, or set specific limits by octave bands.

9.2.2.11 Noise Abatement Methods

The likelihood of a noise complaint is dependent on several factors, mostly related to human perceptions. As a result, the preferred noise abatement method is time-dependent as well as site-specific.

Reduced transformer sound levels: Since power transformers, voltage regulators, and reactors are the primary sources of continuously radiated discrete tones in a substation, careful attention to equipment design can have a significant effect on controlling noise emissions at the substation property line. This equipment can be specified with noise emissions below manufacturer's standard levels, with values as much as 10 dB below those levels being typical.

In severely restrictive cases, transformers can be specified with noise emissions 20 dB less than the manufacturers' standard levels, but usually at a significant increase in cost. Also, inclusion of bid evaluation factor(s) for reduced losses in the specification can impact the noise level of the transformer. Low-loss transformers are generally quieter than standard designs.

Low-impulse noise equipment: Outdoor-type switching equipment is the cause of most impulse noise. Switchgear construction and the use of vacuum or puffer circuit breakers, where possible, are the most effective means of controlling impulse emissions. The use of circuit switchers or air-break switches with whips and/or vacuum bottles for transformer and line switching, may also provide impulse-emission reductions over standard air-break switches.

RF noise and corona-induced audible noise control: Continuously radiated RF noise and corona-induced audible noise can be controlled through the use of corona-free hardware and shielding for high-voltage conductors and equipment connections, and through attention to conductor shapes to avoid sharp corners. Angle and bar conductors have been used successfully up to 138 kV without objectionable corona if corners are rounded at the ends of the conductors and bolts are kept as short as possible.

Tubular shapes are typically required above this voltage. Pronounced edges, extended bolts, and abrupt ends on the conductors can cause significant RF noise to be radiated. The diameter of the conductor also has an effect on the generation of corona, particularly in wet weather. Increasing the size of single grading rings or conductor diameter may not necessarily solve the problem. In some cases it may be better to use multiple, smaller diameter grading rings.

Site location: For new substations to be placed in an area known to be sensitive to noise levels, proper choice of the site location can be effective as a noise abatement strategy. Also, locations in industrial parks or near airports, expressways, or commercial zones that can provide almost continuous background noise levels of 50 dB or higher will minimize the likelihood of a complaint.

Larger yard area: Noise intensity varies inversely with distance. An effective strategy for controlling noise of all types involves increasing the size of the parcel of real estate on which the substation is located.

Equipment placement: Within a given yard size, the effect of noise sources on the surroundings can be mitigated by careful siting of the noise sources within the confines of the substation property. In addition, making provisions for the installation of mobile transformers, emergency generators, etc. near the center of the property, rather than at the edges, will lessen the effect on the neighbors.

Barriers or walls: If adequate space is not available to dissipate the noise energy before it reaches the property line, structural elements might be required. These can consist of walls, sound-absorbing panels, or deflectors. In addition, earth berms or below-grade installation may be effective. It may be possible to deflect audible noises, especially the continuously radiated tones most noticeable to the public, to areas not expected to be troublesome. Foliage, despite the potential aesthetic benefit and psychological effect, is not particularly effective for noise reduction purposes.

Properly constructed sound barriers can provide several decibels of reduction in the noise level. An effective barrier involves a proper application of the basic physics of

1. Transmission loss through masses
2. Sound diffraction around obstacles
3. Standing waves behind reflectors
4. Absorption at surfaces

For a detailed analysis of wall sound barriers, refer to IEEE Std. 1127-1998.

Active noise cancellation techniques: Another solution to the problem of transformer noise involves use of active noise control technology to cancel unwanted noise at the source, and is based on advances in digital controller computer technology. Active noise cancellation systems can be tuned to specific problem frequencies or bands of frequencies achieving noise reduction of up to 20 dB.

9.2.3 Electric and Magnetic Fields (IEEE Std. 644-1994)

Electric substations produce electric and magnetic fields. In a substation, the strongest fields around the perimeter fence come from the transmission and distribution lines entering and leaving the substation. The strength of fields from equipment inside the fence decreases rapidly with distance, reaching very low levels at relatively short distances beyond substation fences.

In response to the public concerns with respect to EMF levels, whether perceived or real, and to governmental regulations, the substation designer may consider design measures to lower EMF levels or public exposure to fields while maintaining safe and reliable electric service.

9.2.3.1 Electric and Magnetic Field Sources in a Substation

Typical sources of electric and magnetic fields in substations include the following:

1. Transmission and distribution lines entering and exiting the substation
2. Buswork
3. Transformers
4. Air core reactors
5. Switchgear and cabling
6. Line traps
7. Circuit breakers
8. Ground grid
9. Capacitors
10. Battery chargers
11. Computers

9.2.3.2 Electric Fields

Electric fields are present whenever voltage exists on a conductor. Electric fields are not dependent on the current. The magnitude of the electric field is a function of the operating voltage and decreases with the square of the distance from the source. The strength of an electric field is measured in volts per meter. The most common unit for this application is kilovolts per meter. The electric field can be easily shielded (the strength can be reduced) by any conducting surface such as trees, fences, walls, buildings, and most structures. In substations, the electric field is extremely variable due to the screening effect provided by the presence of the grounded steel structures used for electric bus and equipment support.

Although the level of the electric fields could reach magnitudes of approximately 13 kV/m in the immediate vicinity of high-voltage apparatus, such as near 500 kV circuit beakers, the level of the electric field decreases significantly toward the fence line. At the fence line, which is at least 6.4 m (21 ft) from the nearest live 500 kV conductor (see the NESC), the level of the electric field approaches zero kV/m. If the incoming or outgoing lines are underground, the level of the electric field at the point of crossing the fence is negligible.

9.2.3.3 Magnetic Fields

Magnetic fields are present whenever current flows in a conductor, and are not voltage dependent. The level of these fields also decreases with distance from the source but these fields are not easily shielded. Unlike electric fields, conducting materials such as the earth, or most metals, have little shielding effect on magnetic fields.

Magnetic fields are measured in Webers per square meter (Tesla) or Maxwells per square centimeter (Gauss). One Gauss = 10^{-4} T. The most common unit for this application is milliGauss (10^{-3} G). Various factors affect the levels of the fields, including the following:

1. Current magnitude
2. Phase spacing
3. Bus height
4. Phase configurations
5. Distance from the source
6. Phase unbalance (magnitude and angle)

Magnetic fields decrease with increasing distance (r) from the source. The rate is an inverse function and is dependent on the type of source. For point sources such as motors and reactors, the function is $1/r^2$; and for single-phase sources such as neutral or ground conductors the function is $1/r$. Besides distance, conductor spacing and phase balance have the largest effect on the magnetic field level because they control the rate at which the field changes.

Magnetic fields can sometimes be shielded by specially engineered enclosures. The application of these shielding techniques in a power system environment is minimal because of the substantial costs involved and the difficulty of obtaining practical designs.

9.2.4 Safety and Security

9.2.4.1 Fences and Walls

The primary means of ensuring public safety at substations is by the erection of a suitable barrier, such as a fence or a wall with warning signs. As a minimum, the barrier should meet the requirements of the NESC and other applicable electrical safety codes. Recommended clearances from substation live parts to the fence are specified in the NESC, and security methods are described in IEEE 1402-2000.

9.2.4.2 Lighting

Yard lighting may be used to enhance security and allow equipment status inspections. A yard-lighting system should provide adequate ground-level lighting intensity around equipment and the control-house area for security purposes without disruption to the surrounding community. High levels of nightly illumination will often result in complaints.

9.2.4.3 Grounding

Grounding should meet the requirement of IEEE Std. 80-2000 to ensure the design of a safe and adequate grounding system. All non-current-carrying metal objects in or exiting from substations should be grounded (generally to a buried metallic grid) to eliminate the possibility of unsafe touch or step potentials, which the general public might experience during fault conditions.

9.2.4.4 Fire Protection

The potential for fires exists throughout all stations. Although not a common occurrence, substation fires are an important concern because of potential for long-term outages, personnel injury or death, extensive property and environmental damage, and rapid uncontrolled spreading. Refer to IEEE Std. 979-1994 for detailed guidance and identification of accepted substation fire-protection design practices and applicable industry standards.

9.2.5 Permitting Process

A variety of permits may be required by the governing bodies before construction of a substation may begin. For the permitting process to be successful, the following factors may have to be considered:

1. Site location
2. Level of ground water
3. Location of wetlands
4. Possibility of existing hazardous materials
5. Need for potable water and sewage
6. Possible noise
7. Aesthetics
8. EMF

Timing for the permit application is a critical factor because the permit application may trigger opposition involvement. If it is determined that the situation requires public involvement, the preparation and implementation of a detailed plan using public participation can reduce the delays and costs associated with political controversy and litigation. In these situations, public involvement prior to permit application can help to build a positive relationship with those affected by the project, identify political and community concerns, obtain an informed consensus from project stakeholders, and provide a basis for the utility to increase its credibility and reputation as a good neighbor.

9.3 Construction

9.3.1 Site Preparation

9.3.1.1 Clearing, Grubbing, Excavation, and Grading

Concerns include the creation of dust, mud, water runoff, erosion, degraded water quality, and sedimentation. The stockpiling of excavated material and the disposal of excess soil, timber, brush, etc. are additional items that should be considered. Protective measures established during the design phase or committed to through the permitting process for ground water, wetlands, flood plains, streams, archeological sites, and endangered flora and fauna should be implemented during this period.

9.3.1.2 Site Access Roads

The preparation and usage of site access roads create concerns that include construction equipment traffic, dust, mud, water runoff, erosion, degraded water quality, and sedimentation. Access roads can also have an impact on agriculture, archaeological features, forest resources, wildlife, and vegetation.

9.3.1.3 Water Drainage

Runoff control is especially important during the construction process. Potential problems include flooding, erosion, sedimentation, and waste and trash carried off the site.

9.3.2 Noise

Noise control is important during construction in areas sensitive to this type of disturbance. An evaluation should be made prior to the start of construction to determine noise restrictions that may be imposed at the construction site.

9.3.3 Safety and Security

Safety and security procedures should be implemented at the outset of the construction process to protect the public and prevent unauthorized access to the site. These procedures should be developed in conformance with governmental agencies. See IEEE 1402-2000 for detailed descriptions of the security methods that can be employed. The safety and security program should be monitored continuously to ensure that it is functioning properly.

The following are suggestions for safety and security at the site:

1. Temporary or permanent fencing
2. Security guards
3. Security monitoring systems
4. Traffic control
5. Warning signs
6. Construction safety procedures
7. Temporary lighting

9.3.4 Site Housekeeping

During construction, debris and refuse should not be allowed to accumulate. Efforts should be made to properly store, remove, and prevent these materials from migrating beyond the construction site. Burning of refuse should be avoided. In many areas this activity is prohibited by law. Portable toilets that are routinely serviced should be provided.

9.3.5 Hazardous Material

The spillage of transformer and pipe cable insulating oils, paints, solvents, acids, fuels, and other similar materials can be detrimental to the environment as well as a disturbance to the neighborhood. Proper care should be taken in the storage and handling of such materials during construction.

9.4 Operations

9.4.1 Site Housekeeping

9.4.1.1 Water and Sediment Control

Routine inspection of control for water flows is important to maintain proper sediment control measures. Inspection should be made for basin failure and for gullies in all slopes. Inspection of all control measures is necessary to be sure that problems are corrected as they develop and should be made a part of regular substation inspection and maintenance.

9.4.1.2 Yard Surface Maintenance

Yard surfacing should be maintained as designed to prevent water runoffs and control dust. If unwanted vegetation is observed on the substation site, approved herbicides may be used with caution to prevent runoff from damaging surrounding vegetation. If runoffs occur, the affected area should be covered with stone to retard water runoff and to control dust.

9.4.1.3 Paint

When material surfaces are protected by paint, a regular inspection and repainting should be performed to maintain a neat appearance and to prevent corrosion damage.

9.4.1.4 Landscaping

Landscaping should be maintained to ensure perpetuation of design integrity and intent.

9.4.1.5 Storage

In some areas, zoning will not permit storage in substations. The local zoning must therefore be reviewed before storing equipment, supplies, etc. The appearance of the substation site should be considered so it will not become visually offensive to the surrounding community.

9.4.1.6 Noise

Inspection of all attributes of equipment designed to limit noise should be performed periodically.

9.4.1.7 Safety and Security

All substations should be inspected regularly, following established and written procedures to ensure the safety and security of the station. Safe and secure operation of the substation requires adequate knowledge and proper use of each company's accident prevention manual. See IEEE 1402-2000 for detailed descriptions of the security methods that can be employed.

Routine inspections of the substation should be performed and recorded, and may include the following:

1. Fences
2. Gates
3. Padlocks
4. Signs
5. Access detection systems
6. Alarm systems
7. Lighting systems
8. Grounding systems
9. Fire-protection equipment
10. All oil-filled equipment
11. Spill-containment systems

9.4.2 Fire Protection

Refer to IEEE Std. 979-1994 for detailed guidance and identification of accepted substation fire-protection practices and applicable industry standards. Any fire-protection prevention system installed in the substation should be properly maintained.

9.4.3 Hazardous Material

A spill-prevention control and counter-measures plan should be in place for the substation site and should meet governmental requirements. For general guidance, see IEEE Std. 980-1994.

9.5 Defining Terms (IEEE, 1998)

A-weighted sound level: The representation of the sound pressure level that has as much as 40 dB of the sound below 100 Hz and a similar amount above 10,000 Hz filtered out. This level best approximates the response of the average young ear when listening to most ordinary, everyday sounds. Generally designated as dBA.

Commercial zone: A zone that includes offices, shops, hotels, motels, service establishments, or other retail/commercial facilities as defined by local ordinances.

Hazardous material: Any material that has been so designated by governmental agencies or adversely impacts human health or the environment.

Industrial zone: A zone that includes manufacturing plants where fabrication or original manufacturing is done, as defined by local ordinances.

Noise: Undesirable sound emissions or undesirable electromagnetic signals/emissions.

Residential zone: A zone that includes single-family and multifamily residential units, as defined by local ordinances.

Wetlands: Any land that has been so designated by governmental agencies. Characteristically, such land contains vegetation associated with saturated types of soil.

For additional definitions, see IEEE Std. 100.

References

Guide for Electric Power Substation Physical and Electronic Security, IEEE 1402-2000.
IEEE Guide for Containment and Control of Oil Spills in Substations, IEEE Std. 980-1994.
IEEE Guide for the Design, Construction, and Operation of Electric Power Substations for Community Acceptance and Environmental Compatibility, IEEE Std. 1127-1998.
IEEE Guide for Safety in AC Substation Grounding, IEEE Std. 80-2000.
IEEE Guide to Specifications for Gas-Insulated, Electric Power Substation Equipment, IEEE Std C37.123-1996.
IEEE Guide for Substation Fire Protection, IEEE Std. 979-1994.
IEEE Standard Procedures for Measurement of Power Frequency Electric and Magnetic Fields from AC Power Lines, IEEE Std. 644-1994.
National Electrical Safety Code® (NESC®) (ANSI), Accredited Standards Committee C2-2007, Institute of Electrical Electronics Engineers, Piscataway, NJ, 2007.
The IEEE Standard Dictionary of Electrical and Electronics Terms, IEEE Std. 100.

10
Animal Deterrents/Security

Mike Stine
TE Energy

10.1 Animal Types 10-2
 Clearance Requirements • Squirrels • Birds • Snakes • Raccoons
10.2 Mitigation Methods 10-4
 Barriers • Deterrents • Insulation • Isolation Devices

The vast majority of electrical utility substations designed to transform transmission voltages to distribution class voltages employ an open-air design. The configurations may vary, but usually consist of equipment that utilizes polymer or porcelain insulators or bushings to create electrically insulated creepage and dry arc distances between the potential voltage carried by the bus or conductor and the grounded portions of the equipment or structure. Although these insulators or bushings provide the proper insulation distance for normal operation voltages (AC, DC, and BIL), they do not provide sufficient distances to eliminate bridging of many animals from potential to ground. This animal bridging situation usually exists at the low side or distribution voltage portion of the substation (12–36 kV), but depending on the size and type of the animal, it can also affect higher voltage equipment. Utilities have reported that animal-caused outages have become a major problem affecting the reliability and continuity of the electrical system and are actively taking steps to prevent it.

The effects of animal bridging range from nuisance trips of the electrical system, which may be a momentary occurrence, to faults that may interrupt power for long periods of time. Aside from the inconvenience and reliability aspects of animal-induced outages, there can be damage to the substation equipment ranging from porcelain bushings and insulators that may cost as little as $20.00 to complete destruction of large transformers running into millions of dollars. There can also be an environmental risk involved with catastrophic failure such as oil spillage from equipment that has ruptured due to electrical faults.

Damage from outages is not limited to the equipment owned by the electrical utility. Many heavy industrial plants such as pulp and paper, petrochemical, and car manufacturers employ processes that are sensitive to interruptions and may result in significant time and money to reestablish production. The proliferation of computers, programmable logic controllers, and other electrically sensitive devices in the workplace is also a reliability concern.

In addition to the concern for protecting assets such as substation equipment, improving the reliability of the system, eliminating environmental risks, and ensuring customer satisfaction and loyalty, the conservation of endangered and protected animal species is an issue. It is important to be educated and informed about the species and types of animals that are protected in each individual area or location.

To evaluate the problem and its possible solutions, several aspects need to be investigated:

- Animal type, size, and tendencies
- Equipment voltage rating and clearance from electrical ground
- Natural surroundings

- Methods by which animals enter substation
- Influences attracting the animals
- Barrier methods available to keep the animal out
- Deterrent methods to repel the animals
- Insulation options

Concern from the U.S. Fish & Wildlife Service (USFWS) and other International Agencies has prompted electrical substation, distribution, and transmission system owners to reevaluate design criteria and retrofit existing equipment to mitigate the risk of electrocution of protected bird species. Laws enacted to protect the bird population include

Migratory Bird Treaty Act

- Provides protection of all avian species except English Sparrow, European Starling, Rock Dove (common Pigeon), Monk Parakeet, and Upland Game Birds
- Unlawful to pursue, wound, or kill
- Strict Liability Law—Intent to harm not needed for prosecution

Bald and Golden Eagle Protection Act

- Provides protection for Eagle species
- Unlawful to pursue, wound, kill, *molest*, or *disturb*
- Criminal Liability Law—Must show knowingly or with wanton disregard for the consequence of the act

Endangered Species Act

- Provides penalties for specific animals listed

The most often enforced for electrocution of avian species is the Migratory Bird Treaty Act. The USFWS provides training for owners in reporting and mitigation programs.

10.1 Animal Types

10.1.1 Clearance Requirements

Table 10.1 has been developed to aid in establishing minimum phase-to-ground and phase-to-phase clearances for the associated animals. This table is for reference only.

10.1.2 Squirrels

In North America, a common culprit causing bridging is the squirrel. Although there are many varieties of squirrels, it can be assumed that the nominal length of a squirrel is 18 in. (450 mm). Using this

TABLE 10.1 Typical Clearance Requirement by Animal

Animal Type	Phase to Phase	Phase to Ground
Squirrel	18 in. (450 mm)	18 in. (450 mm)
Opossum/raccoon	30 in. (750 mm)	30 in. (750 mm)
Snake	36 in. (900 mm)	36 in. (900 mm)
Crow/grackle	24 in. (600 mm)	18 in. (450 mm)
Migratory large bird	36 in. (900 mm)	36 in. (900 mm)
Frog	18 in. (450 mm)	18 in. (450 mm)
Cat	24 in. (600 mm)	24 in. (600 mm)

dimension, you can evaluate equipment and clearances to determine areas where bridging could occur between potential and ground or phase to phase. Clearances for modern substation equipment rated 35 kV and above will normally be sufficient to eliminate squirrel-caused problems; however, distances between phases and between phase and grounded structures should be examined.

There are several schools of thought regarding the reason squirrels often enter substations. One explanation offered is the proximity of trees and vegetation near the substation site that may attract squirrels. Some utilities report that removal of this vegetation had no effect on the squirrel-caused outages. Experts have theorized that the animals' path is predetermined and the construction of a structure will not deter a squirrel from following his intended route. Others believe that the animals are attracted by heat or vibration emitted from the electrical equipment. Regardless of the reason, squirrels are compelled toward intrusion.

The entry into the substation does not always occur over, under, or through the outer fence of the site. Squirrels are very adept at traveling along overhead conductors and often enter the substation in this manner. Because of this fact, perimeter barriers are often ineffectual in preventing squirrel entry.

10.1.3 Birds

Birds create several problems when entering an electrical substation. The first and most obvious is the bridging between phase to ground or phase to phase caused by the wingspan when flying into or exiting the structure. Another problem is the bridging caused by debris used to build nests. Many times material such as strands of conductors or magnetic recording tape may be readily available from the surrounding area and be utilized by the birds. This conductive debris is often dragged across the conductor/busbar and results in flashovers, trips, or faults. The third problem is contamination of insulators caused by regurgitation or defecation of the birds. When this residue is allowed to remain, it can result in flashovers from potential to ground across the surface of the porcelain or polymer insulator by essentially decreasing the insulated creepage distance. The fourth possibility is commonly known as a "streamer outage." Streamers are formed when a bird defecates upon exiting a nest that has been built above an insulator. The streamers may create a path between the structure and conductor/bus, resulting in a flashover. Birds will tend to make nests in substations in an effort to eliminate possible predators from attacking the nest for food. The construction of nests in substations can, in turn, attract other animals such as snakes, cats, and raccoons into the area searching for food.

10.1.4 Snakes

Snakes are a major contributor to substation outages. In some areas, snakes are responsible for virtually all substation wildlife outages. Because of their size and climbing ability, snakes can reach most parts of a substation without difficulty. Snake-proofing substations can sometimes create problems rather than solving them. Snakes typically enter substations hunting birds and eggs. Eliminating these predators can lead to an increase in the bird population inside the substation boundaries. This bird infestation can then lead to bird-induced problems unless additional measures are taken.

10.1.5 Raccoons

Raccoons are excellent climbers and can easily gain access to substations. Unlike snakes, raccoons will occasionally enter substations for no particular reason except curiosity. Because of their large size, raccoons can easily bridge phase-to-phase and phase-to-ground distances on equipment with voltage ratings up to 25 kV.

10.2 Mitigation Methods

10.2.1 Barriers

Some of the barrier methods available include cyclone fences, small mesh wire fences, smooth climbing guards, electric fences, solid wall barriers, and fences with unconventional geometries. Barrier methods can be very effective against certain animals. Some utilities report that the use of small mesh fencing along the lower 3–4 ft (1–1.3 m) of the perimeter has prevented intrusion of certain types of snakes. Several substation owners have incorporated the use of a bare wire attached to a PVC pipe energized with a low-voltage transformer creating an electric fence that surrounds the structure inside the normal property fence. This method has also been proven effective for snakes. Although these barrier designs prevent snakes from entering substations, they do little or nothing to eliminate legged animal intrusions. Smooth climbing guards are also used on structures to prevent some animals from scaling the vertical framework. While these guards work for some legged animals such as dogs and foxes, more agile animals such as squirrels, opossums, and cats can easily circumnavigate the devices.

10.2.2 Deterrents

There are myriad commercially available deterrent devices on the market. Many of the devices have actually come from applications in the household market to repel pests such as squirrels and pigeons from property. Although numerous, most devices have a limited effect on wildlife. Some of these include ultrasonic devices, devices producing loud noises at intermittent periods, chemical repellents, sticky gels, predator urine, plastic owls or snakes, poisons, and spined perching deterrents for birds. Ultrasonic devices tend to have an initial impact on animals but have reportedly become ineffective after a relatively short period of time either due to the animal adapting to the sounds or the need to maintain the devices. Loud noise devices, like ultrasonics, soon lose the ability to repel the animals as they become familiar with the sound and lack of consequence. Chemical repellents, sticky gels, and predator urine have been shown effective against some animals when reapplied at frequent intervals. Poisons have been used to curb infestations of pests such as pigeons but will sometimes result in collateral effects on pets and other animals if the pest is allowed to die outside the substation boundaries. Spined perching deterrents have proven very successful in preventing smaller birds from building nests or congregating above electrically sensitive areas but can sometimes serve as a functional anchor for greater sized birds to secure large nests.

10.2.3 Insulation

Insulating live conductors and hardware can be very effective in eliminating animal outages. Insulation systems are available in several forms:

- Spray on RTV coatings
- Insulating tapes
- Heat-shrinkable tubings, tapes, and sheet materials
- Preformed insulating covers

Insulation systems should be used at locations where animals can possibly make contact phase to ground or phase to phase. Typical applications include

- Equipment bushing hardware (i.e., circuit breakers, reclosers, transformers, potential transformer, capacitors, regulators, etc.)
- Bus support insulator connections to structure or bus

Animal Deterrents/Security

- Hook switch insulator connections to switch base or bus
- Any area where clearance between bus and grounded equipment or structure is insufficient to eliminate bridging
- Busbar and conductors where phase-to-phase spacing is inadequate

Because these products are used as insulation on bus, conductor, or hardware, it is critical that they be of a material that is designed for the rigors of the high-voltage environment. Unlike barriers and deterrents, the insulating materials are subjected to the electric field and are sometimes applied to the leakage path of other insulating materials such as porcelain. Care should be taken to select products that will withstand the outdoor environment as well as the electrical stress to which they may be subjected.

10.2.4 Isolation Devices

Isolation devices are rigid insulating discs that are installed in the leakage path of porcelain insulators. These devices force animals to climb onto them, isolating them from ground. These discs are used on both support insulators as well as switch insulators. As with insulating covers, the insulating material must be designed for the outdoor high-voltage environment.

11
Substation Grounding

11.1 Reasons for Substation Grounding System 11-1
11.2 Accidental Ground Circuit 11-2
     Conditions • Permissible Body Current Limits • Importance
     of High-Speed Fault Clearing • Tolerable Voltages
11.3 Design Criteria 11-9
     Actual Touch and Step Voltages • Soil Resistivity •
     Grid Resistance • Grid Current • Use of the Design
     Equations • Selection of Conductors • Selection
     of Connections • Grounding of Substation Fence •
     Other Design Considerations
References 11-22

Richard P. Keil
*Commonwealth
Associates, Inc.*

11.1 Reasons for Substation Grounding System

The substation grounding system is an essential part of the overall electrical system. The proper grounding of a substation is important for the following two reasons:

1. It provides a means of dissipating electric current into the earth without exceeding the operating limits of the equipment.
2. It provides a safe environment to protect personnel in the vicinity of grounded facilities from the dangers of electric shock under fault conditions.

The grounding system includes all of the interconnected grounding facilities in the substation area, including the ground grid, overhead ground wires, neutral conductors, underground cables, foundations, deep well, etc. The ground grid consists of horizontal interconnected bare conductors (mat) and ground rods. The design of the ground grid to control voltage levels to safe values should consider the total grounding system to provide a safe system at an economical cost.

The following information is mainly concerned with personnel safety. The information regarding the grounding system resistance, grid current, and ground potential rise can also be used to determine if the operating limits of the equipment will be exceeded.

Safe grounding requires the interaction of two grounding systems:

1. Intentional ground, consisting of grounding systems buried at some depth below the earth's surface
2. Accidental ground, temporarily established by a person exposed to a potential gradient in the vicinity of a grounded facility

It is often assumed that any grounded object can be safely touched. A low substation ground resistance is not, in itself, a guarantee of safety. There is no simple relation between the resistance of the grounding system as a whole and the maximum shock current to which a person might be exposed. A substation with relatively low ground resistance might be dangerous, while another substation with very high ground resistance might be safe or could be made safe by careful design.

There are many parameters that have an effect on the voltages in and around the substation area. Since voltages are site-dependent, it is impossible to design one grounding system that is acceptable for all locations. The grid current, fault duration, soil resistivity, surface material, and the size and shape of the grid all have a substantial effect on the voltages in and around the substation area. If the geometry, location of ground electrodes, local soil characteristics, and other factors contribute to an excessive potential gradient at the earth surface, the grounding system may be inadequate from a safety aspect despite its capacity to carry the fault current in magnitudes and durations permitted by protective relays.

During typical ground fault conditions, unless proper precautions are taken in design, the maximum potential gradients along the earth surface may be of sufficient magnitude to endanger a person in the area. Moreover, hazardous voltages may develop between grounded structures or equipment frames and the nearby earth.

The circumstances that make human electric shock accidents possible are

- Relatively high fault current to ground in relation to the area of the grounding system and its resistance to remote earth.
- Soil resistivity and distribution of ground currents such that high potential gradients may occur at points at the earth surface.
- Presence of a person at such a point, time, and position that the body is bridging two points of high potential difference.
- Absence of sufficient contact resistance or other series resistance to limit current through the body to a safe value under the above circumstances.
- Duration of the fault and body contact and, hence, of the flow of current through a human body for a sufficient time to cause harm at the given current intensity.

Relative infrequency of accidents is largely due to the low probability of coincidence of the above unfavorable conditions.

To provide a safe condition for personnel within and around the substation area, the grounding system design limits the potential difference a person can come in contact with to safe levels. IEEE Std. 80, IEEE Guide for Safety in AC Substation Grounding [1], provides general information about substation grounding and the specific design equations necessary to design a safe substation grounding system. The following discussion is a brief description of the information presented in IEEE Std. 80.

The guide's design is based on the permissible body current when a person becomes part of an accidental ground circuit. Permissible body current will not cause ventricular fibrillation, i.e., stoppage of the heart. The design methodology limits the voltages that produce the permissible body current to a safe level.

11.2 Accidental Ground Circuit

11.2.1 Conditions

There are two conditions that a person within or around the substation can experience that can cause them to become part of the ground circuit. One of these conditions, touch voltage, is illustrated in Figures 11.1 and 11.2. The other condition, step voltage, is illustrated in Figures 11.3 and 11.4. Figure 11.1 shows the fault current being discharged to the earth by the substation grounding system and a person touching a grounded metallic structure, H. Figure 11.2 shows the Thevenin equivalent for the person's feet in parallel, Z_{Th}, in series with the body resistance, R_B. V_{Th} is the voltage between terminal H and F when the person is not present. I_B is the body current. When Z_{Th} is equal to the resistance of two feet in parallel, the touch voltage is

$$E_{touch} = I_B(R_B + Z_{Th}) \tag{11.1}$$

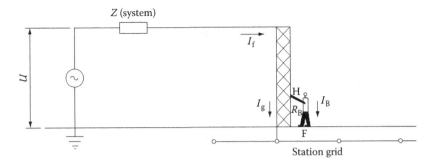

FIGURE 11.1 Exposure to touch voltage.

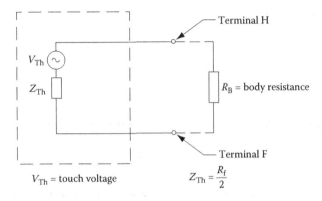

FIGURE 11.2 Touch-voltage circuit.

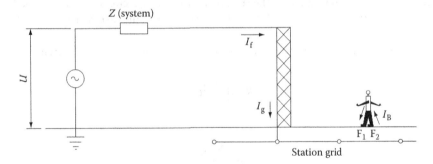

FIGURE 11.3 Exposure to step voltage.

Figures 11.3 and 11.4 show the conditions for step voltage. Z_{Th} is the Thevenin equivalent impedance for the person's feet in series and in series with the body. Based on the Thevenin equivalent impedance, the step voltage is

$$E_{step} = I_B(R_B + Z_{Th}) \tag{11.2}$$

The resistance of the foot in ohms is represented by a metal circular plate of radius b in meters on the surface of homogeneous earth of resistivity ρ (Ω m) and is equal to

$$R_f = \frac{\rho}{4b} \tag{11.3}$$

$$\text{Assuming } b = 0.08, \quad R_f = 3\rho \tag{11.4}$$

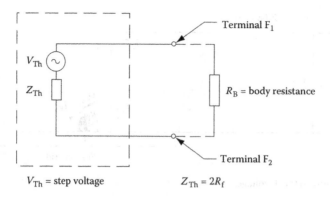

FIGURE 11.4 Step-voltage circuit.

The Thevenin equivalent impedance for two feet in parallel in the touch voltage, E_{touch}, equation is

$$Z_{Th} = \frac{R_f}{2} = 1.5\rho \quad (11.5)$$

The Thevenin equivalent impedance for two feet in series in the step voltage, E_{step}, equation is

$$Z_{Th} = 2R_f = 6\rho \quad (11.6)$$

The above equations assume uniform soil resistivity. In a substation, a thin layer of high-resistivity material is often spread over the earth surface to introduce a high-resistance contact between the soil and the feet, reducing the body current. The surface layer derating factor, C_s, increases the foot resistance and depends on the relative values of the resistivity of the soil, the surface material, and the thickness of the surface material.

The following equations give the ground resistance of the foot on the surface material:

$$R_f = \left[\frac{\rho_s}{4b}\right] C_s \quad (11.7)$$

$$C_s = 1 + \frac{16b}{\rho_s} \sum_{n=1}^{\infty} K^n R_{m(2nh_s)} \quad (11.8)$$

$$K = \frac{\rho - \rho_s}{\rho + \rho_s} \quad (11.9)$$

where
C_s is the surface layer derating factor
K is the reflection factor between different material resistivities
ρ_s is the surface material resistivity in Ω m
ρ is the resistivity of the earth beneath the surface material in Ω m
h_s is the thickness of the surface material in m
b is the radius of the circular metallic disc representing the foot in m
$R_{m(2nh_s)}$ is the mutual ground resistance between the two similar, parallel, coaxial plates, separated by a distance $(2nh_s)$, in an infinite medium of resistivity ρ_s in Ω m

Substation Grounding

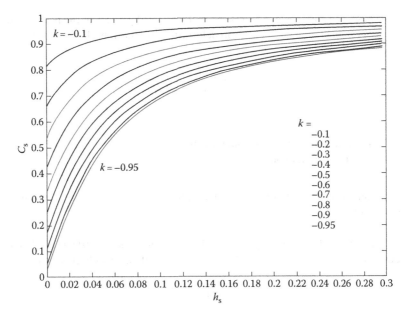

FIGURE 11.5 C_s vs. h_s.

A series of C_s curves has been developed based on Equation 11.8 and $b = 0.08\,\text{m}$, and is shown in Figure 11.5.

The following empirical equation by Sverak [2], and later modified, gives the value of C_s. The values of C_s obtained using Equation 11.10 are within 5% of the values obtained with the analytical method [3]:

$$C_s = 1 - \frac{0.09(1 - (\rho/\rho_s))}{2h_s + 0.09} \qquad (11.10)$$

11.2.2 Permissible Body Current Limits

The duration, magnitude, and frequency of the current affect the human body as the current passes through it. The most dangerous impact on the body is a heart condition known as ventricular fibrillation, a stoppage of the heart resulting in immediate loss of blood circulation. Humans are very susceptible to the effects of electric currents at 50 and 60 Hz. The most common physiological effects as the current increases are perception, muscular contraction, unconsciousness, fibrillation, respiratory nerve blockage, and burning [4]. The threshold of perception, the detection of a slight tingling sensation, is generally recognized as 1 mA. The let-go current, the ability to control the muscles and release the source of current, is recognized as between 1 and 6 mA. The loss of muscular control may be caused by 9–25 mA, making it impossible to release the source of current. At slightly higher currents, breathing may become very difficult, caused by the muscular contractions of the chest muscles. Although very painful, these levels of current do not cause permanent damage to the body. In a range of 60–100 mA, ventricular fibrillation occurs. Ventricular fibrillation can be a fatal electric shock. The only way to restore the normal heartbeat is through another controlled electric shock, called defibrillation. Larger currents will inflict nerve damage and burning, causing other life-threatening conditions.

The substation grounding system design should limit the electric current flow through the body to a value below the fibrillation current. Dalziel [5] published a paper introducing an equation relating the flow of current through the body for a specific time that statistically 99.5% of the population could survive before the onset of fibrillation. This equation determines the allowable body current:

$$I_B = \frac{k}{\sqrt{t_s}} \qquad (11.11)$$

where
- I_B is the rms magnitude of the current through the body, A
- t_s is the duration of the current exposure, s
- $k = \sqrt{S_B}$
- S_B is the empirical constant related to the electric shock energy tolerated by a certain percent of a given population

Dalziel found the value of $k = 0.116$ for persons weighing approximately 50 kg (110 lb) or $k = 0.157$ for a body weight of 70 kg (154 lb) [6]. Based on a 50 kg weight, the tolerable body current is

$$I_B = \frac{0.116}{\sqrt{t_s}} \qquad (11.12)$$

The equation is based on tests limited to values of time in the range of 0.03–3.0 s. It is not valid for other values of time. Other researchers have suggested other limits [7]. Their results have been similar to Dalziel's for the range of 0.03–3.0 s.

11.2.3 Importance of High-Speed Fault Clearing

Considering the significance of fault duration both in terms of Equation 11.11 and implicitly as an accident-exposure factor, high-speed clearing of ground faults is advantageous for two reasons:

1. Probability of exposure to electric shock is greatly reduced by fast fault clearing time, in contrast to situations in which fault currents could persist for several minutes or possibly hours.
2. Both tests and experience show that the chance of severe injury or death is greatly reduced if the duration of a current flow through the body is very brief.

The allowed current value may therefore be based on the clearing time of primary protective devices, or that of the backup protection. A good case could be made for using the primary clearing time because of the low combined probability that relay malfunctions will coincide with all other adverse factors necessary for an accident. It is more conservative to choose the backup relay clearing times in Equation 11.11, because it assures a greater safety margin.

An additional incentive to use switching times less than 0.5 s results from the research done by Biegelmeier and Lee [7]. Their research provides evidence that a human heart becomes increasingly susceptible to ventricular fibrillation when the time of exposure to current is approaching the heartbeat period, but that the danger is much smaller if the time of exposure to current is in the region of 0.06–0.3 s.

In reality, high ground gradients from faults are usually infrequent, and shocks from this cause are even more uncommon. Furthermore, both events are often of very short duration. Thus, it would not be practical to design against shocks that are merely painful and cause no serious injury, i.e., for currents below the fibrillation threshold.

11.2.4 Tolerable Voltages

Figures 11.6 and 11.7 show the five voltages a person can be exposed to in a substation. The following definitions describe the voltages:

Ground potential rise (GPR): The maximum electrical potential that a substation grounding grid may attain relative to a distant grounding point assumed to be at the potential of remote earth. GPR is the product of the magnitude of the grid current, the portion of the fault current conducted to earth by the grounding system, and the ground grid resistance.

Mesh voltage: The maximum touch voltage within a mesh of a ground grid.

Metal-to-metal touch voltage: The difference in potential between metallic objects or structures within the substation site that can be bridged by direct hand-to-hand or hand-to-feet contact.

Note: The metal-to-metal touch voltage between metallic objects or structures bonded to the ground grid is assumed to be negligible in conventional substations. However, the metal-to-metal touch voltage between metallic objects or structures bonded to the ground grid and metallic objects inside the substation site but not bonded to the ground grid, such as an isolated fence, may be substantial. In the case of gas-insulated substations, the metal-to-metal touch voltage between metallic objects or structures bonded to the ground grid may be substantial because of internal faults or induced currents in the enclosures.

Step voltage: The difference in surface potential experienced by a person bridging a distance of 1 m with the feet without contacting any other grounded object.

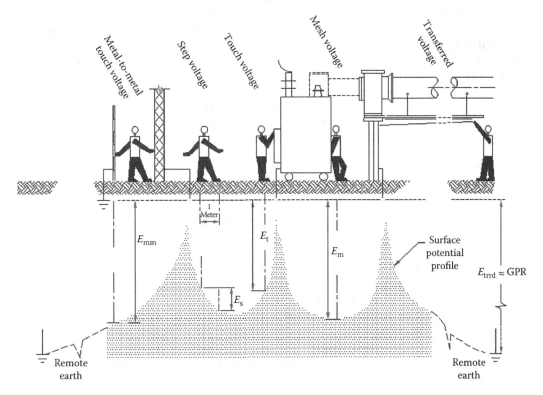

FIGURE 11.6 Basic shock situations.

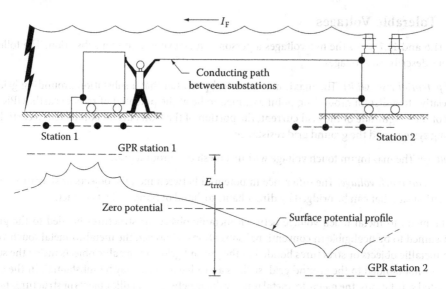

FIGURE 11.7 Typical situation of external transferred potential.

Touch voltage: The potential difference between the GPR and the surface potential at the point where a person is standing while at the same time having a hand in contact with a grounded structure.

Transferred voltage: A special case of the touch voltage where a voltage is transferred into or out of the substation, from or to a remote point external to the substation site. The maximum voltage of any accidental circuit must not exceed the limit that would produce a current flow through the body that could cause fibrillation.

Assuming the more conservative body weight of 50 kg to determine the permissible body current and a body resistance of 1000 Ω, the tolerable touch voltage is

$$E_{\text{touch}50} = (1000 + 1.5 C_s \rho_s) \frac{0.116}{\sqrt{t_s}} \tag{11.13}$$

and the tolerable step voltage is

$$E_{\text{step}50} = (1000 + 6 C_s \rho_s) \frac{0.116}{\sqrt{t_s}} \tag{11.14}$$

where
 E_{step} is the step voltage, V
 E_{touch} is the touch voltage, V
 C_s is determined from Figure 11.5 or Equation 11.10
 ρ_s is the resistivity of the surface material, Ω m
 t_s is the duration of shock current, s

Since the only resistance for the metal-to-metal touch voltage is the body resistance, the voltage limit is

$$E_{\text{mm-touch}50} = \frac{116}{\sqrt{t_s}} \tag{11.15}$$

Substation Grounding

The shock duration is usually assumed to be equal to the fault duration. If re-closing of a circuit is planned, the fault duration time should be the sum of the individual faults and used as the shock duration time t_s.

11.3 Design Criteria

The design criteria for a substation grounding system are to limit the actual step and mesh voltages to levels below the tolerable touch and step voltages as determined by Equations 11.13 and 11.14. The worst-case touch voltage, as shown in Figure 11.6, is the mesh voltage.

11.3.1 Actual Touch and Step Voltages

The following discusses the methodology to determine the actual touch and step voltages.

11.3.1.1 Mesh Voltage (E_m)

The actual mesh voltage, E_m (maximum touch voltage), is the product of the soil resistivity, ρ; the geometrical factor based on the configuration of the grid, K_m; a correction factor, K_i, which accounts for some of the errors introduced by the assumptions made in deriving K_m; and the average current per unit of effective buried length of the conductor that makes up the grounding system (I_G/L_M):

$$E_m = \frac{\rho K_m K_i I_G}{L_M} \tag{11.16}$$

The geometrical factor K_m [2] is as follows:

$$K_m = \frac{1}{2\pi}\left[\ln\left(\frac{D^2}{16hd} + \frac{(D+2h)^2}{8Dd} - \frac{h}{4d}\right) + \frac{K_{ii}}{K_h}\ln\left(\frac{8}{\pi(2n-1)}\right)\right] \tag{11.17}$$

For grids with ground rods along the perimeter, or for grids with ground rods in the grid corners, as well as both along the perimeter and throughout the grid area, $K_{ii} = 1$. For grids with no ground rods or grids with only a few ground rods, none located in the corners or on the perimeter,

$$K_{ii} = \frac{1}{(2n)^{2/n}} \tag{11.18}$$

$$K_h = \sqrt{1+\frac{h}{h_0}} \quad h_0 = 1\,\text{m (grid reference depth)} \tag{11.19}$$

Using four grid-shaped components [8], the effective number of parallel conductors, n, in a given grid can be made applicable to both rectangular and irregularly shaped grids that represent the number of parallel conductors of an equivalent rectangular grid:

$$n = n_a n_b n_c n_d \tag{11.20}$$

where

$$n_a = \frac{2L_C}{L_p} \tag{11.21}$$

$n_b = 1$ for square grids
$n_c = 1$ for square and rectangular grids
$n_d = 1$ for square, rectangular, and L-shaped grids

Otherwise,

$$n_b = \sqrt{\frac{L_p}{4\sqrt{A}}} \qquad (11.22)$$

$$n_c = \left[\frac{L_x L_y}{A}\right]^{0.7 A/L_x L_y} \qquad (11.23)$$

$$n_d = \frac{D_m}{\sqrt{L_x^2 + L_y^2}} \qquad (11.24)$$

where
L_C is the total length of the conductor in the horizontal grid, m
L_p is the peripheral length of the grid, m
A is the area of the grid, m²
L_x is the maximum length of the grid in the x direction, m
L_y is the maximum length of the grid in the y direction, m
D_m is the maximum distance between any two points on the grid, m
D is the spacing between parallel conductors, m
h is the depth of the ground grid conductors, m
d is the diameter of the grid conductor, m
I_G is the maximum grid current, A

The irregularity factor, K_i, used in conjunction with the above-defined n, is

$$K_i = 0.644 + 0.148n \qquad (11.25)$$

For grids with no ground rods, or grids with only a few ground rods scattered throughout the grid, but none located in the corners or along the perimeter of the grid, the effective buried length, L_M, is

$$L_M = L_C + L_R \qquad (11.26)$$

where L_R is the total length of all ground rods, m.

For grids with ground rods in the corners, as well as along the perimeter and throughout the grid, the effective buried length, L_M, is

$$L_M = L_C + \left[1.55 + 1.22\left(\frac{L_r}{\sqrt{L_x^2 + L_y^2}}\right)\right]L_R \qquad (11.27)$$

where L_r is the length of each ground rod, m.

11.3.1.1.1 Geometrical Factor K_m

The equation for K_m has variables of D, the spacing between the conductors; n, the number of conductors; d, the diameter of the conductors; and h, the depth of the grid. Each variable has a different impact on K_m. Figure 11.8 shows how the distance between conductors affects K_m. For this example, changing

Substation Grounding

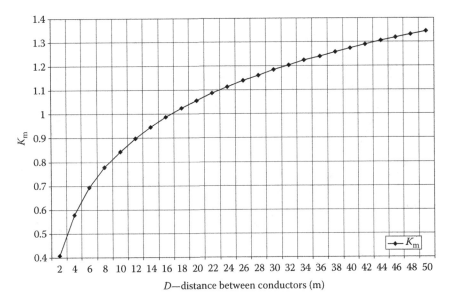

FIGURE 11.8 K_m vs. D.

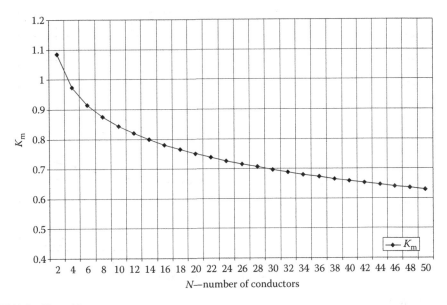

FIGURE 11.9 K_m vs. N.

the spacing from 10 to 40 m only changes K_m from 0.89 to 1.27. The greatest change takes place for relatively small spacings. The closer the spacing, the smaller K_m is. Figure 11.9 shows that as the number of conductors increases and the spacing and depth remain constant, K_m decreases rapidly. The diameter of the ground conductor as shown in Figure 11.10 has very little effect on K_m. Doubling the diameter of the conductor from 0.1 m (2/0) to 0.2 m (500 kcmil) reduces K_m by approximately 12%. D and n are certainly dependent on each other for a specific area of grid. The more conductors are installed, the smaller the distance between the conductors. Physically, there is a limit on how close conductors can be installed and should be a design consideration. Changing the depth as shown in Figure 11.11 also has very little influence on K_m for practical depths.

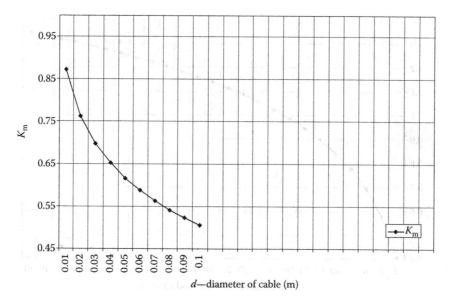

FIGURE 11.10 K_m vs. d.

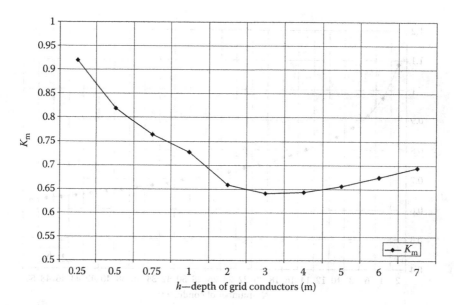

FIGURE 11.11 K_m vs. h.

11.3.1.2 Step Voltage (E_s)

The maximum step voltage is assumed to occur over a distance of 1 m, beginning at and extending outside of the perimeter conductor at the angle bisecting the most extreme corner of the grid. The step-voltage values are obtained as a product of the soil resistivity ρ, the geometrical factor K_s, the corrective factor K_i, and the average current per unit of buried length of grounding system conductor (I_G/L_S):

$$E_s = \frac{\rho K_s K_i I_G}{L_S} \tag{11.28}$$

Substation Grounding

For the usual burial depth of $0.25 < h < 2.5\,\mathrm{m}$ [2], K_s is defined as

$$K_s = \frac{1}{\pi}\left[\frac{1}{2h} + \frac{1}{D+h} + \frac{1}{D}\left(1 - 0.5^{n-2}\right)\right] \quad (11.29)$$

and K_i as defined in Equation 11.25.

For grids with or without ground rods, the effective buried conductor length, L_S, is defined as

$$L_S = 0.75 L_C + 0.85 L_R \quad (11.30)$$

11.3.1.2.1 Geometrical Factor K_s

The equation for K_s also has variables D, n, d, and h. K_s is not affected much by either the distance, D, between or the number, n, of conductors as can be seen in Figures 11.12 and 11.13. This is reasonable since the step voltage lies outside the grid itself. The influence of each conductor as it moves from the edge is reduced. On the other hand, the depth of burial has a drastic affect on K_s. The deeper the

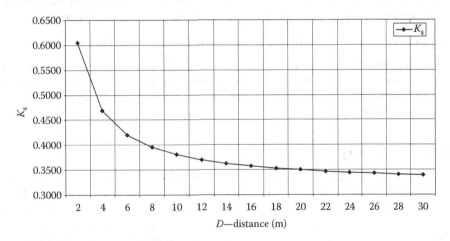

FIGURE 11.12 K_s vs. D.

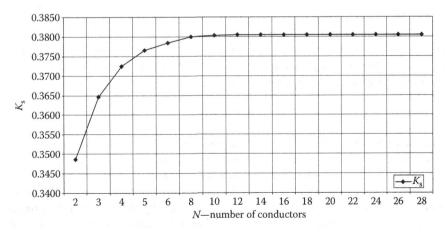

FIGURE 11.13 K_s vs. N.

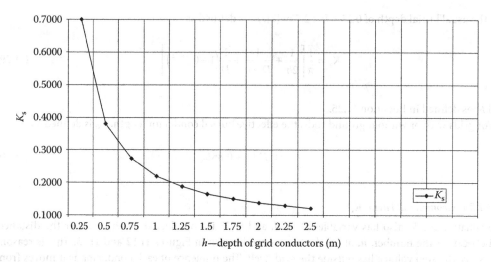

FIGURE 11.14 K_s vs. h.

conductor is buried, the lower the value of K_s as shown in Figure 11.14. This is reasonable since there is a voltage drop as the current passes through the soil reducing the voltage at the surface.

11.3.1.3 Evaluation of the Actual Touch- and Step-Voltage Equations

It is essential to determine the soil resistivity and maximum grid currents to design a substation grounding system. The touch and step voltages are directly proportional to these values. Overly conservative values of soil resistivity and grid current will increase the cost dramatically. Underestimating them may cause the design to be unsafe.

11.3.2 Soil Resistivity

Soil resistivity investigations are necessary to determine the soil structure. There are a number of tables in the literature showing the ranges of resistivity based on soil types (clay, loam, sand, shale, etc.) [9–11]. These tables give only very rough estimates. The soil resistivity can change dramatically with changes in moisture, temperature, and chemical content. To determine the soil resistivity of a particular site, soil resistivity measurements need to be taken. Soil resistivity can vary both horizontally and vertically, making it necessary to take more than one set of measurements. A number of measuring techniques are described in detail in Ref. [12]. The most widely used test for determining soil resistivity data was developed by Wenner and is called either the Wenner or four-pin method. Using four pins or electrodes driven into the earth along a straight line at equal distances of a, to a depth of b, current is passed through the outer pins while a voltage reading is taken with the two inside pins. Based on the resistance, R, as determined by the voltage and current, the apparent resistivity can be calculated using the following equation, assuming b is small compared with a:

$$\rho_a = 2\pi a R \tag{11.31}$$

where it is assumed the apparent resistivity, ρ_a, at depth a is given by the equation.

Interpretation of the apparent soil resistivity based on field measurements is difficult. Uniform and two-layer soil models are the most commonly used soil resistivity models. The objective of the soil model is to provide a good approximation of the actual soil conditions. Interpretation can be done either manually or by the use of computer analysis. There are commercially available computer programs that take the soil data and mathematically calculate the soil resistivity and give a confidence level based on the test. Sunde [10] developed a graphical method to interpret the test results.

Substation Grounding

The equations in IEEE Std. 80 require a uniform soil resistivity. Engineering judgment is required to interpret the soil resistivity measurements to determine the value of the soil resistivity, ρ, to use in the equations. IEEE Std. 80 presents equations to calculate the apparent soil resistivity based on field measurements as well as examples of Sunde's graphical method. Although the graphical method and equations are estimates, they provide the engineer with guidelines of the uniform soil resistivity to use in the ground grid design.

11.3.3 Grid Resistance

The grid resistance, i.e., the resistance of the ground grid to remote earth without other metallic conductors connected, can be calculated based on the following Sverak [2] equation:

$$R_g = \rho \left[\frac{1}{L_T} + \frac{1}{\sqrt{20A}} \left(1 + \frac{1}{1 + h\sqrt{20/A}} \right) \right] \quad (11.32)$$

where
R_g is the substation ground resistance, Ω
ρ is the soil resistivity, Ω m
A is the area occupied by the ground grid, m²
h is the depth of the grid, m
L_T is the total buried length of conductors, m

11.3.3.1 Resistance

The resistance of the grid is mainly determined by the resistivity and the area of the site. Adding more conductors or changing the depth of the grid does little to lower the resistance. The effect of ground rods depends on the location and depth of the ground rod with respect to the soil resistivity. The effects of ground rods on the resistance can be substantial, although it is sometimes difficult to determine the effects. In uniform soil, it is difficult to determine if the addition of more conductors or the addition of ground rods will affect the overall resistance the most. In most cases though, the addition of ground rods has a greater impact because the ground rods discharge current into the earth more efficiently than the grid conductors. Assuming a two-layer soil model with a lower resistivity soil in the lower layer, ground rods can have a substantial impact on the resistance of the grid. The more the ground rods penetrate into the lower resistivity soil, the more the rods will reduce the grid resistance [13–15]. These rods also add stability since the variations in soil resistivity due to moisture and temperature are minimized at lower depths. The effects of moisture and temperature on the soil resistivity can be quite dramatic. Ground rods placed on the outside of the grid have a greater impact than those placed in the interior of the grid because of current density.

The importance of the lower ground grid resistance is reflected in the GPR and actual touch and step voltages. Lowering the resistance of the grid normally reduces the GPR, although not necessarily proportionally. Lowering the resistance may somewhat increase the grid current because the change is the current split between all the ground current return paths. Another way to decrease the resistance is to install counterpoises. This, in effect, results in adding area to the grid. Although IEEE-80 equations cannot take into account these various methods to decrease the resistance, it is important for the engineer to understand there are methods that can be used to lower the resistance of a ground grid.

The following graphs show the effects of the area, number of conductors, and depth for a simple square grid with no ground rods. Figure 11.15 shows conclusively that the area has a great influence on the resistance. The length was not kept constant in the example since more conductor length is needed to cover the area. The number of conductors is related to the change in length and very little decrease in

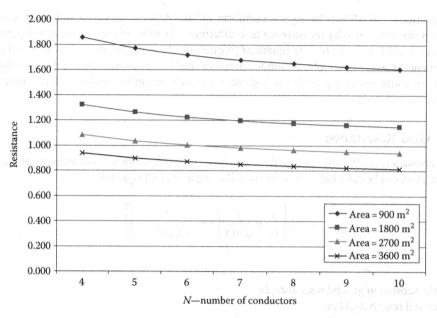

FIGURE 11.15 Resistance vs. N.

resistance takes place when the number of conductors is increased in a constant area. This can be seen by comparing the resistance of a constant area as the number of conductors increases. Since the amount of material is related to the number of conductors, adding more material does not influence the resistance very much.

Figure 11.16 shows the effects of varying the depth of burial of the grid. The area for this example is 900 m². The depth is varied from 0.5 to 2.5 m and the number of conductors from 4 to 10. As can be seen from Figure 11.16, there is very little change in the resistance even if the depth is increased by a factor of 5 and the number of conductors is changed from 4 to 10.

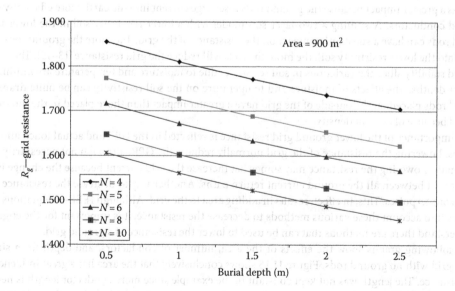

FIGURE 11.16 Resistance vs. depth.

11.3.4 Grid Current

The maximum grid current must be determined, since it is this current that will produce the greatest GPR and the largest local surface potential gradients in and around the substation area. It is the flow of the current from the ground grid system to remote earth that determines the GPR.

There are many types of faults that can occur on an electrical system. Therefore, it is difficult to determine what condition will produce the maximum fault current. In practice, single-line-to-ground and line-to-line-to-ground faults will produce the maximum grid current. Figures 11.17 through 11.19 show the maximum grid current, I_G, for various fault locations and system configurations.

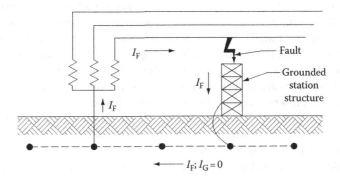

FIGURE 11.17 Fault within local substation, local neutral grounded.

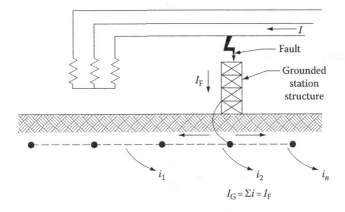

FIGURE 11.18 Fault within local substation, neutral grounded at remote location.

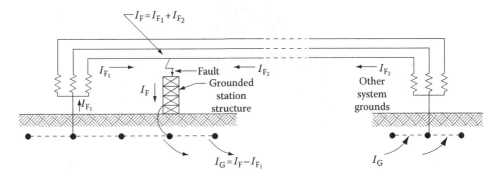

FIGURE 11.19 Fault in substation, system grounded at local station and also at other points.

Overhead ground wires, neutral conductors, and directly buried pipes and cables conduct a portion of the ground fault current away from the substation ground grid and need to be considered when determining the maximum grid current. The effect of these other current paths in parallel with the ground grid is difficult to determine because of the complexities and uncertainties in the current flow.

11.3.4.1 Current Division Consideration

There are many papers that discuss the effects of overhead static wires, neutrals, cables, and other ground paths. As shown in Figure 11.20, the process of computing the current division consists of deriving an equivalent model of the current paths and solving the equivalent circuit to determine what part of the total current flows into the earth and through other ground paths. Endrenyi [16], Sebo [17], Verma and Mukhedkar [18], and Garrett [19] provide approaches to determine the current flows in different current paths for overhead circuits. Dawalibi [20] provides algorithms for deriving simple equations to solve for the currents in the grid and in each tower while Meliopoulos [21] introduces an equivalent conductor to represent the earth using Carson's equations. Sebo [22], Nahman [23], and Sobral [24] provide approaches to determine the current flow when substations are cable fed. Each method can provide insight into the effects of the other current paths on the grid current.

Computer programs are available to determine the split between the various current paths. There are many papers available to determine the effective impedance of a static wire as seen from the fault point. The fault current division factor, or split factor, represents the inverse of a ratio of the symmetrical fault current to that portion of the current that flows between the grounding grid and the surrounding earth.

$$S_f = \frac{I_g}{3I_0} \tag{11.33}$$

where
 S_f is the fault current division factor
 I_g is the rms symmetrical grid current, A
 I_0 is the zero-sequence fault current, A

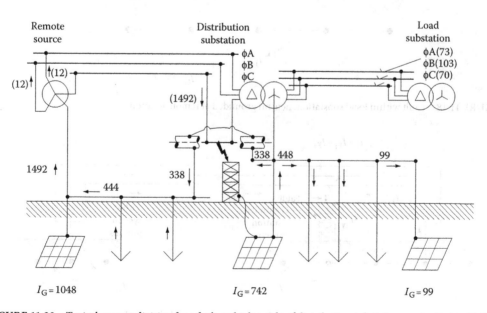

FIGURE 11.20 Typical current division for a fault on higher side of distribution substation.

Substation Grounding

The process of computing the split factor, S_f, consists of deriving an equivalent representation of the overhead ground wires, neutrals, etc., connected to the grid and then solving the equivalent to determine what fraction of the total fault current flows between the grid and earth, and what fraction flows through the ground wires or neutrals. S_f is dependent on many parameters, some of which are

1. Location of the fault
2. Magnitude of substation ground grid resistance
3. Buried pipes and cables in the vicinity of or directly connected to the substation ground system
4. Overhead ground wires, neutrals, or other ground return paths

Because of S_f, the symmetrical grid current I_g and maximum grid current I_G are closely related to the location of the fault. If the additional ground paths of items 3 and 4 above are neglected, the current division ratio (based on remote vs. local current contributions) can be computed using traditional symmetrical components. However, the current I_g computed using such a method may be overly pessimistic, even if the future system expansion is taken into consideration.

IEEE Std. 80 presents a series of curves based on computer simulations for various values of ground grid resistance and system conditions to determine the grid current. These split-current curves can be used to determine the maximum grid current. Using the maximum grid current instead of the maximum fault current will reduce the overall cost of the ground grid system.

11.3.5 Use of the Design Equations

The design equations above are limited to a uniform soil resistivity, equal grid spacing, specific buried depths, and relatively simple geometric layouts of the grid system. The basic requirements for a safe design have not changed through the various revisions of the guide from 1961 to the 2000 edition. The equations in IEEE-80 have changed over the years and will continue to change as better approximate techniques are developed.

It may be necessary to use more sophisticated computer techniques to design a substation ground grid system for nonuniform soils or complex geometric layouts. Commercially available computer programs can be used to optimize the layout and provide for unequal grid spacing and maximum grid current based on the actual system configuration, including overhead wires, neutral conductors, underground facilities, etc. Computer programs can also handle special problems associated with fences, interconnected substation grounding systems at power plants, customer substations, and other unique situations.

11.3.6 Selection of Conductors

11.3.6.1 Materials

Each element of the grounding system, including grid conductors, connections, connecting leads, and all primary electrodes, should be designed so that for the expected design life of the installation, the element will

1. Have sufficient conductivity, so that it will not contribute substantially to local voltage differences
2. Resist fusing and mechanical deterioration under the most adverse combination of a fault current magnitude and duration
3. Be mechanically reliable and rugged to a high degree
4. Be able to maintain its function even when exposed to corrosion or physical abuse

Copper is a common material used for grounding. Copper conductors, in addition to their high conductivity, have the advantage of being resistant to most underground corrosion because copper is cathodic with respect to most other metals that are likely to be buried in the vicinity. Copper-clad steel is usually used for ground rods and occasionally for grid conductors, especially where theft is a problem. Use of copper, or to a lesser degree copper-clad steel, therefore assures that the integrity of an underground network will be maintained for years, so long as the conductors are of an adequate size and not damaged and the soil conditions are not corrosive to the material used. Aluminum is used for ground grids less frequently. Though at first glance the use of aluminum would be a natural choice for GIS equipment with enclosures made of aluminum or aluminum alloys, there are several disadvantages to consider:

- Aluminum can corrode in certain soils. The layer of corroded aluminum material is nonconductive for all practical grounding purposes.
- Gradual corrosion caused by alternating currents can also be a problem under certain conditions.

Thus, aluminum should be used only after full investigation of all circumstances, despite the fact that, like steel, it would alleviate the problem of contributing to the corrosion of other buried objects. However, it is anodic to many other metals, including steel and, if interconnected to one of these metals in the presence of an electrolyte, the aluminum will sacrifice itself to protect the other metal. If aluminum is used, the high-purity electric-conductor grades are recommended as being more suitable than most alloys. Steel can be used for ground grid conductors and rods. Of course, such a design requires that attention be paid to the corrosion of the steel. Use of galvanized or corrosion-resistant steel, in combination with cathodic protection, is typical for steel grounding systems.

A grid of copper or copper-clad steel forms a galvanic cell with buried steel structures, pipes, and any of the lead-based alloys that might be present in cable sheaths. This galvanic cell can hasten corrosion of the latter. Tinning the copper has been tried by some utilities because tinning reduces the cell potential with respect to steel and zinc by about 50% and practically eliminates this potential with respect to lead (tin being slightly sacrificial to lead). The disadvantage of using tinned copper conductor is that it accelerates and concentrates the natural corrosion, caused by the chemicals in the soil, of copper in any small bare area. Other often-used methods are as follows:

- Insulation of the sacrificial metal surfaces with a coating such as plastic tape, asphalt compound, or both.
- Routing of buried metal elements so that any copper-based conductor will cross water pipelines or similar objects made of other uncoated metals as nearly as possible at right angles, and then applying an insulated coating to one metal or the other where they are in proximity. The insulated coating is usually applied to the pipe.
- Cathodic protection using sacrificial anodes or impressed current systems.
- Use of nonmetallic pipes and conduits.

11.3.6.2 Conductor Sizing Factors

Conductor sizing factors include the symmetrical currents, asymmetrical currents, limitation of temperatures to values that will not cause harm to other equipment, mechanical reliability, exposure to corrosive environments, and future growth causing higher grounding-system currents. The following provides information concerning symmetrical and asymmetrical currents.

11.3.6.3 Symmetrical Currents

The short-time temperature rise in a ground conductor, or the required conductor size as a function of conductor current, can be obtained from Equations 11.34 and 11.35, which are taken from the derivation by Sverak [25]. These equations evaluate the ampacity of any conductor for which the material constants are known. Equations 11.34 and 11.35 are derived for symmetrical currents (with no dc offset).

$$I = A_{mm^2} \sqrt{\left(\frac{TCAP \cdot 10^{-4}}{t_c \alpha_r \rho_r}\right) \ln\left(\frac{K_0 + T_m}{K_0 + T_a}\right)} \qquad (11.34)$$

where

- I is the rms current, kA
- A_{mm^2} is the conductor cross section, mm²
- T_m is the maximum allowable temperature, °C
- T_a is the ambient temperature, °C
- T_r is the reference temperature for material constants, °C
- α_0 is the thermal coefficient of resistivity at 0 ∞ C, 1/°C
- α_r is the thermal coefficient of resistivity at reference temperature T_r, 1/°C
- ρ_r is the resistivity of the ground conductor at reference temperature T_r, mΩ cm
- $K_0 = 1/\alpha_0$ or $(1/\alpha_r) - T_r$, °C
- t_c is the duration of current, s
- TCAP is the thermal capacity per unit volume, J/(cm³ °C)

Note that α_r and ρ_r are both to be found at the same reference temperature of T_r °C. If the conductor size is given in kcmils (mm² × 1.974 = kcmils), Equation 11.34 becomes

$$I = 5.07 \cdot 10^{-3} A_{kcmil} \sqrt{\left(\frac{TCAP}{t_c \alpha_r \rho_r}\right) \ln\left(\frac{K_0 + T_m}{K_0 + T_a}\right)} \quad (11.35)$$

11.3.6.4 Asymmetrical Currents: Decrement Factor

In cases where accounting for a possible dc offset component in the fault current is desired, an equivalent value of the symmetrical current, I_F, representing the rms value of an asymmetrical current integrated over the entire fault duration, t_c, can be determined as a function of X/R by using the decrement factor D_f, Equation 11.37, prior to the application of Equations 11.34 and 11.35:

$$I_F = I_f \cdot D_f \quad (11.36)$$

$$D_f = \sqrt{1 + \frac{T_a}{t_f}\left(1 - e^{-2t_f/T_a}\right)} \quad (11.37)$$

where

- t_f is the time duration of fault in s
- T_a is the dc offset time constant in s [$T_a = X/(\omega R)$; for 60 Hz, $T_a = X/(120\pi R)$]

The resulting value of I_F is always larger than I_f because the decrement factor is based on a very conservative assumption that the ac component does not decay with time but remains constant at its initial subtransient value.

The decrement factor is dependent on both the system X/R ratio at the fault location for a given fault type and the duration of the fault. The decrement factor is larger for higher X/R ratios and shorter fault durations. The effects of the dc offset are negligible if the X/R ratio is less than five and the duration of the fault is greater than 1 s.

11.3.7 Selection of Connections

All connections made in a grounding network above and below ground should be evaluated to meet the same general requirements of the conductor used, namely electrical conductivity, corrosion resistance, current-carrying capacity, and mechanical strength. These connections should be massive enough to maintain a temperature rise below that of the conductor and to withstand the effect of heating, be strong enough to withstand the mechanical forces caused by the electromagnetic forces of maximum expected fault currents, and be able to resist corrosion for the intended life of the installation.

IEEE Std. 837 (Qualifying Permanent Connections Used in Substation Grounding) [26] provides detailed information on the application and testing of permanent connections for use in substation grounding. Grounding connections that pass IEEE Std. 837 for a particular conductor size, range, and material should satisfy all the criteria outlined above for that same conductor size, range, and material.

11.3.8 Grounding of Substation Fence

Fence grounding is of major importance, since the fence is usually accessible to the general public, children, and adults. The substation grounding system design should be such that the touch potential on the fence is within the calculated tolerable limit of touch potential. Step potential is usually not a concern at the fence perimeter, but this should be checked to verify that a problem does not exist. There are various ways to ground the substation fence. The fence can be within and attached to the ground grid, outside and attached to the ground grid, outside and not attached to the ground grid, or separately grounded such as through the fence post. IEEE Std. 80 provides a very detailed analysis of the different grounding situations. There are many safety considerations associated with the different fence-grounding options.

11.3.9 Other Design Considerations

There are other elements of substation grounding system design which have not been discussed here. These elements include the refinement of the design, effects of directly buried pipes and cables, special areas of concern including control and power cable grounding, surge arrester grounding, transferred potentials, and installation considerations.

References

1. Institute of Electrical and Electronics Engineers, *IEEE Guide for Safety in AC Substation Grounding*, IEEE Std. 80-2000, IEEE, Piscataway, NJ, 2000.
2. Sverak, J.G., Simplified analysis of electrical gradients above a ground grid: Part I—How good is the present IEEE method? *IEEE Trans. Power Appar. Syst.*, 103, 7–25, 1984.
3. Thapar, B., Gerez, V., and Kejriwal, H., Reduction factor for the ground resistance of the foot in substation yards, *IEEE Trans. Power Delivery*, 9, 360–368, 1994.
4. Dalziel, C.F. and Lee, W.R., Lethal electric currents, *IEEE Spectrum*, 6, 44–50, February 1969.
5. Dalziel, C.F., Threshold 60-cycle fibrillating currents, *AIEE Trans. Power Appar. Syst.*, 79, 667–673, 1960.
6. Dalziel, C.F. and Lee, R.W., Reevaluation of lethal electric currents, *IEEE Trans. Ind. Gen. Appl.*, 4, 467–476, 1968.
7. Biegelmeier, U.G. and Lee, W.R., New considerations on the threshold of ventricular fibrillation for AC shocks at 50–60 Hz, *Proc. IEEE*, 127, 103–110, 1980.
8. Thapar, B., Gerez, V., Balakrishnan, A., and Blank, D., Simplified equations for mesh and step voltages in an AC substation, *IEEE Trans. Power Delivery*, 6, 601–607, 1991.
9. Rüdenberg, R., Basic considerations concerning systems, *Eloktrotech. Z.*, 11–12, 1926.
10. Sunde, E.D., *Earth Conduction Effects in Transmission Systems*, Macmillan, New York, 1968.
11. Wenner, F., A method of measuring earth resistances, Rep. 258, *Bull. Bur. Stand.*, 12, 469–482, 1916.
12. Institute of Electrical and Electronics Engineers, *IEEE Guide for Measuring Earth Resistivity, Ground Impedance, and Earth Surface Potentials of a Ground System*, IEEE Std. 81-1983, IEEE, Piscataway, NJ, 1983.
13. Blattner, C.J., Study of driven ground rods and four point soil resistivity data, *IEEE Trans. Power Appar. Syst.*, PAS-101(8), 2837–2850, August 1982.
14. Dawalibi, F. and Mukhedkar, D., Influence of ground rods on grounding systems, *IEEE Trans. Power Appar. Syst.*, PAS-98(6), 2089–2098, November/December 1979.

15. Tagg, G.F., *Earth Resistances*, Pitman, New York, 1964.
16. Endrenyi, J., Fault current analysis for substation grounding design, *Ont. Hydro Res. Q.*, 2nd Quarter, 1967.
17. Sebo, S.A., Zero sequence current distribution along transmission lines, *IEEE Trans. Power Appar. Syst.*, PAS-88, 910–919, June 1969.
18. Verma, R. and Mukhedkar, D., Ground fault current distribution in substation, towers and ground wire, *IEEE Trans. Power Appar. Syst.*, PAS-98, 724–730, May/June 1979.
19. Garrett, D.L., Determination of maximum ground fault current through substation grounding system considering effects of static wires and feeder neutrals, *Proceedings of Southeastern Electric Exchange*, Atlanta, GA, 1981.
20. Dawalibi, F., Ground fault current distribution between soil and neutral conductors, *IEEE Trans. Power Appar. Syst.*, PAS-99(2), 452–461, March/April 1980.
21. Meliopoulos, A.P., Papalexopoulos, A., and Webb, R.P., Current division in substation grounding system, *Proceedings of the 1982 Protective Relaying Conference*, Georgia Institute of Technology, Atlanta, GA, May 1982.
22. Sebo, S.A. and Guven, A.N., Analysis of ground fault current distribution along underground cables, *IEEE Trans. Power Delivery*, PWRD-1, 9–18, October 1986.
23. Nahman, J. and Salamon, D., Effects of the metal sheathed cables upon the performance of the distribution substation grounding system, *IEEE Trans. Power Delivery*, 7, 1179–1187, July 1992.
24. Sobral, S.T., Costa, V., Campos, M., and Mukhedkar, D., Dimensioning of nearby substations interconnected ground system, *IEEE Trans. Power Delivery*, 3(4), 1605–1614, October 1988.
25. Sverak, J.G., Sizing of ground conductors against fusing, *IEEE Trans. Power Appar. Syst.*, 100, 51–59, 1981.
26. Institute of Electrical and Electronics Engineers, *IEEE Standard for Qualifying Permanent Connections Used in Substation Grounding*, IEEE Std. 837-1989 (reaffirmed 1996), IEEE, Piscataway, NJ, 1996.

12
Direct Lightning Stroke Shielding of Substations*

Robert S. Nowell
(retired)
Commonwealth Associates, Inc.

12.1 Lightning Stroke Protection 12-1
   The Design Problem
12.2 Lightning Parameters 12-2
   Strike Distance • Stroke Current Magnitude • Keraunic Level • Ground Flash Density • Lightning Detection Networks
12.3 Empirical Design Methods 12-5
   Fixed Angles • Empirical Curves
12.4 The Electrogeometric Model (EGM) 12-7
   Whitehead's EGM • Recent Improvements in the EGM • Criticism of the EGM • A Revised EGM • Application of the EGM by the Rolling Sphere Method • Multiple Shielding Electrodes • Changes in Voltage Level • Minimum Stroke Current • Application of Revised EGM by Mousa and Srivastava Method
12.5 Calculation of Failure Probability 12-19
12.6 Active Lightning Terminals 12-19
References 12-19

12.1 Lightning Stroke Protection

Substation design involves more than installing apparatus, protective devices, and equipment. The significant monetary investment and required reliable continuous operation of the facility requires detailed attention to preventing surges (transients) from entering the substation facility. These surges can be switching surges, lightning surges on connected transmission lines, or direct strokes to the substation facility. The origin and mechanics of these surges, including lightning, are discussed in detail in Chapter 10 of *The Electric Power Engineering Handbook* (CRC Press 2001). This section focuses on the design process for providing *effective shielding* (that which permits lightning strokes no greater than those of critical amplitude [less design margin] to reach phase conductors [IEEE Std. 998-1996 (R2002)]) against direct lightning stroke in substations.

---

* A large portion of the text and all of the figures used in this chapter were prepared by the Direct Stroke Shielding of Substations Working Group of the Substations Committee—IEEE Power Engineering Society, and published as IEEE Std. 998-1996 (R2002), *IEEE Guide for Direct Lightning Stroke Shielding of Substrates*, Institute of Electrical and Electronics Engineers, Inc., 1996. The IEEE disclaims any responsibility of liability resulting from the placement or use in the described manner. Information is reprinted with the permission of the IEEE. The author has been a member of the working group since 1987.

12.1.1 The Design Problem

The engineer who seeks to design a direct stroke shielding system for a substation or facility must contend with several elusive factors inherent in lightning phenomena, namely:

- The unpredictable, probabilistic nature of lightning
- The lack of data due to the infrequency of lightning strokes in substations
- The complexity and economics involved in analyzing a system in detail

There is no known method of providing 100% shielding short of enclosing the equipment in a solid metallic enclosure. The uncertainty, complexity, and cost of performing a detailed analysis of a shielding system has historically resulted in simple rules of thumb being utilized in the design of lower voltage facilities. Extra high voltage (EHV) facilities, with their critical and more costly equipment components, usually justify a more sophisticated study to establish the risk vs. cost benefit.

Because of the above factors, it is suggested that a four-step approach be utilized in the design of a protection system:

1. Evaluate the importance and value of the facility being protected.
2. Investigate the severity and frequency of thunderstorms in the area of the substation facility and the exposure of the substation.
3. Select an appropriate design method consistent with the above evaluation and then lay out an appropriate system of protection.
4. Evaluate the effectiveness and cost of the resulting design.

The following paragraphs and references will assist the engineer in performing these steps.

12.2 Lightning Parameters

12.2.1 Strike Distance

Return stroke current magnitude and strike distance (length of the last stepped leader) are interrelated. A number of equations have been proposed for determining the striking distance. The principal ones are as follows:

$$S = 2I + 30(1 - e^{-I/6.8}) \quad \text{Darveniza et al. (1975)} \tag{12.1}$$

$$S = 10I^{0.65} \quad \text{Anderson (1987); IEEE (1993)} \tag{12.2}$$

$$S = 9.4I^{2/3} \quad \text{Whitehead (1974)} \tag{12.3}$$

$$S = 8I^{0.65} \quad \text{IEEE (1985)} \tag{12.4}$$

$$S = 3.3I^{0.78} \quad \text{Suzuki et al. (1981)} \tag{12.5}$$

where
  S is the strike distance in meters
  I is the return stroke current in kiloamperes

It may be disconcerting to note that the above equations vary by as much as a factor of 2:1. However, lightning investigators now tend to favor the shorter strike distances given by Equation 12.4. Anderson, for example, who adopted Equation 12.2 in the 1975 edition of the *Transmission Line Reference Book* (1987), now feels that Equation 12.4 is more accurate. Mousa (1988) also supports this form of the equation. The equation may also be stated as follows:

$$I = 0.041 S^{1.54} \tag{12.6}$$

From this point on, the return stroke current will be referenced as the *stroke current*.

12.2.2 Stroke Current Magnitude

Since the stroke current and striking distance are related, it is of interest to know the distribution of stroke current magnitudes. The median value of strokes to OHGW, conductors, structures, and masts is usually taken to be 31 kA (Anderson, 1987). Anderson (1987) gave the probability that a certain peak current will be exceeded in any stroke as follows:

$$P(I) = \frac{1}{[1+(I/31)^{2.6}]} \tag{12.7}$$

where

$P(I)$ is the probability that the peak current in any stroke will exceed I
I is the specified crest current of the stroke in kiloamperes

Mousa (1989) has shown that a median stroke current of 24 kA for strokes to flat ground produces the best correlation with available field observations to date. Using this median value of stroke current, the probability that a certain peak current will be exceeded in any stroke is given by the following equation:

$$P(I) = \frac{1}{[1+(I/24)^{2.6}]} \tag{12.8}$$

where the symbols have the same meaning as above.

Figure 12.1 is a plot of Equation 12.8, and Figure 12.2 is a plot of the probability that a stroke will be within the ranges shown on the abscissa.

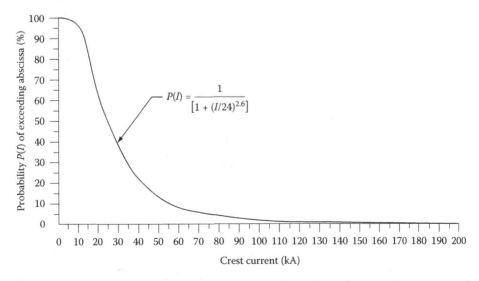

FIGURE 12.1 Probability of stroke current exceeding abscissa for strokes to flat ground. (From IEEE Std. 998-1996 (R2002), IEEE Working Group D5, Substations Committee, *Guide for Direct Lightning Stroke Shielding of Substations*. With permission.)

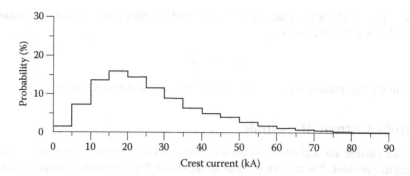

FIGURE 12.2 Stroke current range probability for strokes to flat ground. (From IEEE Std. 998-1996 (R2002), IEEE Working Group D5, Substations Committee, *Guide for Direct Lightning Stroke Shielding of Substations*. With permission.)

12.2.3 Keraunic Level

Keraunic level is defined as the average annual number of thunderstorm days or hours for a given locality. A daily keraunic level is called a thunderstorm-day and is the average number of days per year on which thunder will be heard during a 24 h period. By this definition, it makes no difference how many times thunder is heard during a 24 h period. In other words, if thunder is heard on any one day more than one time, the day is still classified as one thunder day (or thunderstorm day). The average annual keraunic level for locations in the United States can be determined by referring to isokeraunic maps on which lines of equal keraunic level are plotted on a map of the country. Figure 12.3 gives the mean annual thunderstorm days for the United States.

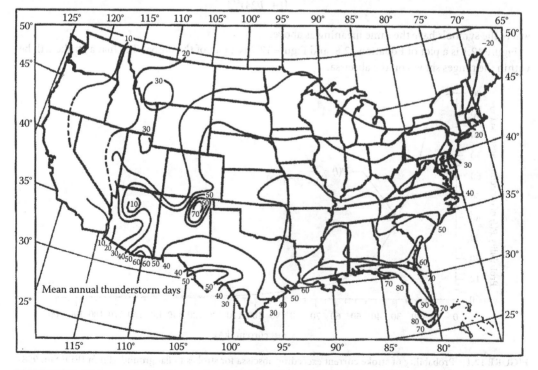

FIGURE 12.3 Mean annual thunderstorm days in the United States. (From IEEE Std. 998-1996 (R2002), IEEE Working Group D5, Substations Committee, *Guide for Direct Lightning Stroke Shielding of Substations*. With permission.)

12.2.4 Ground Flash Density

Ground flash density (GFD) is defined as the average number of strokes per unit area per unit time at a particular location. It is usually assumed that the GFD to earth, a substation, or a transmission or distribution line is roughly proportional to the keraunic level at the locality. If thunderstorm days are to be used as a basis, it is suggested that the following equation be used (Anderson, 1987):

$$N_k = 0.12 T_d \tag{12.9}$$

or

$$N_m = 0.31 T_d \tag{12.10}$$

where
  N_k is the number of flashes to earth per square kilometer per year
  N_m is the number of flashes to earth per square mile per year
  T_d is the average annual keraunic level, thunderstorm days

12.2.5 Lightning Detection Networks

A new technology is now being deployed in Canada and the United States that promises to provide more accurate information about ground flash density and lightning stroke characteristics. Mapping of lightning flashes to the earth has been in progress for over a decade in Europe, Africa, Australia, and Asia. Now a network of direction finding receiving stations has been installed across Canada and the United States. By means of triangulation among the stations, and with computer processing of signals, it is possible to pinpoint the location of each lightning discharge. Hundreds of millions of strokes have been detected and plotted to date.

Ground flash density maps have already been prepared from this data, but with the variability in frequency and paths taken by thunderstorms from year to year, it will take a number of years to develop data that is statistically significant. Some electric utilities are, however, taking advantage of this technology to detect the approach of thunderstorms and to plot the location of strikes on their system. This information is very useful for dispatching crews to trouble spots and can result in shorter outages that result from lightning strikes.

12.3 Empirical Design Methods

Two classical design methods have historically been employed to protect substations from direct lightning strokes:

1. Fixed angles
2. Empirical curves

The two methods have generally provided acceptable protection.

12.3.1 Fixed Angles

The fixed-angle design method uses vertical angles to determine the number, position, and height of shielding wires or masts. Figure 12.4 illustrates the method for shielding wires, and Figure 12.5 illustrates the method for shielding masts. The angles used are determined by the degree of lightning exposure, the importance of the substation being protected, and the physical area occupied by the substation.

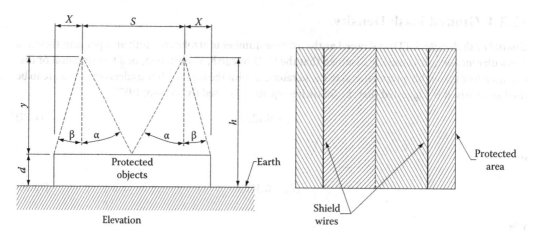

FIGURE 12.4 Fixed angles for shielding wires. (From IEEE Std. 998-1996 (R2002), IEEE Working Group D5, Substations Committee, *Guide for Direct Lightning Stroke Shielding of Substations*. With permission.)

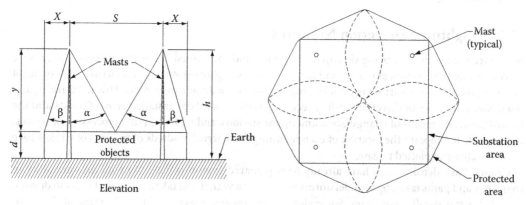

FIGURE 12.5 Fixed angles for masts. (From IEEE Std. 998-1996 (R2002), IEEE Working Group D5, Substations Committee, *Guide for Direct Lightning Stroke Shielding of Substations*. With permission.)

The value of the angle alpha that is commonly used is 45°. Both 30° and 45° are widely used for angle beta. (Sample calculations for low voltage and high voltage substations using fixed angles are given in annex B of IEEE Std. 998-1996 (R2002).)

12.3.2 Empirical Curves

From field studies of lightning and laboratory model tests, empirical curves have been developed to determine the number, position, and height of shielding wires and masts (Wagner et al., 1941a,b; Wagner, 1942). The curves were developed for shielding failure rates of 0.1%, 1.0%, 5.0%, 10%, and 15%. A failure rate of 0.1% is commonly used in design. Figures 12.6 and 12.7 have been developed for a variety of protected object heights, d. The empirical curve method has also been referred to as the Wagner method.

12.3.2.1 Areas Protected by Lightning Masts

Figures 12.8 and 12.9 illustrate the areas that can be protected by two or more shielding masts (Wagner et al., 1942). If two masts are used to protect an area, the data derived from the empirical curves give shielding information only for the point B, midway between the two masts, and for points on the semi-circles drawn about the masts, with radius x, as shown in Figure 12.8a. The locus shown in Figure 12.8a,

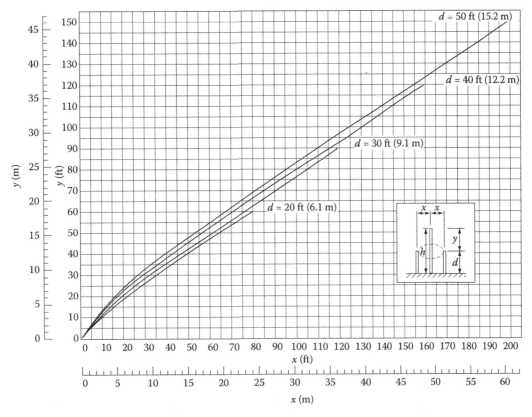

FIGURE 12.6 Single lightning mast protecting single ring of object—0.1% exposure. Height of mast above protected object, y, as a function of horizontal separation, x, and height of protected object, d. (From IEEE Std. 998-1996 (R2002), IEEE Working Group D5, Substations Committee, *Guide for Direct Lightning Stroke Shielding of Substations*. With permission.)

drawn by the semicircles around the masts, with radius x, and connecting the point B, represents an approximate limit for a selected exposure rate. Any single point falling within the cross-hatched area should have <0.1% exposure. Points outside the cross-hatched area will have >0.1% exposure. Figure 12.8b illustrates this phenomenon for four masts spaced at the distance s as in Figure 12.8a.

The protected area can be improved by moving the masts closer together, as illustrated in Figure 12.9. In Figure 12.9a, the protected areas are, at least, as good as the combined areas obtained by superimposing those of Figure 12.8a. In Figure 12.9a, the distance s' is one-half the distance s in Figure 12.8a. To estimate the width of the overlap, x', first obtain a value of y corresponding to twice the distance, s' between the masts. Then use Figure 12.6 to determine x' for this value of y. This value of x is used as an estimate of the width of overlap x' in Figure 12.9. As illustrated in Figure 12.9b, the size of the areas with an exposure greater than 0.1% has been significantly reduced. (Sample calculations for low voltage and high voltage substations using empirical curves are given in annex B of IEEE Std. 998-1996 (R2002).)

12.4 The Electrogeometric Model (EGM)

Shielding systems developed using classical methods (fixed-angle and empirical curves) of determining the necessary shielding for direct stroke protection of substations have historically provided a fair degree of protection. However, as voltage levels (and therefore structure and conductor heights) have increased over the years, the classical methods of shielding design have proven less adequate. This led to the development of the electrogeometric model.

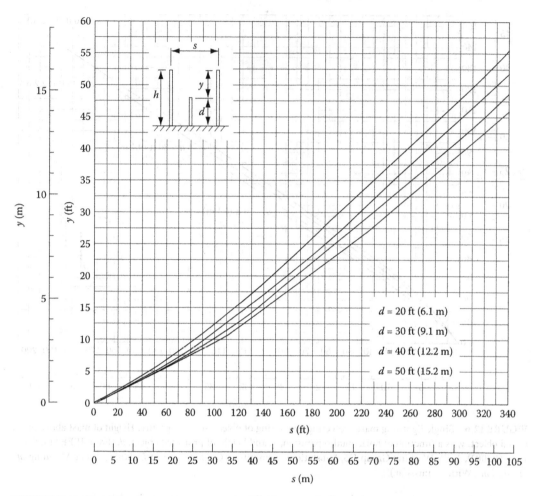

FIGURE 12.7 Two lightning masts protecting single object, no overlap—0.1% exposure. Height of mast above protected object, y, as a function of horizontal separation, s, and height of protected object, d. (From IEEE Std. 998-1996 (R2002), IEEE Working Group D5, Substations Committee. *Guide for Direct Lightning Stroke Shielding of Substations*. With permission.)

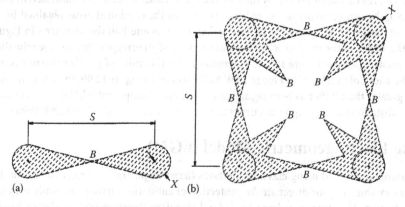

FIGURE 12.8 Areas protected by multiple masts for point exposures shown in Figure 12.5. (a) With two lightning masts; (b) with four lightning masts. (From IEEE Std. 998-1996 (R2002), IEEE Working Group D5, Substations Committee, *Guide for Direct Lightning Stroke Shielding of Substations*. With permission.)

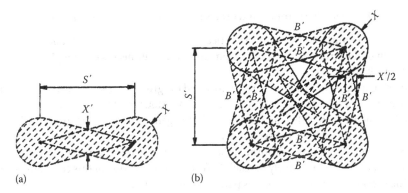

FIGURE 12.9 Areas protected by multiple masts for point exposures shown in Figure 12.5. (a) With two lightning masts; (b) with four lightning masts. (From IEEE Std. 998-1996 (R2002), IEEE Working Group D5, Substations Committee, *Guide for Direct Lightning Stroke Shielding of Substations*. With permission.)

12.4.1 Whitehead's EGM

In 1960, Anderson developed a computer program for calculation of transmission line lightning performance that uses the *Monte Carlo Method* (1961). This method showed good correlation with actual line performance. An early version of the EGM was developed in 1963 by Young et al., but continuing research soon led to new models. One extremely significant research project was performed by Whitehead (1971). Whitehead's work included a theoretical model of a transmission system subject to direct strokes, development of analytical expressions pertaining to performance of the line, and supporting field data that verified the theoretical model and analyses. The final version of this model was published by Gilman and Whitehead in 1973.

12.4.2 Recent Improvements in the EGM

Sargent made an important contribution with the *Monte Carlo Simulation* of lightning performance (1972b) and his work on lightning strokes to tall structures (1972a). Sargent showed that the frequency distribution of the amplitudes of strokes collected by a structure depends on the structure height as well as on its type (mast vs. wire). In 1976, Mousa extended the application of the EGM (which was developed for transmission lines) to substation facilities.

12.4.3 Criticism of the EGM

Work by Eriksson reported in 1978 and later work by Anderson and Eriksson reported in 1980 revealed apparent discrepancies in the EGM that tended to discredit it. Mousa (1988) has shown, however, that explanations do exist for the apparent discrepancies, and that many of them can be eliminated by adopting a revised electrogeometric model. Most investigators now accept the EGM as a valid approach for designing lightning shielding systems.

12.4.4 A Revised EGM

The revised EGM was developed by Mousa and Srivastava (1986, 1988). Two methods of applying the EGM are the modified version of the rolling sphere method (Lee, 1978, 1979; Orell, 1988), and the method given by Mousa and Srivastava (1988, 1991).

The revised EGM model differs from Whitehead's model in the following respects:

1. The stroke is assumed to arrive in a vertical direction. (It has been found that Whitehead's assumption of the stroke arriving at random angles is an unnecessary complication [Mousa and Srivastava, 1988].)
2. The differing striking distances to masts, wires, and the ground plane are taken into consideration.
3. A value of 24 kA is used as the median stroke current (Mousa and Srivastava, 1989). This selection is based on the frequency distribution of the first negative stroke to flat ground. This value best reconciles the EGM with field observations.
4. The model is not tied to a specific form of the striking distance equations (Equations 12.1 through 12.6). Continued research is likely to result in further modification of this equation as it has in the past. The best available estimate of this parameter may be used.

12.4.4.1 Description of the Revised EGM

Previously, the concept that the final striking distance is related to the magnitude of the stroke current was introduced and Equation 12.4 was selected as the best approximation of this relationship. A coefficient k accounts for the different striking distances to a mast, a shield wire, and to the ground. Equation 12.4 is repeated here with this modification:

$$S_m = 8kI^{0.65} \tag{12.11}$$

or

$$S_f = 26.25kI^{0.65} \tag{12.12}$$

where
  S_m is the strike distance in meters
  S_f is the strike distance in feet
  I is the return stroke current in kiloamperes
  k is a coefficient to account for different striking distances to a mast, a shield wire, or the ground plane

Mousa (1988) gives a value of $k = 1$ for strokes to wires or the ground plane and a value of $k = 1.2$ for strokes to a lightning mast.

Lightning strokes have a wide distribution of current magnitudes, as shown in Figure 12.1. The EGM theory shows that the protective area of a shield wire or mast depends on the amplitude of the stroke current. If a shield wire protects a conductor for a stroke current I_s, it may not shield the conductor for a stroke current less than I_s that has a shorter striking distance. Conversely, the same shielding arrangement will provide greater protection against stroke currents greater than I_s that have greater striking distances. Since strokes less than some critical value I_s can penetrate the shield system and terminate on the protected conductor, the insulation system must be able to withstand the resulting voltages without flashover. Stated another way, the shield system should intercept all strokes of magnitude I_s and greater so that flashover of the insulation will not occur.

12.4.4.2 Allowable Stroke Current

Some additional relationships need to be introduced before showing how the EGM is used to design a zone of protection for substation equipment. Bus insulators are usually selected to withstand a *basic lightning impulse level* (BIL). Insulators may also be chosen according to other electrical characteristics, including negative polarity *impulse critical flashover* (C.F.O.) voltage. Flashover occurs if the voltage produced by the lightning stroke current flowing through the surge impedance of the station bus exceeds the withstand value. This may be expressed by the Gilman and Whitehead equation (1973):

$$I_s = \text{BIL} \times \frac{1.1}{(Z_S/2)} = \frac{2.2(\text{BIL})}{Z_S} \tag{12.13}$$

or

$$I_s = 0.94 \times \text{C.F.O.} \times \frac{1.1}{(Z_S/2)} = \frac{2.068(\text{C.F.O.})}{Z_S} \tag{12.14}$$

where
I_s is the allowable stroke current in kiloamperes
BIL is the basic lightning impulse level in kilovolts
C.F.O. is the negative polarity critical flashover voltage of the insulation being considered in kilovolts
Z_S is the surge impedance of the conductor through which the surge is passing in ohms
1.1 is the factor to account for the reduction of stroke current terminating on a conductor as compared to zero impedance earth (Gilman and Whitehead, 1973)

In Equation 12.14, the C.F.O. has been reduced by 6% to produce a withstand level roughly equivalent to the BIL rating for post insulators.

12.4.4.3 Withstand Voltage of Insulator Strings

BIL values of station post insulators can be found in vendor catalogs. A method is given below for calculating the withstand voltage of insulator strings. The withstand voltage in kV at 2 and 6 µs can be calculated as for

$$V_{I2} = 0.94 \times 820w \tag{12.15}$$

$$V_{I6} = 0.94 \times 585w \tag{12.16}$$

where
w is the length of insulator string (or air gap) in meters
0.94 is the ratio of withstand voltage to C.F.O. voltage
V_{I2} is the withstand voltage in kilovolts at 2 µs
V_{I6} is the withstand voltage in kilovolts at 6 µs

Equation 12.16 is recommended for use with the EGM.

12.4.5 Application of the EGM by the Rolling Sphere Method

It was previously stated that it is only necessary to provide shielding for the equipment from all lightning strokes greater than I_s that would result in a flashover of the buswork. Strokes less than I_s are permitted to enter the protected zone since the equipment can withstand voltages below its BIL design level. This will be illustrated by considering three levels of stroke current: I_s, stroke currents greater than I_s, and stroke currents less than I_s. First, let us consider the stroke current I_s.

12.4.5.1 Protection Against Stroke Current I_s

I_s is calculated from Equation 12.13 or 12.14 as the current producing a voltage the insulation will just withstand. Substituting this result in Equation 12.11 or 12.12 gives the striking distance S for this stroke current. In 1977, Lee developed a simplified technique for applying the electromagnetic theory to the shielding of buildings and industrial plants (1978, 1982, 1979). Orrell extended the technique to specifically cover the protection of electric substations (1988). The technique developed by Lee has come to be known as the rolling sphere method. For the following illustration, the *rolling sphere* method will be used. This method employs the simplifying assumption that the striking distances to the ground, a mast, or a wire are the same. With this exception, the rolling sphere method has been updated in accordance with the revised EGM.

Use of the rolling sphere method involves rolling an imaginary sphere of radius S over the surface of a substation. The sphere rolls up and over (and is supported by) lightning masts, shield wires, substation fences, and other grounded metallic objects that can provide lightning shielding. A piece of equipment is said to be protected from a direct stroke if it remains below the curved surface of the sphere by virtue of the sphere being elevated by shield wires or other devices. Equipment that touches the sphere or penetrates its surface is not protected. The basic concept is illustrated in Figure 12.10.

Continuing the discussion of protection against stroke current I_s, consider first a single mast. The geometrical model of a single substation shield mast, the ground plane, the striking distance, and the zone of protection are shown in Figure 12.11. An arc of radius S that touches the shield mast and the ground plane is shown in Figure 12.11. All points below this arc are protected against the stroke current I_s. This is the protected zone. The arc is constructed as follows (see Figure 12.11). A dashed line is drawn parallel to the ground at a distance S (the striking distance as obtained from Equation 12.11 or 12.12) above the ground plane. An arc of radius S, with its center located on the dashed line, is drawn so the radius of the arc just touches the mast. Stepped leaders that result in stroke current I_s and that descend outside of the point where the arc is tangent to the ground will strike the ground. Stepped leaders that result in stroke current I_s and that descend inside the point where the arc is tangent to the ground will strike the shield mast, provided all other objects are within the protected zone. The height of the shield mast that will provide the maximum zone of protection for stroke currents equal to I_s is S. If the mast height is less than S, the zone of protection will be reduced. *Increasing the shield mast height greater than S will provide additional protection in the case of a single mast. This is not necessarily true in the case of multiple masts and shield wires.* The protection zone can be visualized as the surface of a sphere with radius S that is rolled toward the mast until touching the mast. As the sphere is rolled around the mast, a three-dimensional surface of protection is defined. It is this concept that has led to the name *rolling sphere* for simplified applications of the electrogeometric model.

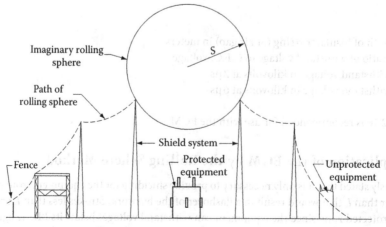

FIGURE 12.10 Principle of the rolling sphere. (From IEEE Std. 998-1996 (R2002), IEEE Working Group D5, Substations Committee, *Guide for Direct Lightning Stroke Shielding of Substations*. With permission.)

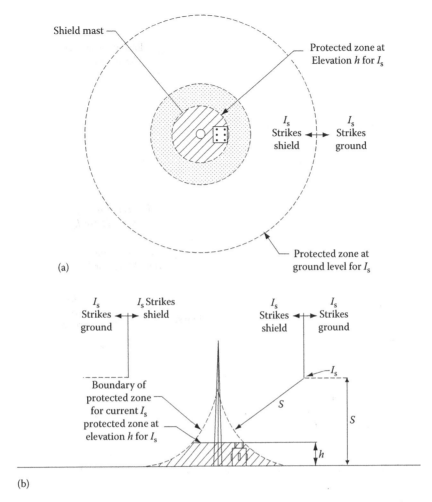

FIGURE 12.11 Shield mast protection for stroke current I_s. (a) Plan view; (b) Elevation view. (From IEEE Std. 998-1996 (R2002), IEEE Working Group D5, Substations Committee, *Guide for Direct Lightning Stroke Shielding of Substations*. With permission.)

12.4.5.2 Protection against Stroke Currents Greater than I_s

A lightning stroke current has an infinite number of possible magnitudes, however, and the substation designer will want to know if the system provides protection at other levels of stroke current magnitude. Consider a stroke current I_{s1} with magnitude greater than I_s. Strike distance, determined from Equation 12.11 or 12.12, is S1. The geometrical model for this condition is shown in Figure 12.12. Arcs of protection for stroke current I_{s1} and for the previously discussed I_s are both shown. The figure shows that the zone of protection provided by the mast for stroke current I_{s1} is greater than the zone of protection provided by the mast for stroke current I_s. Stepped leaders that result in stroke current I_{s1} and that descend outside of the point where the arc is tangent to the ground will strike the ground. Stepped leaders that result in stroke current I_{s1} and that descend inside the point where the arc is tangent to the ground will strike the shield mast, provided all other objects are within the S1 protected zone. Again, the protective zone can be visualized as the surface of a sphere touching the mast. In this case, the sphere has a radius S1.

12.4.5.3 Protection against Stroke Currents Less than I_s

It has been shown that a shielding system that provides protection at the stroke current level I_s provides even better protection for larger stroke currents. The remaining scenario to examine is the

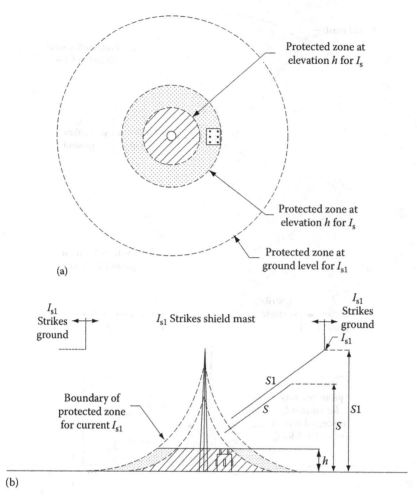

FIGURE 12.12 Shield mast protection for stroke current I_{s1}. (a) Plan view; (b) Elevation view. (From IEEE Std. 998-1996 (R2002), IEEE Working Group D5, Substations Committee, *Guide for Direct Lightning Stroke Shielding of Substations*. With permission.)

protection afforded when stroke currents are less than I_s. Consider a stroke current I_{s0} with magnitude less than I_s. The striking distance, determined from Equation 12.11 or 12.12, is S_0. The geometrical model for this condition is shown in Figure 12.13. Arcs of protection for stroke current I_{s0} and I_s are both shown. The figure shows that the zone of protection provided by the mast for stroke current I_{s0} is less than the zone of protection provided by the mast for stroke current I_s. It is noted that a portion of the equipment protrudes above the dashed arc or zone of protection for stroke current I_{s0}. Stepped leaders that result in stroke current I_{s0} and that descend outside of the point where the arc is tangent to the ground will strike the ground. However, some stepped leaders that result in stroke current I_{s0} and that descend inside the point where the arc is tangent to the ground could strike the equipment. This is best shown by observing the plan view of protective zones shown in Figure 12.13. Stepped leaders for stroke current I_{s0} that descend inside the inner protective zone will strike the mast and protect equipment that is h in height. Stepped leaders for stroke current I_{s0} that descend in the shaded unprotected zone will strike equipment of height h in the area. *If, however, the value of I_s was selected based on the withstand insulation level of equipment used in the substation, stroke current I_{s0} should cause no damage to equipment.*

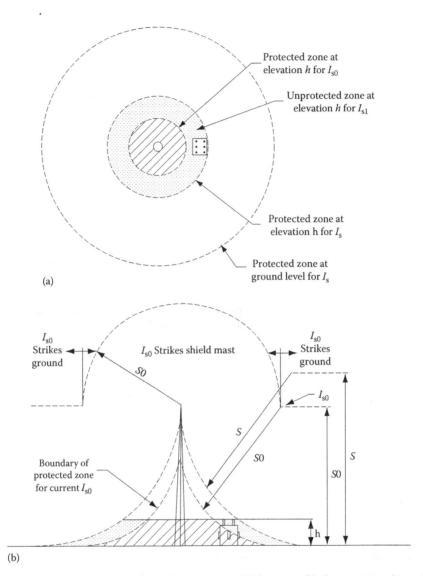

FIGURE 12.13 Shield mast protection for stroke current I_{s0}. (a) Plan view; (b) Elevation view. (From IEEE Std. 998-1996 (R2002), IEEE Working Group D5, Substations Committee, *Guide for Direct Lightning Stroke Shielding of Substations*. With permission.)

12.4.6 Multiple Shielding Electrodes

The electrogeometric modeling concept of direct stroke protection has been demonstrated for a single shield mast. A typical substation, however, is much more complex. It may contain several voltage levels and may utilize a combination of shield wires and lightning masts in a three-dimensional arrangement. The above concept can be applied to multiple shielding masts, horizontal shield wires, or a combination of the two. Figure 12.14 shows this application considering four shield masts in a multiple shield mast arrangement. The arc of protection for stroke current I_s is shown for each set of masts. The dashed arcs represent those points at which a descending stepped leader for stroke current I_s will be attracted to one of the four masts. The protected zone between the masts is defined by an arc of radius S with the center at the intersection of the two dashed arcs. The protective zone can again be visualized as the surface of a sphere with radius S, which is rolled toward a mast until touching the mast, then rolled up and over the

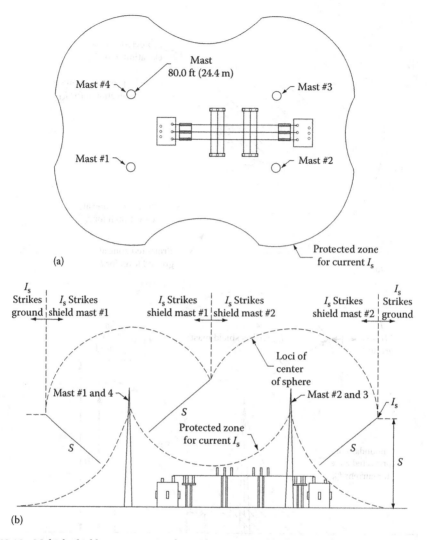

FIGURE 12.14 Multiple shield mast protection for stroke current I_s. (a) Plan view; (b) Elevation view. (From IEEE Std. 998-1996 (R2002), IEEE Working Group D5, Substations Committee, *Guide for Direct Lightning Stroke Shielding of Substations*. With permission.)

mast such that it would be supported by the masts. The dashed lines would be the locus of the center of the sphere as it is rolled across the substation surface. Using the concept of rolling sphere of the proper radius, the protected area of an entire substation can be determined. This can be applied to any group of different height shield masts, shield wires, or a combination of the two. Figure 12.15 shows an application to a combination of masts and shield wires.

12.4.7 Changes in Voltage Level

Protection has been illustrated with the assumption of a single voltage level. Substations, however, have two or more voltage levels. The rolling sphere method is applied in the same manner in such cases, except that the sphere radius would increase or decrease appropriate to the change in voltage at a transformer. (Sample calculations for a substation with two voltage levels are given in annex B of IEEE Std. 998-1996 (R2002).)

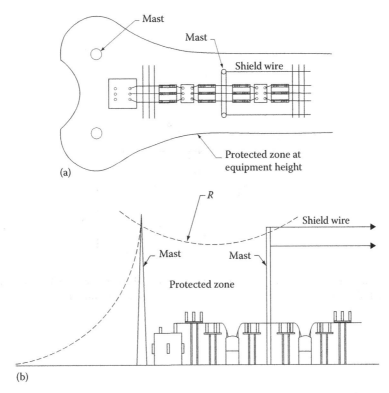

FIGURE 12.15 Protection by shield wires and masts. (a) Plan view; (b) Elevation view. (From IEEE Std. 998-1996 (R2002), IEEE Working Group D5, Substations Committee, *Guide for Direct Lightning Stroke Shielding of Substations*. With permission.)

12.4.8 Minimum Stroke Current

The designer will find that shield spacing becomes quite close at voltages of 69 kV and below. It may be appropriate to select some minimum stroke current, perhaps 2 kA for shielding stations below 115 kV. Such an approach is justified by an examination of Figures 12.1 and 12.2. It will be found that 99.8% of all strokes will exceed 2 kA. Therefore, this limit will result in very little exposure, but will make the shielding system more economical.

12.4.9 Application of Revised EGM by Mousa and Srivastava Method

The rolling sphere method has been used in the preceding paragraphs to illustrate application of the EGM. Mousa describes the application of the revised EGM (1976). Figure 12.16 depicts two shield wires, G1, and G2, providing shielding for three conductors, W1, W2, and W3. S_c is the critical striking distance as determined by Equation 12.11, but reduced by 10% to allow for the statistical distribution of strokes so as to preclude any failures. Arcs of radius S_c are drawn with centers at G1, G2, and W2 to determine if the shield wires are positioned to properly shield the conductors. The factor ψ is the horizontal separation of the outer conductor and shield wire, and b is the distance of the shield wires above the conductors. Figure 12.17 illustrates the shielding provided by four masts. The height h_{mid} at the center of the area is the point of minimum shielding height for the arrangement. For further details in the application of the method, see Mousa (1976). At least two computer programs have been developed

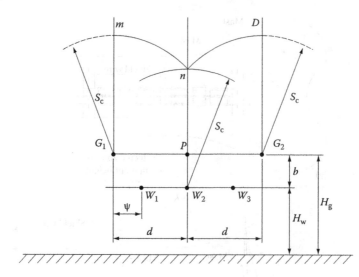

FIGURE 12.16 Shielding requirements regarding the strokes arriving between two shield wires. (From IEEE Std. 998-1996 (R2002), IEEE Working Group D5, Substations Committee, *Guide for Direct Lightning Stroke Shielding of Substations*. With permission.)

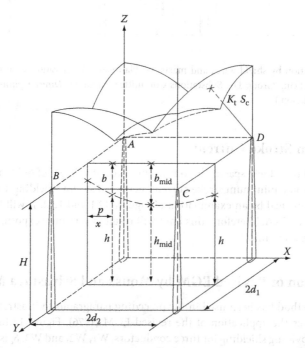

FIGURE 12.17 Shielding of an area bounded by four masts. (From IEEE Std. 998-1996 (R2002), IEEE Working Group D5, Substations Committee, *Guide for Direct Lightning Stroke Shielding of Substations*. With permission.)

that assist in the design of a shielding system. One of these programs (Mousa, 1991) uses the revised EGM to compute the surge impedance, stroke current, and striking distance for a given arrangement of conductors and shield systems, then advises the user whether or not effective shielding is provided. (Sample calculations are provided in annex B of IEEE Std. 998-1996 (R2002) to further illustrate the application.)

12.5 Calculation of Failure Probability

In the revised EGM just presented, striking distance is reduced by a factor of 10% so as to exclude all strokes from the protected area that could cause damage. In the empirical design approach, on the other hand, a small failure rate is permitted, typically 0.1%. Linck (1975) also developed a method to provide partial shielding using statistical methods. It should be pointed out that for the statistical approach to be valid, the size of the sample needs to be large. For power lines that extend over large distances, the total exposure area is large and the above criterion is met. It is questionable, therefore, whether the statistical approach is as meaningful for substations that have very small exposure areas by comparison. Engineers do, however, design substation shielding that permits a small statistical failure rate. Orrell (1988) has developed a method of calculating failure rates for the EGM rolling sphere method. (This method is described with example calculations in annex D of IEEE Std. 998-1996 (R2002).)

12.6 Active Lightning Terminals

In the preceding methods, the lightning terminal is considered to be a *passive* element that intercepts the stroke merely by virtue of its position with respect to the live bus or equipment. Suggestions have been made that lightning protection can be improved by using what may be called *active* lightning terminals. Three types of such devices have been proposed over the years:

- *Lightning rods with radioactive tips* (Golde, 1973). These devices are said to extend the attractive range of the tip through ionization of the air.
- *Early Streamer Emission (ESM) lightning rods* (Berger and Floret, 1991). These devices contain a triggering mechanism that sends high-voltage pulses to the tip of the rod whenever charged clouds appear over the site. This process is said to generate an upward streamer that extends the attractive range of the rod.
- *Lightning prevention devices.* These devices enhance the *point discharge* phenomenon by using an array of needles instead of the single tip of the standard lightning rod. It is said that the space charge generated by the many needles of the array neutralize part of the charge in an approaching cloud and prevent a return stroke to the device, effectively extending the protected area (Carpenter, 1976).

Some of the latter devices have been installed on facilities (usually communications towers) that have experienced severe lightning problems. The owners of these facilities have reported no further lightning problems in many cases.

There has not been sufficient scientific investigation to demonstrate that the above devices are effective; and since these systems are proprietary, detailed design information is not available. It is left to the design engineer to determine the validity of the claimed performance for such systems.

References

This reference list is reprinted in part from IEEE Working Group D5, Substations Committee, *Guide for Direct Lightning Stroke Shielding of Substations*, IEEE Std. 998-1996 (R2002).

Anderson, J.G., Monte Carlo computer calculation of transmission-line lightning performance, *AIEE Transactions*, 80, 414–420, August 1961.

Anderson, J.G., *Transmission Line Reference Book 345 kV and Above*, 2nd edn. Rev. Palo Alto, CA: Electric Power Research Institute, 1987, Chapter 12.

Anderson, R.B. and Eriksson, A.J., Lightning parameters for engineering application, *Electra*, 69, 65–102, March 1980.

Berger, G. and Floret, N., *Collaboration Produces a New Generation of Lightning Rods*, Power Technology International, London, U.K.: Sterling Publications, pp. 185–190, 1991.

Carpenter, R.B. Jr., Lightning elimination. Paper PCI-76-16 given at the *23rd Annual Petroleum and Chemical Industry Conference* 76CH1109-8-IA, 1976.

Darveniza, M., Popolansky, F., and Whitehead, E.R., Lightning protection of UHV transmission lines, *Electra*, 41, 36–69, July 1975.

Eriksson, A.J., Lightning and tall structures, *Transaction South African IEE*, 69(8), 238–252, August 1978. Discussion and closure published May 1979, 70(5), 12 pp.

Gilman, D.W. and Whitehead, E.R., The mechanism of lightning flashover on high voltage and extra-high voltage transmission lines, *Electra*, 27, 65–96, March 1973.

Golde, R.H., Radio-active lightning conductors, in *Lightning Protection*, London, U.K.: Edward Arnold Publishing Co., pp. 37–40, 196–197, 1973.

IEEE Std. 998-1996 (R2002), IEEE Working Group D5, Substations Committee. *Guide for Direct Lightning Stroke Shielding of Substations*.

IEEE Working Group, A simplified method for estimating lightning performance of transmission lines, *IEEE Transactions on Power Apparatus and Systems*, PAS-104(4), 919–932, 1985.

IEEE Working Group, Estimating lightning performance of transmission lines. II. Updates to analytic models, *IEEE Transactions on Power Delivery*, 8(3), 1254–1267, July 1993.

Lee, R.H., Protection zone for buildings against lightning strokes using transmission line protection practice, *IEEE Transactions on Industry Applications*, 1A-14(6), 465–470, 1978.

Lee, R.H., Lightning protection of buildings, *IEEE Transactions on Industry Applications*, IA-15(3), 236–240, May/June 1979.

Lee, R.H., Protect your plant against lightning, *Instruments and Control Systems*, 55(2), 31–34, February 1982.

Linck, H., Shielding of modern substations against direct lightning strokes, *IEEE Transactions on Power Apparatus and Systems*, PAS-90(5), 1674–1679, September/October 1975.

Mousa, A.M., Shielding of high-voltage and extra-high-voltage substations, *IEEE Transactions on Power Apparatus and Systems*, PAS-95(4), 1303–1310, 1976.

Mousa, A.M., A study of the engineering model of lightning strokes and its application to unshielded transmission lines, PhD thesis, University of British Columbia, Vancouver, Canada, August 1986.

Mousa, A.M., A computer program for designing the lightning shielding systems of substations, *IEEE Transactions on Power Delivery*, 6(1), 143–152, 1991.

Mousa, A.M. and Srivastava, K.D., A revised electrogeometric model for the termination of lightning strokes on ground objects, *Proceedings of International Aerospace and Ground Conference on Lightning and Static Electricity*, Oklahoma City, OK, April 1988, pp. 342–352.

Mousa, A.M. and Srivastava, K.D., The implications of the electrogeometric model regarding effect of height of structure on the median amplitudes of collected lightning strokes, *IEEE Transactions on Power Delivery*, 4(2), 1450–1460, 1989.

Orrell, J.T., Direct stroke lightning protection, Paper presented at *EEI Electrical System and Equipment Committee Meeting*, Washington, DC, 1988.

Sargent, M.A., Monte Carlo simulation of the lightning performance of overhead shielding networks of high voltage stations, *IEEE Transactions on Power Apparatus and Systems*, PAS-91(4), 1651–1656, 1972a.

Sargent, M.A., The frequency distribution of current magnitudes of lightning strokes to tall structures, *IEEE Transactions on Power Apparatus and Systems*, PAS-91(5), 2224–2229, 1972b.

Suzuki, T., Miyake, K., and Shindo, T., Discharge path model in model test of lightning strokes to tall mast, *IEEE Transactions on Power Apparatus and Systems*, PAS-100(7), 3553–3562, 1981.

Wagner, C.F., McCann, G.D., and Beck, E., Field investigations of lightning, *AIEE Transactions*, 60, 1222–1230, 1941a.

Wagner C.F., McCann, G.D., and Lear, C.M., Shielding of substations, *AIEE Transactions*, 61, 96–100, 313, 448, February 1942.

Wagner, C.F., McCann, G.D., and MacLane, G.L., Shielding of transmission lines, *AIEE Transactions*, 60, 313–328, 612–614, 1941b.

Whitehead, E.R., Mechanism of lightning flashover, EEI Research Project RP 50, Illinois Institute of Technology, Pub 72-900, February 1971.

Whitehead, E.R., CIGRE survey of the lightning performance of extra-high-voltage transmission lines, *Electra*, 63–89, March 1974.

Young, E.S., Clayton, J.M., and Hileman, A.R., Shielding of transmission lines, *IEEE Transactions on Power Apparatus and Systems*, S82, 132–154, 1963.

13
Seismic Considerations

13.1	Earthquakes and Substations	13-1
13.2	Seismic Design Standards for Substations	13-4
	Importance of Standards • IEEE 693 and Complementary Standards	
13.3	Seismic Design Process	13-6
13.4	Seismic Qualification	13-7
	Qualification Levels and Design Earthquake • Selection of Qualification Level • Methods of Qualification • Equipment Supports • Performance Level and Projected Performance Level	
13.5	Installation	13-14
	References	13-15

Eric Fujisaki
*Pacific Gas and
Electric Company*

13.1 Earthquakes and Substations

This chapter discusses the current seismic design standards and practices employed in high seismic hazard areas of North America. The Institute of Electrical and Electronic Engineers (IEEE) Standard 693-2005, Seismic Design of Substations [1] has been adopted as the seismic design standard for substations by many utilities serving these areas of North America as well as other locations worldwide. Consequently, this chapter refers to provisions and examples particularly related to IEEE 693. This chapter is intended to provide general guidance to substation designers on seismic design concepts and the application of IEEE 693 for equipment qualification. Basic steps for securing and protecting components within a given substation are illustrated. The discussion is not intended to be all-inclusive or to provide all the necessary details to undertake such work. For further details and information on this topic, the reader should review the references listed at the end of this chapter.

Prior to 1970, little guidance existed for the seismic design of electric substation equipment. Even in earthquake-prone regions, typical design practices consisted of applying low static lateral force coefficients modeled after building codes for the seismic design of high-voltage equipment and their supports. It was common for utilities to each develop its own set of seismic qualification requirements, which varied from one another depending on experience, site-specific hazards, engineering practices, and other considerations. In the 1970s through 1990s, several large-magnitude earthquakes struck California, causing millions of dollars in damage to substation components (see Figures 13.1 through 13.4), consequent losses of revenue, and disruption of service for a large number of customers. High-voltage equipment was often damaged due to the use of massive and brittle porcelain components and the absence of recognized seismic design standards and technical guidance. As a result of these losses and the increasing importance of reliable electric service, it became apparent to owners and operators of substation facilities in seismically active areas that the existing seismic design practices for substation components were inadequate.

Earthquake hazards in the western United States, principally in California, are well known; however, significant seismic hazards also exist in other parts of North America (e.g., eastern and western parts of Canada, the Pacific Northwest and New Madrid Fault zones in the United States, and western Mexico)

FIGURE 13.1 500 kV live tank circuit breaker. (Photo courtesy of E. Matsuda, Copyright 1989–2011 Pacific Gas and Electric Company, Oakland, CA. All rights reserved.)

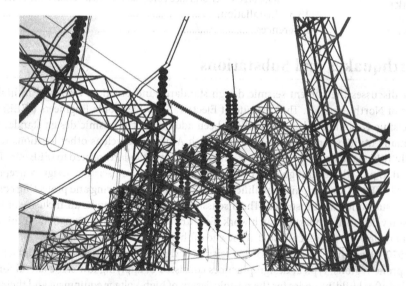

FIGURE 13.2 230 kV air disconnect switch. (Photo courtesy of E. Matsuda, Copyright 1989–2011 Pacific Gas and Electric Company, Oakland, CA. All rights reserved.)

and throughout the world. Large earthquakes have occurred around the Pacific Rim, Indian Ocean, central Asia, eastern Mediterranean, and the Middle East, as well as other regions.

Earthquakes may result from volcanic activity or when there is a sudden slip along a geological fault, resulting in the release of strain energy as shock waves that radiate from the focus or hypocenter, causing ground shaking at the earth's surface. Linear elastic theory provides the basis for understanding these waves that have several types with different characteristics. In general, these waves are attenuated over distance from the hypocenter but may be filtered, amplified, or otherwise modified by the geological structures through which they pass. The effects of the various waves combine to form a complex set of multifrequency vibratory ground motions at a particular site. For the purposes of earthquake engineering, these ground motions are typically described by acceleration time histories or response spectra in the horizontal and vertical directions.

FIGURE 13.3 Porcelain bushing. (Photo courtesy of E. Matsuda, Copyright 1996–2011 Pacific Gas and Electric Company, Oakland, CA. All rights reserved.)

FIGURE 13.4 230 kV circuit breaker and air disconnect switch. (Photo courtesy of E. Matsuda, Copyright 1996–2011 Pacific Gas and Electric Company, Oakland, CA. All rights reserved.)

The response of structures and buildings to this ground motion depends on their dynamic properties, ductility, and design and details of construction. Lightly damped structures that have natural modes of vibration within the frequency band of the ground motion excitation can experience considerable movement, which can generate forces and deflections that the structures may not have been designed to accommodate.

Electrical substation equipment are complex machines whose primary functions require characteristics that are often at odds with those of a robust seismic design. This equipment commonly includes components of relatively low-strength, lightly damped, and brittle materials such as porcelain. The fundamental frequencies of vibration of substation equipment often lie within the frequency band of the earthquake having highly amplified accelerations. Commonly damaged equipment and components include large porcelain bushings, surge arresters, live tank circuit breakers, air disconnect switches, transformer radiators and piping, transformers mounted on rails (due to lack of restraint), instrument transformers, and rigid buswork. In general, equipment of higher voltage classes have suffered more damage due to their height, large porcelain insulators, and flexibility that tends to amplify seismic response. In order for the equipment to be capable of performing its intended functions after the earthquake, the various structural load-carrying components must not be overstressed, and electrical functionality must be maintained.

A further complication occurs when two or more structures or pieces of equipment are mechanically linked by electrical conductors. In such a case, they will interact with one another, thus producing a modified response. If they are either not linked or linked in such a way that the two pieces can move independently—an ideal situation—then the two components are uncoupled, and no forces are transferred between them. However, recent research has shown that even a well-designed link may influence the response of the connected equipment or structures during a seismic event. Inadequate conductor slack was likely a significant contributor to the damage depicted in Figure 13.4.

The vital role of electric power in modern society continues to increase. Other lifelines such as those delivering communications, water supplies, wastewater treatment, and critical emergency services depend on reliable electric power. Economic losses to businesses and other utility customers resulting from interruption of electric power far outweigh the utility's direct losses associated with electric system damage and broken revenue streams. Implementation of effective seismic design practices can significantly improve the reliability and post-earthquake recovery of electric power systems as well as the regional economy.

13.2 Seismic Design Standards for Substations

13.2.1 Importance of Standards

Standards governing the electrical performance of substation equipment are well known and have been applied to good effect. In an analogous manner, seismic design standards for substation equipment are intended to provide assurance that the equipment will perform acceptably during and following a large earthquake. The process demonstrating that equipment and components satisfy the seismic performance requirements specified by a user or purchaser is known as seismic qualification. The outcome of the qualification process is a binary condition: the component or equipment is either qualified or not. In order for equipment to be qualified, it must meet all of the requirements of the given standard.

Seismic design standards also provide a number of other important benefits including the following:

- *Management of assets and risks*—Standards simplify the management of the installed and spare inventory of equipment. Standards also provide for a better understanding of the expected seismic performance of the utility's equipment, which is critical for understanding seismic risks.
- *Transparency for manufacturers*—Clearly defined requirements for seismic qualification allow the manufacturer to build in seismic-resistant features during the design phase rather than as an afterthought.
- *Improvement in the efficiency of the seismic qualification process*—Application of standards reduces cost. It allows all users or purchasers who specify the same standard for a given model or type of equipment to make use of the same qualification.
- *Guidance for users*—Standards provide technical guidance for users who may have limited familiarity or awareness of seismic design practices.

Substation seismic design standards generally specify the seismic demand (usually given in terms of a design earthquake), methods for performing the engineering evaluation (test or analysis) of the equipment, and the acceptance criteria (e.g., allowable stresses). Given the random nature of earthquakes, the mechanical/electrical complexity of such equipment, and the many uncertainties involved, both the demands (loads due to earthquake) and capacities (strength of load-carrying components) must be set appropriately. Most seismic design standards thus explicitly or implicitly require that the structural integrity of the equipment is preserved, that is, there is no fracture or excessive permanent deformation of any part of the equipment. In addition, displacements due to seismic loading must be reasonable, the equipment must maintain the correct operational state, and electrical functions must remain unimpaired following shaking. When applying a seismic design standard, the user should have an understanding of the definition of the design earthquake as specified by the standard, seismic hazard at the site of installation, and seismic margin provided by the qualification against such earthquakes.

IEEE 693 has been adopted by many utilities in high seismic hazard areas of North America and elsewhere. A corresponding collection of seismic design standards developed by the International Electrotechnical Commission (IEC) covering a scope similar to IEEE 693 is also available and has been adopted primarily in Europe. Some countries have national design specifications or codes with their own specific requirements. IEEE 693 and its counterparts of other standards-developing organizations share the common goal of specifying equipment that will continue to function during and following a large earthquake. They differ, however, in the degree of prescription, techniques and methodologies, seismic input, and resultant seismic margin.

IEEE 693-1997 [2] represented a major change in the seismic design practices for substation equipment. This standard and its 2005 successor are based upon theoretical as well as empirical considerations and provide a more prescriptive approach compared to other similar standards. IEEE 693 specifies the seismic inputs for three defined levels of qualification, recommendations for selecting the appropriate level of qualification based upon site seismic hazard, and the acceptable methods of qualification. IEEE 693 requires more stringent methods of qualification for equipment having vulnerable characteristics observed in past earthquakes. The prescriptive or single approach of IEEE 693 gives clarity to the requirements as well as the capabilities of the qualified product, which encourages the equipment purchaser to specify the standard. The equipment manufacturer benefits by having a well-defined set of requirements and, in concept, the ability to serve multiple customers with the same qualification.

IEEE 693 continues to be a living standard, incorporating important improvements based upon user experience; research advancements; earthquake experience; and the continued contributions of utilities, manufacturers, consulting engineers, and the academic community.

13.2.2 IEEE 693 and Complementary Standards

The seismic design of substations is governed by three primary standards, each addressing a particular aspect of the design. Taken together, these standards are intended to provide the requirements for the seismic design of the entire substation.

IEEE 693 provides requirements for the design and qualification of equipment, components, and their "first support," which is the primary above-ground support structure. The first support consists of the entire structure in the case of stand-alone supports or the portion supporting the equipment in the case of structures carrying multiple pieces of equipment or conductor pull-off loads. IEEE 1527 [3] addresses the design of flexible buswork, and the American Society of Civil Engineers (ASCE) *Manual of Practice 113, Guide to the Structural Design of Substations* [4] provides guidance for the design of all other substation structures (A-frames, racks, bus supports, and other structures beyond the first support) and anchorages. Each of these three primary standards also refers to material-based standards and specifications for the detailed design requirements related to structural steel, aluminum, and other materials.

The design of foundations may be performed in accordance with other standards such as ASCE *Minimum Design Loads for Buildings and Other Structures* (ASCE 7) [5] and the American Concrete Institute (ACI) *Building Code Requirements for Structural Concrete* (ACI 318). The seismic design of substation buildings is usually governed by a model building code such as the International Building Code in the United States or a modified version of such a code that may have been adopted by the local jurisdiction. IEEE 693 does not apply to Class 1E nuclear safety-related equipment, which is addressed by IEEE 344-2004 [6].

IEEE 693 is organized into various clauses that discuss the general requirements and recommendations for the qualification of equipment and design of substations, and instructions for implementation. Standard annexes define input motions, the various methods of qualification, acceptance criteria, and documentation requirements. Equipment-specific annexes (e.g., circuit breaker, transformer, or air disconnect switch) state the detailed qualification requirements for different types of equipment and refer to the standard annexes for common provisions. The equipment-specific annexes specify the seismic inputs for different levels of qualification, permissible qualification method for each class of equipment (typically by voltage rating or other key characteristic related to its seismic vulnerability), the qualification procedure, monitoring required during testing, and specific acceptance criteria.

It should be noted that IEEE 693, IEEE 1527, and the ASCE *Substation Structure Design Guide* were written primarily for new installations. These standards may be used to assist designers in the evaluation of the seismic acceptability for existing equipment and structures, but in general, it is difficult to apply such criteria after the fact to an existing facility without encountering the need for reinforcement or modifications. Seismic qualification may be particularly difficult to apply to older existing equipment, since the retrofit or replacement of nonconforming components and details may be costly or unfeasible.

13.3 Seismic Design Process

The seismic design process for substation equipment depends upon standards such as the ones discussed in the previous section. The design process has the following steps, with roles for both the utility or substation owner and the equipment manufacturer:

1. Utility specifies the seismic qualification standard for equipment, noting the equipment type, such as surge arresters or circuit breakers, and schedule for completion of the qualification. Annex U of IEEE 693 provides a template with suggested wording and details for use in specifications.
2. Utility determines and specifies the required seismic qualification level (low, moderate, or high) based on the seismic hazard at the installation site and the equipment in situ configuration, such as mounting information. In some cases, the equipment manufacturer may select the qualification level and structure mounting details in order to prequalify equipment. Typically, this latter case occurs for commonly specified or cataloged equipment, such as air disconnect switches, surge arresters, or instrument transformers.
3. Equipment manufacturer designs the equipment and executes the seismic qualification in accordance with the procedures, methods, and seismic inputs of the standard. If necessary, modifications are made to the equipment in order to satisfy the qualification requirements.
4. Equipment manufacturer prepares the qualification report, which documents the qualification activities and results, and submits it to the utility. IEEE 693 specifies standard report templates, which prescribe the format and content of the report.
5. Utility reviews and accepts the qualification report.
6. Utility receives the equipment and installs it in accordance with the conditions of the seismic qualification.

The details of these steps are further discussed in the following sections.

13.4 Seismic Qualification

Seismic qualification depends upon the following key elements:

- *Design earthquake*—the earthquake input motion used for the seismic qualification
- *Seismic hazard*—the size and likelihood of earthquakes at a particular site
- *Qualification protocol*—the methods and rules by which the qualification is executed
- *Acceptance criteria*—the conditions defining acceptable performance when the equipment is subjected to the design earthquake

IEEE 693 defines the design earthquake, qualification protocols, and the acceptance criteria. The seismic hazard depends upon the characteristics and location of the substation site where the equipment is to be installed. IEEE 693 prescriptively treats all of these aspects of qualification. When equipment is qualified in accordance with the standard, the user is presented with a well-defined picture of the expectations for seismic performance.

13.4.1 Qualification Levels and Design Earthquake

IEEE 693 provides three levels of qualification termed low, moderate, and high, which are associated with different levels of ground shaking that are intended to satisfy the needs of most users. Buildings are usually unique structures and are therefore designed according to site-specific seismic parameters. In contrast, equipment and components are often mass produced, installed at a variety of sites, and are thus best qualified by establishing a few discrete qualification levels. This is intended to reduce the number of qualifications required of equipment manufacturers and simplifies interchangeability of equipment by a utility.

Each of the three qualification levels is associated with a required response spectrum (RRS) or other description of ground shaking that defines the seismic loading environment and may be thought of as the design earthquake for the purpose of qualification. The RRS is a broadband response spectrum that envelops the effects of earthquakes in different areas for site conditions ranging from soft soils to rock as described in the Recommended Provisions for Seismic Regulations for New Buildings [7]. (This data is now found in ASCE 7.)

In particular, the RRS provide longer period coverage for soft sites; however, sites with very soft soils, sites located on moderate to steep slopes, or sites located in the upper floors of buildings may not be adequately covered by these spectral shapes. In such cases, the user may find it necessary to develop a site-specific response spectrum for use in qualification activities. The site-specific spectra, however, must envelop the RRS given in IEEE 693 in order to conform to the standard. This is intended to ensure that the site-specific qualification also results in qualification to one of the standard levels, thus preserving its validity for future users. Implementation of site-specific spectra may also require retesting of previously qualified equipment, at additional costs. The high RRS is anchored to $0.5\,g$, peak ground acceleration (pga), which is the acceleration at the high-frequency end of the spectra. The moderate RRS has the same spectral shape, but is anchored to $0.25\,g$, pga. Equipment that is qualified according to IEEE 693 using the high RRS is said to be seismically qualified to the high seismic level. Similarly, equipment that is qualified using the moderate RRS is said to be qualified to the moderate seismic level. The high RRS and moderate RRS are shown in Figures 13.5 and 13.6, respectively, for different damping ratios.

Finally, equipment that has been qualified to the low-level requirements of IEEE 693 is said to be qualified to the low seismic level. The low seismic qualification level is associated with a pga of $0.1\,g$ or less. Equipment that has no special seismic design features, but has been installed using good seismic installation and construction practices, automatically earns a qualification to the low seismic level. In general, it is expected that the majority of equipment will be qualified to the low seismic level.

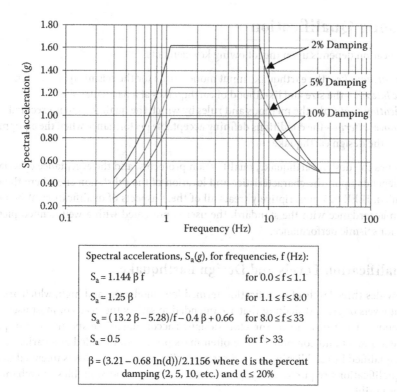

FIGURE 13.5 High required response spectra. (IEEE Std. 693-2005, *IEEE Recommended Practice for Seismic Design of Substations*, Copyright 2006, New York, IEEE. All rights reserved.)

13.4.2 Selection of Qualification Level

The appropriate level of seismic qualification for equipment to be installed at a particular substation site depends upon the seismic hazard present at the site. IEEE 693 recommends the use of either the earthquake hazard or seismic exposure map approach for this assessment.

The earthquake hazard method is the preferred approach and can be used at any site. It consists of the following steps:

1. Establish the probabilistic earthquake hazard exposure of the site where the equipment will be placed. Use the site-specific pga and response spectra developed in a study of the site's seismic hazard, selected at a 2% probability of exceedance in 50 years, modified for site soil conditions.
2. Compare the resulting site-specific pga value and response spectra with one of the three seismic levels—high, moderate, or low—that best accommodates the expected ground motions. If the pga is less than or equal to $0.1\,g$, the site is classified as low. If the pga is greater than $0.1\,g$ but less than or equal to $0.5\,g$, the site is classified as moderate. If the pga is greater than $0.5\,g$, the site is classified as high. If the site-specific spectra exceed the RRS selected on the basis of pga, in the frequency range expected for the equipment, it may be necessary to select a higher level of qualification. This level then defines the seismic qualification level used for procurement.

The seismic exposure map method is an alternative to the earthquake hazard method and may be undertaken utilizing ASCE 7 ground motion maps in the United States, the National Building Code of Canada (NBCC) maps for Canada, or the Manual de Disseno de Obras/de la Comision Federal de Electricidad

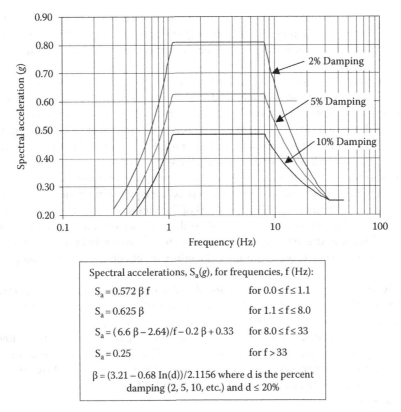

FIGURE 13.6 Moderate required response spectra. (IEEE Std. 693-2005, *IEEE Recommended Practice for Seismic Design of Substations*, Copyright 2006, New York, IEEE. All rights reserved.)

(MDOC/CFE) maps for Mexico. The reader should consult current maps for Canada and Mexico and the current procedures for their use. Other countries should use corresponding country-specific maps. In the United States, ASCE 7 ground motion maps may be applied, using the steps outlined. In the following, the cited chapters, sections, and terminology refer to ASCE 7:

1. Determine the soil classification of the site (A, B, C, D, E, or F) from Section 20.3 Site class definitions.
2. Locate the site on the maps (Section 11.4 and Chapter 22) for the maximum considered earthquake ground motion 0.2-s spectral response acceleration (5% of critical damping).
3. Estimate the site 0.2 s spectral acceleration, "S_s," from this map.
4. Determine the value of the site "F_a" from Section 11.4, as a function of site class and mapped spectral response acceleration at short periods (S_s).
5. Determine the pga, equal to $(F_a S_s)/2.5$.
6. Use the pga to select the seismic qualification level. If the pga is less than or equal to $0.1\,g$, the low qualification level should be used. If the pga is greater than $0.1\,g$ but less than or equal to $0.5\,g$, the moderate qualification level should be used. If the pga is greater than $0.5\,g$, the high qualification level should be used.

Use of one of the three qualification levels given in IEEE 693 and the corresponding RRS is encouraged. Although a utility has the prerogative to specify a customized set of criteria, this practice will likely lead to higher cost and the benefits of the standardized seismic design criteria may be lost or diluted.

Similar methods for establishing the seismic qualification levels from seismic zone maps used in Canada and Mexico are also given in IEEE 693, with appropriate country-specific references and maps as required. Users applying the standard in other countries may use a method similar to that described in IEEE 693.

Judgment and experience must be exercised when selecting the seismic level for qualification as the site hazard may not fall directly on the high, moderate, or low seismic level. In this case, a strategy of accepting more or less risk will be required. A utility may divide its service territory into large geographic areas based upon anticipated levels of shaking corresponding to the qualification levels of IEEE 693 and use a common seismic specification for all equipment to be installed in a given area. For existing facilities requiring upgrade or repair design work, increased efficiencies may be realized. Additional operational requirements, substation network design, emergency response capabilities, equipment spares, and other factors deserve consideration when selecting equipment for an active inventory of an operating utility. The utility may wish to evaluate all of the sites in the service territory and establish a master plan, designating the required (or desired, as the case may be) qualification level of each site and prioritizing those sites that need to be upgraded to meet current standards. Likewise, after a site for new electrical equipment has been identified, the utility should determine the appropriate seismic qualification level.

If the seismic response spectra for a specific site significantly exceed the response spectra depicted in Figures 13.5 and 13.6, then the utility may develop a more appropriate response spectrum for that specific site. Even in this case, the basic procedure outlined in this chapter and IEEE 693 may still be used. However, the high, moderate, and low levels specified in IEEE 693 should be used without deviation unless it is very clear that one of the qualification levels will not adequately represent the site or sites. Note that if the utility elects to reduce, not meet, or remove any part of the IEEE 693 requirements, then the equipment may not be claimed to be in compliance with IEEE 693 and the benefits of standardization would be forfeited.

13.4.3 Methods of Qualification

Different methods of qualification are available for electrical substation equipment. These methods fall into two major categories:

- *Analytical methods*—static, static coefficient, and dynamic analysis
- *Test methods*—static pull, time history shake table, and sine beat tests

Various seismic qualification standards may employ the same basic methods of qualification; however, IEEE 693 also specifies the acceptable methods depending on the type of equipment being considered. Because of the semiempirical basis of IEEE 693, the most vulnerable equipment, which tend to be those in higher voltage classes (most massive, with the tallest insulators), are usually required to undergo the most stringent qualification procedures, often including shake table testing. For other equipment or components, analysis of different types, or, perhaps, static pull testing are sufficient to satisfy the requirements of the standard.

IEEE 693 also utilizes a third category of qualification methods that permit simplified means of demonstrating acceptability based on equipment type, or if a qualification has already been performed via a different standard. Most power equipment whose voltage class is less than 35,000 V is classified in IEEE 693 as "inherently acceptable," meaning that this equipment has performed well in earthquakes without imposing additional requirements. Certain types of equipment can be qualified using the experience-based qualification method of Annex Q of IEEE 693. If the equipment was qualified to a previous version of IEEE 693, the "grandfathering" provisions of the IEEE 693-2005 may apply. The grandfathering provisions require that the equipment manufacturer provides a justification reconciling the differences between the current version of the standard and the version to which the equipment was previously qualified. In certain instances, the grandfathering provisions do not apply as indicated in the standard.

In the 2005 version, only the shed seal test for composite polymer insulators is specifically excluded from the grandfathering clauses.

Analytical methods are generally specified in IEEE 693 for equipment that has been observed to be less vulnerable in past earthquakes. Relatively rigid or simple equipment with favorable earthquake performance are permitted to be qualified by static analysis, which uses the zero-period acceleration (i.e., the acceleration at the high frequency end of the spectrum in Figure 13.5 or 13.6). Amplification factors are specified for the qualification of power transformer appendages (radiators and conservator tanks) due to their flexibility and poor performance in past earthquakes. Lower voltage flexible equipment may be qualified by static coefficient analysis, which uses the plateau of the RRS multiplied by a factor of up to 1.5, depending on the expected complexity of its dynamic behavior. Dynamic analysis is specified for higher voltage or complex equipment that usually exhibits multimode behavior. The appropriate 2%-damped RRS is used as the basis for all analytical methods unless a higher damping can be justified. The reliability and quality of the analysis depend greatly upon modeling details and the judgment of the analyst.

Test methods are specified in IEEE 693 for complex components or equipment, or those at the highest voltage ratings. Static pull tests are used for components whose historic earthquake performance has been generally good, such as transformer bushings and cable terminations of medium voltage ratings. Static pull tests consist of the application of a specified terminal force to a component fixed to a test stand.

Time history shake table testing is required for equipment types and classes that have sustained damage in past earthquakes or have unfavorable characteristics. The time history test is intended to simulate the equipment response during an earthquake. IEEE 693 provides detailed requirements for development of the input motions to be used in the time history test, as well as sets of preapproved standard motions. These tests require more extensive planning to ensure that the proper test article configuration, input motions, monitoring instrumentation (strain gages, accelerometers, displacement transducers, and other measuring devices), test sequences, and procedures are utilized. Shake tables are often limited in their ability to deliver the required input motions to the test article particularly in the low frequency range; this requires careful pretest planning with the laboratory operator. Electrical functionality of the equipment is demonstrated by subjecting the test article to routine electrical tests following shaking. In addition, other functions such as continuity or relay state may be required to be monitored during shaking. Care should be taken to gather all of the data required for qualification since such tests entail significant time and expense and are not easily repeated.

Sine beat tests are single-frequency tests that apply only to high-voltage circuit breaker and gas-insulated switchgear qualifications under IEEE 693. The standard specifies the detailed requirements of the sine beat test, including the number of cycles in each beat, number of beats, and frequencies at which the equipment should be tested.

Table 13.1 summarizes the methods of qualification required by IEEE 693 for different types of equipment, based upon voltage class or other characteristics.

Equipment that meets the qualification requirements of a given level is expected to be undamaged and functional after being subjected to the shaking described by the corresponding RRS. Further, it is anticipated that the equipment will continue to function acceptably after ground shaking up to two times the RRS with little or no structural damage (see Section 13.4.5). This expectation is based upon the rather conservative acceptance criteria specified by the standard. These acceptance criteria follow allowable strength design (ASD) or load resistance factor design (LRFD) with applicable load factors. In ASD, the internal stresses caused by the applied loads (in this case, earthquake forces) in the various structural load-carrying parts are measured or assessed, and compared with allowable stresses. Allowable stresses are usually defined as a percentage (typically 50%–60%) of the yield strength, breaking strength, or load-carrying capacity of a material or element. In the evaluation of a component's strength using IEEE 693, no reduction of seismic load due to inelastic energy absorption or overstrength is permitted. Through this combination of conservative acceptance criteria and ground

TABLE 13.1 Required Methods of Qualification in IEEE 693-2005 for Different Equipment Types

Equipment Type	Voltage Class (kV) or Characteristic and Required Method of Qualification						
	Inherently Acceptable	Load Path	Static Analysis	Static Coefficient Analysis	Dynamic Analysis	Static Pull Test	Time History Test
Circuit breaker	<35	—	—	35 to <121	121 to <169	—	≥169[a]
Transformer	<35	35 to <115	≥115	—	—	—	—
Transformer bushing	<35	—	—	—	—	35 to <161	≥161
Air disconnect switch	<35	—	—	35 to <121	121 to <169	—	≥169
Instrument transformer	<35[b]	—	—	35 to <69[b]	69 to <230	—	≥230[b]
Air core reactor	<35	—	—	35 to <115	≥115	—	—
Circuit switcher	<35	—	—	35 to <121	121 to <169	—	≥169
Surge arrester	<35[c]	—	—	35 to <54[c]	54 to <90[c]	—	≥90[c]
Metalclad switchgear	<35	—	—	—	≥35	—	—
Cable termination	<35	—	—	—	—	35 to <220	≥220
Capacitor	<38	—	—	38 to <230	≥230	—	—
Gas-insulated switchgear	<35	—	—	35 to <121	121 to <169	—	≥169[a]
Battery racks	—	—	Rigid or any 1 stack	—	Non-rigid with 2 stacks	—	Non-rigid with ≥3 stacks
Batteries	All units installed per Annex J	—	—	—	—	—	—

[a] Also requires sine beat test.
[b] Time history shake table test is required when total height of equipment on support is ≥20 ft.
[c] Duty cycle voltage shown; apply amplification factor per Annex D.4.6.

motion, IEEE 693 seeks to provide a degree of assurance that qualified equipment possesses a significant seismic safety margin beyond the loading levels of the RRS.

13.4.4 Equipment Supports

A well-designed support structure must satisfy two important requirements as related to seismic design. It must have sufficient strength to resist the effects of the earthquake, and it must have the appropriate dynamic properties such that the response (displacements, accelerations, and stresses) of the equipment is within acceptable limits. In most cases, seismic response of the supported equipment is the governing condition.

During an earthquake, a piece of equipment and the support upon which it is mounted are subjected to ground motions as a mechanical system. The seismic response of the equipment may be amplified or significantly altered by the support structure, depending on its characteristics. Very stiff support structures will usually pass the ground motions to the equipment with limited amplification unless a resonance occurs, while more flexible support structures will tend to amplify the ground motion. Equipment mounted on tall, flexible support structures may experience amplification of responses by factors of two or more compared to the case in which the same equipment is mounted on a rigid support or attached directly to a foundation at ground level.

The seismic qualification of equipment should be performed with the same support structure that is to be used in the field installation. In some cases, this approach is impractical because the support structure is too large to test, may not have been designed at the time of qualification of the equipment, or perhaps a variety of different support structures will be used. In those cases, the adverse effects or uncertainties associated with the use of the support structure must be considered in the seismic qualification.

This is relatively straightforward for analytical qualification methods but raises difficulties when time history shake table tests are required. For shake table testing, amplified input motions may be used, provided that the shake table is capable of delivering the required motions. The required amplification factors for cases in which the support structure parameters are unknown or when equipment is qualified without a support structure are given in IEEE 693.

The substation designer should be cognizant of the structure being used to support a piece of equipment when considering its seismic design. Support structures are sometimes furnished by the manufacturer with the equipment. In some cases, the utility will use a support structure of its own design. IEEE 693 uses the term "dynamically equivalent" or better to describe a structure that will cause the equipment to experience equal or lower response than the original structure used during the qualification of the equipment. In order for a seismic qualification to remain valid, the same or equivalent support structure must be used.

In IEEE 693, support structures are designed for adequate strength by applying the appropriate material-based code or standard. Just as for load-carrying elements of the equipment, no credit is taken for inelastic energy absorption or anticipated overstrength. For this reason, ductile seismic detailing of the structural elements of supports is not required by the standard.

13.4.5 Performance Level and Projected Performance Level

The reader may have noted that during the selection process for the appropriate qualification level for equipment at a particular location, sites having pga between $0.1\,g$ up to $0.5\,g$ required the use of the moderate-level RRS ($0.25\,g$ pga, see Figure 13.6). In a similar manner, sites with a peak ground accelerations exceeding $0.5\,g$ required the use of the high-level RRS ($0.5\,g$ pga, see Figure 13.5). This implies that equipment qualified to the high- or moderate-level RRS may be installed at sites subjected to, albeit infrequently, shaking levels higher than those used in its qualification. This is indeed the case, and the reasons for such an approach are discussed in this section.

As discussed earlier, IEEE 693 specifies a conservative set of acceptance criteria coupled with the ground motion defined by the RRS to provide some assurance that a seismic safety margin beyond the qualification level is provided. IEEE 693 intends that equipment qualified to the requirements of the standard remains functional after an earthquake that imposes levels of shaking twice that actually tested. The level of shaking defined by twice the qualification level (RRS) is known as the performance level (PL). At the high seismic level, the RRS is anchored to a pga of $0.5\,g$, and the high PL spectrum would be defined by the same spectral shape anchored to a pga equal to twice $0.5\,g$ or $1\,g$. That is, the entire $0.5\,g$ RRS is scaled by a factor of two to produce the PL spectrum. In a similar manner, the moderate PL spectrum is anchored to a pga of twice $0.25\,g$ or $0.5\,g$. The low seismic level is considered to be the same as the low PL and is set at $0.1\,g$ pga. Since substation site seismic hazards defining the high and moderate seismic levels are based upon earthquakes having a 2% probability of exceedance in 50 years or the maximum considered earthquake, it is clear that the earthquakes corresponding to the PL spectra are indeed very infrequent events. IEEE 693 permits seismic qualification to be demonstrated by testing to the PL, and due to the extreme nature of the load levels imposed on the equipment, the acceptance criteria for the PL are set at a level much closer to failure of the equipment. In the ideal case, it would be desirable to test every piece of equipment to the appropriate PL spectrum. It is, however, often neither practical nor cost effective to do so. Some of these considerations are as follows:

- Test laboratory shake tables may not be able to attain the required acceleration levels, especially at low frequencies.
- Testing at loading levels close to the theoretical fracture strength of brittle materials such as porcelain may create safety hazards.
- Permanent deformation of ductile elements in a test, which may be permissible under the PL acceptance criteria, would require a costly test article to be scrapped, resulting in a significant financial loss.

For these reasons, the primary means of qualification described in IEEE 693 are based upon testing or analysis using the qualification, or RRS levels rather than the PL. However, since testing to the appropriate PL is the most comprehensive and reliable means of demonstrating performance, IEEE 693 does not preclude nor discourage the testing of equipment to the PL. For components such as base isolators and certain oil-filled transformer bushings, PL testing is required.

When applying one of the qualification levels in IEEE 693, the equipment is tested or analyzed to a level of shaking that is only one-half of the PL. As discussed previously, such equipment is anticipated to perform acceptably at twice the qualification level due to the application of an ASD basis or the LRFD method with appropriately factored loads as part of the qualification procedure. In the absence of tests performed at levels exceeding those at which the equipment was qualified, its performance at higher levels of shaking must be projected. If a reasonable understanding of the modes of failure of the equipment has been achieved, projecting the performance of the equipment beyond the qualification level may be justified.

It is cautioned that the validity of this approach is dependent upon identifying the most critically stressed locations within an individual piece of equipment and then monitoring the stresses at these locations during testing or analysis. If the testing or analysis is not carried out in this manner, the critical locations within the equipment may fail prematurely—at levels below the PL. In addition to these considerations, the response of the equipment to the dynamic load may change between the RRS level and the PL. If such effects are not properly accounted for, premature failures may occur.

The previous discussion pertains to the structural performance of the equipment. Qualification by analysis provides no assurance of electrical function. Shake table testing provides assurance for only those electrical functions verified by electrical testing and only to the levels applied in the test. Shake table testing may be required for equipment that was qualified by dynamic analysis in accordance with other standards but performed poorly during past earthquakes. However, static or static-coefficient analysis may still be specified if past seismic performance of equipment qualified by such methods has consistently led to acceptable performance.

A rigorous seismic qualification, such as that required to meet the high and moderate PLs, is not required for equipment qualified to the low PL. That is, no RRS or seismic report is required. However, the following criteria should be satisfied:

1. Anchorage for the low seismic PL shall be capable of withstanding at least 0.2 times the equipment weight applied in one horizontal direction and combined with 0.16 times the weight applied in the vertical direction at the center of gravity of the equipment and support. The resultant load should be combined with the maximum normal operating load and dead load to develop the greatest stress on the anchorage. The anchorage should be designed using the requirements specified in IEEE 693 and ASCE Manual 113.
2. Equipment and its support structure should have a well-defined load path. The determination of the load path should be established so that it describes the transfer of loads generated by, or transmitted to, the equipment from the point of origin of the load to the anchorage of the supported equipment. Among the forces that should be considered are seismic (simultaneous triaxial loading—two horizontal and one vertical), gravitational, and normal operating loads. The load path should not include the following:
   a. Sacrificial collapse members
   b. Materials or elements that will deform inelastically or undergo unrestrained translation or rotation
   c. Solely friction-dependent restraint (controlled energy-dissipating devices excepted)

13.5 Installation

A good deal of the preceding discussion has focused on the qualification of equipment for which the manufacturer assumes much of the responsibility. The utility or substation owner, however, must install the equipment in a manner consistent with the seismic qualification in order to maintain its validity. These installation practices include use of the appropriate support structure, anchorages, and conductor slack.

The equipment should be installed on the same support structure used in the qualification or one that is dynamically equivalent as discussed in Section 13.4.4.

The equipment or support structure anchorages should be capable of resisting the design loads from the seismic qualification. Anchorage of large equipment is usually accomplished by welding the equipment supports or baseplates to steel plates or shapes embedded in the concrete foundation. The embedded steel should be well connected to the concrete by using welded studs, reinforcing steel, or other shear connectors. Anchor bolts are also frequently used to anchor equipment and structures. Such anchors should be sized for adequate strength and designed with sufficient embedment and reinforcing steel detailing to permit yielding of the anchor bolts before the occurrence of a brittle failure mode such as concrete breakout. ASCE Manual 113 provides detailed guidance on the design of anchorages.

Equipment subjected to a qualification procedure is generally tested or analyzed in an unconnected or stand-alone condition. When connected by conductors, equipment responses to an earthquake may be altered, particularly if insufficient slack or flexibility is provided in the conductor or rigid bus work. Slack or flexibility is introduced by providing additional length in cable conductors or expansion connectors (such as thermal joints) in rigid bus work. Interaction effects have likely caused damage in past earthquakes, and these effects have been observed in sophisticated analyses as well as various experiments.

The most obvious instance of interaction between two connected pieces of equipment occurs when the differential displacement between the equipment items exceeds the available slack. In that case, large forces may be developed in the conductor. Less obvious is the case in which sufficient conductor slack is provided to fully accommodate the displacement of two equipment items based upon their stand-alone displacements. In this case, some degree of interaction will still occur because of the presence of the conductor that has non-zero mass and stiffness.

The design of flexible bus work utilizes the terminal displacements computed or measured in the seismic qualification of the stand-alone equipment. IEEE 693 provides guidance in determining the minimum length of flexible bus to accommodate the displacement demand, while IEEE 1527 provides a more comprehensive design procedure. When specifying conductor slack, electrical clearance requirements must also be considered. At present, seismic design standards provide little guidance related to interaction of equipment connected by rigid buses. A discussion of both types of bus work may be found in the *Application Guide for Design of Flexible and Rigid Buswork* [8].

References

1. Institute of Electrical and Electronics Engineers, *IEEE Recommended Practice for Seismic Design of Substations*, IEEE Std. 693-2005, IEEE, Piscataway, NJ, 2006.
2. Institute of Electrical and Electronics Engineers, *IEEE Recommended Practice for Seismic Design of Substations*, IEEE Std. 693-1997, IEEE, Piscataway, NJ, 1998.
3. Institute of Electrical and Electronics Engineers, *IEEE Recommended Practice for the Design of Flexible Buswork Located in Seismically Active Areas*, IEEE Std. 1527-2006, IEEE, Piscataway, NJ, 2006.
4. American Society of Civil Engineers, *Guide for the Design of Substation Structures, Manual of Practice 113*, ASCE, Reston, VA, 2007.
5. American Society of Civil Engineers, *Minimum Design Loads for Buildings and Other Structures*, ASCE 7-10, ASCE, Reston, VA, 2010.
6. Institute of Electrical and Electronics Engineers, *IEEE Recommended Practice for Seismic Qualifications of Class 1E Equipment for Nuclear Power Generating Stations*, IEEE Std. 344-2004, IEEE, Piscataway, NJ, 2004.

7. Federal Emergency Management Agency, *Recommended Provisions for Seismic Regulations for New Buildings*, NEHRP-2000 (National Earthquake Hazards Reduction Program), FEMA, Washington, DC, 2000.
8. Dastous, J.-B. and Der Kiureghian, A., *Application Guide for the Design of Flexible and Rigid Bus Connections between Substation Equipment Subjected to Earthquakes*, PEER 2010/04, Pacific Earthquake Engineering Research Center, Berkeley, CA.

14

Substation Fire Protection

14.1 Fire Protection Objectives 14-2
 Electrical Supply Reliability • Operational Safety • Revenue
 and Asset Preservation • Compliance • Risk Management
14.2 Fire Protection Philosophies 14-4
 Fire Prevention/Safety by Design • Automatic Fire
 Protection • Manual Fire Suppression • Fire Recovery •
 Incident Management and Preparedness
14.3 Fire Hazards 14-6
 Control of Fuels • Control of Ignition Sources •
 Substation Hazards • Specific Substation Building Hazards •
 Switchyard Hazards
14.4 Fire Mitigation Measures 14-9
 Substation Site–Related Fire Protection • Substation Building
 Fire Protection Measures • Exit Facilities • Passive Measures •
 Active Fire Protection Measures • Manual Measures •
 Substation Switchyard Mitigation Measures • Fire Protection
 Selection Criterion • Economic Risk Analysis Example
14.5 Fire Incident Management and Preparedness 14-19
 Fire Safety Plan • Operations Plan • Fire Training •
 Environmental Preparedness • Fire Recovery
14.6 Conclusion 14-22
14.A Appendix A: Control Building Fire Protection
 Assessment Checklist 14-22
14.B Appendix B: Switchyard Fire Protection
 Assessment Checklist 14-23
References 14-24

Don Delcourt
BC Hydro
and
Glotek Consultants Ltd.

The risk of fire in substations has been historically low, but the possible impact of a substation fire can be catastrophic. Fires in substations can severely impact the supply of power to customers and the utility company's revenue and assets. These fires can also create fire hazards to utility personnel, emergency personnel, and the general public. The recognition of the fire hazards, the risks involved, and the appropriate fire protection mitigation measures are some of the key considerations for the design and operation of new or existing substations.

This chapter provides an overview to help substation designers to identify fire protection objectives, fire hazards within the substation, and remedial measures and to evaluate the value of incorporating these measures. It is only an overview and is not intended to be all-inclusive or to provide all the necessary details to carry out a project. For further details and information on this topic, it is recommended that the designer refers to IEEE 979 Guide for Substation Fire Protection 2010 [1].

14.1 Fire Protection Objectives

A substation designer must clearly understand the substation owner's fire protection objectives before evaluating or selecting fire protection for a new and existing substation. The utility company or substation owner may have objectives of maintaining electrical supply reliability, operational safety, preservation of the revenue and assets, or compliance with regulatory standards. Each of these objectives may determine or influence the type and level of fire protection that should be selected. The following is a simplified explanation of these objectives.

14.1.1 Electrical Supply Reliability

Electrical supply customers and regulatory agencies are demanding higher levels of reliability for electrical services. In order to meet these demands, the utility companies or substation owners must understand all of the hazards that may impact their ability to provide a reliable electrical supply. Fires are one of the hazards that must be included in any reliability analysis. Some utility companies are setting goals of having operational reliability of 99.96% or greater. A transformer fire at a major substation or a fire in a substation control building or control house can result in a severe and long-term outage that will directly impact the utilities' reliability. In addition to the operational reliability, there is a growing concern that outages caused by fires can have significant societal impacts. In the past, it has been difficult to analyze and quantify the societal impacts but recent research programs have identified methods to quantify the impacts. These studies have shown that the societal impacts can range from $1 to $10 per kW h depending on the customer mix and the outage duration. It then becomes easier to estimate the benefits of increased reliability provided by fire protection, in terms of load loss and the projected societal loss costs.

14.1.2 Operational Safety

Fires in substations can directly and indirectly impact the public, emergency response services, and employee safety. The following are a number of the fire-related direct impacts that the substation designer should consider.

14.1.2.1 Direct Public Impacts

- The explosive failure of any oil-insulated electrical equipment can cause shrapnel to hit adjacent properties at distances of 76 m or more.
- The resulting blast from a transformer explosion can result in blast impacts or oil sprays on to adjacent public properties.
- Insulating oil pool fires from a transformer fire can create levels of thermal radiation beyond the perimeter of the station fence and may ignite combustible structures and vegetation in surrounding public properties.
- Insulating oil fires can also create very large fire plumes with flame heights of 33 m or more. During periods of high winds, the flames and the smoke plume can tilt significantly and impact adjacent buildings and structures. This can result in the ignition of the adjacent structure or the contamination of the structures in the area with soot or other particulate from the fire plume.
- In substations without appropriate fire spill containment, burning oil spill fires can spread beyond station boundary and impact adjacent public building structures and other public facilities.

14.1.2.2 Indirect Fire-Related Impacts

A fire in the substation can directly impact the electrical power supply to the public. Some of the indirect consequences of this interruption of the electrical supply are the loss of heating and cooling during inclement weather, the loss of the interior lighting, the impaired operation of elevators and large high-rise buildings, the impacts of loss of computer communication on stores and businesses, and the resulting business revenue losses related to the outage with the loss of wages during those outage periods. The extent of these electrical power supply losses can be estimated and included in any analysis or justification of substation fire protection.

14.1.2.3 Emergency Response Services

Fire department personnel responding to substation fires can be exposed to significant fire and electrical safety hazards that they may not be trained to deal with. The types of fire hazards found in indoor and outdoor substations are significantly different than the typical hazards most firefighters are normally exposed to. The most significant hazards that the fire department personnel are exposed to in substations are electrical safety hazards. Fire department personnel are trained to take active and aggressive roles in suppressing fires. In the case of a fire in electrical substation, there may be long delays before the station operating personnel can arrive on site and make the station electrically safe. In some cases, it may take up to an hour for electrical personnel to arrive and make the station safe. Therefore, the fire department personnel have to be patient and trained in the hazards to be expected in a working substation. These delays or periods of inactivity while the fire department wait to gain access to the station can create conditions where they are unable to respond to other critical alarms. The specialized nature of the equipment and systems within substations are also very different to the normal environments that a lot of fire departments work in. The delays in suppression and the safety issues are significant factors for the justification for automatic fire protection in a substation.

14.1.2.4 Employee Safety

The occurrence of a fire at a substation can create abnormal risks for the utility employees. This is especially true if these employees try to manually suppress a fire involving an indoor substation or very large transformer. Utility employees normally do not get the same level of training as professional firefighters and are not always aware of the physical stress that they may be put under during firefighting. The National Fire Protection Association (NFPA) standard on industrial fire brigades [2], along with any applicable health and safety regulations, should be considered before there is any decision made to use employees as a primary means of fire suppression.

14.1.3 Revenue and Asset Preservation

One of the major objectives of any utility is to generate revenue or at least generate sufficient cash flow to cover their operating costs. Private sector utilities have a strong objective to generate profit in order to continue to be a viable enterprise. The occurrence of a major substation fire can result in revenue losses or asset losses that can greatly impact the utilities' financial future. These fire-related risks can be controlled through the provision of fire protection or offset through the provision of insurance coverage. When considering revenue and asset preservation, it is very important that the substation designer analyze the critical elements of the station and their need for fire protection. Typically, transformers can create a very large catastrophic fire, but in a considerable number of cases, these transformers can be replaced without significant outages. A fire in a substation control building or control room can create a situation that may result in an outage that could last as long as 2 months. Therefore, the protection of control buildings and control rooms should be considered a high priority. Revenue in asset losses can be fairly easily estimated for various substation fire scenarios and can, therefore, be used in a benefit–cost analysis to justify the provision of fire protection to mitigate those critical risks.

14.1.4 Compliance

Some utilities operate in jurisdictions with mandated fire protection requirements for electrical substations. In these cases, the mandatory compliance with the appropriate codes and standards is a critical fire protection objective. For example, the "Guide for Fire Protection in Substations JEAG 5002-2001" [3] has been adopted in Japan. There are a number of other quasi-compliance objectives that the utility may have to use in the design of a new substation or changes to an existing substation. The IEEE 979 Substation Fire Protection guide [1] is one such document that is used by regulatory bodies to ensure that the utilities are providing a level of fire protection that meets the definition of good engineering practice. Utilities may also be exposed to some pressure to meet these general standards and guidelines from a due diligence perspective or they may be subject to litigation from customers, ratepayers, or regulatory organizations. The NFPA [4], CIGRE [5], and IEC [6] substation fire protection documents should also be considered as reference standards.

14.1.5 Risk Management

Utilities are starting to recognize that a risk strategy is important in maintaining their operations and controlling their asset losses. Risks are composed of two elements, the frequency or likelihood of the event and the consequences of the event. In the past, utilities have not directly set fire-risk objectives for their operations. There have been significant fire-risk strategies embedded in their insurance coverage and strategies; these strategies have not normally been quantified into frequencies or consequences. Risk practices are being adopted within the fire protection industry, based on the use of performance-based criteria from the petrochemical industry and nuclear industry. New performance-based building codes are being adopted in Europe and North America. These new codes and standards have embedded probabilistic criteria to evaluate acceptable levels for fire fatalities or, in the nuclear industry, acceptable levels of reactor shutdown frequencies.

There are no commonly adopted criteria for acceptable fire risks within utility substations. Published data on the maximum tolerable risk criteria has been published in various countries that could be easily adapted to form employee safety objectives. The typical accepted ranges for individual maximum tolerable risk criteria are from 10^{-4} to 10^{-6} fatalities/year. The Health and Safety Executive in the United Kingdom has published individual risk criteria for workers and has recommended that the maximum tolerable risk per year for industry workers to be 10^{-3} fatalities/year [7].

Although the probabilistic methodology is one that could be adapted to substation design, more commonly the qualitative methodology is applied in a formal and informal process. The NFPA and Society of Fire Protection Engineering (SFPE) have available publications on these risk management methods.

14.2 Fire Protection Philosophies

14.2.1 Fire Prevention/Safety by Design

Fire prevention is a philosophy of the selection of equipment, materials, and processes that will eliminate or lower the risk of a fire. Eliminating or lessening the hazards created by substation equipment and design is the most effective method of fire protection. The removal or isolation of fuel and ignition sources is some of the key fire prevention strategies. The following are a number of typical fire prevention measures used in substations:

- Use SF_6 circuit breakers in place of oil-insulated breakers.
- Install lightning protection for the station.
- Use high flashpoint insulating oils for electrical equipment insulating in cooling fluids.

14.2.2 Automatic Fire Protection

Automatic fire protection philosophies are measures that are incorporated into the substation design to mitigate the hazards that are inherent in the substation design or equipment selection. The typical types of fire protection measures are discussed in greater detail in Section 14.4. These measures are designed to automatically detect a fire, activate the system, and suppress or control the fire. The following are a number of typical automatic fire protection measures used in substations:

- Installation of water spray deluge on large transformers
- Installation of gaseous fire protection systems in substation control rooms
- Installation of wet sprinkler protection in cable spreading rooms

14.2.3 Manual Fire Suppression

If fire prevention and automatic fire protection measures are not incorporated into the substation design or equipment selection, there may be an expectation that the local fire department or fire brigade will suppress the fire. Manual fire suppression can only be considered a valid fire protection philosophy if the substation designer has evaluated the fire department and fire brigade response. There are many conditions that have to be in place in order for manual fire suppression to be successful. The following are a number of manual fire suppression conditions that should be considered:

- Early notification to the fire department fire brigade of the occurrence of the fire.
- Can they arrive at the site in a reasonable length of time?
- The speed at which the substation be made electrically safe for firefighting when they arrive.
- There is a suitable water supply system available.
- Have sufficient manpower and training to effectively suppress the fire.

The manual fire suppression philosophy may have a very low initial costs, but it also has a very low effectiveness. The result is that any reliance on this type of measure may result in significant losses of assets, revenue, and service to customers. The substation designer should clearly document this type of philosophy because of the significant impacts that can be expected with a fire occurrence. The following are a number of typical manual suppression measures:

- Trained station fire brigade
- A water supply system with hydrants throughout the station
- Station equipped with 160 kg dry chemical wheeled units and AFFF foam units located onsite

14.2.4 Fire Recovery

Another parallel philosophy to the manual fire suppression philosophy is to provide measures to recover the substation operation in the event of a catastrophic fire; some examples are utilities that have temporary mobile substations and an inventory of spare breakers and transformers. There are components of a substation where a replacement or work-around philosophy will not work, such as large substation control buildings and control equipment. It can be applied to specific portions of the substation such as the provision of two transformers that are each capable of carrying the total of substation capacity. In the event of the loss of a single transformer, a substation is still operational. This philosophy can work for specific scenarios if a critical element analysis has been done and multiple critical elements are not at risk from a single event. The substation control building is a critical element where the fire recovery philosophy may not be practical to implement.

14.2.5 Incident Management and Preparedness

An incident management and preparedness philosophy is one in which the utility formulates a detailed plan of the operational, safety, and firefighting actions that should be taken and how the incident can be coordinated with the fire department. This type of philosophy avoids the need to delay fire suppression in order to plan and coordinate a fire attack during a fire. More detail on this philosophy is provided in Section 14.4.

14.3 Fire Hazards

One of the key steps in the design of new substations and the assessment of existing substations is to identify conditions that are a fire hazard. If fire hazards are identified during the design stage, measures can be incorporated to eliminate, lessen, or mitigate the hazards.

Fire hazards are the conditions that create the potential for a fire. Fire hazards normally have the following attributes:

- A probability that a fire will actually occurr during a specified time interval
- An initial fire size and a probability of growth or spread
- Consequences that can be characterized in terms of life lost or injuries, asset loss, revenue loss, and operational losses

In order for fire hazards to exist, there must be fuel and some sort of ignition source available. The following are some of the generic methods of controlling typical fire hazards.

14.3.1 Control of Fuels

This type of fire prevention measure eliminates or limits the fuel available to a fire, thereby controlling the size and duration of the fire. An example would be a management control requiring that all flammable liquids in approved safety cans be stored in approved flammable liquid cabinets or vaults. Another example would be a designed containment basin around an indoor transformer for the capture of transformer oil. The control of fuels measures also covers hazards such as

1. The storage of rags (clean and soiled)
2. The storage of combustible material (wood and electrical cable)
3. Use and storage of combustible liquids (transformers, reactors, and oil-filled cable)
4. Storage and handling of refuse
5. Use of combustible cable (PVC jacketed)

14.3.2 Control of Ignition Sources

This type of fire prevention measure is intended to control or eliminate ignition sources so that there are no easily ignitable fuel sources. An example would be a management control prohibiting smoking in all areas other than those with a low fuel load (i.e., lunchroom). Another example would be a management control requiring a permit and fire watch when any welding or hot work is carried out. The control of ignition sources measures also covers hazards such as

1. Electrical apparatus
2. Electrical cable
3. Hot work
4. Fuel-fired heating equipment
5. Friction producing equipment

Fire codes mandate prevention measures for common hazards relating to the use, storage, and handling of flammable and combustible liquids and hazardous processes such as compressed gases, paint spray booths, welding and cutting, general fire safety and fire prevention design, and maintenance standards.

The NFPA standards set out the minimum fire prevention and protection standards for various occupancies (i.e., Nuclear Generating Plants, Fossil Fuel Fired Generating Plants, LNG, and LPG Plants), various hazardous processes or equipment (i.e., gas-fired boilers, electrical equipment maintenance), and the standards for the fire protection to prevent against the hazards.

14.3.3 Substation Hazards

There are a wide range of types and causes of fires in substations and these are good indicators of typical fire hazards. The types of fires are based on the equipment and systems used in the stations. The following are some types of substation equipment that have caused fires:

- High-voltage DC valves
- Outdoor or indoor oil-insulated equipment
- Oil-insulated cable
- Hydrogen-cooled synchronous condensers
- Polychlorinated biphenyls (PCB)-insulated equipment

There are a number of other substation specific types of fires that are not as well documented. IEEE 979 [1], Factory Mutual Data Sheets, NFPA 850 [8] Recommended Practice for Fire Protection for Current Converter Stations, and CIGRE TF 14.01.04 Report on Fire Aspects of HVDC Valves and Valve Halls [10] give guidance on these types of fires. Also, the Edison Electric Institute's "Suggested Guidelines for Completing a Fire Hazards Analysis for Electric Utility Facilities (Existing or in Design) 1981" [11] provides reference guidelines for the fire hazard analysis process.

Energized electrical cables with combustible insulation and jacketing can be a major hazard because they are a combination of fuel supply and ignition source. A cable failure can result in sufficient heat to ignite the cable insulation that could continue to burn, produce high heat and large quantities of toxic smoke. Oil-insulated cables are an even greater hazard since the oil increases the fuel load and spill potential.

The hazard created by mineral oil-insulated equipment such as transformers, reactors, and circuit breakers is that the oil is a significant fuel supply that can be ignited by an electrical failure within the equipment. Infiltration of water, failure of core insulation, exterior fault currents, and tap changer failures are some of the causes of internal arcing within the mineral insulating oil that can result in fire. This arcing can produce breakdown gases such as acetylene and hydrogen. Depending on the type of failure and its severity, the gases can build up sufficient pressure to cause the external shell of the transformer tank or ceramic bushings to fail or rupture. Once the tank or bushing fails, there is a strong likelihood that a fire or explosion will occur and the possible explosion could create blast damage. This blast can result in an oil spill fire that can spread to form a large pool fire depending on the volume of oil, spill containment, slope of the surrounding area, and the type of the surrounding ground cover (i.e., gravel or soil). The oil spill fire can also create thermal radiation damage to surrounding structures and convective heat damage to structures above the fire area.

Substations can contain common industrial hazards such as the use and storage of flammable compressed gases, hot work, storage and handling of flammable liquid, refuse storage, presence of heating equipment, and dangerous goods storage. The local fire codes or NFPA codes can provide assistance in the recognition of common fire hazards.

14.3.4 Specific Substation Building Hazards

The basic problem with major fires in indoor stations is that the building contains the blast pressure, heat, and smoke, which can result in the following types of damage or impacts:

- Oil-insulated equipment explosions could cause blast damage to the building structure, possible structural failure and thermal damage to the building structure and possible structural failure.
- Smoke damage to critical substation equipment (corrosion damage).
- Smoke and thermal exposure can also create severe impacts on building occupants and emergency personnel trying to safely evacuate the facility or to carry out subsequent firefighting.

FIGURE 14.1 Shows large distribution transformer in an explosion containment tank inerted with nitrogen.

Some of the specific fire hazards in substation buildings are

- Use of exposed combustible construction
- Use of combustible finishes
- Presence of emergency generators, shops, office, and other noncritical facilities in the buildings
- Batteries and charger systems
- System protection equipment
- Communication equipment
- SCADA equipment
- Computer equipment
- Cable spreading areas, cable trenches, cable tunnels, and cable vaults
- HVDC equipment
- Switchyard cable openings that have not been fire stopped
- Adjacent oil-insulated transformers, breakers, and other electrical apparatus
- High-voltage equipment
- Dry transformers

Fires in any of the above components can result in damage or the destruction of critical operation control or protection equipment. Damages can result in a long-duration outage to customers as well as significant revenue losses (Figure 14.1).

14.3.5 Switchyard Hazards

Some of the specific components in substation switchyards that are a potential fire hazard are

- Oil-insulated transformers and breakers
- Voltage regulators
- Oil-insulated cable
- Pipe type cable
- Oil-insulated potheads
- Capacitors
- Cable oil pump houses

- Oil processing plants
- Hydrogen-cooled synchronous condensers
- Gasoline storage or dispensing facilities
- Vegetation
- Combustible service building
- Pesticide or dangerous goods storage
- Storage warehouses
- Stand-by diesel generator buildings

The failure of some of the critical components such as transformers and breakers directly results in losses of revenue or assets. Other switchyard components can create a fire exposure hazard to critical operational components (i.e., combustible service buildings located close to bus support structures or transmission lines). Also, see the Appendix 14.B.

14.4 Fire Mitigation Measures

14.4.1 Substation Site–Related Fire Protection

When reviewing a new site or considering changes to an existing site, the substation designer should consider the impacts the location of the site and the site have on the station fire protection. The following are a number of location and site-related considerations:

1. External exposures
   a. External exposures are fire hazards that exist external to the substation but may impact the substation. These external exposures are no different than the typical internal exposures except that the utility may not have any control over the exposing property. Examples could be large combustible buildings or structures immediately adjacent to the substation (i.e., chemical plants, power plants, LNG storage facilities, and other high-risk industrial operations). These exposures can also be in the form of natural hazards such as heavily forested areas and grasslands immediately adjacent to the station. The possible impact of a fire on the station, the level of risk involved, and whether measures can be taken within the substation to mitigate the risk of this exposure should be evaluated.
2. Site grading
   a. The concern is that heavily sloping sites may exacerbate the spread of mineral oil spill fires unless spill containment is incorporated in the design. The preferred design option for heavily sloping sites would be to locate mineral oil equipment down the slope from other critical building structures and equipment.
3. Prevailing winds
   a. The substation designer should research any data on the prevailing wind direction and strength in the area of the substation site. An analysis of this data may indicate that the major transformers should be located in a different orientation to other transformers or control buildings, if the prevailing wind direction and strength warrant it. If the substation designer determines that the site is exposed to very strong frequent prevailing winds from the north, then lying out the equipment in a north–south direction may not be appropriate. The substation designer should then consider orienting equipment in an east–west direction or north–south in a chevron configuration.
4. Fire emergency response capabilities
   a. One of the most critical aspects of the substation site is availability of public fire protection to fight a fire within the station. It is a very important element of the fire safety and protection for a substation. If automatic or passive fire protection measures are not integrated into the design of the substation, it would normally be left up to the public fire department or a station fire brigade to fight the fire. The designer should also check on the availability of the firefighting water supply adjacent to the substation since this would be critical for manual firefighting.

14.4.2 Substation Building Fire Protection Measures

The measures to mitigate or lessen fire hazards are normally called fire protection measures. The NFPA standards and local building codes set the standards for application and design of fire protection. The type of measures can be broken down as follows:

1. Fire safety
2. Passive fire protection
3. Active fire protection
4. Manual fire protection

The following is a brief description of the previous measures.

14.4.2.1 Fire Safety

Fire safety measures generally include the fire protection measures required under the building, fire, or life safety codes. The main objective of these codes is to ensure that

- The occupants are able to leave the station without being subject to hazardous or untenable conditions (thermal exposure, carbon monoxide, carbon dioxide, soot, and other gases)
- Firefighters are safely able to effect a rescue and prevent the spread of fire
- Building collapse does not endanger people (including firefighters) who are likely to be in or near the building

To meet these objectives, fire safety systems provide the following performance elements:

- Detect a fire at its earliest stage.
- Signal the building occupants and the fire department of a fire.
- Provide adequate illumination to an exit.
- Provide illuminated exit signs.
- Provide fire-separated exits within reasonable travel distances from all areas of a building. These exits shall terminate at the exterior of the building.
- Provide fire separations between building floors and high-hazard rooms to prevent the spread of fire.
- Provide passive protection to structural components to prevent their failure due to fire.

14.4.2.1.1 Fire Alarm System

The primary function of the fire alarm and detection system is to provide a means of detecting a fire at an early stage so that personnel life safety can be signaled, and manual and automatic fire suppression activities can be initiated. The fire alarm systems can report to a remote monitoring station that would then, in turn, dispatch the local fire department. The fire alarm system will also provide information to the fire department on the location of the fire.

A fire alarm system comprises a control panel, power supplies, signal-initiating or fire detection devices (such as water flow, smoke, heat, or flame detectors), audible signal appliances (such as bells, horns, or sirens), and annunciators, all of which is interconnected via electrical wiring. In addition, the fire alarm system also interfaces various types of automatic suppression systems (Sprinkler Systems, Water Deluge, or Gaseous Extinguishing Systems).

Electrical substations contain significant quantities of materials that produce dense, toxic, and corrosive smoke when involved in a fire. In order to prevent the propagation of smoke under fire conditions, air duct smoke detectors should be installed at all supply and return air systems, and the fire alarm system designed to shut down all ventilation systems on the operation of any type of fire detector (smoke, heat, flame, etc.) or an automatic suppression system.

The provision of an approved fire alarm system for early detection, combined with rapid fire suppression (whether by manual or automatic means), can reduce and minimize both direct and indirect losses incurred as a result of fire.

The design, installation, verification, inspection, and testing of fire alarm system are governed by various Codes and Standards, such as

- The local fire and building codes
- NFPA-72 National Fire Alarm and Signaling Code [12]

Typically, fire alarm systems for substations have a fire alarm panel located in the entry vestibule or control room. In control, relay, and telecommunications rooms, spot type photoelectric detectors are installed at the ceiling and below the sub floor area, and stations where a rapid response is required, air sampling detection can be installed. In feeder, cable spreading, and switchgear areas, spot type photoelectric detection, linear beam smoke detection, and air sampling detection are installed. In other areas of the substation, photoelectric detection, thermal detection, or sprinklers can be used as the detection means. Commonly, horns strobe type signaling devices are installed throughout station, and the fire alarm system is monitored by the local monitoring company or by the utility control center.

14.4.3 Exit Facilities

One of the most critical fire safety facilities in the building are the exit facilities. These facilities allow the occupants to evacuate the building while being protected against exposure to smoke and fire. The exit facilities also allow the firefighting personnel to enter and gain access to the floor areas of the building without being exposed to smoke and fire until they reach the fire floor.

A building egress system consists of vertical or horizontal exits and various components to compliment these facilities. Vertical exit facilities comprise fire separated stairs or ramps, and horizontal exits are composed of fire-rated exit corridors. Both the horizontal and vertical exit facilities extend to the various floors of the building so that no part of the building area exceeds the maximum allowable travel distance. At the start of the exit facility, an approved type exit door is provided that allows a fire separated opening to the exit corridor. The travel distance is measured from this door in the most remote area of a room or area served by the door. The vertical or horizontal exits are constructed of a fire-rated separation of at least 3/4 of an hour to ensure that its integrity is in place for at least 3/4 of an hour of fire exposure. The exit is then provided with an emergency lighting system providing a minimum of 10 lx of lighting to ensure that the occupants of the building can escape during periods of a power outage. The exit facility then extends to the exterior of the building or a public thoroughfare outside the structure of the building.

The requirements for exits are governed by the local building and fire codes, and the Life Safety Code [13] NFPA 101. These codes base their recommendations for exit facilities on the occupancy, height, size, and floor area of the particular building involved. Typically, the requirements for the exit facilities would be as follows:

1. The maximum travel distance to an exit allowable would be 30 m.
2. The fire separation for corridors and stairs would have to be a minimum of 3/4 of an hour or at least equal to the required slab through which the exit is puncturing.
3. The minimum width of the exit would be 1 m.
4. Exit signs would have to be provided at each exit door from the floor area.

The exit and the access to exit path would have to be illuminated with an emergency lighting system capable of at least 10 lx at floor level.

Typical exit facilities for substations consist of minimum of 1 h rated fire separated stairs and corridors and exterior doors.

14.4.4 Passive Measures

Passive measures are static measures that are designed to control the spread of fire and withstand the effects of fire. These measures are the most frequently used methods of protecting life and property in buildings from a fire. This protection confines a fire to a limited area or ensures that the structure remains sound for a designate period of fire exposure. Its popularity is based on the reliability of this type of protection, since it does not require human intervention or equipment operation. Common types of passive protection include the following:

- Fire stopping
- Fire separations
- Use of noncombustible construction materials
- Use of low flame spread or smoke developed rated materials

The degree of passive protection would be based on the occupancy of the area and the required structural integrity. Of the two criteria, the structural integrity is the most critical in order to preserve life and property, since the premature failure of the building structure could trap occupants, severely hamper firefighting, and possibly cause the total destruction of the building by collapse.

Another important passive life safety facility is a fire separated exit corridor. The protection of a corridor or exit allows all the building occupants to exit the building without being exposed to fire/smoke and allows firefighters to safely reach the fire involved floor area without exposure to the fire. The actual requirements for the passive protection of an exit facility is based on the group classification for the building since the exit facilities must have the same rating as the floor assemblies which it is piercing, but no less than 3/4 of an hour.

Typically, substation buildings should be of noncombustible construction. The following are some of the typical fire separation ratings that are recommended by IEEE 979:

- Two hour fire resistance rating for control rooms, battery rooms, switchgear rooms, cable spreading rooms or tunnels, telecommunication rooms, shops, offices, warehouse areas, emergency diesel generator, and flammable and combustible storage rooms
- A minimum of a 3 h fire resistance rating for transformer vaults and indoor oil breaker vaults

14.4.5 Active Fire Protection Measures

Active fire protection measures are automatic fire protection measures that warn occupants of the existence of fire and extinguish or control the fire. These measures are designed to automatically extinguish or control a fire at its earliest stage, without risking life or sacrificing property. The benefits of these systems have been universally identified and accepted by building and insurance authorities. Insurance companies have found significant reductions in losses when automatic suppression systems have been installed.

An automatic suppression system consists of an extinguishing agent supply, control valves, a delivery system, and fire detection and control equipment. The agent supply may be virtually unlimited (such as with a city water supply for a sprinkler system) or of limited quantity (such as with water tank supply for a sprinkler system). Typical examples of agent control valves are deluge valves and sprinkler valves. The agent delivery systems are a configuration of piping, nozzles, or generators that apply the agent in a suitable form and quantity to the hazard. Fire detection and control equipment may be either mechanical or electrical in operation. These systems incorporate a fire detection means such as sprinkler heads or they use a separate fire and detection system as part of their operation. These detection systems detect a fire condition, signal its occurrence, and activate the system.

Active systems include wet, dry, and pre-action sprinklers, deluge systems, foam systems, and gaseous systems. Detailed descriptions of each of these systems, code references, and recommendations on application are covered in IEEE 979.

14.4.6 Manual Measures

The major provisions for manual firefighting are portable fire extinguishers, interior fire hose facilities, and smoke/heat venting and control. The provision of portable firefighting equipment is normally required by the local fire codes. The objective is to provide readily accessible portable fire equipment for the extinguishment of incipient stage fires by the building occupants. Since the majority of all fires start small, it is an advantage to extinguish them during their incipient stage to ensure that the losses are minimized.

Interior fire hose facilities are used as firefighting water supplies for occupants and the fire department. The interior fire hose facilities or standpipes are installed in order to provide effective fire streams in the shortest possible time in places such as the upper storeys of a high building or a building having a large area, using either 30 m of 3.8 cm diameter hose or 64 cm diameter hose valve stations. The provision of interior fire hose facilities is normally required by the local building code when the building exceeds height and area limits. The actual design of the standpipe system is governed by the requirements of the NFPA [14]. Typically, the number of standpipes required is based on the requirement that the standpipe hose station be capable of reaching all areas of the building with a 9 m spray from a nozzle on a 30 m hose.

The provision of smoke/heat venting and control is important in occupancies such as substations where there are significant quantities of materials that produce large amounts of toxic, corrosive, dense smoke, and high heat while usually having little means of ventilation (windows or other openings). One of the most important objectives of a smoke/heat venting system is to increase the visibility to allow the building occupants to safely evacuate the building, and so the fire department can enter the building and quickly find the fire without encountering the visibility problems encountered with the smoke. Another important use of these systems is to control and release the heat being built up in the fire area to prevent it spreading to other areas of the building. This system can also control and safely release the corrosive combustion by-product of PVC or Neoprene cable jacketing or insulation fire (i.e., hydrochloric acid). The normal methods of heat/smoke control are as follows:

1. To provide for the shutdown of the building ventilation systems on the detection of a fire by the fire alarm system
2. To provide a seal for ducts piercing the hazard area using smoke dampers controlled by the fire alarm system
3. To provide controls at the main fire control panel to allow the fire department to selectively operate the building ventilation to evacuate the smoke
4. To provide special smoke evacuation system to vent high-hazard areas

14.4.7 Substation Switchyard Mitigation Measures

Substation switchyard passive, active, and manual mitigation measures are very similar to those discussed for substation buildings. The only difference being these types of systems target specific hazards that are commonly found within switchyards.

14.4.7.1 Passive Measures

14.4.7.1.1 Spatial Separation

The spatial separation of oil-insulated equipment is one of the most important aspects of substation switchyard fire mitigation. A thorough analysis should be done of the equipment within the switchyard and the possible exposure that this equipment would create to overhead structures, transmission lines, adjacent oil-insulated equipment, and critical substation buildings.

Various codes and standards recommend specific spacing between oil-insulated equipment and critical structures. This criterion is very generic and may not provide a suitable level of protection for

FIGURE 14.2 Shows damage to the wall of substation control building from a transformer fire. The flame reached above the roof and caused the roof structure to collapse into the control room.

specific applications. In the past, documents and standards have recommended spacing criteria based on the voltage of the equipment or the number of gallons of mineral oil that an equipment contains. From a practical fire protection perspective the most important issue regarding oil-insulated equipment spacing is how close will an oil spill fire get to the exposed equipment or structure. If a transformer is located 30 m away from a control building, but the transformer is located up a deeply sloping site and has no fire spill containment, a separation distance of 30 m is not adequate. If the transformer is located 30 m away from the control building and the site slopes from the control building to the transformer and the transformer has fire spill containment, a 15 m separation may be adequate. The substation designer must carefully analyze the proposed design to determine the possible spill fire consequences, instead of merely selecting separation distances from generic tables covering equipment oil volumes or voltages (Figure 14.2).

Conservative spacing criteria outlined in the following table should be considered as appropriate to layout station equipment exposing substation buildings for the conceptual stage of design:

Oil-Insulated Equipment Separation Distances to Buildings			
Equipment Oil Volume (L)	2 h Rated Construction (m)[a]	Non-Combustible Construction (m)[a]	Combustible Construction (m)[a]
<1,893	1.5	4.6	8
1,893–18,927	4.6	8	15
>18,927	8	15	30

[a] The distance from the edge of the flame front of the postulated spill to the exposed structure.

IEEE 979 and NFPA 850 guides provide some generic spacing guidelines for the spacing of oil-insulated equipment, but these generic distances must be adjusted based on the type of insulating oil used, slope of the site, adequacy of the fire spill containment, type of bushing material, and the expected fire response. A detailed radiant heat flux should be carried out for all situations where the separation criteria cannot be met. IEEE 979 provides information on how to use heat flux nomographs and U.S. Nuclear Regulatory

FIGURE 14.3 Shows closely spaced transformer without any fire spill containment or flame suppressing stone. A transformer fire at this substation damaged three additional transformers before the fire was suppressed.

Commission (NUREG) Fire Dynamics Tools [15] spreadsheets can be to carry out a detailed engineering analysis of the required spatial separations for exposed oil-insulated equipment and critical buildings.

Where the aforementioned separation criteria cannot be met, the substation designer should look at the provision of fire spill containment, provision of fire barriers, or the provision of water spray deluge protection (Figure 14.3).

14.4.7.1.2 Ground Cover

Ground cover can play an important role in controlling the impact of a spreading oil spill fire originating in oil-insulated equipment. When burning mineral oil spreads to an area with a minimum thickness of 15 cm of 1.9 cm diameter stone, the flames will be suppressed. Also, where fire spill containment is used, a common practice is to backfill the containment with stone to also act as a flame suppressant.

14.4.7.1.3 Fire Spill Containment

Fire spill containment is another of the common protection measures used in substation switchyards to prevent a spill fire from exposing adjacent critical equipment and structures. A burning oil spill can spread for significant distances if the substation site is sloped and this spill is not contained. Fire spill containment has some similar objectives to oil spill containment except that it is focused on minimizing the impacts of a mineral oil spill fire. The following are some of the typical types of spill containment:

- Reinforced concrete containment basins backfilled with flame suppressing stone or using cover grading with a 15 cm depth of stone
- Excavated pits that have been backfilled with stone, which surround oil-insulated equipment
- Curbed or heavily sloped areas around transformers leading to remote burn off pits

14.4.7.1.4 Fire Barriers

In situations, where oil-insulated equipment cannot be spaced sufficiently far apart from each other or from critical buildings such as control buildings, fire barriers are used to act as thermal barriers and as shrapnel shields. These barriers can be constructed of heavy gauge sheet metal, reinforced concrete,

FIGURE 14.4 Fire barriers between individual single-phase transformers that are not high enough to protect the 230 kV bushings on adjacent transformers.

masonry, and specially designed heat-resistant material. These barriers should extend to the outermost boundary of any oil containment area or the postulated spill. The minimum barrier height should be determined based on an analysis of the expected flame height of the fire in the piece of equipment it is exposing the other critical components of the station. The height of the barrier should be no less than 1 m higher than the height of the highest oil-filled component of equipment that is exposed. See IEEE 979 for specific fire barrier recommendations (Figure 14.4).

14.4.7.1.5 Cable Systems

Switchyard cable trenches located within 6 m of large oil-insulated electrical equipment should be designed so that the trench walls are liquid tight and the trench covers are both noncombustible and liquid tight. The objective is to prevent burning oil from entering critical cable trenches and damaging control and power cable. In addition, any cable entries from switchyard cable trenches into control buildings should be fire stopped to a minimum of a 2 h rating to prevent fire extension into the control building. In stations with extensive cable trench systems, consideration should be given to installing sand fill or sandbags at major transitions or intersections of the system to prevent a fire from spreading along the trenches.

14.4.7.2 Active Systems

14.4.7.2.1 Automatic Fire Protection Systems

Automatic fire protection systems such as water spray deluge and foam systems are used to protect against fires in large pieces of oil-insulated equipment that are either closely spaced or spaced too close to critical station facilities such as control buildings. Where adequate spatial separation, fire barriers or other passive measures cannot be installed, then active fire protection system should be installed. Water spray deluge systems are extremely effective in suppressing fires in large oil-insulated transformers and breakers but will require a large volume water supply. Automatic foam systems require smaller volumes of water to be effective but create additional environmental concerns that have to be addressed as part of the design. IEEE 979 provides additional details on available options (Figure 14.5).

FIGURE 14.5 Transformer deluge system in operation.

14.4.7.2.2 Explosion Suppression

Internal faults within oil-insulated equipment can create accumulations of hydrogen and acetylene gas that can result in a rupture of the equipment tank and a subsequent explosion. Systems are currently available that are able to detect very small internal pressure changes within the transformer tanks, vent and inert the transformer tanks before an explosion can occur. As a result of these types of explosion suppression, equipment can also prevent a fire from occurring.

Some transformer manufacturers are now building transformer tanks that are designed to withstand internal over pressurization. These explosion resistive designs can also help to prevent the occurrence of a transformer fire.

14.4.7.3 Manual Systems

14.4.7.3.1 Fire Extinguishers

Portable fire extinguishers and wheeled dry chemical units are the most common types of manual fire protection measures used in a substation switchyard. Commonly, 14 kg ABC dry chemical extinguishers are located in weather-resistant boxes adjacent to oil-insulated equipment or other switchyard hazards. At stations with very large transformers, and no public fire department response; utilities sometimes provide 160 kg ABC dry chemical wheeled extinguishers in weatherproof sheds or shelters. These large capacity dry chemical extinguishers can provide significant fire suppression. In addition, it is also common for utilities to provide an on-site supply of firefighting foam or small foam trailers to help suppress fires in a large transformers or breakers.

14.4.7.3.2 Water Supplies

Water supplies are critical for fire suppression of oil-insulated equipment fires. The substation designer should work with the on-site staff and the local fire department to determine whether there are any suitable firefighting water supplies available. This analysis should consider the available water supply flow rates, pressures, and duration. Municipal piped water supplies should have available flow rates of no less than 500 L/min at 133 kpa to be effective water supplies. At some substations, there will be no available municipal supply, and the designer will have to look at alternate available supplies such as creeks, rivers, and ponds. The local fire department also may be able to provide a tanker shuttle water supply from remote hydrants.

14.4.8 Fire Protection Selection Criterion

Generally, fire protection measures can be subdivided into fire safety and investment categories. Fire safety measures are considered to be mandatory by fire codes, building codes, and safety codes. Investment-related fire protection is provided to protect assets, conserve revenue, and help to maintain service to customers. This type of fire protection is not commonly mandated by legislation but is driven by economic reasons such as asset losses, revenue losses, and the possible loss of customers. Therefore, there is considerable flexibility in how investment fire risks are mitigated, the types of fire protection measures used, whether the risk is offset by purchasing insurance, or whether the risk accepted as a cost of doing business (the "do nothing" option).

The selection of investment-related fire protection can be done based on company policies and standards, insurance engineering recommendations, industry practices, specific codes and standards (IEEE 979 and NFPA 850), or by an economic risk analysis.

The economic risk analysis is the evaluation of the investment measures in relation to the probability of fire, the potential losses due to fire, and the cost of the fire protection measures. This analysis requires a reasonable database of the probability of fires for the different hazard areas or types, an assessment of the effectiveness of the proposed fire protection measures, an estimate of the fire loss costs, and a fair degree of engineering judgment. The potential losses usually include the equipment loss as well as an assessment of the lost revenue due to the outage resulting from the loss of equipment.

One of the most common economic risk analysis measures is a benefit/cost analysis. This analysis is calculated from the following equation:

$$\text{Benefit/cost ratio} = \frac{(\text{Annual frequency of fire} \times \text{Fire loss costs [assets and revenue]})}{(\text{Cost of the fire protection} \times (1/\text{Effectiveness of the fire protection measure})}$$

Normally, this ratio should be greater than 1 and preferably greater than 2. A benefit/cost ratio of 2 means that the avoided fire loss cost or benefit is twice the cost of the fire protection. Therefore, it is a good investment.

One of the greatest difficulties is to estimate the frequency of fire for the specific hazards. Some companies have extensive fire loss histories and loss databases. These databases can be used to estimate specific fire frequencies, but the results may be poor due to the small statistical sample size based on the companies' records. There are a number of other databases and reports that are in the public domain that provide useful data (i.e., NFPA data shop [16], EPRI Fire Induced Vulnerability Evaluation Methodology [17], and IEEE 979 Transformer Fire Survey). The IEEE 979 Transformer Fire Survey's [18] estimated probability of fire is given in IEEE 979.

Once the potential financial loss due to a fire has been calculated, the designer should input costs and effectiveness of any proposed fire protection measure into the benefit/cost equation and determine the B/C ratio. If the B/C ratio is less than 1, the provision of the fire protection measure is not an acceptable investment.

14.4.9 Economic Risk Analysis Example

The following is a simplified example of an economic risk analysis:

- Substation has four 138 kV single-phase oil-insulated transformers. One of these transformers is a spare and is located remote from the others. The load supplied by these transformers is 25 MW. A water spray deluge system is being considered to suppress or control a fire in the transformers. The deluge system is expected to protect the adjacent transformers. The estimated cost of a deluge system for all three transformers is $60,000. The individual transformers have a replacement value of $300,000.
- Utility's chief financial officer questions whether this is a good investment.

Substation Fire Protection

- Company uses a discount rate of 10% and requires that all investments have a benefit/cost ratio of greater than 2. The assigned value of energy is $25/MW. The standard amortization period is 25 years.
- Annual frequency of fire for a single 138 kV transformer is estimated to be 0.00025/year and the combined frequency for the three transformers is 0.00075 fires/year.
- Estimated effectiveness of the deluge system protecting the adjacent transformers is 0.9. The deluge system will not save the transformer in which the fire originates; it is assumed to be a total loss.
- Fire is assumed to originate in the center transformer in the bank of three single-phase transformers. It is assumed that in the absence of the suppression, the fire will spread to destroy the two adjacent transformers. The spare transformer is not affected because it is remote from the other transformers.
- Estimated station outage period for this scenario is the difference between the outage time to replace all three transformers (a fire in the center transformer could destroy all three transformers) and the outage time to replace the center transformer (assuming the deluge system will protect the adjacent transformers). The outage time to replace a single unit is 5 days and to replace three units is 40 days. Therefore, the expected outage loss period is 35 days.
- Expected lost revenue is 35 days × 24 h/day × 25 MW/h × $25/MW = $525,000.
- Estimated annual revenue and equipment loss costs = (Composite annual fire frequency) × (Revenue loss for the station outage period + Replacement value of the adjacent transformers) = (0.00075 fires/year) × [$525,000 + (2 × $300,000)] = $843.75/year.
- Net present value of the annual revenue and equipment losses for the 25 year amortization period at a discount rate of 10% = $7659.
- Benefit/cost ratio = $7,659/ ($60,000 × (1.0/0.9)) = 0.115.

14.4.9.1 Example Conclusion

The calculated benefit/cost ratio of 0.115 is considerably less than the minimum required ratio of 2. The proposal to install deluge protection should be rejected since it is not economical. Other fire protection measures could be considered or the risk could be transferred by purchasing insurance to cover the possible loss of the assets (transformers) and the revenue. These other measures can also be analyzed using this economic risk analysis methodology.

It should be noted that the previous example does not include societal costs, loss of reputation, and possible litigation.

14.5 Fire Incident Management and Preparedness

The purpose of fire incident management is to provide a process to coordinate and manage substation fire incidents with private or public emergency responders. This management process covers the following components:

- Substation fire safety plans
- Fire operational plans
- Fire incident management training
- Environmental preparedness
- Fire recovery

14.5.1 Fire Safety Plan

Fire safety plans are generally required by the fire codes to ensure that the building owner describes the fire safety components of the building, the building response organization, and emergency response procedures for building occupants.

A fire safety plan should be prepared in cooperation with the fire department and other applicable regulatory authorities and shall include

1. The emergency procedures to be used in case of fire, including.
   a. Sounding the fire alarm.
   b. Notifying the fire department.
   c. Instructing occupants on procedures to be followed when the fire alarm sounds.
   d. Evacuating occupants, including special provisions for persons requiring assistance.
   e. Confining, controlling, and extinguishing the fire.
2. The appointment and organization of designated supervisory staff to carry out fire safety duties.
3. The training of supervisory staff and other occupants in their responsibilities for fire safety.
4. Documents, including diagrams, showing the type, location, and operation of the building fire emergency systems.
5. The holding of fire drills.
6. The control of fire hazards in the building.
7. The inspection and maintenance of building facilities provided for the safety of occupants.
8. *Posting of fire emergency plans*—The emergency plans from the fire safety plan shall be posted throughout the building to inform and educate the building occupants in the building emergency procedures. At least one copy of the fire emergency procedures shall be prominently posted on each floor area.

14.5.2 Operations Plan

A written and accessible fire operations plan consisting of the following elements is required to assist emergency responders in their incident management activities and to assure their safety. This cannot be done during the emergency and should include the following:

A. Fire operations plan drawings that provide
   - Site drawings
   - Access routes
   - Building locations
   - Floor plans
   - Ventilation capabilities and controls to allow for smoke and heat venting systems
   - Identification of major hazards
   - Locations of energized oil-filled equipment
   - Description of the nature and location of electrical hazards
   - Location of detection/fixed protection
   - Fire alarm system
   - Fire alarm panel location, type, instruction on the use of the panel, and any required passwords or keys
   - Drawings showing the location of all system devices
   - Site municipal water supply information and connections' locations
   - Location of fire hydrants, fire dept connections, fire pumps, water supply tanks
   - Location and types of standpipes
   - Sprinkler system types, coverage areas, and valve locations
   - Location and types of fire extinguishers
   - Location and types of firefighting foam systems and foam supplies
   - Location and types of gaseous systems
   - Electrical single line drawings
   - Electrical isolation plans/switching orders for all major electrical equipment and systems in battery systems

- Location, type, and installed controls for the elevators
- Location and types of fuel supplies (i.e., propane and natural gas)
- Inventory, UN number, and location of all hazardous materials

B. The fire operations plan should include the following:
- Pre-established lines of communication.
- Identify who is going to be the emergency response organization (private or public) and who will be the incident commander. They will have overall authority for the management of the incident.
- Site owner/person in charge (PIC).
- Preestablish protocols so that authority is with most senior, knowledgeable, individual at all times. This person acts as liaison with the incident commander in charge of the emergency response.
- Develop written organization charts that identify positions and reporting relationships.
- Conduct periodic site walk downs/drills to familiarize emergency responders and owner representatives.
- Predetermine role of site owner.
- Participant role in the isolation of affected equipment/electrical circuits (local or remote).
- Team members' responsibilities in assisting emergency responders in size up of incident.
- Identification of SCBA qualified individuals.

14.5.3 Fire Training

All new employees, contractors, and visitors should be trained in the facility evacuation procedures, station fire hazards, and fire protection provisions. All station electrical workers should annually review the station fire-related local operating orders and the station fire operations plan. Also, annually the local fire department should be given an orientation tour of all major substations.

14.5.4 Environmental Preparedness

Environmental exposure will involve the containment of oil runoff from the equipment involved. The runoff can involve burning oil from the fire as well as oil floating on top of the water that is used to suppress the fire. This will need to be contained and prevented from running into nearby streams, rivers, lakes, etc. The following should be considered in developing spill mitigation plans:

- Slope of site
- Drainage
- Cable trenches and control building
- Oil containment
- Value of station stone

14.5.5 Fire Recovery

The primary purpose of a post-fire recovery plan is to put in place procedures to promote the continued safety of emergency responders as well as owner personnel responsible for damage assessment and equipment restoration. The secondary purpose is to expedite the restoration of service to customers and to manage the potential ongoing environmental exposure of oily/contaminated water and fire suppression agents. Issues for consideration include the following:

- Air quality (particularly in buildings) that may be contaminated with PCBs and asbestos.
- Qualified personnel need to conduct air quality assessments; public fire departments do not normally provide this service.

- If air quality is not acceptable, and restoration has to be expedited, owner personnel will be required to wear self-contained breathing apparatus or respirators depending on what contaminant is present.
- Fire protection systems that have operated need to be restored to service as well as the recharging of fire extinguishers.
- A detailed damage assessment needs to be conducted to establish what needs to be isolated prior to restoration of service.
- A root cause analysis to minimize the potential for reoccurrence.

14.6 Conclusion

The assessment of the hazards involved with an existing or planned substation and the selection of the most appropriate fire protection are the best ways to ensure that the power supply to customers, company revenue, and assets are protected from fire. Substation, switchyard, and control building fire protection review checklists are enclosed in Appendices 14.A and 14.B to aid in the assessment process. The IEEE 979 "Guide for Substation Fire Protection" provides an excellent guide to the assessment process.

14.A Appendix A: Control Building Fire Protection Assessment Checklist

14.A.1 Risk Assessment

- Review the criticality of the control room and building fire loss to the substation operation and asset base.
- Review the historical frequency of fire in control buildings.

14.A.2 Life Safety Assessment

- Review the control room layout to ensure that the room has a minimum of two outward swinging exit doors.
- Ensure that the travel distance from any area within the control building to an exit does not exceed 30 m.
- Ensure that exit signs are installed at each exit door.
- Review that emergency lighting is provided that will provide a minimum lighting level of 10 lx at the floor, along the exit paths.
- Review the size and number of stories of the building to ensure that proper exits are provided and also ensure that maximum travel distances to the exits do not exceed 30 m.
- Determine if there are any building or fire code requirements for the installation of a fire detection system.

14.A.3 Fire Protection Assessment

- Review the availability of a fire department response to the site.
- Review the availability of firefighting water supply at or adjacent to the site.
- Review the adequacy of any existing control building fire protection.
- Review criticality of control building equipment, hazards involved, and response time of station personnel and the fire department.

Substation Fire Protection

- Determine the type of detection that will provide an acceptable very early detection (air sampling detection) to detect a fire at a very early stage (small electronic component failure—arcing) or at an early stage with smoke detection (photoelectric detection) to detect a fire at a smoldering or small flame stage.
- Determine the type of fire suppression system that will provide an acceptable equipment losses and outages (i.e., gaseous suppression systems to suppress a fire at an early stage [component loss] or sprinkler protection to suppress a fire at the stage where the loss would be restricted to a single control cabinet).
- Review the occupied hours of the building and ability of site personnel to safely extinguish a fire with portable fire equipment. Determine the levels of portable fire equipment required by the local fire code and that is suitable for safe staff operation.

14.A.4 Hazard Assessment

- Review the other uses (shops, offices, storage, etc.) within the control building and their exposure to the critical substation equipment.
- Review the use of combustible construction in the control building (i.e., exterior surfaces and roofs).
- Review the use of combustible interior surface finishes in the control room and ensure that the surface finishes have a flame spread rating of less than 25.
- Review the combustibility of any exposed cable used in the building to ensure that it meets the requirements of IEEE 383 [18].
- Review the control room separation walls to other occupancies to ensure that the walls have a fire resistance rating of a minimum of 1 h.

14.B Appendix B: Switchyard Fire Protection Assessment Checklist

Determine the initial electric equipment layout and equipment types.

14.B.1 Risk Assessment

1. Review the criticality of the various pieces of equipment.
2. Review types of insulating fluid used and their flammability.
3. Review the historical frequency of fire for the various types of equipment.
4. Review the availability of a fire department response to the site.
5. Review the availability of a firefighting water supply at or adjacent to the site.
6. Review the adequacy of any existing substation fire protection.

14.B.2 Radiant Exposure Assessment

1. Review the spacing between individual single-phase transformers and breakers with IEEE 979.
2. Review the spacing between large three-phase transformers, banks of single-phase transformers, or groups of breakers with IEEE 979.
3. Review the spacing of oil-filled equipment with respect to substation buildings with IEEE 979. Note: the presence of combustible surfaces and unprotected windows on exposed surfaces of the buildings may require detailed thermal radiation calculations or the application of safety factors

to the table distances. The Society of Fire Protection Engineers "Engineering Guide for Assessing Flame Radiation to External Targets from Pool Fires [19]" can be used as a reference for detailed thermal radiation calculations.
4. Review the distances between oil-filled equipment and the property line. Note: combustible vegetation and building structures beyond the property line of the substation may be exposed to high enough heat fluxes to ignite combustible surfaces. Detailed thermal radiation calculations should be considered.
5. Review the use of the various methods of fire protection discussed in IEEE 979 that will address the hazard determined in the radiant exposure assessment such as changing the type of equipment and insulating fluid used, increased spacing, provision of gravel ground cover, oil containment, fire barriers, and automatic water deluge fire protection.

14.B.3 Fire Spread Assessment

1. Is the surface around oil-filled equipment pervious (gravel) or impervious? Use of 31 cm thick gravel ground covers will suppress the flames from a burning oil spill fire. Impervious surfaces can allow the burning oil to form a large pool fire, which will increase the heat flux to adjacent equipment and structures.
2. Is there any oil containment in place around the oil-filled equipment? Oil containment can contain pool fires and prevent their spread.
3. Does the grade surrounding the oil-filled equipment slope toward the equipment or away from the oil-filled equipment toward adjacent oil-filled equipment, cable trenches, drainage facilities, or buildings? The burning oil released from ruptured oil-filled equipment can spread for significant distances if the ground surrounding the equipment has a slope greater than 1%.
4. Review the use of the various methods of fire protection discussed in IEEE 979 that will address the hazard determined in the fire spread assessment. These methods include the following:
    a. Changing the type of equipment and insulating fluid used.
    b. Increasing the spacing, use of gravel ground cover.
    c. Provision of oil containment.
    d. Changing the grade surrounding the equipment.
    e. Use of liquid-tight noncombustible cable trench cover adjacent to oil-filled equipment.
    f. Fire stopping of cable trench entries into control buildings.
    g. Use of automatic water deluge fire protection.

References

1. IEEE (Institute of Electrical and Electronics Engineers) Std. 979, *IEEE Guide for Substation Fire Protection*.
2. NFPA Std. 600, *Standard on Industrial Fire Brigades*.
3. JEAG 5002-2001, *Japan Electric Association Electrical Technology Guides—Substation Fire Protection Guideline*.
4. National Fire Protection Association (NFPA), Codes and standards. Available at www.nfpa.org
5. CIGRE, International Council on Large Electric Systems publications. Available at www.cigre.org
6. IEC (International Electrotechnical Commission) Std. 61936, *Power Installations Exceeding 1 kV AC*. Available at www.iec.ch
7. Health & Safety Executive—UK, publication R2P2, Reducing risks, protecting people. Available at the HSE Website www.hse.gov.uk
8. FM global property loss prevention data sheets. Available at www.fmglobal.com

9. NFPA Std. 850, *Recommended Practice for Fire Protection for Electric Generating Plants and High Voltage Direct Current Converter Stations*, 2010.
10. CIGRE Std TF 14.01.04, *Report on Fire Aspects of HVDC Valves and Valve Halls*.
11. Edison Electric Institute, *Suggested Guidelines fire Completing a Fire Hazard Analysis for Electric Utility Facilities* (existing or in design), 1981.
12. NFPA Std. 72, *National Fire Alarm and Signalling Code*, 2010.
13. NFPA Std. 101, *Life Safety Code*, 2012.
14. NFPA Std. 14, *Installation of Standpipe and Hose Systems*.
15. NUREG 1805, *Fire Dynamics Tools Quantitative Fire Hazard Analysis Methods for the U.S. Nuclear Regulatory Commission Fire Protection Inspection Program*. US Nuclear Regulatory Commission Final Report. Available at www.nrc.gov/reading-rm/doc-collections/nuregs/staff/sr1805/final-report
16. NFPA, *One Stop Data Shop Reports*. Available at www.nfpa.org
17. EPRI TR-100370, *Fire-Induced Vulnerability Evaluation (FIVE) Report*. Electric Power Research Institute, Palo Alto, CA.
18. IEEE (Institute of Electrical and Electronics Engineers) Std. 383, *Qualification of Electric Cables and Splices for Nuclear Plants*.
19. SFPE (Society of Fire Protection Engineers), *Engineering Guide for Assessing Flame Radiation to External Targets from Pool Fires*. Available at www.sfpe.org

15
Substation Communications

15.1	Introduction	15-2
15.2	Supervisory Control and Data Acquisition: Historical Perspective	15-2
15.3	SCADA Functional Requirements	15-4
15.4	SCADA Communication Requirements	15-5
15.5	Relay Communication Requirements	15-5
	Communications and the Smart Grid • Time Distribution • Phasor Measurements • Quality of Service	
15.6	Components of a SCADA System	15-8
15.7	Structure of a SCADA Communication Protocol	15-9
15.8	SCADA Communication Protocols: Past, Present, and Future	15-11
	Distributed Network Protocol • IEC 60870-5 • UCA 1.0 • ICCP • UCA 2.0 • IEC 61850 • Transmission Control Protocol/Internet Protocol • Continuing Work	
15.9	Security for Substation Communications	15-15
	SCADA Security Attacks • Security by Obscurity • SCADA Message Data Integrity Checking • Encryption • Denial of Service	
15.10	Electromagnetic Environment	15-17
15.11	Communication Media	15-18
	Advanced Radio Data Information Service • Cellular Telephone Data Services • Digital Microwave • Fiber Optics • Hybrid Fiber/Coax • Multiple Address Radio • Mobile Computing Infrastructure • Mobile Radio • Mobitex Packet Radio • Paging Systems • Power Line Carrier • Satellite Systems • Short Message System • Spread Spectrum Radio and Wireless LANs	
15.12	Telephone-Based Systems	15-24
	Telephone Lines: Leased and Dial-Up • Integrated Services Digital Network • Digital Subscriber Loop • T1 and Fractional T1 • Frame Relay • Asynchronous Transfer Mode • SONET • Multiprotocol Label Switching	
15.13	For More Information	15-28
	Useful Websites • Power System Relay Communication References • Relevant Standards	

Daniel E. Nordell
Xcel Energy

15.1 Introduction

Modern electric power systems have been dubbed "the largest machine made by mankind" because they are both physically large—literally thousands of miles in dimension—and operate in precise synchronism. In North America, for example, the entire West Coast, everything east of the Rocky Mountains, and the state of Texas operate as three autonomous interconnected "machines." The task of keeping such a large machine functioning without breaking itself apart is not trivial. The fact that power systems work as reliably as they do is a tribute to the level of sophistication that is built into them. Substation communication plays a vital role in power system operation. This chapter provides a brief historical overview of substation communication, followed by sections that

- Review functional and communication requirements
- Examine the components of both traditional and emerging supervisory control and data acquisition (SCADA) systems
- Review the characteristics of past, present, and future substation communication protocols
- Review the role of standards for substation communication
- Discuss the electromagnetic environment which substation communication devices must withstand
- Discuss security aspects of substation communications
- Discuss communication media options for substation communications

15.2 Supervisory Control and Data Acquisition: Historical Perspective

Electric power systems as we know them began developing in the early twentieth century. Initially, generating plants were associated only with local loads that typically consisted of lighting and electric transportation. If anything in the system failed—generating plant, power lines, or connections—the lights would quite literally be "out." Customers had not yet learned to depend on electricity being nearly 100% reliable, so outages, whether routine or emergency, were taken as a matter of course.

As reliance on electric power grew, so did the need to find ways to improve reliability. Generating stations and power lines were interconnected to provide redundancy and higher voltages were used for longer distance transportation of electricity. Points where power lines came together or where voltages were transformed came to be known as "substations." Substations often employed protective devices to allow system failures to be isolated so that faults would not bring down the entire system and operating personnel were often stationed at these important points in the electrical system so that they could monitor and quickly respond to any problems that might arise. They would communicate with central system dispatchers by any means available—often by telephone—to keep them apprised of the condition of the system. Such "manned" substations were normative throughout the first half of the twentieth century.

As the demands for reliable electric power became greater and as labor became a more significant part of the cost of providing electric power, technologies known as "supervisory control and data acquisition," or SCADA for short, were developed that would allow remote monitoring and even control of key system parameters. SCADA systems began to reduce and even eliminate the need for personnel to be on hand at substations.

Early SCADA systems provided remote indication and control of substation parameters using technology borrowed from automatic telephone switching systems. As early as 1932, Automatic Electric was advertising "remote-control" products based on its successful line of "Strowger" telephone switching apparatus (see Figure 15.1). Another example (used as late as the 1960s) was an early Westinghouse REDAC system that used telephone-type electromechanical relay equipment at both ends of a conventional "twisted-pair" telephone circuit. Data rates on these early systems were slow—data were

FIGURE 15.1 October 31, 1932 electrical world advertisement.

sent in the same manner as rotary-dial telephone commands—10 bit/s—so only a limited amount of information could be passed using this technology.

Early SCADA systems were built on the notion of replicating remote controls, lamps, and analog indications at the functional equivalent of pushbuttons, often placed on a mapboard for easy operator interface. The SCADA masters simply replicated, point-for-point, control circuits connected to the remote, or slave, unit.

During the same time frame as SCADA systems were developing, a second technology—remote teleprinting, or "Teletype"—was coming of age, and by the 1960s had gone through several generations of development. The invention of a second device—the "modem" (MOdulator/DEModulator)—allowed digital information to be sent over wire pairs that had been engineered to only carry the electronic equivalent of human voice communication. With the introduction of digital electronics, it was possible to use faster data streams to provide remote indication and control of system parameters. This marriage of Teletype technology with digital electronics gave birth to remote terminal units (RTUs), which were typically built with discrete solid-state electronics and which could provide remote indication and control of both discrete events and analog voltage and current quantities.

Beginning also in the late 1960s and early 1970s, technology leaders began exploring the use of small computers (minicomputers at that time) in substations to provide advanced functional and communication capability. But early application of computers in electric substations met with industry resistance because of perceived and real reliability issues.

The introduction of the microprocessor with the Intel 4004 in 1971 (see http://www.intel4004.com for a fascinating history) opened the door for increasing sophistication in RTU design that is still continuing today. Traditional point-oriented RTUs that reported discrete events and analog quantities could be built in a fraction of the physical size required by previous discrete designs. More intelligence could be introduced into the device to increase its functionality. For the first time RTUs could be built, which reported quantities in engineering units rather than as raw binary values. One early design developed at Northern States Power Company in 1972 used the Intel 4004 as the basis for a standardized environmental data acquisition and retrieval (SEDAR) system, which collected, logged, and reported environmental information in engineering units using only 4 kB of program memory and 512 nibbles (half-bytes) of data memory.

While the microprocessor offered the potential for greatly increased functionality at lower cost, the industry also demanded very high reliability and long service life measured in decades that were difficult to achieve with early devices. Thus, the industry was slow to accept the use of microprocessor technology in mission-critical applications. By the late 1970s and early 1980s, integrated microprocessor-based devices were introduced, which came to be known as "intelligent electronic devices" or IEDs.

Early IEDs simply replicated the functionality of their predecessors—remotely reporting and controlling contact closures and analog quantities using proprietary communication protocols. Increasingly, IEDs are being used also to convert data into engineering unit values in the field and to participate in field-based local control algorithms. Many IEDs are being built with programmable logic controller (PLC) capability and, indeed, PLCs are being used as RTUs and IEDs to the point that the distinction between these different types of smart field devices is rapidly blurring.

Early SCADA communication protocols were usually proprietary in nature and were also often kept secret from the industry. A trend beginning in the mid-1980s has been to minimize the number of proprietary communication practices and to drive field practices toward open, standards-based specifications. Two noteworthy pieces of work in this respect are the International Electrotechnical Commission (IEC) 60870-5 family of standards and the IEC 61850 standard. The IEC 60870-5 work represents the pinnacle of the traditional point-list-oriented SCADA protocols, while the IEC 61850 standard is the first of an emerging approach to networkable, object-oriented SCADA protocols based on work started in the mid-1980s by the Electric Power Research Institute, which became known as the utility communication architecture (UCA).

15.3 SCADA Functional Requirements

Design of any system should always be preceded by a formal determination of the business and corresponding technical requirements that drive the design. Such a formal statement is known as a "functional requirements specification." Functional requirements capture the intended behavior of the system. This behavior may be expressed as services, tasks, or functions the system is required to perform.

In the case of SCADA, it will contain such information as system status points to be monitored, desired control points, and analog quantities to be monitored. It will also include identification of acceptable delays between when an event happens and when it is reported, required precision for analog quantities, and acceptable reliability levels. The functional requirements analysis will also include a determination of the number of remote points to be monitored and controlled. It should also include identification of communication stakeholders other than the control center, such as maintenance engineers and system planners who may need communication with the substation for reasons other than real-time operating functionality.

Substation Communications

The functional requirements analysis should also include a formal recognition of the physical, electrical, communications, and security environment in which the communication is expected to operate. Considerations here include recognizing the possible (likely) existence of electromagnetic interference from nearby power systems, identifying available communication facilities, identifying functionally the locations between which communication is expected to take place, and identifying communication security threats that might be presented to the system.

It is sometimes difficult to identify all of the items to be included in the functional requirements, and a technique that has been found useful in the industry is to construct a number of example "use cases," which detail particular individual sets of requirements. Aggregate use cases can form a basis for a more formal collection of requirements.

15.4 SCADA Communication Requirements

After the functional requirements have been articulated, the corresponding architectural design for the communication system can be set forth. Communication requirements include those elements that must be included in order to meet the functional requirements.

Some elements of the communication requirements include

- Identification of communication traffic flows—source/destination/quantity
- Overall system topology—for example, star, mesh
- Identification of end system locations
- Device/processor capabilities
- Communication session/dialog characteristics
- Device addressing schemes
- Communication network traffic characteristics
- Performance requirements
- Timing issues
- Reliability/backup/failover
- Application service requirements
- Application data formats
- Operational requirements (directory, security, and management of the network)
- Quantification of electromagnetic interference withstand requirements

15.5 Relay Communication Requirements

Relay systems perform vital functions to isolate local failures in generation, transmission, and distribution systems so that they will not spread to other parts of an interconnected power system.

Communication systems are a vital component of wide area power system relaying. They provide the information links needed for the relay and control systems to operate. Because of potential loss of communication, relay systems must be designed to detect and tolerate failures in the communication system and must be independent of the communication system and not subject to the same failure modes.

The communication system must be designed for fast, robust, and reliable operation. Factors that can influence the performance of the relay system include communication system bandwidth, delay, latency, jitter, reliability, and error handling. Relay systems frequently require communication delays not to exceed a few milliseconds and often cannot tolerate timing jitter.

The following paragraphs discuss several data communication systems, which can be used for either relay or SCADA communications. It should be noted that frequently relay applications have more stringent requirements for speed, latency, and jitter than do SCADA applications.

Electric utilities use a combination of analog and digital communication systems for their operations consisting of power line carrier (PLC), radio, microwave, leased phone lines, satellite systems, and fiber optics. Each of these systems has characteristics that make them well suited to particular applications. The advantages and disadvantages of each are briefly summarized in the following:

- Transmission PLC is usually an economic choice for relay communication but has limited distance of coverage and low bandwidth. It is best suited to station-to-station protection as well as communications to small stations that are difficult or costly to access otherwise.
- Company-owned microwave is cost-effective and reliable but is maintenance intensive. Microwave is useful for general communications for all types of applications.
- Radio systems provide narrower bandwidths but are nonetheless useful for mobile applications or communication to locations difficult to reach otherwise.
- Satellite systems likewise are effective for reaching difficult-to-access locations but are not good where the long delay is a problem. They also tend to be expensive.
- Leased phone lines are very effective where a solid link is needed to a site served by standard telephone service. They tend to be expensive in the long term, so are usually not the best solution where many channels are required.
- Fiber-optic systems are a new option. They are expensive to install and provision but are expected to be very cost-effective. They have the advantage of using existing rights-of-way and delivering communications directly between points of use. In addition, they have the very high bandwidth needed for modern data communications.
- Spread spectrum radio is a new option, which can provide affordable solutions using unlicensed services. Advances in this field are appearing rapidly and they should be examined closely to determine their usability to satisfy relaying requirements.

The increasing speed of wide area networks combined with the availability of high-precision satellite time references is increasingly making it possible to share communication facilities between corporate information, SCADA, and relaying applications. The IEC 61850 standard (based on the UCA) provides specifications for communication facilities using wide area network techniques, which will not only serve SCADA needs but also provide functionality for relay systems.

15.5.1 Communications and the Smart Grid

While the electricity grid has been increasingly "Smart" from its beginnings more than 100 years ago, the "Energy Independence and Security Act of 2007" put the term "Smart Grid" into the public vocabulary. Title XIII Section 1301 of the Act calls for "... modernization of the electric T&D system to maintain a reliable and secure infrastructure that can meet future demand growth and to achieve each of the following, which together characterize a Smart Grid":

1. Increased use of digital technology
2. Dynamic optimization of grid operations and resources, with full cyber security
3. Deployment and integration of distributed resources and generation
4. Development and incorporation of demand response
5. Deployment of "smart" technologies (real-time, automated, interactive technologies that optimize the physical operation of appliances, and consumer devices) for metering, communications concerning grid operations and status, and distribution automation (DA)
6. Integration of "smart" appliances and consumer devices
7. Deployment and integration of advanced electricity storage and peak-shaving technologies, including plug-in electric and hybrid electric vehicles, and thermal-storage air-conditioning
8. Provision to consumers of timely information and control options
9. Development of standards for communication and interoperability of appliances and equipment connected to the electric grid, including the infrastructure serving the grid
10. Identification and lowering of unreasonable or unnecessary barriers to adoption of smart grid technologies, practices, and services

Section 1305 of the Act calls for the National Institute for Standards and Technology (NIST) to be given primary responsibility to develop a framework including protocols and model standards for the Smart Grid. The traditional NIST role is to maintain measurement standards (time, length, temperature, etc.)—not commercial standards but, with initial work beginning in 2008 and in recognition that there exist nearly 100 Standards-Development Organizations (SDOs) working on commercial standards, NIST chose to establish the Smart Grid Interoperability Panel (SGIP) to cooperatively develop the mandated framework (reference URL: www.nist.gov/smartgrid). The first two significant documents issued by NIST with work being done by the SGIP include the "Framework and Roadmap for Smart Grid Interoperability Standards, Release 1.0" and the companion document "NISTIR 7628, Smart Grid Cyber Security Strategy and Requirements." Release 2.0 of the *Framework* document has been prepared and will soon be released for public review. These documents, together with the related Priority Action Plans, which are being directed to the several SDOs, and the *Catalog of Standards*, which will list recommended standards for use in implementing the mandated Smart Grid, promise to have a significant impact on future utility communication practices.

15.5.2 Time Distribution

A number of Smart Grid functions require accurate knowledge of time. Depending on the application, accuracy requirements range from a few seconds to a few microseconds. A number of time synchronization techniques have been developed to meet those needs. They include the Internet Network Time Protocol (www.ntp.org), the use of IRIG for substation synchronization (www.irigb.com), and IEEE 1588 "Standard for a Precision Clock Synchronization Protocol for Networked Measurement and Control Systems" (www.nist.gov/el/isd/ieee/ieee1588.cfm).

15.5.3 Phasor Measurements

An important function in power system management is knowledge of instantaneous voltage and current phase angles. Synchronous measurement of these parameters is known as "synchrophasor measurements" and can only be achieved if disparate locations within the power system have a precise knowledge of absolute time so that the measurements can be appropriately time-tagged. Traditional time distribution has been achieved using the IRIG time protocols, but more recently standards have been developed that make use of geographic position satellite (GPS) time, which can be distributed using IEEE 1588 "Standard for a Precision Clock Synchronization Protocol for Networked Measurement and Control Systems." The time-tagged phasor data can be transmitted using specifications found in C37.118 "IEEE Standard for Synchrophasors for Power Systems" and in IEC 61850. A useful discussion can be found at http://www.nist.gov/el/isd/ieee/ieee1588.cfm

15.5.4 Quality of Service

Unless otherwise managed, messages that share a common communication network will compete for service based on "best effort" by the network. In fact, some messages (e.g., power system relay messages) are deemed to be more time-critical than are others (e.g., business and accounting information). These have been traditionally separated by the use of physically separate communication infrastructures but they might share the same wire if suitable priority mechanisms are employed.

The mechanism by which priority can be given to different messages—either configured by network address or on an individual message basis—is called "quality of service" (QoS) (reference: "en.wikipedia.org/wiki/Quality_of_service"). With the coming adoption of IPv6 (see separate discussion) utilities will have a mechanism by which individual messages can be tagged and treated with appropriate QoS.

15.6 Components of a SCADA System

Traditional SCADA systems grew up with the notion of a SCADA "master" and a SCADA "slave" or "remote." The implicit topology was that of a "star" or "spoke and hub," with the master in charge. In the historical context, the master was a hardwired device with the functional equivalent of indicator lamps and pushbuttons (see Figure 15.2).

Modern SCADA systems employ a computerized SCADA master in which the remote information is either displayed on an operator's computer terminal or made available to a larger "energy management system" through networked connections. The substation RTU is either hardwired to digital, analog, and control points or frequently acts as a "submaster" or "data concentrator" in which connections to intelligent devices inside the substation are made using communication links. Most interfaces in these systems are proprietary, although in recent years standards-based communication protocols to the RTUs have become popular. In these systems if other stakeholders such as engineers or system planners need access to the substation for configuration or diagnostic information, separate, often ad hoc, provision is usually made using technologies such as dial-up telephone circuits.

With the introduction of networkable communication protocols, typified by the IEC 61850 series of standards, it is now possible to simultaneously support communication with multiple clients located at multiple remote locations. Figure 15.3 shows how such a network might look. This configuration will support clients located at multiple sites simultaneously accessing substation devices for applications as diverse as SCADA, device administration, system fault analysis, metering, and system load studies.

SCADA systems as traditionally conceived report only real-time information, but interfaces built according to standards IEC 61968 and IEC 61970 will allow integration of both control center and enterprise information systems as shown in Figure 15.3. A feature that may be included in a modern SCADA system is that of a historian, which time-tags each change of state of selected status parameters or each change (beyond a chosen deadband) of analog parameters and then stores this information in an efficient data store, which can be used to rebuild the system state at any selected time for system performance analyses.

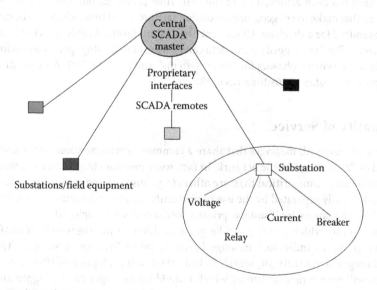

FIGURE 15.2 Traditional SCADA system topology.

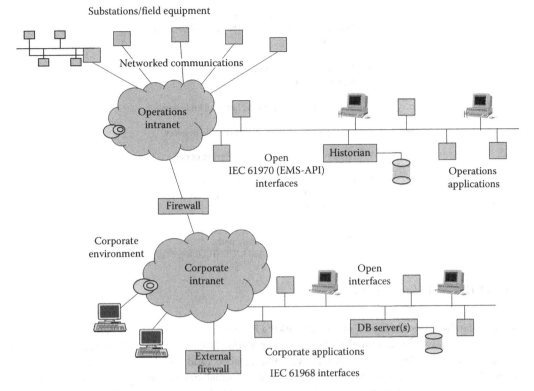

FIGURE 15.3 Networked SCADA communications.

15.7 Structure of a SCADA Communication Protocol

The fundamental task of a SCADA communication protocol is to transport a "payload" of information (both digital and analog) from the substation to the control center and to allow remote control in the substation of selected operating parameters from the control center. Other functions that are required but usually not included in traditional SCADA protocols include the ability to access and download detailed event files and oscillography and the ability to remotely access substation devices for administrative purposes. These functions are often provided using ancillary dial-up telephone-based communication channels. Newer, networkable, communication practices such as IEC 61850 make provision for all of the aforementioned functionality and more using a single wide area network connection to the substation.

From a communication perspective, all communication protocols have at their core a payload of information that is to be transported. That payload is then wrapped in either a simple addressing and error detection envelope and sent over a communication channel (traditional protocols) or is wrapped in additional layers of application layer and networking protocols, which allow transport over wide area networks (routable object-oriented protocols like IEC 61850).

In order to help bring clarity to the several parts of protocol functionality, in 1984 the International Organization for Standardization (ISO) issued Standard ISO/IEC 7498 entitled "Reference Model of Open Systems Interconnection" or, simply, the "OSI Reference Model." The model was updated with a 1994 issue date, with the current reference being "ISO/IEC 7498-1:1994," and available from "http://www.iso.org."

The OSI reference model breaks the communication task into seven logical pieces as shown in Figure 15.4. All communication links have a data source (application layer 7 information) and a physical path (layer 1). Most links also have a data link layer (layer 2) to provide message integrity protection.

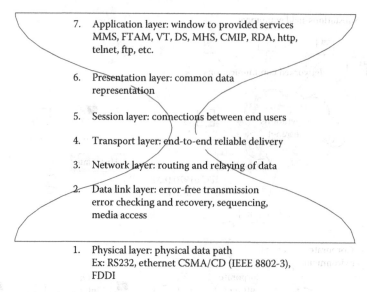

FIGURE 15.4 OSI reference model.

Security can be applied at layer 1 or 2 if networking is not required but must be applied at or above the network layer (layer 3) and is often applied at the application layer (layer 7) to allow packets to be routed through a network. More sophisticated, networkable, protocols add one or more of layers 3–6 to provide networking, session management, and sometimes data format conversion services. Note that the OSI reference model is not, in and of itself, a communication standard. It is just a useful model showing the functionality that might be included in a coordinated set of communication standards.

Also note that Figure 15.3 as drawn shows a superimposed "hourglass." The hourglass represents the fact that it is possible to transport the same information over multiple physical layers—radio, fiber, twisted pair, etc.—and that it is possible to use a multiplicity of application layers for different functions. In the middle—the networking—layers, interoperability over a common network can be achieved if all applications agree on common networking protocols. For example, the growing common use of the Internet protocols TCP/IP represents a worldwide agreement to use common networking practices (common middle layers) to route messages of multiple types (application layer) over multiple physical media (physical layer—twisted pair, Ethernet, fiber, radio) in order to achieve interoperability over a common network (the Internet).

Figure 15.5 shows how device information is encapsulated (starting at the top of the diagram) in each of the lower layers in order to finally form the data packet at the data link layer, which is sent over the physical medium. The encapsulating packet—the header and trailer and each layer's payload—provides the added functionality at each level of the model, including routing information and message integrity protection. Typically, the overhead requirements added by these wrappers are small compared with the size of the device information being transported. Figure 15.6 shows how a message can travel through multiple intermediate systems when networking protocols are used.

Traditional SCADA protocols, including all of the proprietary legacy protocols, distributed network protocol (DNP), and IEC 60870-5-101, use layers 1, 2, and 7 of the reference model in order to minimize overheads imposed by the intermediate layers. IEC 60870-5-104 and recent work being done with DNP add networking and transport information (layers 3 and 4) so that these protocols can be routed over a wide area network. IEC 61850 is built using a "profile" of other standards at each of the reference model layers so that it is applicable to a variety of physical media (lower layers), is routable (middle layers), and provides mature application layer services based on ISO 9506—the manufacturing message specification (MMS).

Layered protocols encapsulate each successive layer

Device information	Device data/model	Device information
Application	Application protocol data unit:	(hdrDev Infotlr)
Presentation	Presentation packet:	(hdrPres(hdrDev Infotlr)tlr)
Session	Session packet:	(hdrSess(hdrPres(hdrDev Infotlr)tlr)tlr)
Transport	Transport packet:	hdrTrns(hdrSess(hdrPres(hdrDev Infotlr)tlr)tlr)tlr
Network	Network packet:	(hdrNtwk(hdrTrns(hdrSess(hdrPres(hdrDev Infotlr)tlr)tlr)tlr)tlr
Data link	Data link packet:	(hdrDlnk(hdrNtwk(hdrTrns(hdrSess(hdrPres(hdrDev Infotlr)tlr)tlr)tlr)tlr)tlr
Physical	Electrical signals:	(hdrDlnk(hdrNtwk(hdrTrns(hdrSess(hdrPres(hdrDev Infotlr)tlr)tlr)tlr)tlr)tlr

FIGURE 15.5 Layered message structure.

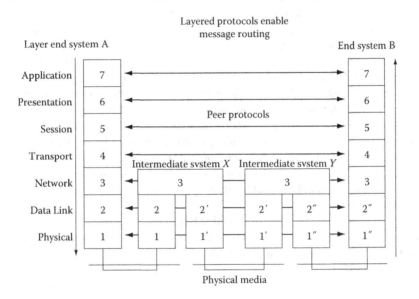

FIGURE 15.6 End-to-end messaging in OSI model.

15.8 SCADA Communication Protocols: Past, Present, and Future

As noted in the section on SCADA history, early SCADA protocols were built on electromechanical telephone switching technology. Signaling was usually done using pulsed direct-current signals at a data rate on the order of 10 pulses/s. Analog information could be sent using "current loops," which are able to communicate over large distances (thousands of feet) without loss of signal quality. Control and status points were indexed using assigned positions in the pulse train. Analog information was sent using current loops, which could provide constant current independent of circuit impedance. Communication security was assured by means of repetition of commands or mechanisms such as "arm" and "execute" for control.

With the advent of digital communications (still precomputer), higher data rates were possible. Analog values could be sent in digital form using analog-to-digital converters, and errors could be detected using parity bits and block checksums. Control and status points were assigned positions in the data blocks that needed to be synchronized between the remote and master devices. Changes of status were detected by means of repetitive "scans" of remote devices, with the "scan rate" being a critical system design factor. Communication integrity was assured by the use of more sophisticated block ciphers including the "cyclical redundancy check," which could detect both single- and multiple-bit errors in communications. Control integrity was ensured by the use of end-to-end "select-check-operate" procedures. Each manufacturer (and sometimes user) of these early SCADA systems would typically define their own communication protocol and the industry became known for the large number of competing practices.

Computer-based SCADA master stations, followed by microprocessor-based RTUs, continued the traditions set by the early systems of using points-list-based representations of control and status information. Newer, still-proprietary, communication protocols became increasingly sophisticated in the types of control and status information that could be passed. The notion of "report by exception" was introduced in which a remote terminal could report "no change" in response to a master station poll, thus conserving communication resources and reducing average poll times.

By the early 1980s, the electric utility industry enjoyed the marketplace confusion brought by on the order of 100 competing proprietary SCADA protocols and their variants. With the rising understanding of the value of building on open practices, a number of groups began to approach the task of bringing standard practices to bear on utility SCADA practices.

As shown in Figure 15.7, a number of different groups are often involved in the process of reaching consensus on standard practices. The process reads from the bottom to the top, with the "International Standards" level the most sought-after and also often the most difficult to achieve. Often the process starts with practices which have been found useful in the marketplace but which are, at least initially, defined and controlled by a particular vendor or, sometimes, end user. The list of vendor-specific SCADA protocols is long and usually references particular vendors. One such list (from a vendor's list of supported protocols) reads like a "who's who" of SCADA protocols and includes Conitel, CDC Type I and Type II, Harris 5000, Modicon MODBus, PG&E 2179, PMS-91, QUICS IV, SES-92, TeleGyr 8979, PSE Quad 4 Meter, Cooper 2179, JEM 1, Quantum Qdip, Schweitzer Relay Protocol (221, 251, 351), and Transdata Mark V Meter.

Groups at the Institute of Electrical and Electronics Engineers (IEEE), the International Electrotechnical Commission (IEC), and the Electric Power Research Institute (EPRI) all started in the mid-1980s to look at the problem of the proliferation of SCADA protocols. IEC Technical Committee 57 (IEC TC57) working group 3 (WG3) began work on its "60870" series of telecontrol standards. Groups within the IEEE substations and relay committees began examining the need for consensus for SCADA

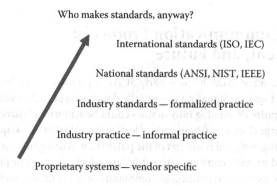

FIGURE 15.7 The standard process.

protocols, and EPRI began a project that became known as the "utility communication architecture" in an effort to specify an enterprise-wide, networkable, communication architecture, which would serve business applications, control centers, power plants, substations, distribution systems, transmission systems, and metering systems.

15.8.1 Distributed Network Protocol

With the IEC work partially completed, a North American manufacturer adapted the IEC 60870-5-3 and 60870-5-4 draft documents plus additional North American requirements to draft a new DNP, which was released to the DNP Users Group (www.dnp.org) in November 1993. DNP3 was subsequently selected as a recommended practice by the IEEE C.2 task force for an RTU to IED communication protocol (IEEE Std 1379-1997 IEEE Trial-Use Recommended Practice for Data Communications between Intelligent Electronic Devices and Remote Terminal Units in a Substation). DNP has enjoyed considerable success in the marketplace and represents the pinnacle of traditional points-list-oriented SCADA protocols.

15.8.2 IEC 60870-5

The IEC TC57 WG3 continued work on its telecontrol protocol and has issued several standards in the IEC 60870-5 series (www.iec.ch), which collectively define an international consensus standard for telecontrol. IEC 60870-5 has recently issued a new transport profile (104), which can be used over wide area networks. 60870-5 represents the best international consensus for traditional control center-to-substation telecommunication and, as noted earlier, is closely related to the North American DNP protocol.

15.8.3 UCA 1.0

The EPRI UCA project published its initial results in December 1991. The UCA 1.0 specification outlines a communication architecture based on existing international standards. It specifies the use of the MMS (MMS: ISO 9506) in the application layer for substation communications. The SCADA concepts outlined in UCA 1.0 have been codified in the IEC 61850 series of standards. Other parts of the UCA specification have been used in the development of other IEC standards discussed elsewhere in this chapter. The timeline for the UCA work from its inception through standardization as an IEC standard is shown in Figure 15.8.

1. 1986 (December): EPRI workshop
2. 1987 (December): Assessment
3. 1988 (December): Projects
4. 1991 (December): UCA documents published by EPRI
5. 1992 (May): MMS Forum begins
6. 1993: Demonstration projects started
7. 1994: ICCP released
8. UCA 2.0 demo projects include
    a. "AEP Initiative"—Substation LAN
    b. City public service distribution automation
9. 1997: UCA 2.0 completed
10. 1998: IEEE SCC36 formed
11. 1998: IEC TC57 61850 standards started
12. 1999: IEEE TR1550 published
13. 2005: IEC 61850 completed

FIGURE 15.8 UCA timeline.

15.8.4 ICCP

The UCA 1.0 work became the basis for IEC 60870-6-503 (2002–04), entitled "Telecontrol equipment and systems—Part 6-503: Telecontrol protocols compatible with ISO standards and ITU-T recommendations—TASE.2 Services and protocol." Also known as ICCP, this specification calls for the use of MMS and was designed to provide standardized communication services between control centers but has also been used to provide communication services between a control center and its associated substations.

15.8.5 UCA 2.0

Continuing work to develop the substation and IED communication portions of UCA was conducted in the MMS Forum (sponsored by Northern States Power Company) beginning in 1992. This work resulted in the issuance of a UCA 2.0 report, which was published as IEEE Technical Report 1550 (TR1550) (www.ieee.org) in December 1998. The concepts outlined in UCA 2.0 have been codified in the IEC 61850 series of standards.

15.8.6 IEC 61850

IEEE TR1550 became the basis for the new generation of IEC 61850 standards for communication between substation devices. The feature, which distinguishes UCA and its IEC 61850 successor from traditional SCADA protocols, is that they are networkable and are object oriented so that a device can describe its attributes when asked. This allows the possibility of self-discovery and "pick-list" configuration of SCADA systems rather than the labor-intensive and more error-prone points-list systems associated with earlier SCADA protocols.

Work is currently underway to extend the use of IEC 61850 between substations and for SCADA communication between substations and control centers.

15.8.7 Transmission Control Protocol/Internet Protocol

The transmission control protocol (TCP) and the Internet protocol (IP), commonly referred to as TCP/IP, are transport layer (TCP) and network layer (IP) protocols, which are the essential protocols for transporting messages in the Internet and are maintained by the Internet Engineering Task Force (IETF). The UCA and the IEC 61850 standards employ TCP/IP to transport messages for relaying and SCADA applications.

It was recognized a number of years ago that IP version 4, in use since 1981, with its 32 bit address would soon run out of available space. Continued development work has resulted in specification of IP version 6 (IPv6), which uses a 128 bit address space and also incorporates a number of other advanced features, including support security and quality of service (QoS), which is important for successful sharing of infrastructure between disparate applications.

15.8.8 Continuing Work

Work is continuing in IEC TC57 WG13 and WG14 to define object-oriented presentation of real-time operations information to the business enterprise environment using best networking practices. Working Group 13 is producing Standard IEC 61970: Energy management system application program interface (EMS-API). Working Group 14 is producing Standard IEC 61968: System interfaces for distribution management (SIDM).

TC57 Working Group 15 has developed security practices for the IEC protocols as Standard IEC 62351: Data and Communication Security. Section 15.9 discusses the fundamental principles of data security, which are being codified by WG15.

Section 15.13 contains a detailed listing of the several standards being produced by Working Groups 13, 14, and 15.

15.9 Security for Substation Communications

Until recently the term "security" when applied to SCADA communication systems meant only the process of ensuring message integrity in the face of electrical noise and other disturbances to the communications. But, in fact, security also has a much broader meaning, as discussed in depth in Chapters 16 and 17. Security, in the broader sense, is concerned with anything that threatens to interfere with the integrity of the business. Our focus here will be to examine issues related more narrowly to SCADA security.

In an earlier section, we discussed the role of the OSI Reference Model (ISO 7498-1) in defining a communication architecture. In similar fashion, ISO 7498-2, "Information Processing Systems—Open Systems Interconnection—Basic Reference Model—Part 2: Security Architecture," issued in 1989, provides a general description of security services and related mechanisms that fit into the reference model and defines the positions within the reference model where they may be provided. It also provides useful standard definitions for security terms.

ISO 7498-2 defines the following five categories of security service:

- *Authentication*: the corroboration that an entity is the one claimed
- *Access control*: the prevention of unauthorized use of a resource
- *Data confidentiality*: the property that information is not made available or disclosed to unauthorized individuals, entities, or processes
- *Data integrity*: the property that data have not been altered or destroyed in an unauthorized manner
- *Nonrepudiation*: data appended to, or a cryptographic transformation of, a data unit that allows a recipient of the data unit to prove the source and integrity of the unit and protect against forgery, for example, by the recipient

Note that ISO 7498-2 provides standard definitions and an architecture for security services but leaves it to other standards to define the details of such services. It also provides recommendations on where the requisite security services should fit in the seven-layer reference model to achieve successful secure interoperability between open systems.

Security functions can generally be provided alternately at more than one layer of the OSI model. Communication channels, which are strictly point to point and for which no externally visible device addresses need to be observable, can employ encryption and other security techniques at the physical and data link layers. If the packets need to be routable, either messages need to be encrypted at or above the network layer (the OSI recommendation) or the security wrapper needs to be applied and removed at each node of the interconnected network. This is a bad idea because of the resultant complexities of security key management and the resultant probability of security leaks.

15.9.1 SCADA Security Attacks

A number of types of security challenges to which SCADA systems may be vulnerable are recognized in the industry. The list includes the following:

- *Authorization violation*: An authorized user performing functions beyond his level of authority.
- *Eavesdropping*: Gleaning unauthorized information by listening to unprotected communications.
- *Information leakage*: Authorized users sharing information with unauthorized parties.
- *Intercept/alter*: An attacker inserting himself (either logically or physically) into a data connection and then intercepting and modifying messages for his own purposes.

- *Masquerade (Spoofing)*: An intruder pretending to be an authorized entity and thereby gaining access to a system.
- *Replay*: An intruder recording a legitimate message and replaying it back at an inopportune time. An often-quoted example is recording the radio transmission used to activate public safety warning sirens during a test transmission and then replaying the message sometime later. An attack of this type does not require more than very rudimentary understanding of the communication protocol.
- *Denial-of-service attack*: An intruder attacking a system by consuming a critical system resource such that legitimate users are never or infrequently serviced.

15.9.2 Security by Obscurity

The electric utility industry frequently believes that the multiplicity and obscurity of its SCADA communication protocols make them immune to malicious interference. While this argument may have some (small) merit, it is not considered a valid assumption when security is required. An often-quoted axiom states that "security by obscurity is no security at all." In the same way that the operation of door locks is well understood but the particular key is kept private on a key ring, it is better to have well-documented and tested approaches to security in which there is broad understanding of the mechanisms but in which the keys themselves are kept private.

15.9.3 SCADA Message Data Integrity Checking

Early SCADA protocols based on telephone switching technology did not have message integrity checking built into the protocols. Incoming (status) information integrity was not considered mission-critical on a per-message basis, and errors would be corrected in the course of repeat transmissions. Control message integrity was provided by redundant messages and by select-check-operate sequences built into the operation.

Traditional packet-based SCADA protocols provide message integrity checking at the data link layer through the use of various checksum or cyclic redundancy check (CRC) codes applied to each data packet. These codes can detect single- and many multiple-bit errors in the transmission of the data packet and are extremely useful for detecting errors caused by electrical noise and other transmission errors. The selection of the particular frame checking algorithm has been the subject of a great deal of study in the development of the several existing SCADA protocols. Usually, the frame check sequence is applied once to the entire packet. In the case of IEC 60870-5 and DNP, however, a CRC is applied to both the header of a message and every 16 octets within the message to ensure message integrity in the face of potentially noisy communication channels.

The OSI reference model prescribes data link integrity checking as a function to be provided by the link layer (layer 2), so all protocols built on this model (e.g., IEC 61850) will have CRC-based frame check sequences built into their lower layers, although they may not be optimized for performance in very noisy communication channels as is the case with the IEC 60870-5 family of protocols.

Since the link layer integrity checks discussed earlier do not include encryption technology and they use well-documented algorithms, they provide protection only against inadvertent packet corruption caused by hardware or data channel failures. They do not provide, nor do they attempt to provide, encryption that can protect against malicious interference with data flow.

15.9.4 Encryption

Security techniques discussed in this section are effective against several of the attacks discussed earlier, including eavesdropping, intercept/alter, and masquerade (spoofing). They can also be effective against replay if they are designed with a key that changes based upon some independent entity such as packet sequence number or time.

The OSI Reference Model separates the function of data link integrity checking (checking for transmission errors) from the function of protecting against malicious attacks to the message contents. Protection from transmission errors is best done as close to the physical medium as possible (data link layer), while protection from message content alteration is best done as close to the application layer as possible (network layer or above). An example of this approach is the IP security protocol (IPsec), which is inserted at the IP level (network layer) in the protocol stack of an Internet-type network.

For those instances where packet routing is not required, it is possible to combine error checking and encryption in the physical or data link layer. Commercial products are being built, which intercept the data stream at the physical (or sometimes data link) layer, add encryption and/or error detection to the message, and send it to a matching unit at the other end of the physical connection, where it is unwrapped and passed to the end terminal equipment. This approach is particularly useful in those situations where it is required to add information security to existing legacy systems. If such devices are employed in a network where message addressing must be visible, they must be intelligent enough to encrypt only the message payload while keeping the address information in the clear.

For systems in which the packets must be routed through a wide area network, the addition of a physical-layer device, which does not recognize the packet structure, is unusable and it is more appropriate to employ network layer or above security protection to the message. This can be accomplished using either proprietary (e.g., many virtual private network [VPN] schemes) or standards based (e.g., IPsec), which operate at the network layer or above in the OSI model.

15.9.5 Denial of Service

Denial-of-service attacks are attacks in which an intruder consumes a critical system resource with the result that legitimate users are denied service. This can happen on a wide area network by flooding the network with packets or requests for service, on a telephone network by simultaneously going "off-hook" with a large number of telephone sets, or on a radio network by jamming the frequency used by radio modems. Defense against such attacks varies depending on the type of communication facility being protected.

Denial of service is usually not an issue on networks that are physically isolated. The exception is defending against system failures, which might arise under peak load conditions or when system components fail.

Defense against denial-of-service attacks in an interconnected wide area network is difficult and can only be accomplished using techniques such as packet traffic management and QoS controls in routers. Denial of service during normal system peak loading is a consideration that must be made when the system is designed.

Defense on a telephone system might include managing "QoS" by giving preferential dial tone to "critical" users while denying peak-load service to "ordinary" users.

Defense on a radio system might include the use of spread spectrum techniques that are designed to be robust in the face of cochannel interference.

15.10 Electromagnetic Environment

The electromagnetic environment in which substation communication systems are asked to operate is very unfriendly to wired communication technologies. It is not unusual to expose communication circuits to several thousands of volts during system faults or switching as a result of electromagnetic induction between high-voltage power apparatus and both internal and external (e.g., telephone) communication facilities.

IEEE Standard 487-2000 states the following:

Wire-line telecommunication facilities serving electric supply locations often require special high-voltage protection against the effects of fault-produced ground potential rise or induced voltages, or both. Some of the telecommunication services are used for control and protective relaying purposes and may be called upon to perform critical operations at times of power system faults. This presents a major challenge in the design and protection of the telecommunication system because power system faults can result in the introduction of interfering voltages and currents into the telecommunication circuit at the very time when the circuit is most urgently required to perform its function. Even when critical services are not involved, special high-voltage protection may be required for both personnel safety and plant protection at times of power system faults. Effective protection of any wire-line telecommunication circuit requires coordinated protection on all circuits provided over the same telecommunication cable.

Tools that can be used to respond to this challenge include the use of isolation and neutralizing transformers for metallic telephone circuits, protection (and qualification testing) of connections to communication apparatus, and proper shielding and grounding of wired circuits.

The use of fiber-optic communication systems for both local networking (e.g., fiber Ethernet) and for telecommunication circuits is a valuable tool for use in hazardous electromagnetic environments.

IEEE and IEC standards that have been issued to deal with electromagnetic interference issues include the following (www.standards.ieee.org):

- IEC Technical Committee No. 65: Industrial-Process Measurement and Control, Electromagnetic Compatibility for Industrial-Process Measurement and Control Equipment, Part 3: Radiated
- IEEE Std C37.90-1994 Standard for Relays and Relay Systems Associated with Electric Power Apparatus
- IEEE Std C37.90.1-2002 Surge Withstand Capability (SWC) Tests for Protective Relays and Relay Systems
- IEEE Std C37.90.2-2001 Withstand Capability of Relay Systems to Radiated Electromagnetic Interference from Transceivers
- IEEE Std C37.90.3-2001 Electrostatic Discharge Tests for Protective Relays
- IEEE Std 487-2000 IEEE Recommended Practice for the Protection of Wire-Line Communication Facilities Serving Electric Supply Locations
- IEEE Std 1613 Environmental Requirements for Communications Networking Devices Installed in Electric Power Substations

15.11 Communication Media

This section discusses each of several communication media that might be used for SCADA communications and reviews their merits in light of the considerations discussed earlier.

15.11.1 Advanced Radio Data Information Service

Advanced radio data information service (ARDIS) was originally developed jointly by Motorola and IBM in the early 1980s for IBM customer service engineers and is owned by Motorola. Service is now available to subscribers throughout the United States with an estimated 65,000 users, mostly using the network in vertical market applications. Many of these users are IBM customer engineers.

ARDIS is optimized for short message applications, which are relatively insensitive to transmission delay. ARDIS uses connection-oriented protocols that are well suited for host/terminal applications. With typical response times exceeding 4 s, interactive sessions generally are not practical over ARDIS. As a radio-based service, ARDIS can be expected to be immune to most of the electromagnetic compatibility (EMC) issues associated with substations. It provides either 4800 bps or 19.2 kbps service using a 25 kHz channel in the 800 MHz band.

15.11.2 Cellular Telephone Data Services

Several different common-carrier services that are associated with cell phone technologies are being offered in the marketplace. Space here permits only cursory mention of the several technologies and their general characteristics.

Advanced mobile phone system (AMPS) is the analog mobile phone system standard, introduced in the Americas during the early 1980s. Though analog is no longer considered advanced at all, the relatively seamless cellular switching technology AMPS introduced was what made the original mobile radiotelephone practical and was considered quite advanced at the time.

Cellular digital packet data (CDPD) is a digital service that can be provided as an adjunct to existing conventional AMPS 800 MHz analog cellular telephone systems. It is available in many major markets but often unavailable in rural areas. CDPD systems use the same frequencies and have the same coverage as analog cellular phones. CDPD provides IP-based packet data service at 19.2 kbps and has been available for a number of years. Service pricing on a use basis has made it prohibitively costly for polling applications, although recent pricing decreases have put a cap in the range of $50 per month for unlimited service. As a radio-based common-carrier service, it is immune to most EMC issues introduced by substations. CDPD is nearing the end of its commercial life cycle and will be decommissioned in the relatively near future by major carriers.

AMPS includes a supervisory data channel used to provide service and connection management to cell phone users. This data channel has been adapted and offered for sale to the electric utility community to support low-data-requirement SCADA functions for simple utility RTU and remote monitoring applications. Refer the discussion of short message system (SMS) in Section 15.11.13 for more information on control channel signaling.

Although AMPS represents the only "universal" standard practice in the North American cell phone industry, it does use technology now regarded as obsolete and is scheduled for retirement in the not-too-distant future.

New applications should consider the use of other common-carrier digital systems such as personal communication service (PCS), time division multiple access (TDMA), global system for mobile communications (GSM), or code division multiple access (CDMA). A third generation of cell phone technology is currently under development, using new technologies called "wideband," including enhanced data rates for GSM evolution (EDGE), also called enhanced GPRS (EGPRS), W-CDMA, CDMA2000, and W-TDMA. The marketplace competition among these technologies can be expected to be lively. While these technologies can be expected to play a dominant role in the future of wireless communications, because of the rapidly changing marketplace, it remains unclear what the long-term availability or pricing of any particular one of these technologies will be.

Common-carrier services are increasingly being offered by cellular telephone providers. These have traditionally been "data on top of voice," but the emerging practice is to put "voice on top of data," in which the network is engineered primarily to support high-speed data with QoS, in which traditional voice services are but one (albeit an important one) of many services supported. Such networks have the potential of providing very cost-effective utility communication services. But utility users should beware that if they do not have both contractual and technical support for QoS, service quality may degrade to unacceptable levels with increasing network loading.

15.11.3 Digital Microwave

Digital microwave systems are licensed systems operating in several bands ranging from 900 MHz to 38 GHz. They have wide bandwidths ranging up to 40 MHz per channel and are designed to interface directly to wired and fiber data channels such as asynchronous transfer mode (ATM), Ethernet, synchronous optical networking (SONET), and T1 derived from high-speed networking and telephony practice.

As a licensed radio system, the Federal Communications Commission (FCC) allocates available frequencies to users to avoid interference. Application of these systems requires path analysis to avoid obstructions and interconnection of multiple repeater stations to cover long routes. Each link requires a line-of-sight path.

Digital microwave systems can provide support for large numbers of both data and voice circuits. This can be provided either as multiples of DS3 ($1 \times DS3 = 672$ voice circuits) signals or DS1 ($1 \times DS1 = 24$ voice circuits) signals, where each voice circuit is equivalent to 64 kbps of data or (increasingly) as ATM or 100 Mbps Fast Ethernet, with direct RJ-45, Category 5 cable connections. They can also link directly into fiber-optic networks using SONET/SDH.

Digital microwave is costly for individual substation installations but might be considered as a high-performance medium for establishing a backbone communication infrastructure that can meet the utility's operational needs.

See also the discussion in Section 15.11.14 for future directions.

15.11.4 Fiber Optics

Fiber-optic cables offer at the same time high bandwidth and inherent immunity from electromagnetic interference. Large amounts of data as high as gigabytes per second can be transmitted over the fiber.

The fiber cable is made up of varying numbers of either single- or multimode fibers, with a strength member in the center of the cable and additional outer layers to provide support and protection against physical damage to the cable during installation and to protect against effects of the elements over long periods of time. The fiber cable is connected to terminal equipment that allows slower speed data streams to be combined and then transmitted over the optical cable as a high-speed data stream. Fiber cables can be connected in intersecting rings to provide self-healing capabilities to protect against equipment damage or failure.

Two types of cables are commonly used by utility companies: optical power ground wire (OPGW), which replaces a transmission line's shield wire, and all dielectric self-supporting (ADSS). ADSS is not as strong as OPGW but enjoys complete immunity to electromagnetic hazards, so it can be attached directly to phase conductors.

Although it is very costly to build an infrastructure, fiber networks are highly resistant to undetected physical intrusion associated with the security concerns outlined earlier. Some of the costs of a fiber network can be shared by joint ventures with other users or by selling bandwidth to other entities such as common carriers. Optical fiber networks can provide a robust communication backbone for meeting a utility's present and future needs.

15.11.5 Hybrid Fiber/Coax

Cable television systems distribute signals to residences primarily using one-way coaxial cable, which is built using an "inverted tree" topology to serve large numbers of customers over a common cable, using (analog) intermediate amplifiers to maintain signal level. This design is adequate for one-way television signals but does not provide the reverse channel required for data services. Cable systems are being upgraded to provide Internet service by converting the coaxial cables to provide two-way communications and adding cable modems to serve customers. The resulting communication data rate is usually asymmetrical, in which a larger bandwidth is assigned downstream (toward the user), with a much smaller bandwidth for upstream service.

Typically the system is built with fiber-optic cables providing the high-speed connection to cable head-ends. Since coaxial cables are easier to tap and to splice, they are preferred for delivery of the signals to the end user. The highest quality, but also most costly, service would be provided by running the fiber cable directly to the end user. Because of the high cost of fiber, variations on this theme employ fiber to the node (FTTN) (neighborhood fiber), fiber to the curb (FTTC), and fiber to the home (FTTH).

Because of the difficulty in creating undetected taps in either a coaxial line or a fiber-optic cable, these systems are resistant to many security threats. However, the fact that they typically provide Internet services makes them vulnerable to many of the cyber attacks discussed earlier, and appropriate security measures should be taken to ensure integrity of service if this service is chosen for utility applications.

15.11.6 Multiple Address Radio

Multiple address (MAS) radio is popular due to its flexibility, reliability, and small size. An MAS radio link consists of a master transceiver (transmitter/receiver) and multiple remote transceivers operating on paired transmit/receive frequencies in the 900 MHz band. The master radio is often arranged to transmit continuously, with remote transmitters coming up to respond to a poll request. Units are typically polled in a "round-robin" fashion, although some work has been done to demonstrate the use of MAS radios in a contention-based network to support asynchronous remote device transmissions.

The frequency pairs used by MAS must be licensed by the FCC and can be reused elsewhere in the system with enough space diversity (physical separation). Master station throughput is limited by radio carrier stabilization times and data rates limited to a maximum of 9.6 kbps. The maximum radius of operation without a special repeater is approximately 15 km, so multiple master radios will be required for a large service territory.

MAS radio is a popular communication medium and has been used widely by utilities for SCADA systems and DA systems.

MAS radio is susceptible to many of the security threats discussed earlier, including denial of service (radio jamming), spoof, replay, and eavesdropping. In addition, the licensed frequencies used by these systems are published and easily available in the public domain. For this reason, it is important that systems using MAS radio be protected against intrusion using the techniques discussed earlier.

15.11.7 Mobile Computing Infrastructure

Systems and personal devices that allow "on-the-go" communications, often including Internet access, are rapidly emerging in the marketplace. These systems offer opportunities to provide communications for IP-based utility applications, often with easy setup and low service costs. New wireless technologies can be expected to provide data rates in excess of 100 kbps. Applications built on these technologies should include network level or above security protection similar to that required of other networked communication systems. For additional discussion on these emerging technologies, refer also to the discussion in Section 15.11.14.

15.11.8 Mobile Radio

Mobile radio systems operating in the VHF, UHF, and 800 MHz radio bands have sometimes been pressed into shared data service along with their primary voice applications. Such use is problematic due to the facts that the systems are designed for analog (voice) rather than digital (data) applications and that they are shared with voice users. It is difficult to license new applications on these channels and their use for digital applications should be discouraged. The emerging "mobile computing" technologies are much more attractive for these applications.

15.11.9 Mobitex Packet Radio

Mobitex is an open, international standard for wireless data communications developed by Ericsson. It provides remote access for data and two-way messaging for mobile computing applications.

The technology is similar to that used in ARDIS and cellular telephone systems. Like mobile telephone systems, the Mobitex networks are based on radio cells. Base stations allocate digital channels to active terminals within limited geographic areas. Data are divided into small packets that can be transmitted individually and as traffic permits. Setup time is eliminated and network connections are instantaneous. Packet switching makes more efficient use of channel capacity. Area and main exchanges handle switching and routing in the network, thereby providing transparent and seamless roaming within the United States. A modest data rate of 8 or 16 kbps makes it useful for small amounts of data or control but not for large file transfers. Service is offered to large portions of the U.S. population (primarily in the East), but rural service may be lacking. As part of a public network, applications should employ end-to-end application-layer security.

15.11.10 Paging Systems

Paging systems have been used very effectively for certain utility applications that typically require only one-way command operation. Paging networks are built using carefully engineered sets of system controllers, transmitters, and data links designed to make sure the system has optimal coverage and response while minimizing interference. Some systems use satellite channels to provide wide area coverage. Most paging systems use simulcast techniques and multiple transmitters to give continuous coverage over their service areas. Typical systems provide publicly accessible interfaces using dial-up, modem, and/or Internet access. The over-the-air protocol is the British Post Office Code Standardisation Advisory Group (POCSAG) standard operating in the 900 MHz band. Most systems are one way (outbound), but a few also offer inbound messaging services. Systems have large capacities but are subject to intolerable delays when overloaded. Service cost is typically very low, making this system very attractive for certain applications.

As part of a public network, application layer security to protect from masquerading attacks is appropriate. A coordinated denial-of-service attack may be possible but is unlikely to occur in the types of applications for which this system is suited.

15.11.11 Power Line Carrier

PLC systems operating on narrow channels between 30 and 500 kHz are frequently used for high-voltage line protective relaying applications. Messages carried by PLC systems are typically simple status indicators such as 1 bit messages which use either amplitude- or frequency-shift keying to tell the other end of the dedicated PLC circuit to trip or to inhibit the tripping of a protective relay.

Other PLC systems have been developed for specialized distribution feeder applications such as remote meter reading and DA. Early in the development of PLC systems, it was observed that signals below approximately 10 kHz would propagate on typical distribution feeders, with the primary impediments coming from shunt power factor correction capacitors and from series impedances of distribution transformers. These two components work together as a low-pass filter to make it difficult to transmit higher frequency signals. In addition, signaling at lower frequencies approaching the power line frequency is difficult because of harmonic interference from the fundamental power line itself.

One successful system uses frequency shift keying (FSK) signals in the 10 kHz range to provide communications for DA.

Two systems, the two-way automatic communication system (TWACS) and the Turtle, use communications based on modification of the 60 Hz waveform itself. Both systems use disturbances of the voltage waveform for outbound communication and of the current waveform for inbound communication. The primary difference between the two systems is that TWACS uses relatively higher power and data rates of 60 bit/s, while the Turtle system uses extremely narrow bandwidth signaling—on the order of 1/1000 bit/s—and massively parallel communications in which each remote device has its own logical channel. The TWACS is used for both automatic meter reading and DA, while the Turtle system is used mostly for meter reading.

With the proper equipment, both of these systems would be subject to both eavesdropping and masquerading types of security threats, so security measures are appropriate. With the limited data rates of these systems, only simple encryption techniques using secret keys are appropriate.

Recent and much-publicized work has been conducted to develop high-speed data services, which claim to deliver data rates as high (in one case) as a gigabit per second. Known in the industry as broadband over power lines (BPL), these systems use spread spectrum techniques to deliver data rates previously unattainable. But fundamental physical constraints make it unlikely that successful data rates will be delivered much above several megabits per second. A technical issue with these technologies is the fact that they occupy the same spectrum as do licensed users in the high-frequency space (3–30 MHz) and can cause interference to, as well as receive interference from, these other services. FCC Part 15 rules require any unlicensed users (such as BPL) to not interfere with existing uses and to vacate the frequency if found to be interfering, so utility application of this technology should only be done with caution.

PLC systems are exposed to public access, so encryption techniques are appropriate to protect any sensitive information or control communications.

15.11.12 Satellite Systems

Satellite systems, which offer high-speed data service, have been deployed in two different forms, broadly categorized by their orbits.

Hughes built the first geosynchronous orbit (GEO) communication satellite in the early 1960s under a NASA contract to demonstrate the viability of such satellites operating in an Earth orbit 22,300 miles (35,900 km) above the ground. The attractive feature of these satellites is that they appear fixed in the sky and therefore do not require costly tracking antennas. Such satellites are commonly used today to distribute radio and television programming and are useful for certain data applications.

Because of the large distances to the satellite, GEO systems require relatively large parabolic antennas to keep satellite transponder power levels to a manageable level. Because of the distances involved, each trip from Earth to satellite and back requires 1/4 s of time. Some satellite configurations require all data to pass through an Earth station on each hop to or from the end user, thereby doubling this time before a packet is delivered to the end device. If a communication protocol is used, which requires link-layer acknowledgements for each packet (typical of most legacy SCADA protocols), this can add as much as 1 s to each poll/response cycle. This can be unacceptably large and have a significant impact on system throughput, so careful protocol matching is appropriate if a GEO satellite link is being considered. This long delay characteristic also makes GEO satellites undesirable for two-way telephone links.

A second satellite technology that is gaining popularity is the "low Earth orbit" (LEO) satellite. LEOs operate at much lower altitudes of 500–2000 km. Because of the lower altitude, the satellites are in constant motion (think of a swarm of bees), so a fixed highly directional antenna cannot be used. But compensating for this is the fact that the smaller distances require lower power levels, so if there are a sufficient number of satellites in orbit and if their operation is properly coordinated, LEOs can provide ubiquitous high-speed data or quality voice service anywhere on the face of the Earth. LEO systems can be quickly deployed using relatively small Earth stations. There are a number of competing service providers providing several varieties of LEO service: "Little LEOs" for data only; "Big LEOs" for voice plus limited data; and "Broadband LEOs" for high-speed data plus voice. Search "LEO satellite" on the Internet for more information.

All satellite systems are subject to eavesdropping, so the use of appropriate security measures is indicated to avoid loss of confidential information.

15.11.13 Short Message System

SMS (also known as "text messaging") uses the forward and reverse control channels (FOCC and RECC, respectively) of cell phone systems to provide two-way communication service for very short telemetry messages. The FOCC and RECC are the facilities normally used to authorize and set up cell phone calls.

Since the messages are short and the channel is unused during a voice call, there is surplus unused bandwidth available in all existing analog cell phone systems that can be used for this service. SMSs send information in short bursts of 10 bit in the forward (outbound) direction and 32 bits in the reverse (inbound) direction, making them well suited for control and status messaging from simple RTUs. Message integrity is enhanced through the use of three out of five voting algorithms. A number of companies are offering packaged products and services, which can be very economic for simple status and control functions. Utility interface to the system is provided using various Internet, telephone, and pager services. Search the web for "SMS telemetry" for more information.

15.11.14 Spread Spectrum Radio and Wireless LANs

New radio technologies are being developed as successors to traditional MAS and microwave radio systems, which can operate unlicensed in the 900 MHz, 2.4 GHz, and 5.6 GHz bands or licensed in other nearby bands. These systems typically use one of several variants of spread spectrum technology and offer robust, high-speed, point-to-point, or point-to-multipoint service. Interfaces can be provided ranging from 19.2 kbps RS232 to Ethernet. Line-of-sight distances ranging from 1 to 20 miles are possible, depending on antenna and frequency band choices and transmitter power. Higher-powered devices require operation in licensed bands.

This technology has been successfully used both for communication within the substation fence as well as communication over larger distances between the enterprise and the substation or between substations. An example of communication within the substation is adding new functionality, such as transformer condition monitoring, to an existing substation. An internal substation radio connection can make such installations very cost-effective while at the same time providing immunity to electromagnetic interference, which might otherwise arise from the high electric and magnetic fields found in a substation environment.

As contrasted to traditional radio systems, spread spectrum radio transmits information spread over a band of frequencies either sequentially (frequency hopping spread spectrum [FHSS]) or in a so-called chirp (direct sequence spread spectrum [DSSS]). Other closely related but distinct modulation techniques include orthogonal frequency division multiplexing (OFDM), which sends data in parallel over a number of subchannels. The objective in all of these systems is to allow operation of multiple systems concurrently without interference and with maximum information security. The existence of multiple systems in proximity to each other increases the apparent noise background but is not immediately fatal to successful communications. Knowledge of the frequency hopping or spreading "key" is necessary for the recovery of data, thus at the same time rendering the system resistant to jamming (denial of service) and eavesdropping attacks.

Variants of DSSS, FHSS, and OFDM are being offered in commercial products and are being adopted in emerging wireless LAN standards such as the several parts of IEEE 802.11 (Wireless LAN) and 802.16 (Broadband Wireless Access).

This is a rapidly changing technology. Search the web for "spread spectrum," "DSSS," "FHSS," and "OFDM" for more information and to discover a current list of vendors.

15.12 Telephone-Based Systems

The following paragraphs present short discussions of systems based on traditional "wired" telephone technology. They range from low-speed leased and dial-up circuits through very high-speed optical connections.

15.12.1 Telephone Lines: Leased and Dial-Up

Dedicated so-called "leased" or "private" voice-grade lines with standard 3 kHz voice bandwidth can be provided by the telephone company. Dial-up telephone lines provide similar technical characteristics, with the key difference being the manner of access (dial-up) and the fact that the connection is "temporary."

Commonly thought of as providing a "private twisted pair," leased lines are seldom built in this manner. Rather, these circuits are routed, along with switched lines, through standard telephone switches. Unless otherwise ordered, routing (and performance characteristics) of such circuits may change without warning to the end user. Dedicated circuits, known in the industry as 3002 circuits, can support modem data rates up to 19.2 kbps and up to 56 kbps with data compression. High-performance so-called digital data service (DDS) circuits can support modem communications up to 64 kbps with special line conditioning.

Security issues for all telephone circuits include the fact that they are easily tapped in an unobtrusive manner, which makes them vulnerable to many of the security attacks discussed earlier. In addition, they can be rerouted in the telephone switch by a malicious intruder, and dial-up lines are easily accessed by dialing their phone numbers from the public telephone network. Thus it is important that these circuits be protected by the appropriate physical, data-link, or network layer measures as discussed earlier. In the case of IED interfaces accessible by dial-up phone lines, they must at a minimum be protected by enabling sign-on passwords (and changing of the system default passwords), with the possibility of using other systems such as dial-back modems or physical-layer encryption as discussed in Chapter 17.

Telephone circuits are susceptible to all of the electromagnetic interference issues discussed earlier and should be protected by appropriate isolation devices.

15.12.2 Integrated Services Digital Network

Integrated services digital network (ISDN) is a switched, end-to-end wide area network designed to combine digital telephony and data transport services. ISDN was defined by the International Telecommunications Union (ITU) in 1976. Two types of service are available: ISDN basic access (192 kbps), sold as ISDN2, 2B + D, or ISDN BRI; and ISDN primary access (1.544 Mbps), sold as ISDN23, 23B + D, or ISDN PRI. The total bandwidth can be broken into either multiple 64 kbps voice channels or from one to several data channels. ISDN is often used by larger businesses to network geographically dispersed sites.

Broadband ISDN (B-ISDN) provides the next generation of ISDN, with data rates of either 155.520 or 622.080 Mbps.

ISDN can be configured to provide private network service, thereby sidestepping many of the security issues associated with public networks. However, it is still subject to security issues that arise from the possibility of an intruder breaking into the telephone company equipment and rerouting private services.

As a wired service, it is also subject to the electromagnetic interference issues, which substations create. The high-speed digital signals will not successfully propagate through isolation and neutralizing transformers and will require isolation using back-to-back optical isolators at the substation.

15.12.3 Digital Subscriber Loop

Digital subscriber loop (DSL) transmits data over a standard analog subscriber line. Built upon ISDN technology, DSL offers an economical means to deliver moderately high bandwidth to residences and small offices. DSL comes in many varieties known as xDSL, where x is used to denote the varieties. Commonly sold to end users, ADSL (asymmetric DSL) sends combined data and voice over ordinary copper pairs between the customer's premises and the telephone company's central office. ADSL can provide data rates ranging from 1.5 to 8 Mbps downstream (depending on phone line characteristics) and 16 to 640 kbps upstream. The digital and analog streams are separated at both the central office and the customer's site using filters, and an ADSL modem connects the data application to the subscriber line.

Telephone companies use high-speed DSL (HDSL) for point-to-point T1 connections and symmetric or single line DSL (SDSL) to carry T1 on a single pair. HDSL can carry T1 (1.544 Mbps) and FT1 (fractional T1) data in both directions. The highest speed implementation to date is very high-speed DSL (VDSL) that can support up to 52 Mbps in the downstream data over short ranges. ADSL can operate up to 6000 m, whereas VDSL can only attain full speed up to about 300 m.

A key advantage of DSL is its competitive pricing and wide availability. A disadvantage is that service is limited to circuit lengths of less than 3.6 km without repeaters.

As a wired service, DSL has the same security and EMC issues as ISDN.

15.12.4 T1 and Fractional T1

T1 is a high-speed digital network (1.544 Mbps) developed by AT&T in 1957 and implemented in the early 1960s to support long-haul pulse-code modulation (PCM) voice transmission. The primary innovation of T1 was to introduce "digitized" voice and to create a network fully capable of digitally representing what was, up until then, a fully analog telephone system. T1 is part of a family of related digital channels used by the telephone industry that can be delivered to an end user in any combination desired.

The T1 channel data rates are shown in Table 15.1.

T1 is a well-proven technology for delivering high-speed data or multiple voice channels. Depending on the proximity of the utility facility to telephone company facilities, the cost can be modest or high. See also the discussion of DSL for additional options.

As a wired facility, T1 is subject to the electromagnetic interference issues discussed earlier unless it is offered using fiber-optic facilities (see discussion of fiber optic).

Since T1 was originally designed to serve voice users, delivery of data bits with a minimum of latency and jitter is important, but occasional discarded data are not considered a problem. Therefore, equipment using T1 links should provide link error checking and retransmission.

A T1 link is point to point and interfacing to a T1 facility requires sophisticated equipment, so a T1 facility is resistant to casual eavesdropping security attacks. But since it is part of a system exposed to outside entities and with the possibility that an intruder to the telephone facility could eavesdrop or redirect communications, it is important that systems using T1 facilities employ end-to-end security measures at the network layer or above as discussed in Section 15.9.

15.12.5 Frame Relay

Frame relay is a service designed for cost-efficient intermittent data transmission between local area networks and between end points in a wide area network. Frame relay puts data in a variable-size unit called a frame and leaves error correction up to the end points. This speeds up overall data transmission. Usually, the network provides a permanent virtual circuit (PVC), allowing the customer to see a continuous, dedicated connection without paying for a full-time leased line. The provider routes each frame

TABLE 15.1 DS Data Rates

Name	Date Rate	Number of T1s	Number of Voice Channels
DS0	64 kbps	1/24	1
DS1	1.544 Mbps	1	24
DS1C	3.152 Mbps	2	48
DS2	6.312 Mbps	4	96
DS3	44.736 Mbps	28	672
DS3C	88.472 Mbps	56	1344
DS4	274.176 Mbps	168	4032

to its destination and can charge based on usage. Frame relay makes provision to select a level of service quality, prioritizing some frames. Frame relay is provided on fractional T1 and full T-carrier. It provides a mid-range service between ISDN (128 kbps, Section 15.12.2) and ATM (155.520 or 622.080 Mbps, Section 15.12.6). Based on the older X.25 packet-switching technology, frame relay is being displaced by ATM and native IP-based protocols (see discussion of MPLS).

15.12.6 Asynchronous Transfer Mode

ATM is a cell relay protocol that encodes data traffic into small fixed-size (53 byte with 48 bytes of data and 5 bytes of header) "cells" instead of variable-sized packets as in packet-switched networks. ATM was originally intended to provide a unified networking solution that could support both synchronous channel networking and packet-based networking along with multiple levels of service quality for packet traffic. ATM was designed to serve both circuit-switched networks and packet-switched networks by mapping both bit streams and packet streams onto a stream of small fixed-size "cells" tagged with virtual circuit identifiers. These cells would be sent on demand within a synchronous time slot in a synchronous bit stream. ATM was originally designed by the telecommunications industry and intended to be the enabling technology for broadband integrated services digital network (B-ISDN), replacing the existing switched telephone network. Because ATM came from the telecommunications industry, it has complex features to support applications ranging from global telephone networks to private local area computer networks.

ATM has enjoyed widespread deployment but only partial success. It is often used as a technology to transport IP traffic but suffers significant overhead for IP traffic because of its short cell size. Its goal of providing a single integrated technology for LANs, public networks, and user services has largely failed (see discussion of MPLS).

15.12.7 SONET

SONET is a standard for sending digital information over optical fiber. It was developed for the transport of large amounts of telephone and data traffic. The more recent synchronous digital hierarchy (SDH) standard developed by the ITU is built on experience gained in the development of SONET. SONET is used primarily in North America and SDH in the rest of the world. SONET can be used to encapsulate earlier digital transmission standards or used directly to support ATM. The basic SONET signal operates at 51.840 Mbps and is designated synchronous transport signal one (STS-1). The STS-1 frame is the basic unit of transmission in SONET. SONET supports multiples of STS-1 up to STS-3072 (159.252480 Gbps).

15.12.8 Multiprotocol Label Switching

Multiprotocol label switching (MPLS) is a data-carrying mechanism that operates in the OSI model one layer below protocols such as IP. Designed to provide a unified service for both circuit-based and packet-switching clients, it provides a datagram service model. It can carry many different kinds of traffic, including voice telephone traffic and IP packets. Previously, a number of different systems, including frame relay and ATM, were deployed with essentially identical goals. MPLS is now replacing these technologies in the marketplace because it is better aligned with current and future needs. MPLS abandons the cell-switching and signaling protocol of ATM and recognizes that small ATM cells are not needed in modern networks, since they are so fast (10 Gbps and above) that even full-length 1500 byte packets do not incur significant delays. At the same time, MPLS preserves the traffic engineering and out-of-band control needed for deploying large-scale networks. Originally created by CISCO as a proprietary protocol, it was renamed when it was handed over to the IETF for open standardization.

15.13 For More Information

As discussed earlier, a number of standards organizations have produced standards that can be used as guidelines when designing substation communication systems, and there are Internet resources that can be studied for further information. References to some of these standards and websites are provided in the following to help the reader find more detailed information.

15.13.1 Useful Websites

American National Standards Institute (ANSI): www.ansi.org
Institute of Electrical and Electronics Engineers (IEEE): www.ieee.org
International Electrotechnical Commission (IEC): www.iec.ch
Internet Engineering Task Force (IETF): www.ietf.org
International Standards Organization (ISO): www.iso.ch
National Institute of Standards and Technology (NIST): www.nist.gov
International Telecommunications Union (ITU): www.itu.int
DNP User's Group: www.dnp.org
UCA User's Group: www.ucausersgroup.org
Information on Systems Engineering: http://www.bredemeyer.com
Publicly available ISO Standards: http://www.acm.org/sigcomm/standards/ Useful overview discussions on emerging data communication technologies can be found by searching Wikipedia at "www.wikipedia.org"

15.13.2 Power System Relay Communication References

The following references are useful when designing communication facilities to support power system relaying. The IEEE-PES Power System Relay Committee web page http://www.pes-psrc.org/ contains many of these reports. The radio engineering references are valuable in the design of wireless services for either SCADA or relaying.

1. Application of peer-to-peer communication for protective relaying, *IEEE Transactions on Power Delivery*, 17(2), April 2002.
2. Digital communications for relay protection, Working Group H9 of the IEEE Power System Relaying Committee.
3. D. Holstein, J. Tengdin, and E. Udren. IEEE C37.115-2003. Standard test method for use in the evaluation of message communications between intelligent electronic devices in integrated substation protection, control, and data acquisition systems.
4. Special considerations in applying power line carrier for protective relaying, IEEE Power Systems Relaying Committee Special paper, Relaying Communications Subcommittee, Working Group H9.
5. Application considerations of IEC 61850/UCA 2 for substation ethernet local area network communication for protection and control, IEEE PSRC H6 Special Report.
6. Using spread spectrum radio communication for power system protection relaying applications: A report to the IEEE Power System Relaying Committee, Prepared by IEEE/PSRC Working Group H2.
7. Wide area protection and emergency control, Working Group C-6, System Protection Subcommittee, IEEE PES Power System Relaying Committee, Final Report.
8. Application of peer-to-peer communication for protective relaying, IEEE PES PSRC Working Group H5 Report to the Relay Communications Subcommittee, *IEEE Transactions on Power Delivery*, 17(2), April 2002, 446.
9. Application of peer-to-peer communications for protective relaying, IEEE PES PSRC H5 Working Group Report to the Relay Communications Subcommittee, March 2001 [Online]. Available: http://www.pes-psrc.org/h/H5doc.zip

10. R.F. White. *Engineering Considerations for Microwave Communications Systems*, Edited by GTE Network Systems, Northlake, IL. Copyright 1975 By GTE Lenkurt Incorporated. Third Printing, 1983.
11. R.U. Laine. Digital microwave systems applications seminar (notes). Volume 1, Digital Microwave Link Engineering, Issue 7, December. A.R. Lunan and W. Quan (contributors).
12. IEC 60834-1, Ed. 2.0, b: 1999, Teleprotection equipment of power systems—Performance and testing—Part 1: Command systems.
13. *The ARRL Antenna Book*. 944 pp., 20th edn., © 2003, The American Radio Relay League (ISBN: 0-87259-904-3) #9043.
14. *The ARRL Handbook*. Softcover. 1216 pp., 2004 edn. (81st edn.), © 2003, The American Radio Relay League, Inc. (ISBN: 0-87259-196-4) #1964. This handbook is updated annually.
15. J.J. Kumm, M.S. Weber, E.O. Schweitzer, D. Hou, Assessing the effectiveness of self-tests and other monitoring means in protective relays, *Western Protective Relay Conference*, Spokane, WA, October 1994.
16. NERC Planning Standards, North American Electric Reliability Council Engineering Committee, September 1997.
17. Proposed reliability management system (RMS), Western Systems Coordinating Council, FERC Declaratory Filing, July 23, 1998.
18. Reliability criteria for transmission system planning, Western Systems Coordinating Council, March 1999.
19. WSCC Telecommunications Work Group Design guidelines for critical communications circuits, Western Systems Coordinating Council Telecommunications Work Group.
20. Communications systems performance guide for protective relaying applications, WSCC Telecommunications and Relay Work Groups, November 21, 2001. http://www.wecc.biz/documents/library/procedures/Comm_Perf_Guide_for_Relays.pdf

15.13.3 Relevant Standards

15.13.3.1 IEEE 802.x Networking Standards

IEEE 802.x standards are available from www.standards.ieee.org
802 Standards are currently available in electronic form at no cost 6 months after publication.

15.13.3.2 IEEE Electromagnetic Interference Standards

IEEE Std C37.90-1994 Standard for relays and relay systems associated with electric power apparatus.
IEEE Std C37.90.1-2002 surge withstand capability (SWC) tests for protective relays and relay systems.
IEEE Std C37.90.2-2001 Withstand capability of relay systems to radiated electromagnetic interference from transceivers.
IEEE Std C37.90.3-2001 Electrostatic discharge tests for protective relays.
IEEE Std 487-2000 IEEE recommended practice for the protection of wire-line communication facilities serving electric supply locations.
IEEE Std 1613 Environmental requirements for communications networking devices installed in electric power substations.

15.13.3.3 IEC 60870 Standards for Telecommunication

IEC/TR 60870-1-1 {Ed.1.0}Telecontrol equipment and systems. Part 1: General considerations. Section 1: General principles.
IEC 60870-1-2 {Ed.1.0} Telecontrol equipment and systems. Part 1: General considerations. Section 2: Guide for specifications.
IEC/TR 60870-1-3 {Ed.2.0} Telecontrol equipment and systems—Part 1: General considerations—Section 3: Glossary.

IEC/TR 60870-1-4 {Ed.1.0} Telecontrol equipment and systems—Part 1: General considerations—Section 4: Basic aspects of telecontrol data transmission and organization of standards IEC 870-5 and IEC 870-6.

IEC/TR 60870-1-5 {Ed.1.0} Telecontrol equipment and systems—Part 1–5: General considerations—Influence of modem transmission procedures with scramblers on the data integrity of transmission systems using the protocol IEC 60870-5.

IEC 60870-2-1 {Ed.2.0} Telecontrol equipment and systems—Part 2: Operating conditions—Section 1: Power supply and electromagnetic compatibility.

IEC 60870-2-2 {Ed.1.0} Telecontrol equipment and systems—Part 2: Operating conditions—Section 2: Environmental conditions (climatic, mechanical, and other nonelectrical influences).

IEC 60870-3 {Ed.1.0} Telecontrol equipment and systems. Part 3: Interfaces (electrical characteristics).

IEC 60870-4 {Ed.1.0} Telecontrol equipment and systems. Part 4: Performance requirements.

IEC 60870-5-1 {Ed.1.0} Telecontrol equipment and systems. Part 5: Transmission protocols—Section 1: Transmission frame formats.

IEC 60870-5-2 {Ed.1.0} Telecontrol equipment and systems—Part 5: Transmission protocols—Section 2: Link transmission procedures.

IEC 60870-5-3 {Ed.1.0} Telecontrol equipment and systems—Part 5: Transmission protocols—Section 3: General structure of application data.

IEC 60870-5-4 {Ed.1.0} Telecontrol equipment and systems—Part 5: Transmission protocols—Section 4: Definition and coding of application information elements.

IEC 60870-5-5 {Ed.1.0} Telecontrol equipment and systems—Part 5: Transmission protocols—Section 5: Basic application functions.

IEC 60870-5-101 {Ed.2.0} Telecontrol equipment and systems—Part 5–101: Transmission protocols—Companion standard for basic telecontrol tasks.

IEC 60870-5-102 {Ed.1.0} Telecontrol equipment and systems—Part 5: Transmission protocols—Section 102: Companion standard for the transmission of integrated totals in electric power systems.

IEC 60870-5-103 {Ed.1.0} Telecontrol equipment and systems—Part 5–103: Transmission protocols—Companion standard for the informative interface of protection equipment.

IEC 60870-5-104 {Ed.1.0} Telecontrol equipment and systems—Part 5–104: Transmission protocols—Network access for IEC 60870-5-101 using standard transport profiles.

15.13.3.4 DNP3 Specifications for Device Communication

IEEE Std 1379-1997 IEEE Trial-Use Recommended Practice for Data Communications between Intelligent Electronic Devices and Remote Terminal Units in a Substation DNP 3.0 specifications, available from www.dnp.org. Available on-line: "A DNP3 Protocol Primer." Specifications in four documents available to Users Group members: DNP V3.00 Data Link Layer Protocol Description; DNP V3:00 Transport Functions; DNP V3.00 Application Layer Protocol Description; DNP V3:00 Data Object Library.

15.13.3.5 IEC 60870-6 TASE.2 (UCA/ICCP) Control System Communications

IEC/TR 60870-6-1 {Ed.1.0} Telecontrol equipment and systems—Part 6: Telecontrol protocols compatible with ISO standards and ITU-T recommendations—Section 1: Application context and organization of standards.

IEC 60870-6-2 {Ed.1.0} Telecontrol equipment and systems—Part 6: Telecontrol protocols compatible with ISO standards and ITU-T recommendations—Section 2: Use of basic standards (OSI layers 1–4).

IEC 60870-6-501 {Ed.1.0} Telecontrol equipment and systems—Part 6: Telecontrol protocols compatible with ISO standards and ITU-T recommendations—Section 501: TASE.1 Service definitions.

IEC 60870-6-502 {Ed.1.0} Telecontrol equipment and systems—Part 6: Telecontrol protocols compatible with ISO standards and ITU-T recommendations—Section 502: TASE.1 Protocol definitions.

IEC 60870-6-503 {Ed.2.0} Telecontrol equipment and systems—Part 6-503: Telecontrol protocols compatible with ISO standards and ITU-T recommendations—TASE.2 Services and protocol.

IEC/TS 60870-6-504 {Ed.1.0} Telecontrol equipment and systems—Part 6-504: Telecontrol protocols compatible with ISO standards and ITU-T recommendations—TASE.1 User conventions.

IEC/TR 60870-6-505 {Ed.1.0} Telecontrol equipment and systems—Part 6-505: Telecontrol protocols compatible with ISO standards and ITU-T recommendations—TASE.2 User guide.

IEC/TR 60870-6-505-am1 {Ed.1.0} Amendment 1—Telecontrol equipment and systems—Part 6-505: Telecontrol protocols compatible with ISO standards and ITU-T recommendations—TASE.2 User guide.

IEC 60870-6-601 {Ed.1.0} Telecontrol equipment and systems—Part 6: Telecontrol protocols compatible with ISO standards and ITU-T recommendations—Section 601: Functional profile for providing the connection-oriented transport service in an end system connected via permanent access to a packet switched data network.

IEC/TS 60870-6-602 {Ed.1.0} Telecontrol equipment and systems—Part 6-602: Telecontrol protocols compatible with ISO standards and ITU-T recommendations—TASE transport profiles.

IEC 60870-6-701 {Ed.1.0} Telecontrol equipment and systems—Part 6-701: Telecontrol protocols compatible with ISO standards and ITU-T recommendations—Functional profile for providing the TASE.1 application service in end systems.

IEC 60870-6-702 {Ed.1.0} Telecontrol equipment and systems—Part 6-702: Telecontrol protocols compatible with ISO standards and ITU-T recommendations—Functional profile for providing the TASE.2 application service in end systems.

IEC 60870-6-802-am1 {Ed.2.0} Amendment 1—Telecontrol equipment and systems—Part 6-802: Telecontrol protocols compatible with ISO standards and ITU-T recommendations—TASE.2 Object models.

IEC 60870-6-802 {Ed.2.1} Telecontrol equipment and systems—Part 6-802: Telecontrol protocols compatible with ISO standards and ITU-T recommendations—TASE.2 Object models.

15.13.3.6 IEC 61850/UCA Standards for Substation Systems

IEEE TR 1550-1999 EPRI/UCA Utility Communications Architecture (UCA) Version 2.0 1999, IEEE Product No: SS1117-TBR IEEE Standard No: TR 1550-1999.

IEC/TR 61850-1 {Ed.1.0} Communication networks and systems in substations—Part 1: Introduction and overview.

IEC/TS 61850-2 {Ed.1.0} Communication networks and systems in substations—Part 2: Glossary.

IEC 61850-3 {Ed.1.0} Communication networks and systems in substations—Part 3: General requirements.

IEC 61850-4 {Ed.1.0} Communication networks and systems in substations—Part 4: System and project management.

IEC 61850-5 {Ed.1.0} Communication networks and systems in substations—Part 5: Communication requirements for functions and device models.

IEC 61850-6 {Ed.1.0} Communication networks and systems in substations—Part 6: Configuration description language for communication in electrical substations related to IEDs.

IEC 61850-7-1 {Ed.1.0} Communication networks and systems in substations—Part 7-1: Basic communication structure for substation and feeder equipment—Principles and models.

IEC 61850-7-2 {Ed.1.0} Communication networks and systems in substations—Part 7-2: Basic communication structure for substation and feeder equipment—Abstract communication service interface (ACSI).

IEC 61850-7-3 {Ed.1.0} Communication networks and systems in substations—Part 7-3: Basic communication structure for substation and feeder equipment—Common data classes.

IEC 61850-7-4 {Ed.1.0} Communication networks and systems in substations—Part 7-4: Basic communication structure for substation and feeder equipment—Compatible logical node classes and data classes.

IEC 61850-8-1 {Ed.1.0} Communication networks and systems in substations—Part 8–1: Specific Communication Service Mapping (SCSM)—Mappings to MMS (ISO 9506-1 and ISO 9506-2) and to ISO/IEC 8802-3.

IEC 61850-9-1 {Ed.1.0} Communication networks and systems in substations—Part 9–1: Specific Communication Service Mapping (SCSM)—Sampled values over serial unidirectional multidrop point to point link.

IEC 61850-9-2 {Ed.1.0} Communication networks and systems in substations—Part 9–2: Specific communication service mapping (SCSM)—Sampled values over ISO/IEC 8802-3.

IEC 61850-10 {Ed.1.0} Communication networks and systems in substations—Part 10: Conformance testing.

IEC 61850-SER {Ed.1.0} Communication networks and systems in substations—All parts.

IEC 61850-90-1 Use of IEC 61850 for the communication between substations.

IEC 61850-90-2 Use of IEC 61850 for the communication between control centers and substations.

IEC 61850-7-410 Hydroelectric power plants—Communication for monitoring and control.

IEC 61850-7-420 Communications systems for Distributed Energy Resources (DER)—Logical nodes.

IEC 61850-7-500 Use of logical nodes to model functions of a substation automation system.

IEC 61850-7-510 Use of logical nodes to model functions of a hydro power plant.

IEC 61850-90-1 Use of IEC 61850 for the communication between substations.

IEC 61850-90-2 Use of IEC 61850 for the communication between control centers and substations.

IEC 61850-90-3 Using IEC 61850 for condition monitoring.

IEC 61850-90-4 IEC 61850—Network Engineering Guidelines.

IEC 61850-90-5 Use of IEC 61850 to transmit synchrophasor information according to IEEE C37.118.

IEC 61850-80-1 Guideline to exchanging information from a CDC-based data model using IEC 60870-5-101 or IEC 60870-5-104.

IEC 61400-25 IEC 61850 Adaptation for wind turbines.

IEC 61400-25-1 Wind turbines—Part 25-1: Communications for monitoring and control of wind power plants—Overall description of principles and models.

IEC 61400-25-2 Wind turbines—Part 25-2: Communications for monitoring and control of wind power plants—Information models.

IEC 61400-25-3 Wind turbines—Part 25-3: Communications for monitoring and control of wind power plants—Information exchange models.

IEC 61400-25-4 Wind turbines—Part 25-4: Communications for monitoring and control of wind power plants

    Mapping to communication profile
    Mapping to web services
    Mapping to MMS [Refer to IEC 61850-8-1]
    Mapping to OPC XML DA
    Mapping to IEC 60870-5-104 [Refer to IEC 62445-3]
    Mapping to DNP3

IEC 61400-25-5 Wind turbines—Part 25-5: Communications for monitoring and control of wind power plants—Conformance testing.

IEC 61400-25-6 Wind turbines—Part 25-6: Communications for monitoring and control of wind power plants—Logical node classes and data classes for condition monitoring.

IEC 62271-3 Communications for monitoring and control of high-voltage switchgear.

15.13.3.7 IEC 61968 Standards for Distribution Application Integration

IEC 61968-1 {Ed.1.0} Application integration at electric utilities—System interfaces for distribution management—Part 1: Interface architecture and general requirements.

IEC/TS 61968-2 {Ed.1.0} Application integration at electric utilities—System interfaces for distribution management—Part 2: Glossary.

IEC 61968-3 {Ed.1.0} Application integration at electric utilities—System interfaces for distribution management—Part 3: Interface for network operations.
IEC 61968-4 Application integration at electric utilities—System interfaces for distribution management—Part 4: Interface standard for records and asset management.
IEC 61968-5 Application integration at electric utilities—System interfaces for distribution management—Part 5: Interface standard for operational planning and optimization.
IEC 61968-6 Application integration at electric utilities—System interfaces for distribution management—Part 6: Interface standard for maintenance and construction.
IEC 61968-7 Application integration at electric utilities—System interfaces for distribution management—Part 7: Interface standard for network extension planning.
IEC 61968-8 Application integration at electric utilities—System interfaces for distribution management—Part 8: Interface standard for customer inquiry.
IEC 61968-9 Application integration at electric utilities—System interfaces for distribution management—Part 9: Interface standard for meter reading and control.
IEC 61968-10 Application integration at electric utilities—System interfaces for distribution management—Part 10: Interface standard for systems external to, but supportive of, distribution management.
IEC 61968-11 Application integration at electric utilities—System interfaces for distribution management—Part 11: Distribution information exchange model (DIEM).
IEC 61968-12 Application integration at electric utilities—System interfaces for distribution management—Part 12: Use case examples.
IEC 61968-13 Application integration at electric utilities—System interfaces for distribution management—Part 13: CIM RDF model exchange format for distribution.

15.13.3.8 IEC 61970 Standards for Energy Management System Integration

IEC 61970-1 Energy management system application program interface (EMS-API)—Part 1: Guidelines and general requirements.
IEC/TS 61970-2 Energy management system application program interface (EMS-API)—Part 2: Glossary.
IEC 61970-301 {Ed.1.0} Energy management system application program interface (EMS-API)—Part 301: Common information model (CIM) base.
IEC 61970-302 Energy management system application program interface (EMS-API)—Part 302: Common information model (CIM) financial, energy scheduling, and reservations.
IEC/TS 61970-401 Energy management system application program interface (EMS-API)—Part 401: Component interface specification (CIS) framework.
IEC 61970-402 Energy management system application program interface (EMS-API)—Part 402: Component Interface Specification (CIS)—Common services.
IEC 61970-403 Energy management system application program interface (EMS-API)—Part 403: Component interface specification (CIS)—Generic data access.
IEC 61970-404 Energy management system application program interface (EMS-API)—Part 404: Component interface specification (CIS)—High speed data access.
IEC 61970-405 Energy management system application program interface (EMS-API)—Part 405: Component interface specification (CIS)—Generic eventing and subscription.
IEC 61970-407 Energy management system application program interface (EMS-API)—Part 407: Component interface specification (CIS)—Time series data access.
IEC 61970-453 Energy management system application program interface (EMS-API)—Part 453: Exchange of graphics schematics definitions (common graphics exchange).
IEC 61970-501 Energy management system application program interface (EMS-API)—Part 501: Common information model (CIM) XML codification for programmable reference and model data exchange.

15.13.3.9 Other IEC Standards

IEC 62351-1 Data and communication security—Part 1: Introduction and overview.
IEC 62351-2 Data and communication security—Part 2: Glossary of terms.
IEC 62351-3 Data and communication security—Part 3: Profiles including TCP/IP.
IEC 62351-4 Data and communication security—Part 4: Profiles including MMS.
IEC 62351-5 Data and communication security—Part 5: Security for IEC 60870-5 and derivatives.
IEC 62351-6 Data and communication security—Part 6: Security for IEC 61850 profiles.
IEC 62351-7 Data and communication security—Part 7: Management information base (MIB) requirements for end-to-end network management.

15.13.3.10 ISO Reference Models (Available from www.iso.ch)

ISO/IEC 7498-1:1994 2nd edn.: Information technology—Open systems interconnection—Basic Reference Model: The Basic Model.

ISO/IEC 7498 Security Architecture, Part 2 (superseded by ISO/IEC 10745 and ITU-T X.803 "Upper Layers Security Model," ISO/IEC 13594 and ITU-T X.802 "Lower Layers Security Model," and ISO/IEC 10181-1 and ITU-T X.810 "Security Frameworks, Part 1: Overview").

ISO/IEC 7498-3:1997 2nd edn.: Information technology—Open systems interconnection—Basic Reference Model: Naming and Addressing.

ISO/IEC 7498-4:1989 1st edn.: Information processing systems—Open systems interconnection—Basic Reference Model—Part 4: Management Framework.

16
Physical Security of Substations

	16.1 Introduction	16-1
	Definitions	
	16.2 Electric System Today	16-3
	Size • As an Essential Service • Structure • Need for Security	
	16.3 Threat Assessment	16-5
	Presence, Capability, and Intent • Intruders	
	16.4 System Analysis	16-8
	Criticality Assessment • Vulnerability Assessment • Risk Assessment	
John Oglevie		
POWER Engineers, Inc.	16.5 Risk Management	16-10
	16.6 Responsibility for Security	16-11
W. Bruce Dietzman	Owner/Operator • Federal Government • NERC Security Guidelines	
Oncor Electric Delivery Company	16.7 Implementation (Methods)	16-12
	Physical Methods • System Methods • Contractual Methods • Management/Organizational Methods	
Cale Smith Oncor Electric Delivery Company		
	References	16-20

16.1 Introduction

Electric substations exist in almost every neighborhood and in every city. They are adjacent to or within the town proper of every small town. They exist in isolated, remote, and rural areas. More often than not, they exist next to every major commercial or industrial facility in the country. They range in size from fused cutouts that separate a couple of feeders to huge substations covering tens of acres. They exist at voltages up to 765 kV AC and ±500 kV DC.

They are also essential to our everyday life. As the nerve centers of our electricity supply system, substations are essential components of its reliable operation—hence the need to protect them from all types of threats: internal and external and cyber and physical.

But how do you protect the thousands of substations spread out around the country? Who is a threat? What needs to be protected? How do you protect it? How much does it cost? Who is responsible?

A threat can be either man made or natural. Events such as September 11, 2001, the Northeast Blackout of 2003, and hurricane Katrina of 2010 have raised the bar for security and reliability. In addition, now that terrorism is a part of the energy lexicon, the federal government has established organizations to audit the electric power industry. Not only are they collecting information about the loss of power, but they have also mandated the reporting of these events in real time to allow the detection of a coordinated attack on the system.

As a result, the protection of the electric system now requires that an owner/operator use a systematic, comprehensive approach to the development and design of a physical substation security system.

The intent of this chapter is to help shed light on this complicated topic and to answer the most pressing questions with regard to the physical security of substations. Its intent is not to answer all questions and provide definitive answers but to be a primer for the development and implementation of a physical security program.

A security program can be designed and implemented using a systematic approach. This chapter discusses the basic elements of such an approach that will determine the type of physical security most appropriate for a particular electric system. This includes the following critical steps:

1. Threat assessment
2. System analysis (which includes)
    a. Criticality assessment
    b. Vulnerability assessment
    c. Risk assessment
3. Risk management
4. Implementation

It is important to note that this chapter only discusses physical substation security, specifically security measures to prevent human intrusion. The chapter does not discuss security programs and methods for power plants or the physical security methods for substations attached to nuclear power plants since the substation security methods for these stations are generally included in the plant's security program. Cyber network security, the prevention of intrusion into a substation using electronic methods (hacking), will only be discussed in so far as it relates to assuring electronic equipment in a substation is secure from intrusion once other physical security systems have been breached.

16.1.1 Definitions

This chapter uses several key terms with specific definitions. The following table outlines these key definitions:

Term	Definition
Asset	Any substation or substation component
Critical asset	Substation or substation components, which, if damaged, destroyed, or rendered unavailable, would affect the performance of business, the reliability or the operability of the electric supply system
ESISAC	Electric Sector Information Sharing and Analysis Center, the organization responsible for communicating with other security organizations and the government about threat indications, vulnerabilities, and protective strategies
Intruder	Any individual or organization performing unauthorized activity within a substation
Intrusion	Unauthorized entry into a substation. Intrusion can be a person or organization, a vehicle, a projectile, or it can be electronic (hacking)
NERC	North American Electric Reliability Council, the organization whose principal mission is to set standards for the reliable operation and planning of the bulk electric system
Owner/operator	Anyone, or entity, that has the responsibility for the security and safety of an electric supply substation
Risk	The probability that a particular threat will exploit a particular vulnerability of a substation
Risk management	Decisions to accept exposure or to reduce vulnerabilities by either mitigating the risks or applying cost-effective controls
Substation	Meant to be all inclusive for this chapter. The term substation includes all stations classified as switching, collector bus, transmission, and distribution substations
Terrorist attack	A specific type of threat that involves the use of violence and force, if necessary, to deliberately cause the destruction or functional loss of a substation
Threat	Any event or circumstance with the potential to damage or destroy a substation or a component of a substation
Vulnerability	The degree to which a component of a substation is open to a threat

16.2 Electric System Today

In order to put the need for physical security into its proper context, we must understand the size and importance of the electric system to our current way of life.

16.2.1 Size

According to U.S. Energy Information Administration (EIA) the electric system today is growing at a significant rate. In 2009, the total electricity demand for the United States was 3745 billion kW h. The EIA projects that the demand on the electric system will grow another 31% by 2035, totaling a demand of 4908 billion kW h [1]. In 2002, the total installed generation capacity for the United States was 813 GW [2]. By 2009, installed capacity had reached 1025 GW [3]. Due to these high projected growth rates, utilities will have to increase their investment in the electric system in the coming years to support consumer demand. This will include the construction of new substation facilities.

16.2.2 As an Essential Service

The system size and growth rate are directly tied to its need within our society. Every facet of our life and our economy is dependent on the presence of electricity. It is the key essential service (see Figure 16.1). Paul Gilbert, the Director Emeritus of Parson's Brinckerhoff, Inc. in testimony before Congress for a hearing on the "Implications of Power Blackouts on America's Cyber Networks and Critical Infrastructure," outlined the importance of this system: "Our way of life is dependant on a highly utilized set of infrastructure components that provide our communities and way of life with vitally needed services and support. Only the electric supply system has the unique ability to seriously impact, or cause the complete loss of all of the others" [4].

As the *key* essential service, our electric system comes with a daunting requirement: it must never fail. One measure of the increased need for system reliability is availability. Over the past decade, the total time the system is allowed to be unavailable per year has dropped from hours to fractions of a minute. *IEEE Power and Energy* published availability standards in its September 2005 issue (see Table 16.1), which demonstrate the high degree of availability required, even for "managed" system types (99% available).

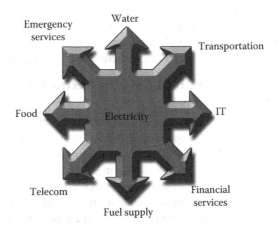

FIGURE 16.1 Electricity is the key essential service within our society. (From Gent, M., *IEEE Power Energ. Mag.*, 1(1), 46, 2003.)

TABLE 16.1 Availability Standards Require a High Degree of Availability

System Type	Unavailability (min/year)	Availability (%)	Availability Class
Unmanaged	50,000	90	1
Managed	5,000	99	2
Well managed	500	99.9	3
Fault tolerant	50	99.99	4
Highly available	0	99.999	5
Very highly available	0.5	99.9999	6
Ultra highly available	0.05	99.99999	7

Source: Gellings, C. et al., *IEEE Power Energ. Mag.*, 2(5), 41, 2004.
Even managed systems need to be operating 99% of the time.

16.2.3 Structure

The overall structure of the electric utility system has changed dramatically in the past 50 years. The electric system today has evolved away from the post–World War II, vertically organized system of public and privately owned utilities. At that time, these companies owned their own generation, transmission, and distribution systems. Today, the general trend is for the electric system to be horizontally organized into separate generation, transmission, and distribution companies. It is comprised of owner/operators that consist of public and private entities, independent power producers, and industrial companies. To allow equal access to the electric system for all, independent system operators (ISO) have been formed to schedule and manage power flows across the system.

With the Energy Policy Act of 2005, Congress has now mandated "the development of a new mandatory system of reliability standards and compliance that would be backstopped in the United States by the Federal Energy Regulatory Commission (FERC). On August 8, 2005, President Bush signed into law the Energy Policy Act of 2005, which authorizes the creation of an electric reliability organization (ERO) with the statutory authority to enforce compliance with reliability standards among all market participants" [5].

16.2.4 Need for Security

Many believe that the size of the system, the rate of growth, the lack of investment in recent years, the demand on reliability, and the changes in business structure make the system more vulnerable to failure than in the past. These vulnerabilities coupled with the electric system's role as the key essential service mean our way of life cannot tolerate the loss of the system due to natural disasters, human-made disasters, or terrorist attacks. Hence, the need for protection of these facilities is more important now than ever.

16.2.4.1 IEEE Standard 1402-2000, IEEE Guide for Electric Power Substation Physical and Electronic Security

In the mid-1990s, a group of utility engineers and private consultants worked together to develop a physical security standard for electric supply substations. The result was IEEE Standard 1402-2000, IEEE Guide for Electric Power Substation Physical and Electronic Security. This standard was completed prior to the events of September 11, 2001. The term "intrusion" was coined at that time to define any unauthorized entry into a substation whether it was by a person, a vehicle, a projectile, or electronic means (hacking). The focus was on the most prevalent problems all owner/operators were facing at the time. The guide looked at ways intrusion could occur; the stages of property ownership and use; and geographical, economic, social, and industrial conditions that could influence when an intrusion would occur.

Physical Security of Substations

The group conducted a nationwide survey to collect data on the effectiveness of the most common security systems in use at the time. The balance of the guide discussed the various security methods available and weighed their associated advantages and disadvantages. A final section was added that outlined the structure of a security program.

Since IEEE Standard 1402 was last published, national events have brought security and reliability to the forefront. The protection of the electric system now requires that the owner/operator use a systematic, comprehensive approach to the development and design of their physical substation security system. These changes will likely result in significant revisions to IEEE Standard 1402 in coming years.

16.3 Threat Assessment

The first step in a comprehensive security plan is the threat assessment. The threat assessment identifies the outside elements that can cause harm to a substation or its component equipment. A threat is any event, circumstance, or sequence of steps that leads to substation property being stolen, damaged, or destroyed. A threat can be a single act such as a gunshot at the equipment located in a substation or it can be a series of acts such as an intruder cutting or climbing a fence to enter a substation, crossing the yard, breaking into the control building, and then operating circuit breakers. In both cases listed earlier, the motives of the person or group that creates a threat to the system may be different. A gunshot can be an act of frustration during hunting season or it can be a deliberate attempt to damage or destroy the substation or its equipment.

Regardless of intent, an owner/operator must choose the appropriate physical security methods to protect the substation from as many threats as possible. For the purposes of the threat assessment, the owner/operator must be able to identify potential threats and potential vulnerabilities.

Intruders create potential threats. Intrusion is the unauthorized entry onto the substation property or into the fenced area of the substation. An intruder then can be a person or a group, or it can be an object such as a bullet or a rock. If a person or object is not authorized to be there, then he, she, or it is defined as an intruder. The owner/operator must be able to identify the persons or groups that pose the threat and the circumstances or conditions that may allow them to gain unauthorized access. But the selection of individuals or groups with the potential to cause harm is not a random process based on speculation. It must be a methodical investigation of the potential intruders active in the area and the type of activities they are likely to perform.

16.3.1 Presence, Capability, and Intent

Any individual or group to be watched should meet each of these three criteria. They should have

- *Presence*—Although nationally or internationally active, has the group or individual exhibited local activity?
- *Capability*—Does the group or individual have the ability to attack the substation?
- *Intent*—Does the group or individual have the interest and desire, that is, the intent, to attack or harm the substation in any way?

Be aware of the activities of those groups specifically identified to be operating in your area. Look for and be vigilant for signs of activity and take extra precautions, if warranted. This can include general threats to the electric industry, its customers, or special interest groups, which have been threatened. Look for warning signs that activity might be imminent. This can include people seen at sites with no apparent reason for being there, missing documents or credentials, etc., which would be necessary to gain knowledge or access to critical sites and components. In 2003, the Department of Homeland Security and the Federal Bureau of Investigation issued a memorandum on suspicious activity.

It is important to make the distinction between perceived and actual threats to your system. For example, the International Council on Large Electric Systems (CIGRE) conducted a survey in 2004 that assessed the perceived threats to electric power substations. Thirty-five percent of the respondents believed terrorism was the largest threat, followed by theft and vandalism (both around 20%). However, actual historical information reported by the survey respondents in the same report was in sharp contrast to these perceived threats. This information showed the highest percentage of intruder events involved theft (32%). Terrorism was not on the list.

A comprehensive threat assessment must include a review of historical data, that is, the threats faced as well as the potential threats. The physical methods to be used must take both into consideration. But, when looking at the potential threats, it is vitally important to apply the test of presence, capability, and intent in order to assure that the physical security methods chosen are focused on the probability of a threat to your substations rather than speculative possibilities based on national/international events. For instance, the focus today is on international terrorist groups. However, national terrorist organizations such as the Environmental Liberation Front (ELF) may pose a more immediate and larger threat. In another instance, if a terrorist organization is active and making threats against the federal government, what is the probability that this organization will become a threat in a particular area to an owner/operator of an electric supply system?

16.3.2 Intruders

For the purpose of this chapter, all potential threats to a substation involve intrusion. They are initiated by outside sources or influences. To reiterate, intrusion is defined as the unauthorized access to a substation property. The gunshot in the aforementioned example is an intrusion even though no one entered the yard. The second example is more easily understood as an intrusion because it involves the act of someone entering the substation. However, in both cases the acts are equal under the definition of intrusion. Intrusion can be accidental, deliberate but nonspecific (vandalism), or deliberate with a premeditated purpose (malicious destruction or a terrorist act). Intruders can be classified as follows:

- General public
- Thieves
- Vandals
- Disgruntled employees
- Terrorists

16.3.2.1 General Public

The general public is the group that most frequently comes into contact with these facilities. On a daily basis we drive by and/or work next to some portion of the transmission system. In the event that we come into contact with substations, it is generally by accident. Unplanned, "accidental" intrusion is the category of least concern. The threat to the system is minimal. The risk to the intruder is the largest concern.

16.3.2.2 Thieves

Theft is driven by economic conditions. Theft will rise and fall with the cost of copper, aluminum, and other common metals. In one incident, a utility used concertina (razor) wire on the substation fence to prevent intrusion only to have a thief steal the wire instead. Thieves will enter a site for its copper ground wire, copper control and power cable, or any aluminum material stored on site. Thieves will break into a control building in hopes of finding tools, components, or hardware that may be of value. Attempts to track down thieves and recover stolen material have had mixed results. In most cases, the materials can be found but it is very difficult to prove who the original owner is. Without proof of ownership, you cannot prove it was stolen.

Preventing theft is the most common reason for many of the physical security methods discussed later in this chapter and the most difficult to prevent. Thieves can be of any age and of any economic background.

Although their purpose is premeditated, their choice of targets is random and unpredictable. Any substation that is deemed to have the requisite metals or materials will be of interest to the thief.

16.3.2.3 Vandals

Individuals, couples, groups of teenagers, vagrants, drug dealers, and urban gangs all have been known to use a substation property as a gathering place to have parties, to reside, or to conduct unlawful activities. The damage these groups do to a substation may or may not be deliberate and malicious, but these actions generally are considered random. They have not targeted the system or the substation for a political or social purpose and, therefore, their actions can be classified as acts of vandalism. Some actions practiced by these groups that can lead to a forced outage of the equipment include the discharging of firearms, which could, in turn, damage circuit breaker and transformer bushings, the throwing of chains over an energized bus, and the damaging of control panels in control buildings.

Communities want substations to be invisible. Today's communities view substations as detractors to the value of their property. The owner/operator will try to make them more acceptable by planting trees and shrubbery around them to give them an aesthetic, environmentally pleasing appearance. However, hiding the substation can increase the probability of vandalism by making the grounds more attractive as a meeting place where their actions can be hidden from the public eye.

In many cases, especially in urban locations, the property can be part of a gang's territory. To indicate this, they will mark their territory with the type of graffiti called "tagging." By tagging the property, they have warned rival gangs of their claim to the property, which can lead to "turf wars" between these gangs on the property that, again, increases the threat of destruction to the property.

16.3.2.4 Disgruntled Employees

There are numerous instances on record of disgruntled employees causing damage within a substation as the result of a grievance with the owner/operator. They pose a higher threat than the previous three because they represent an internal threat to the system. Their acts are not only deliberate; they possess specialized knowledge that can be used to increase the impact to the owner/operator and to the community.

There are many cases of disgruntled employees opening the oil valves on transformers. There is a case on record of a disgruntled employee attaching an improvised explosive device (IED) to a piece of electrical equipment. One disgruntled employee during the 2002 Winter Olympics went into the substation control building and opened several specific breakers, which caused an area power outage of 1 h affecting 33,000 residential customers [6].

Disgruntled employees usually work alone. They prefer to remain anonymous. Their anger is focused solely at their employer. Their acts are not intended to cause regional or national outages.

16.3.2.5 Terrorists

Terrorism is defined in the United States by the Code of Federal Regulations as "… the unlawful use of force and violence against persons or property to intimidate or coerce a government, the civilian population, or any segment thereof, in furtherance of political or social objectives" (28 C.F.R. Section 0.85).

There is no generally accepted definition or classification in use today for terrorist groups. They can be classified as either domestic or internationally based. They can be grouped into such general categories as social or political groups, environmental and religious extremists, rogue states or state-sponsored groups, and nationalist groups. They can act for social, ideological, religious, or apocalyptic reasons. In today's world, rather than being structured groups like those of the 1960s and 1970s, they now can be ad hoc groups that coalesce, come together for an attack, and then disband and disperse. In today's world, they practice a form of leaderless resistance. That is, the specific tactics are left up to the individuals so that compromise of a part does not compromise the whole [7].

Regardless of the method used to classify them, terrorists differ from all of the other groups discussed in several ways. They are determined. Their intent is to interfere with some process or the completion of some project. Their actions are planned and carefully organized; their intent is to bring public attention to their cause. They wish to create chaos, panic, and terror within the community or to create a diversion as a cover for some other primary but unrelated target. They can be well funded; they can have access to weapons and intelligence.

16.3.2.6 Resources

Aside from an owner/operator's own historical database of incidents, there are numerous other resources available to assist with gathering information about potential intruders that may be active in a particular area. The most critical step is for an owner/operator to establish a relationship with local and state police agencies to get a sense of the threat conditions in their area. In addition to these, other essential sources include the FBI's InfraGard program, the U.S. Bureau of Alcohol, Tobacco, and Firearms (ATF), U.S. Marshals, and state and local law enforcement organizations.

Online Internet sources include the National Memorial Institute for the Prevention of Terrorism (www.mipt.org), the Antidefamation League (www.adl.org), and LexisNexis (www.lexisnexis.com). In addition, there are numerous private security consulting companies that can be retained to assist.

16.4 System Analysis

The second step in a comprehensive security program is to determine in what way the system may be affected by the threats identified previously. System research includes determining the criticality of the assets an owner/operator owns (stations and components) and the vulnerabilities they possess. Once these two are known, then the risk to loss of these assets can be quantified.

A comprehensive system analysis must take into account two important ideas: a substation's "criticality" and a substation's "vulnerability." There is a distinct difference between the criticality of a substation or its components and the vulnerability of the same substation or its components. Criticality is a measure of the importance of the substation or equipment to the performance of the system. A substation or substation's components are critical if their loss affects system reliability and system operations or represents an unacceptable loss to the owner/operator. The more unacceptable a loss is and the more impact it has on the electric system, the more critical the asset becomes.

Vulnerability, on the other hand, identifies individual weaknesses or ways in which the substation or its components can be compromised. All substations have some form of vulnerability. These two terms are explored further in the following sections.

16.4.1 Criticality Assessment

All substations require some level of physical security but not all substations are critical or have critical assets. Factors that would raise or lower the criticality of a substation would include its importance to generation, switching or load, voltage levels, the type of bus arrangement, overall age, whether it is enclosed in a building or outdoors, above or below ground, or the effect the loss of the station would have on public image. Criticality would be assigned an increasing value as the effect of its loss moves from local, to regional, to the overall system. Its value would also be increased in relation to the cost of replacement power while the station is out of service or the cleanup costs if a spill occurs and the impact to public perception. The criticality of individual components would be raised or lowered based on replacement costs, delivery time, location of manufacturing (domestic or foreign), type of communications and EMS equipment they use, or if they have highly specialized technology, such as a high voltage, solid state, electronically controlled device, for example, a static VAR compensator (SVC) or DC/AC inverter.

For the owner/operator, a critical asset can be a substation with only local importance. However, at the federal level, ESISAC would classify a critical asset as only those assets that would affect the reliability or operability of the bulk electric system. Each owner/operator must decide what constitutes a critical asset for themselves. However, in general, the criticality of a substation or its components can be determined by the following conditions:

- *Unable to provide service to customers.* Does the substation serve critical or sensitive loads such as emergency or other essential services, military installations, or major industries that have been the target of threats themselves? Does critical load served by this substation have alternate sources of power available during emergencies? What is the maximum length of time required to restore or replace damaged critical equipment in this substation? Is this length of time acceptable? Can the system or its critical and sensitive loads tolerate loss of power for this length of time?
- *Unable to maintain system integrity.* Is the substation designed to provide at least double contingency protection for all critical loads and lines?
- *Unable to maintain system reliability.* Will loss of this equipment cause area wide voltage fluctuation, frequency, stability, or reliability problems?

16.4.2 Vulnerability Assessment

As stated earlier, vulnerability is the individual weaknesses or ways in which a substation or its components can be compromised. All station equipment is vulnerable to damage or destruction from an external threat, such as a gunshot, but there are other vulnerabilities as well: the operation of breakers and control equipment by someone in the control building; damage or destruction of equipment operator mechanisms; the loss of control and protection schemes by cutting control wiring in the yard, in marshalling cabinets or in the back of relay panels in the control building; the loss of primary protection communication channels by switching them off; bus faults caused by throwing chains over air-insulated buses; structure failure caused by the removal of structural bolts or destruction of insulator supports; or the loss of system control by switching equipment from remote to local control using the manual switches in the control building. These vulnerabilities are exacerbated by other environmental factors (which are vulnerabilities in and of themselves). These include the location of a substation (urban, suburban, rural, or industrial), access to a substation by the public, and whether it is manned or unmanned. The stages of development of a substation (preconstruction, storage, construction, operation, and when decommissioned) are also vulnerabilities that affect the risk of a threat.

16.4.3 Risk Assessment

The third component of the system analysis is the risk assessment. To manage risk, it must first be measured. Once measured, the risk to substations and components can be prioritized. To perform a risk assessment, the owner/operator must first have a completed threat, vulnerability, and criticality assessment.

To measure risk is to quantify the probability that a threat to a specific vulnerability will occur, resulting in loss of a critical component or substation. Typical scenarios would include

- What is the risk of a gunshot penetrating the wall of a critical transformer, causing failure of the transformer?
- What is the risk that the loss of the aforementioned transformer will cause a severe capacity overload of the electric system resulting in a system cascade?
- What is the risk that an intruder breaking into a control building will destroy or disable the protection schemes for several critical transmission lines? What is the risk this will occur without detection by the owner/operator?
- What is the risk that failure to one of these lines will lead to major damage of the line before repair or replacement of the damaged protection equipment?
- What is the risk that an intruder will break into the control building of a substation and gain control of the substation via the local control switches?

- What is the risk that an intruder will break into the control building of a substation and gain control of the substation using a laptop to access the SCADA RTU or protective relays? What is the risk that the intruder will use this access to gain access to databases or gain control of equipment located in the operations center or other substations?
- What is the risk that theft of a portion of the substation ground grid will cause a severe disturbance to the system?
- What is the risk that someone will be injured or killed during the attempt to steal portions of the ground grid?

A substation or a substation's components must be susceptible to the identified threat for there to be a risk of failure. In addition, the risk must be ranked in importance to the owner/operator and to the system as a whole. For instance, in the case of the transformer described earlier, its importance to the system may be critical and risk of intrusion to the substation quite high. But, if the only identified threat to the substation is entry to steal the ground grid, then the overall risk associated with the transformer, the critical component within the substation, and the operability of the substation itself may be very low. Hence, extra effort to protect the transformer from damage or destruction from this type of threat may not be warranted. Only the owner/operator of the electric system can make these decisions.

A number of risk assessment guides along with sample work sheets have been created to assist with the aforementioned process. Four, in particular, are located on ESISAC's website, http://www.esisac.com/library-assessments.htm. They are

- Vulnerability Assessment Methodology, September 2002
- Energy Infrastructure Risk Management Checklists for Small and Medium Sized Energy Facilities, August 19, 2002
- Energy Infrastructure Vulnerability Survey Checklists, February 2002
- Vulnerability Assessment Survey Lessons Learned/Best Practices, September 2001

16.5 Risk Management

Now, in order to take appropriate action for the risks that have been identified, an owner/operator must review the data collected during the system analysis (discussed in Section 16.4) and determine the best physical security methods. Together, the analysis and decision process is called "risk management."

The dictionary definition of risk management is "a decision to accept exposure or to reduce vulnerabilities by either mitigating the risks or applying cost effective controls" [8]. In the case of physical security of the electric system, there are four possible avenues to manage the risk associated with a threat. The owner/operator can

- Accept the risk by acknowledging that all risk cannot be eliminated.
- Avoid the risk by preventing the incident from happening.
- Mitigate the risk by reducing the consequences.
- Transfer the risk by moving it to another party such as insurance or contract [9].

Risk management is practiced all the time in the design of a station, its layout, and protection for system-type failures. Consider a power transformer. Manufacturing defects, age, and operating conditions create vulnerabilities that can individually—or in concert—cause a transformer to fail under through-fault conditions. Since it is recognized that these types of failures can happen, owner/operators employ various bus arrangements, parallel transformers, switching schemes, and protection schemes to protect the transformer and the supply of electricity to customers from the consequences of this type of vulnerability.

In this example, the fault is the threat, while the age and operating condition of the transformer is the vulnerability. Risk management is exercised in several ways when dealing with the consequences of this type of failure. The risk can be eliminated by replacing the transformer before age and operating conditions make the risk of loss too high. The risk can be mitigated by implementing protection schemes designed to clear the fault before it can damage the transformer. The risk can be transferred by employing bus arrangements and switching operations to route the power to other equipment if the fault persists and the transformer bus must be cleared. Finally, insurance may be purchased to reduce the cost of replacement of the transformer if it is damaged or destroyed.

16.6 Responsibility for Security

Security concerns for the electric system are no longer the responsibility of the owner/operator alone. Recent legislative changes and federal agency rule implementations have shifted some of the substation owner/operator's responsibilities to a shared responsibility between the owner/operator and the federal government.

16.6.1 Owner/Operator

For an effective physical substation security program, it is important that an owner/operator identify the person (or persons) or department who will have responsibility for security implementation and administration. To be successful, a security plan should include a policy, procedures, and guidelines in order to be effective across the organization. To establish a successful security plan, the policy must be created and mandated at the corporate level, then monitored at all levels to assure its implementation. Defined levels of responsibility, along with procedures and guidelines that define specific tasks, are required throughout the organization.

Each company should have someone specifically in charge of substation security. This individual should be responsible for assuring that a security plan is developed, implemented, regularly reviewed, and updated. The regular inspection of each substation to assure that security measures are in effect should be part of the security plan. In addition, employee training and the development of methods that enable employees to report irregularities or breaches of security are also necessary. Management must accept a high level of responsibility to make sure that those security systems installed are maintained at all times.

16.6.2 Federal Government

The federal government has taken a number of actions in recent years to both legislate and mandate the auditing of the level of preparedness for incidents, monitoring the operation of the system, and reporting in real time the occurrence of outages. The most significant recent acts include

- The federal government has established security agencies specifically required to monitor, collect data, and perform analyses on the probability, and probable location, of a terrorist strike on the electric supply system.
- The federal government has initiated security audits of major utilities.
- The Energy Policy Act of 2005, which establishes the creation of an ERO, has been enacted. The ERO will have the ability to levy fines for outages that affect the stability and reliability of the electric system.
- The federal government has set up guidelines defining a significant power outage and is now requiring the mandatory, real-time reporting of these occurrences to the Department of Energy (DOE).
- The federal government is requiring owner/operators to verify the ratings of their existing lines and equipment and publish on one-line drawings the "Limiting Series Element."

16.6.2.1 Reportable Incidents

The Department of Energy has prepared a guideline that defines a reportable incident. Reportable incidents include every kind of outage that can occur in a system including those that can be caused by vandalism. They include

- Loss of firm power
- Minimum 3% voltage reductions
- Voluntary load reductions if needed to maintain power
- Vulnerability action that includes any incident that degrades reliability
- Fuel supply emergencies
- Other events that result in continuous 3 h or longer interruption

16.6.2.2 Reporting Procedures

In the event of an outage, a full technical report may be required by the DOE from the utility including restoration procedures utilized. The report is required on a timely basis. During the event, the DOE encourages interim notification via email on a "heads up" basis to be followed by a full report later.

Government form OE-417 covers the reporting requirements and the format for the report. The form can be found on the EIA website at http://www.eia.gov/cneaf/electricity/page/forms.html.

16.6.3 NERC Security Guidelines

To assist the owner/operator with the development of security management departments, security reporting and methods, NERC has developed a comprehensive set of guidelines [10]. These guidelines are being developed in cooperation with the industry and cover in more detail the topics we have covered in this primer. The guidelines can be obtained at http://www.esisac.com/library-guidelines.htm. They cover the following areas:

- Physical security
- Vulnerability and threat assessment
- Threat response
- Emergency plans
- Continuity of business processes
- Communications
- Employment background screening
- Protecting potentially sensitive information

16.7 Implementation (Methods)

Upon completion of a vulnerability analysis and risk assessment, a program needs to be implemented that addresses the findings of those studies. This section provides many methods that can be used in such a program. These methods have been used to prevent or deter access to a substation—or failing that, provide notification that access has occurred. Each of these methods has a cost/benefit question that varies by application and should be reviewed by the user prior to application. All of the methods are listed for their security value only.

While a number of methods listed are in common use today, many others are still in the developmental stage. Some of these methods are meant to be suggestive, imaginative, or thought-provoking in an attempt to encourage new methods to be developed by the reader.

Security requirements should be identified in the early design stages of a substation project. Generally, it is more economical to anticipate and incorporate security measures into the initial design rather than retrofit substations at a later date. The type of security used should take into account the type of intrusion it will likely see over the course of its life.

Physical Security of Substations   16-13

The types of methods listed here have been divided into four categories: physical, system, contractual, and organizational/management. Each of these categories represents a different approach. While they may be used alone, it is highly recommended that security systems be layered by using a combination of methods.

In utilizing these methods, the user must take into consideration whether the intended measures are counterproductive. Do the measures actually aid the intruder? Do they interfere with visual patrols putting maintenance personnel at risk? Do they interfere with the operation and maintenance of the substation? For example, solid walls are ideal for preventing equipment damage due to projectiles, but they also prevent drive-by inspection of the interior by patrols from the outside. Once inside, an intruder can, if detection equipment is not present, roam the substation property without detection. There are reported incidents of maintenance crews being attacked during daylight hours by individuals that have entered a substation with solid walls. Locking gates while a crew is inside is one solution, but some owner/operators have instituted rules that prohibit locking gates so the crews inside have an exit route during emergencies.

16.7.1 Physical Methods

16.7.1.1 Fences and Walls

As a minimum, every substation should have a perimeter fence around that portion of the property used for energized equipment. It is the first line of defense against intrusion. Internal fences provide an additional safety barrier for trained personnel working within the yard. Fences may also be used to prevent vehicular intrusion into the substation. Two or more layers of fencing might also be employed for additional protection.

The IEEE/ANSI Standard C2 of the 2012 National Electric Safety Code (NESC) can provide basic fence requirements. The NESC sets minimum requirements for height and material. It also specifies a minimum distance from internal energized equipment to the fence. In addition, there are numerous IEEE standards and guides detailing the fencing requirements for various applications such as shunt and series capacitor banks and shunt reactors.

Fences can be constructed of various materials. A fabric fence will stop the casual intruder from entering the substation yard. They usually consist of a chain link fabric attached to metal posts with three strands of barbed wire at the top to discourage an intruder from climbing the fence to gain entry. This barbed wire may be angled to the inside or the outside of the substation, subject to local ordinances.

Most applications use a commercial grade, galvanized fabric, consisting of either 3.0 mm (11 gauge) or 3.8 mm (9 gauge) wire. The mesh opening size should preferably be 50 mm (2 in.) but not larger than 60 mm (2.4 in.) to resist climbing. The addition of fiberglass or wooden slats to the fence fabric can also provide visual screening.

The height of the fence should be a minimum of 2.1 m (7 ft) aboveground line. In areas of the country that experience snowfall, it is desirable for the fence to be of sufficient height such that 2.1 m (7 ft) of the fence projects above the maximum snow accumulation. In higher risk areas consider using razor wire, coiled in a cylinder along the fence or angled toward the outside away from the facility. Areas vulnerable to vehicle penetration may require 19 mm or larger aircraft cable mounted to the fence supports, inside the mesh fabric, at a height of approximately 0.76 m (30 in.) aboveground level to reinforce the fence. The U.S. Department of State provides various crash rated designs, which can be accessed for further information. Please refer to "Specification for Vehicle Crash Test of Perimeter Barriers and Gates, SD-STD-02.01, Revision A," dated March 2003." This specification provides for various "K" ratings, which are based on vehicle weight and speed.

Metal fabric fences should be grounded for safety. Further information is available in Chapter 11. Solid walls, which can be of either masonry or metal construction, are an alternative to fencing. Solid walls have the advantage of providing screening of the substation and its equipment. In addition, solid walls may prevent external vandalism such as gunshot damage. The walls should be constructed to provide a barrier similar in height to the fencing discussed earlier and in a manner that does not provide hand or footholds.

16.7.1.2 Gates and Locks

Access to a substation is generally through a swinging gate of various materials. The gate should provide at least the same level of protection as the substation fence or wall. Crash-proof gates of an acceptable rating might be considered. (See U.S. Department of State specification mentioned earlier.) No openings that would allow a small child to enter the substation should be allowed.

All entrances to a substation should be locked. Control building doors should have locked metal doors. All outdoor electrical equipment should have a provision for locking cabinets and operating handles. Padlocks should utilize nonreproducible keys.

It is strongly recommended that a key control program be implemented to control the distribution and return of all keys. This will ensure that each person in possession of a key is accountable for the location and control of the key.

16.7.1.3 Landscaping

While landscaping can be utilized to screen the substation and its electrical equipment from view and may be required to obtain public acceptance or approval, this section may be best described more as what *not* to do. Any landscaping treatment around substations should be carefully designed so as not to create hidden areas that are attractive to people and groups as a meeting place. Landscaping must be regularly maintained. Trees and shrubbery must be pruned to avoid concealing intrusion and illegal activity. A "pride in ownership" appearance, along with regular personnel visits, will keep many would-be intruders away from the facility. No tree should be planted in a location that will allow the tree to be used as an access or climbing aid into the substation. The final growth dimensions of the tree should be considered in determining this location.

Landscaping, in the form of berms, can be used to prevent direct line of site with equipment similar to solid walls or fences with fabric. Although they require additional land, berms may be less expensive to construct than walls or fences with fabric screens. The additional land can provide a clear buffer zone around the substation. Berms may also be a deterrent to vehicular intrusion into a substation. A series of small nonlandscaping-type berms can be installed around a substation outside the fence so that a vehicle attempting to penetrate the substation property will become high centered. Keep in mind that these berms need to be placed so that a vehicle would not be able to navigate between them.

16.7.1.4 Barriers

Access to energized equipment and bus work may be of concern if the perimeter security measures are breached. Polycarbonate or other barriers on ladders and structure legs can be used to provide additional barriers. Refer to the NESC and Occupational Safety and Health Administration (OSHA) requirements. These barriers may also function as animal deterrents.

All sewer and storm drains that are located inside the substation perimeter, with access from the outside, should be fitted with a vertical grillwork or similar barrier to prevent entry. Manhole covers or openings should be located on the inside of the substation perimeter fence and locked down.

Driveway barriers (gates, guardrails, ditches, etc.) at the property line for long driveways can help limit vehicular access to the substation property. Additionally, use boulders, jersey barriers, or impassable ditches in locations where vehicle or off-road vehicles can gain access to the perimeter fence.

16.7.1.5 Grounding and Ground Mats

Theft of metals, particularly copper, is one reason for intruders to enter a substation. To reduce the availability of copper, the use of a copper-clad steel conductor for substation ground grid system construction is recommended. Additional considerations should include

- Making connections from the grounding grid to the fence fabric and fence posts below grade
- Covering exposed grounding connections with conduit or other material to hide from view
- Adding a hardening, ground enhancing material to the soil used to backfill the grounding grid trench. This material will increase the diameter of the ground grid conductor making it more difficult to pull from the ground
- Placing a notification sign on the fence stating the use of copper-clad steel conductors
- Forming a partnership program with local police officials and scrap yards to identify and recover stolen materials

16.7.1.6 Lighting

The exterior and interior of the substation may be designed with dusk-to-dawn lighting. A typical minimum lighting level of approximately 20 lx (2 fc) is suggested. However, it is recommended that the use of sodium vapor lighting be avoided if the lighting is intended to assist with the identification of intruders or if lighting is intended to be used in conjunction with video surveillance equipment. This type of lighting produces a yellow or orange cast, which will interfere with attempts to identify the person, his clothing, and the description of the person's vehicle.

Wiring to the lighting posts should be in conduit or concealed to minimize tampering by an intruder. Areas outside the fence, but within the facility property, should also be considered for lighting to deter loitering near the substation. Community acceptance of this level of lighting may prohibit its use. Light fixture covers and lenses should be damage resistant.

16.7.1.7 Control Building Design

In general, most building materials provide adequate security protection. However, the type of building construction should be suitable for the level of security risk. This construction could include hardened walls, armor plating of outside building walls, and impenetrable ceilings. Additional features that should be included are steel doors with tamper-proof hinges and roof-mounted heating/air-conditioning units. Any wall openings (i.e., wall air conditioners) should have security bars over and around the unit. A building that is part of the perimeter fence line should be at least as secure as the fence.

16.7.1.8 Security Patrols

At critical substations or in areas where vandalism has been a chronic problem, the judicious use of a security patrol service should be considered. A partnership can be established with local law-enforcement agencies to facilitate these patrols. Specific security procedures should be established that identify who handles security alarms and what the response notification procedures should be within the company and with local law enforcement agencies. These procedures may be augmented to include rapid response by local authorities to substation sites when alerted by the owner/operator, posted security guards, and required identity checks during unusual occasions—labor disputes, major events, or visiting dignitaries.

16.7.1.9 Signs

Signs should be installed on the perimeter fence to warn the public that

- There is a danger of electrical shock inside.
- Entry is not permitted.
- Alarm systems are providing security for the substation.

Please refer to the NESC, as there are additional requirements.

16.7.1.10 Clear Areas and Safety Zones

In addition, structures and poles should be kept a sufficient distance from the fence perimeter to minimize the use of structures as a climbing aid.

Where practical, a 6 m (20 ft) to 9 m (30 ft) clear zone around the exterior of the perimeter fence should be considered. This zone will provide a clear field of view that will make it easier to detect someone from illegal entry.

16.7.1.11 Site Maintenance

Frequent and routine inspection of the site should be provided to ensure a minimum level of care. Maintaining the substation in a clean and orderly fashion can help discourage various groups from using the property as a meeting place. Any deficiencies should be rapidly reported.

16.7.1.12 Intrusion Detection Systems

Numerous systems are now on the market to detect unlawful entry to a substation. These include video, motion, sound, and seismic detection systems that can distinguish, among other things, two-legged versus four-legged intruders along with their speed of movement, sound, and vibration. Before employing any of these systems, the owner/operator must determine the objectives for their use. For instance, if the purpose is to stop an intrusion in progress, the anticipated response to a detected intrusion by local authorities must be determined. If no response or a delayed response is all that is possible, then the use of these systems will still act as deterrent but may not provide their original intention.

Sophisticated motion detection systems, video camera surveillance equipment, and building security systems are becoming more common. The software associated with these systems can be customized to filter out some nuisance detections. The systems use local processors for detection and only send data to the operations center when a change is detected. All of these systems are designed to provide early detection of an intruder.

Perimeter systems using photoelectric, laser-sensing, fiber optics, or microwave may be utilized to provide perimeter security. Generally, these sensors are mounted on a fence and alarm upon detection of either movement of the fence or the characteristic sound that a fence will make when it is moved. These can be effective to detect intruders that either climb the fence or pry the fence up at the bottom. Other technologies use a fiber-optic cable embedded in the fence fabric that look at the speckle pattern and received level of light, then alarm at a predefined threshold. In both cases, the systems are able to locate the source of the alarm. Other perimeter systems use microwave technology to set a volume (three-dimensional space) of detection around the exterior of the station, immediately inside the station, or around critical equipment. Defeating this type of detection system is much more difficult than the fence-mounted systems because of the size of the three-dimensional space within the detection zone.

Video systems can be deployed to monitor the perimeter of the substation, the entire substation area, or building interiors. They can have zoom lenses added to them that allow reading gauges located in the yard. They can be programmed to move to a specific combination of angle and zoom to provide a clear reading of a level or temperature gauge. They can also be moved to view any specific area of the substation yard where illegal entry is suspected. Video systems are available that use microwave and infrared to activate a slow-scan video camera, which can be alarmed and monitored remotely and automatically videotaped.

Control building door alarm systems are quite common. These systems include, at a minimum, magnetic contacts on all the doors and have provisions to communicate to the operations center through the existing telephone network or SCADA system. They can also include a local siren or strobe light located on the outside of the building that is activated under an alarm condition. The systems should be capable of being activated or deactivated using an alphanumeric keypad, keyed switch, or card reader system located inside the building. All siren boxes and telephone connections should have contacts to initiate an alarm if they are tampered with.

16.7.1.13 Video Motion Detection Systems

A completely new family of technologies is becoming commercially available that utilizes a CCTV coupled with motion sensing equipment for intrusion detection. The concept of video motion detection is not new. When it was first introduced three decades ago, an alarm was generated when one or more pixels in a video scene changed. While this provided a very sensitive way to detect motion, it turned out to be very difficult to implement in an exterior situation because of the many ways an exterior scene changes that does not necessarily represent a potential intrusion. For example, the passage of a cloud over the sun would darken all or a portion of the field of view, generating an alarm.

With the advent of digital cameras and the increase in available on board computing power, a number of products are now on the market that hold promise for cost-effective video-based motion detection in the exterior environment. These particular products offer a number of attractive features that allow the end user to tailor the performance of the system to their specific site and security needs. These features include the ability to define multiple detection areas in a camera's field of view. The software also uses various decision algorithms to distinguish between normal and threatening activity. These algorithms can include shape filters to identify the source of the motion. Other filters include direction of motion, speed of motion, and consistency of internal motion of the object. They can also provide the ability to perform various camera control functions upon detection of motion. This would include zooming in on the source of the motion upon detection and, in some cases, tracking the moving object subsequent to detection.

16.7.1.14 Substation Service

Two or more substation/station service power sources are recommended. A third source based on generation from a local fuel supply may be warranted.

16.7.1.15 Personnel Access

As described in Section 16.7.1.1, substation access keys should only be provided to personnel on an "as-needed" basis. For critical substations, this group may be limited even further. Electronic access systems using electronic identification cards and touch pad password readers should be considered. The identification cards should be used only for substation access and the passwords changed periodically. A higher level of security can be provided by biometric devices, if required, such as fingerprint and retinal pattern readers. Smart badges (utilizing RFI devices), which can follow the location of personnel within the substation, are also available.

16.7.1.16 Drawings and Information Books

The storage of information at a substation facility, which could allow a knowledgeable intruder to cause damage or increase the damage that is done, should be avoided. Serious consideration should be given before making the following information available on site:

- Station construction drawings
- One-line and three-line drawings
- Relay functional drawings
- Relay and SCADA manuals
- Electrical equipment manuals
- Anything containing password information
- Easy access to this information from an onsite Internet device

16.7.1.17 SCADA/Communication Equipment

Special attention should be paid to SCADA and communication systems. These systems should be provided with at least two independent power sources. Installing these systems in a separate, hardened location—accessed with a separate entry control—is also recommended. This location might be shielded to restrict electronic intrusion.

16.7.1.18 Relay and Control Equipment

The proper operation of relaying and control equipment systems is critical. Unauthorized personnel must be restricted from access and easy operation of these systems. The following is a list of possible methods to prevent unauthorized access to relay and control equipment:

1. Primary and backup relaying for each line is not the same or on adjacent panels.
2. No panel, protective equipment, or switch legends. All equipment is unidentified on the panels.
3. No visible control switches or the switches are remote mounted in a hardened cabinet (patrolman carries switch handles with them).
   a. Dual switch handles one on the panel and one master switch remotely located, so it takes two people to operate a switched device such as a breaker.
   b. Specially keyed shafts similar to antitheft lugs used on car tire lugs. Unique shafts would prevent use of common switch handles taken from other substations or ordered from a manufacturer being used in a critical substation.
   c. Control switches with a socket instead of shaft protruding from the panel to prevent turning the switch with pliers. The shaft is part of the switch handle instead of part of the switch.
4. Each panel covered with tamper-proof doors front and back with bullet-resistant glass on the front doors, the breaking of which should set off alarms.
5. Each relay requires a password to access its display panel and activate function keys. Two passwords are required: one created by a random number generator and issued by the system operator within an hour of the request to access the relay; the other is linked to the Personnel ID system.
6. Relay communication ports use intrusion detection cables.
7. Encryption software on each intelligent electric device within the substation should be activated.
8. Remote access to the intelligent electric device through the use of a dial-up phone line can only be accomplished by closing a SCADA operated contact from the operations center. This access can be automatically limited to a specific time interval.

16.7.2 System Methods

Some security measures are much more effective than others. None are 100% effective, either singly or working in combination. With this in mind, the security of a substation must be viewed in the larger context of the security of the overall electric system. To provide this greater security, the following should be considered:

- Perform load flow, stability, and reliability studies that measure how the system will react if key equipment or substations are destroyed or put offline. Focus the upgrade and design of new substations with built-in contingency capacity and equipment that mitigates the effect of the loss of key substations and equipment.
- Build a redundant network. The transmission system should remain stable during the outage of any two lines or substations anywhere in the system. This is a double contingency (N-2) planning criteria.
- Maintain spares, for critical hard-to-replace pieces of equipment. This can include transformers, breakers, and even substation structures.
- Partial assembly and storage of all the equipment necessary to build an entire substation. Partial assembly would allow a more rapid response during an emergency.
- Earmark spare equipment that is critical to the region and enter into agreements with the utilities in the region for the storage, availability, and use of the equipment.

Physical Security of Substations

- Design substation yards to include wide internal roadways that provide easy access for mobile transformer equipment. Provide attachment points for mobile transformer equipment to minimize the time required to restore partial power to critical loads.
- Use underground feeders for critical loads.
- Consider distributed generation resources near critical loads or near critical substations that do not meet the N-2 double contingency criteria.
- Consider placing new substations indoors using gas-insulated equipment and use underground feeders.
- Use single-phase units for very large power transformers. Single-phase units are easier to move in emergency situations than three-phase units. Also, they can be shared more easily with other utilities in the immediate area.
- Consider alternate bus designs that provide higher levels of line protection and source-load redundancy. Space equipment farther apart, if possible, to prevent collateral damage.

16.7.3 Contractual Methods

If the detection and deterrent methods described earlier are unsuccessful, the replacement or restoration of damaged equipment may be required. The cost of these actions can, of course, be covered by an insurance contract, but, to expedite this work, the following contractual arrangements or plans should be considered:

1. Establish agreements with adjacent utilities in regard to common design standards, which will facilitate the use of spare equipment for emergencies.
2. Establish construction contracts with specific contractors for the emergency repair and replacement of failed equipment.
3. Negotiate agreements with major equipment manufacturers for the emergency production and delivery of major, long lead, or special design components that cannot be or are not readily available from storage. Such equipment should include
    a. Large power transformers
    b. Solid-state switching equipment associated with SVC and AC/DC inverters
    c. Specialized cooling systems for critical components
    d. Circuit breakers if their voltage class or fault duty present long lead time delivery
    e. Switches when manufacturing lead times exceed acceptable limits
    f. Steel structures when manufacturing lead times exceed acceptable limits
    g. Electrical bus, insulators, and connectors
    h. Control cable
4. Establish agreements with local law enforcement agencies, pubic or private, to provide rapid response during emergencies or when called to a substation site by the operations center. Provide their personnel with substation safety training to assure their conduct and safety within an energized substation.

16.7.4 Management/Organizational Methods

The methods used for substation security should be modified as needed to address current and changing needs. The following are suggested to determine those needs:

- Continually review past events and determine how to upgrade or configure the system to prevent deliberate and similar events.
- Establish alliances with other utilities to develop and share information about the history of intrusion and the results of security analysis.
- Engage the services of a security consultant to help develop profiles and perform risk assessments.

The following additional organization and procedural methods that will limit access to sensitive information are also provided:

- Do not publish design guides and design standards that provide installation and operation details about security devices to be adopted for substations.
- Develop security design details on separate drawings and place these drawings in separate secure file directories.

References

1. U.S. Energy Information Agency. 2011. *Annual Energy Outlook 2011*, p. 73.
2. IEEE-USA, Position Paper on *Electric Power Reliability Organization*. November 13, 2002; www.ieee.usa.org
3. U.S. Energy Information Agency. 2011. Existing Capacity by Generation Source, 2009.
4. Gilbert, P., Testimony at joint hearing on implications of power blackouts on America's cyber networks and critical infrastructure. *108th Congress, 2nd Session*. Congress, National Academies of Engineering, 2003.; www.nae.edu
5. NERC Web site, introductory comments about NERC.
6. The Salt Lake Tribune, Friday, September 20, 2002.
7. NA, Field Manual Section 1: Principles Justifying the Arming and Organizing of a Militia, 1994, Wisconsin, The Free Militia, p. 78. *IA/NE Conference*, October 04, 2005, David Cid.
8. www.utmb.edu/is/security/glossary.htm
9. 71 ESISAC Risk-Assessment Methodologies for Use in the Electric Utility Industry.
10. Security Guidelines for the Electricity Sector DOC43.
11. Gent, M., Reflections on security. *IEEE Power and Energy Magazine*, 1(1), 46–52, 2003.
12. Gellings, C., Samotyj, M., Howe, B. The future's smart delivery system. *IEEE Power and Energy Magazine*, 2(5), 41, 2004.

17
Cyber Security of Substation Control and Diagnostic Systems

17.1 Introduction **17**-1
17.2 Definitions and Terminology.. **17**-4
17.3 Threats to the Security of Substation Systems.................... **17**-7
SA System Security Misconceptions • SA System Threat Actors
17.4 Substation Automation System Security Challenges **17**-11
Slow Processors with Stringent Real-Time Constraints •
Legacy Real-Time Operating Systems That Preclude Security •
Insecure Communication Media • Open Protocols • Lack of
Authentication • Organizational Issues • Lack of Centralized
System Administration • Large Numbers of Remote
Devices • Substation Diagnostic Systems
17.5 Measures to Enhance SA Cyber Security.................................**17**-14
Protecting Substation Systems against Cyber Intrusion •
Detecting Cyber Intrusion • Responding to Cyber Intrusion
17.6 Devising a Security Program **17**-24
17.7 Future Measures... **17**-25
Authentication and Encryption • Secure Real-Time
Operating Systems • Functional Cyber Security Testing
and Certification • Test Beds • Incident Reporting Sites •
Intrusion Prevention/Detection Systems and Firewalls •
Developed and Emerging Technical Standards and Guidelines

Daniel Thanos*
GE Energy–Digital Energy

References.. **17**-28

17.1 Introduction

There have been significant cyber security developments that affect the substation automation (SA) industry and broader critical infrastructure with the advent of four defining events; the development of the North American Electric Reliability Corporation (NERC) Critical Infrastructure Protection (CIP) standards [1], the U.S. Department of Energy (DoE) stimulus funding of Smart Grid projects, the development of National Institute of Standards Technology (NIST) Smart Grid Guidelines [2], and the first verified automated and targeted cyber attack on control systems known as Stuxnet [3,4] that is believed to have had significant physical impacts on the assets which it was programmed to strike. The NERC CIP standards that are mandated on the bulk power system (BPS) [5] have created a tidal shift in how cyber security is viewed by asset-owning organizations and it is having significant impacts on operations and

---

* 2nd edition author and editor.

associated technologies both positive and negative as all stakeholders try to wrestle with what security means in this new technical domain. For the first time, noncompliance to the CIPs carries the levying of significant fines of up to one million dollars per day. The advent of the U.S. DoE-backed Smart Grid has meant an unprecedented infusion of government investment matched by utility organizations to modernize the power grid and introduce a wide range of control system automation, communications, and information technologies to enable general modernization, smart meters (for demand response), distribution automation, large introduction of clean distributed generation (DG), large-scale support of electric vehicles, wide area situational awareness, and many other applications.

These Smart Grid paradigm changing technologies have introduced three fundamental shifts to the physical and logical underpinnings of the power grid, namely, two-way communication, control, and power flows. All of these supported by new information technology (IT) as encapsulated by general computing and network systems (e.g., general corporate systems for enterprise applications, network management, and general workstations), industrial control system (ICS) automation technologies and supporting IT elements (e.g., PC software for HMIs, SCADA, SA engineering, and configuration tools) as encapsulated in operational technologies (OT), and the communication technologies that connect IT and OT systems within themselves and to each other. These three shifts are a fundamental challenge to the engineering principles that have kept the power grid reliable for nearly a century and it is introducing cyber risks that have reliability impacts that did not previously exist on a wide scale. In this regard, the NERC Smart Grid Task Force (SGTF) has produced a report that discusses possible reliability impacts of the Smart Grid to the traditional BPS [6]. As industry and government have begun to realize the cyber risks interconnected and distributed automation (DA) technologies will have on the new emerging Smart Grid, they have necessitated the development of cyber security guidelines by the U.S. NIST in order to give common grounding for building secure and interoperable systems especially for U.S. DoE stimulus funded projects. Those guidelines are known as NISTIR 7628 [2] that are freely available in three volumes plus an introduction. The guidelines are comprehensive and cover the following: high-level requirements to inform a complete security program (Volume 1), consumer privacy issues (Volume 2), and, of particular note, the bottom-up-oriented technical analysis of Smart Grid cyber security issues (Volume 3). The latter gives detailed content on cyber security design considerations and analysis for OT systems found throughout the Smart Grid (including SA) that are legacy to emerging in nature. In Volume 3, there is also a section on future security R&D that will be needed as the scope, function, and complexity of the Smart Grid begin to be more fully realized in new types of automation technologies.

On a very real and technical level, Stuxnet has forever ended the debate of whether esoteric and so-called "proprietary" or "obscure" ICS technologies can be targeted and successfully cyber attacked (without needing an attacker to be physically present or being a privileged insider) in practice. It has done so using a complex composition of advanced computer malware and virus technologies, system penetration, social engineering, ICS scanning, and exploitation, and all in an automated and remote fashion with a level of stealth previously thought to only exist in creative works of fiction. Stuxnet represents what is effectively a new breed of advanced cyber weaponry (malicious computer software/hardware with the ability to target and cause material harm to physical equipment and people), in all likelihood engineered with the resources of a nation state, and with evidence of the operational planning and precision commonly found in many military and intelligence organizations. However, what is most concerning is that Stuxnet is not a closely held state secret, only to be used in instances of military or covert intervention by hopefully responsible and reasonable actors. Rather its malicious technical genius has now been made public on the Internet through the Github social code sharing site, and without doubt has been downloaded and analyzed by countless individuals and groups [7]. Stuxnet code can now be further learnt from and adapted to retarget originally affected ICSs and perhaps new ones with further innovation by attackers and their sponsoring hostile nation states. This sobering reality provides a compelling reason to address cyber security in SA systems with a vigor that perhaps has not existed in industry before this event. While Stuxnet was extremely sophisticated, there are practical procedures, adaptive methods, and technologies that can be deployed and continually improved to provide levels of defenses against the threats that it represents, as well as the ones not yet known. We will begin to discuss some of these in this chapter.

All of these factors of new cyber threats, regulations, standards, and guidelines have created substantial market incentive for security innovation and adoption for the power and automation systems industry. This has led to a sizable increase of product and service offerings that seek to help utility and supporting vendor organizations become "compliant" and "secure." However, it has been clear for some time to those that are professionals in the field of critical infrastructure cyber security that compliance does not necessarily mean security, and procurement of security solutions targeted at IT systems does not necessarily give any real assurance of security for OT systems. In fact the converse can be true as power and automation system reliability risks can increase at a greater rate than the cyber risks one perceives to be mitigating by introducing poorly designed and configured security technologies that are not meant to fulfill OT requirements. This runs contrary to the fundamental goal of cyber security in power and automation systems, which is to ensure cyber elements do not at worst degrade reliability and at best help enhance it. Examples of this abound, but consider the implementation of a common lockout security access control in which after n number of failed authentication attempts a system may block subsequent access for a certain defined period of time, or in more restrictive cases may even lockdown the entire system until certain procedures are executed to restore access. The motivating factor behind such a control is based on denying an attacker the ability to execute a dictionary attack (e.g., use a program to automatically guess many classes of alpha-numeric passwords) and correctly guessing a password for a crucial system account or operation (these attacks can be remarkably effective). In many classical IT system contexts, this is a sound control and can be implemented as is, especially if the primary security goal is protecting access to certain sensitive data or information, as in this case, the confidentiality of the information is a higher security priority than the availability of the system. An unauthorized disclosure would be catastrophic to the business and hence it is the governing risk that needs to be mitigated above all others, or in other words all things being equal we can accept loss of availably over loss of confidentiality. Now let us consider an SA system perhaps used to trigger a breaker (for the protection of physical power system assets) on certain detected power system events. If we were to implement this security control without any detailed system analysis and to its highest protection level, we could impede reliability and create a denial of service (DoS) vulnerability condition that is trivial for an attacker to exploit. The attacker could purposely lock down the automation system (effectively taking it offline) with knowledge of the protection it is providing by purposely triggering the number of failed authentications and then execute the primary attack through another control/communications channel that is meant to cause damage to the targeted power system asset. Similarly, if the SA system is providing visibility into the power system through certain measured values, or other critical device statuses, the attacker may want to "blind" the monitoring and controlling processes at the exact point of an attack as to degrade the ability to detect and respond to it, thus making containment unlikely and wide spread damages certain. Poorly designed and implemented security controls in an OT environment carry dire impacts and can give enhanced capabilities to attackers while giving system operators a false sense of security. It is clear that within the context of SA systems availability is the governing cyber risk that must be mitigated above all others, and this means having a much more nuanced understanding of how to apply classical IT security controls that fit the context of OT systems. Further to this, one also has to understand that security models in any critical control automation system are more weighted on the idea of gradient failure and resilience, as opposed to the binary nature of many security controls that are either satisfied or not and often dictate a denied access to a system when certain conditions are not met. It is important to realize that the power grid and its supporting automation systems at their very foundation are designed to accept failures of equipment and still operate by switching over to other systems or detaching other systems or loads/generation to keep grid wide stability and reliability while allowing for recovery of affected systems. Any cyber security must thus be precisely aligned to this type of gradient failure and resilience model while having a deep contextual knowledge of underlying power and SA systems to make truly intelligent decisions for mitigating all the known risks.

TABLE 17.1 Differences between the Traditional Threats to Utility Substation Assets and Contemporary Threats

Traditional Threats	Contemporary Threats
The threat is direct damage to the physical assets of the utility	The threat is damage to utility computer systems, which may lead to damage to the physical assets
The threat is local	The threat originates from local or distant sources
The threat is from an individual	The threat may come from individuals, competitors, or well-funded and highly motivated organizations
An attack occurs at a single site	An attack may be unleashed simultaneously at many sites within many utilities, and may be coordinated with cyber or physical attacks on other elements of key infrastructure
A successful attack causes immediate and obvious damage	A successful attack may be undetected. It may result in changes to utility software that lie dormant and are triggered to operate at some future time
A successful attack causes obvious damage	A utility may not know the nature of the damage to software caused by a successful attack
An attack is a single episode	As a result of an attack, software may be modified to cause continued damage
Restoration can take place safely after the attack	Since the attacker may still have access to the systems, restoration plans can be impacted

In this chapter, we will attempt to give some introductory guidance on the topic of cyber security as they apply to SA systems and present some of the challenges that must be addressed as well as how to start the process of finding the right solutions for that to be accomplished. What will be discussed will in part be based on knowledge of existing standards, guidelines, and best practices as referenced throughout the chapter. This chapter should not be treated as a guide for achieving security compliance with NERC CIP [1] or be viewed to be comprehensive enough to build a security program or plan as informed by NISTIR 7268 [2]; the scope of such a work product would require its own dedicated book. The contents of this chapter are only meant to provide an introduction to the SA/ICS security field and provide the reader a starting point to become further acquainted with all the deeper technical issues and literature that is available for their further research and education. For further exploration of issues in ICS cyber security, there is a dedicated text by one of the coauthors of the first edition of this chapter, which is recommended reading [8]. Lastly, the guidance and examples presented here are by no means meant to be exhaustive or a replacement for the sound thinking and experience of cyber security and SA/ICS engineering professionals working together to practically identify and mitigate all system risks as they uniquely apply to the diversity of operations that exist in the industry. If there is one central piece of guidance, we can leave the reader with it is to work with true SA/ICS focused cyber security professionals from the planning/design phase of their SA systems to their implementation and deployment. These professionals can often be found in utility, vendor, and consulting organizations; it is best to have each of them as stakeholders for any SA project. Lastly, to give a practical and initial lay understanding of cyber security in SA systems, please refer Table 17.1. It summarizes the differences between the traditional threats to utility substation assets and contemporary cyber threats. (The traditional threats have by no means evaporated; the new threats have to be seen as an addition to, and not as a replacement of, the traditional threats.)

17.2 Definitions and Terminology

AAA   Authentication, Authorization, and Accounting is typically an authentication server infrastructure that is responsible for centrally authenticating user and device accesses to applications and networks. This is the authentication component. The server stores a user and/or their group memberships that specify the permissions

	they have in target applications, devices, or systems. This is the authorization component. Many AAA servers implement user activity accounting or logging as commonly found in remote authentication dial in user service (RADIUS) implementations [9]. This is the accounting component. Another common example of functional AAA is lightweight directory access protocol (LDAP) [10] based servers.
BPS	"Bulk-Power System means facilities and control systems necessary for operating an interconnected electric energy transmission network (or any portion thereof) and electric energy from generating facilities needed to maintain transmission system reliability. The term does not include facilities used in the local distribution of electric energy" [5].
CIGRE	CIGRE is the International Conference on Large High Voltage Electric Systems. CIGRE is recognized as a permanent nongovernmental and nonprofit-making international association based in France. It focuses on issues related to the planning and operation of power systems, as well as the design, construction, maintenance, and disposal of high-voltage equipment and plants.
Cyber Security	Security (q.v.) from threats conveyed by computer or computer terminals; also, the protection of other physical assets from modification or damage from accidental or malicious misuse of computer-based control facilities.
Default Password	A "password" is a sequence of characters that one must input to gain access to a file, application, or computer system. A "default password" is the password that was implemented by the supplier of the application or system.
DNP3	Distributed Network Protocol is a nonproprietary communication protocol (q.v.) designed to optimize the transmission of data acquisition information and control commands from one computer to another.
Firewall	A device that implements security policies to keep a network safe from unwanted data traffic. It may operate by simply filtering out unauthorized data packets based on their addresses (source and destination) and ports/service types, or it may involve more complex inspection of the sequence of messages to determine whether the communications are legitimate according to their connection state, as is commonly found in stateful firewalls that are capable of stateful packet inspection (SPI). A firewall may also be used as a relay between two networks, breaking the direct connection to outside parties. Simple firewalls are typically found integrated into Router or VPN equipment and can provide port/service control as well as SPI. Firewalls or similar network filtering technology can also be found in endpoint devices such as PCs running many generic operating system (OS) platforms, or even many newer forms of SA embedded systems. Basic firewalls typically provide filtering of packets based on layers 2 and 3 (e.g., IP and TCP/UDP) of a network. Typically, firewalls provide the first level of defense in a network from external threats. They are commonly used to form the outer layer of what is known as the electronic security perimeter (ESP) as defined in the NERC CIP standards [1] with deeper forms of protection further into the ESP being provided by intrusion detection/prevention systems (see the following).
IDS	Intrusion Detection System is a hardware device or software application that monitors the traffic on a communication line or operations on endpoint devices themselves (e.g., file access patterns on PC host device) with the aim of detecting and reporting preidentified unauthorized activities (rule set–based detection) or anomalous behaviors (anomaly-based detection) by any communication stream entering the network. A network-based IDS is also known as a network intrusion detection system (NIDS). An endpoint device–based IDS is also known as a host-based intrusion detection system (HIDS). At their most fundamental level, IDSs are

programmed to identify, track, and classify (as types of attacks) patterns of activity by monitoring specific data in network packets or other sources of information found on endpoint devices (e.g., executing process and memory activity and file accesses) with the intent of altering an operator so that action can be taken to respond to a possible attack that is detected and prevent widespread impact. Modern NIDSs typically accomplish their detection functions through deep packet inspection (DPI). That is the ability to inspect and trigger alerts based on the contents of a packet (header and payload) (this can be from layer 1 to 7, or simply the ability to detect application layer intrusions down to the link layer of the network) and the connection state (SPI). An NIDS can be configured to monitor both incoming and outgoing network traffic.

IEC
International Electrotechnical Commission is an international organization whose mission it is to prepare and publish standards for all electrical, electronic, and related technologies.

IED
Intelligent electronic device is any device incorporating one or more processors with the capability to receive or send data and control from or to an external source (e.g., electronic multifunction meters, digital relays, and controllers) [11].

IPS
Intrusion Prevention System is the same as an IDS with the added ability to take predetermined and automated actions on the detection of an intrusion attempt and hence its ability to prevent attack. Examples can include dynamically writing a firewall rule that blocks the source address of a detected attack attempt and pushing it to a firewall for automatic protection, or the IPS itself functioning as the firewall and dynamically blocking the communication at the source address. IPS also applies to HIDS in that anti-malware technologies can prevent file infections, network communication, and other actions on an endpoint device upon a threat agent being detected. Most modern IDSs are really IPSs with the automatic protection disabled and having the hardware or software providing only alerts or logging that is monitored by a higher order security monitoring system known as a security event manager (see the following). Throughout this text, IDS and IPS can be used interchangeably expect when making specific note of one over the other. Also when referring to the endpoint or the host-based version, we will make specific mention of the HIDS acronym.

NAC
Network Access Control is method and technology commonly implemented on networking infrastructure or as software agents on an endpoint that ensures a device meets a network's security policy before it is admitted to the network. Examples include running scans to ensure the device has the latest: antivirus/anti-malware installed with updated signatures, operating system patches, and other critical configurations. NAC is a preventative security technology that is designed to limit risks from entering secure networks by ensuring the endpoint devices that associate to them meet all the security requirements of the network.

NIST
National Institute of Standards and Technology.

Port
A communication pathway into or out of a computer or networked device such as a server. Ports are often numbered and associated with specific application programs. Well-known applications have standard port numbers; for example, port 80 is used for HTTP traffic (web traffic).

Protocol
A formal set of conventions governing the format and relative timing of message exchange between two communication terminals; a strict procedure required to initiate and maintain communication [12].

Remote Access
Access to a control system or IED by a user whose operations terminal is not directly connected to the control systems or IED. Applications using remote access

	include Telnet, SSH, and remote desktop software such as PC Anywhere, Exceed, DameWare, and VNC. Transport mechanisms typical of remote access include dial-up modem, frame relay, ISDN, Internet (TCP/IP), and wireless technologies.
RTU	Remote Terminal Unit is the entire complement of devices, functional modules, and assemblies that are electrically interconnected to effect the remote station supervisory functions. The equipment includes the interface with the communication channel but does not include the interconnecting channel [13].
Security	The protection of computer hardware and software from accidental or malicious access, use, modification, destruction, or disclosure [14].
SEM	Security Event Manager is typically enterprise class software or a hardware-based appliance that centrally aggregates all event logging (both dedicated security logs and general logs) found in devices and systems throughout the network and does further analysis (through event detection rule sets and/or correlation analysis) to provide alerts to event and incident responders. These systems are typically designed to process a very large volume of events per/second in order to provide as close to real-time performance as possible for detecting possible malicious behavior over a very large network that can be hundreds to thousands of devices. Their data store can usually be queried and any number of reports built on monitored event logs. This makes SEMs useful for supporting security audit and forensic functions beyond the centralized monitoring they are meant to provide. It is also important to note that SEMs fulfill a critical security event monitoring and analysis gap that a network of IPSs cannot provide on their own and that is the centralized log processing and event correlation. This is because what may look to be a benign/authorized access pattern on single point in the network may actually be a broader attack if the same pattern is occurring simultaneously at many points in the network. The same can be true of certain events recorded on endpoint devices. One example could be the detection of a single user account logging into hundreds of devices simultaneously through the analysis of access logs. A security analyst would often write a rule with some thresholds to detect this type of activity and provide an alert as it is usually indicative of a compromised account and a possible automated attack. Finally, industry has also come to call an SEM, a security event information managers (SEIM), these terms can be used interchangeably, and we will use SEM throughout this chapter.

17.3 Threats to the Security of Substation Systems

17.3.1 SA System Security Misconceptions

There are some fundamental differences with SA systems with their associated SCADA and IEDs as compared to classical business IT systems. Hence, the scope of how classical IT security applies to substation environments is also not always well understood, especially when considering cases where various security controls can degrade reliability as discussed in Section 17.1. There are also misconceptions about security technologies that can often be found in both the SA and classical IT domain as it is easy to try and trivialize security as a few technologies and concepts. We will discuss a few of these.

Electronic security perimeters as implemented by Firewalls are sufficient to secure a SA network against all threats. Classical SPI firewalls (and their lesser stateless predecessors) are a tool to provide basic network segregation and control (early firewall technologies are ineffective against most/all modern attacks) from threats originating from an external source. If configured properly basic firewalls can also enhance the reliability properties of the SA system by ensuring improper network traffic does not accidently enter/leave the SA network through port/service control. The term accidental is key because for intelligent attackers of malicious intent classical firewalls are insufficient and more advanced IDS/IPS technologies tailored and configured for SA are needed to form a more secure ESP.

Virtual private network (VPN) and other forms of network link encryption or "tunneling" provide the highest degree of security. Firstly, there are many kinds and grades of VPNs and they are not all equally secure. It is enough to say that VPN has become an often abused term in industry and it can mean securely encrypted networking with a full suite of strong cryptography and accepted ciphers (see the cryptography and key management system requirements in NISTIR 7628 [2]) or it can mean nothing more than providing a means for disparate remote systems to share a logical network over a public network, but with no encryption and security. IP Security (IPSec) [15] and transport layer security (TLS) [16] based VPNs are good examples of encrypted and secure technologies; they are also a benchmark with which to compare the security of other solutions. However, it is important to understand that an encrypted channel alone is not a guarantee of security, in fact if not implemented correctly VPNs can provide cover for attacks. For example, if encrypted traffic is going through an IDS/IPS, it renders DPI useless as the packets cannot be inspected and analyzed for malicious or unauthorized activity. Therefore, it is necessary that VPN connections are terminated and the outer layer of the ESP before moving into the inner layer for IDS/IPS filtering. The common security error of assuming VPNs infer trust on endpoint devices can introduce significant risks. The classical case of this error is when end users run corporate VPN clients on improperly controlled/personally owned computers and connect into a critical corporate network, or worse yet an SA network and their devices are infected with viruses or malware. In this case, just the scanning and infection attempts that many viruses and malware generate can flood a network or critically impede control devices to the point of causing a persistent failure that requires its return to manufacturing for repair/replacement. Lastly, there are performance issues that must be critically evaluated with many VPN and network link encryption technologies that may not make them suitably reliable for many hard-real-time classes of messaging that are required in SA systems (protection class messaging is an important example) because of the possible latencies and non-deterministic behavior that they introduce. This does not rule out their use, only that rigorous engineering study and analysis should be taken to evaluate and empirically understand the performance impacts that are introduced to ensure the worst-case expected performance of the system is preserved. Otherwise this becomes another example of cyber risk being mitigated at the cost of reliability, which is unacceptable as system reliability (equivalent to availability) is the governing risk to be mitigated.

Intrusion prevention systems (IPSs) can identify all ICS/SA system attacks and prevent them from ever getting to a control device/asset; thus, they completely solve the security problem. IPS (and IDS) technologies have application in SA systems; however, they are not a panacea that provides a universal solution against all threats, in fact they are not currently designed out-of-the-box to address SA devices and systems. They are primarily targeted for IT-based threats (which are not be discounted as IT attacks can impact ICS/SA systems) and currently need to be adapted and configured to address ICS/SA systems with considerable skill and understanding by ICS/SA and cyber security professionals. Thus, their, in the field, performance will vary considerably based on many factors (including the skill of the implementers) and face high probabilities of being removed as they are more likely to create a flood of false alarms (with insufficient design and tuning of rule sets), which become unacceptable to operational personnel as this becomes a distraction that impacts power system reliability. The latter can also mean that a poorly designed and implemented IPS can more easily enable an attack as a malicious agent can target a false attack to trigger a flood of alarms with the intent of distracting system operators or causing the IPS to be disabled long enough to execute their primary mission.

The network is secure because it is "private" and its architecture does not allow two-way command/data flows or routing to a public network such as the Internet. Most modern utility networks are consequently based on the IP protocol or use the IP protocol at some point in the infrastructure; hence, this makes them two-way and routable to outside networks by nature. Smart Grid applications through DA and DG, as well as wide area monitoring protection and control (WAMPAC) require two-way data/control by design. A Wide Area Network (WAN) may indeed be designed and deployed by a utility as a private network versus public or leased/shared networks; however, devices (other network devices or host endpoints with multiple network connections/interfaces) connect to "private" networks and routinely act as bridges to outside and untrusted networks. This often occurs without the knowledge of system managers and operators and can be the result of poor security policy and enforcement, erroneous configuration, insider attacks, or social engineering that makes use of insiders and their system access to the private network. One example of this reality is that many users bring their mobile smart phones to work (of particular note are the ones they personally own); these devices are almost always connected to the Internet through their 3G/4G cellular networks; it is not uncommon for these users to want to "share" this connection with their organizationally assigned computer (perhaps even a workstation or laptop managing an SA system: engineering tools, monitoring, or HMI) that can be accomplished through Bluetooth, USB, or WiFi wireless networking as many smart phones have embedded access points. In the latter case, it can also be offered up to their colleagues as well to offer a quick view of the news, a check of personal email, a social network status, etc. The point here is unless there are specific and technical network and end point device security controls that preclude this type of activity, the idea of a private network exists only in theory but not in practice. In other words, network architecture (private, segregated, and otherwise) as a security method is only as good as how it is technically enforced throughout the entire network and on every device that joins the network. Fortunately, methods such as NAC and associated endpoint security controls, Port Network Access Control (PNAC) (The IEEE 802.1X Standard), and physical network composition can provide the necessary technical enforcement. Lastly and very importantly, regular network vulnerability scans (especially tuned as to not impact fragile control system assets) can help periodically verify if all the devices (especially high-risk IT ones) on the network are correctly configured, controlled, and patched to meet the security policy and architecture of the network.

ICS/SA devices are obscure and highly proprietary and closed systems; thus, hackers do not understand how to attack them. IEDs, RTUs, PLCs, and other field-level control devices have in past been penetrated and attacked, regardless of the obscurity they enjoyed. As mentioned earlier, Stuxnet brought this entire line of thinking into a very public and spectacular end. Most, if not all, field-level devices can be disassembled physically and logically. Their hardware examined and firmware dumped often with a simple EEPROM reader. Hex editors and advanced reverse engineering tools that target embedded platforms that most of these field devices share can provide remarkable insight into their system internals and workings, often at a level that is greater than what the vendor understands as they are often building on top of other platforms and software development kits or have lost knowledge with many deployed legacy systems. Field devices are still controlled and monitored through applications that run on generic platform operating systems (often with a large number of unknown and exploitable vulnerabilities), regularly ones that are dated and have lost all security patching support. This makes attack and penetration of ICS/SA devices a simpler affair than many well-understood and open IT systems. There can be no substitute for evolving more modern security controls into ICS/SA devices themselves and treating the IT components that communicate with them as less trusted. In particular, this requires support of strong cryptographically based authentication measures using AAA services for engineering access and configuration, especially for firmware upgrade processes, as well as enhanced security event logging that can be integrated with SEM systems. In the future, as SA field devices become more autonomous in their control actions and peer-to-peer oriented in their communications, it will be necessary to embedded IPS-like and VPN-like functionality into them as well.

Our SA systems include the latest security technologies and we cannot be attacked. What is such common and accepted truth in the security field that it has become cliché is that there is no such thing as perfect security, and any adversary with enough time and money can successfully attack your system. Secondly, it is not technology alone (no matter how elegant and advanced) that makes any system secure; it is its people and their knowledge and training in concert with the technology and tools they have to work with that provides real security; this is especially true of SA systems. Many industries are rife with security "shelf-ware," that is, solutions that are procured, but are inadequate, or never deployed because of insufficient training to operate and manage them. The best guarantee of security is intelligence and agility and repeated exercises and training to handle actual cyber security incidents and their response. This can only be attained with a well-educated and cyber-trained work force that understands security as it applies to their role and responsibilities from the CEO to every worker in the organization. Most importantly, frontline SA system operators and engineers working together with cyber security personnel need to have a deep understanding of how to operate, manage, and continually evolve/improve their security systems and program to counter new and emerging threats. This requires constant education and training for all stakeholders with regular testing and cyber exercise that simulate real and plausible attacks and measure how effective an integrated ICS/SA and cyber security team is at detecting the attack and responding by stopping/containing it. This is often accomplished through what are known as red and blue team exercises. The red team being the simulated attacker and blue team the defender, or, in this case, the SA system operational team in concert with supporting cyber security personnel. It is highly recommended that skilled adversaries are hired in the red team, experts that are often known as penetration testers that understand and use all the latest cyber offensive technologies and methods, the same that a real and sophisticated attacker would use. Of course we also recommend that such exercises not be run on operational in the field systems, but in laboratory environments that would have identical systems and configurations to test and do exercises against.

17.3.2 SA System Threat Actors

Even though the general public may not be aware of these systems, the hardware, software, architecture, and communication protocols for substations are well known to the utilities, equipment suppliers, contractors, and consultants throughout the industry. Often, the suppliers of hardware, software, and services to the utility industry share the same level of trust and access as the utility individuals themselves. Consequently, the concept of an insider is more encompassing. A utility employee knows how to access the utility's computer systems to gather information or cause damage and also has the necessary access rights (keys and passwords). The utility has to protect itself against disgruntled employees who seek to cause damage as well as against employees who are motivated by the prospect of financial gain. Computer-based systems at substations have data of value to a utility's competitors as well as data of value to the competitors of utility customers (e.g., the electric load of an industrial plant). Corporate employees have been bribed in the past to provide interested parties with valuable information; we have to expect that this situation will also apply to utility employees with access to substation systems. Furthermore, we cannot rule out the possibility of an employee being bribed or blackmailed to cause physical damage or to disclose secrets that will allow other parties to cause damage.

A second potential threat comes from employees of suppliers of substation equipment. These employees also have the knowledge that enables them to access or damage substation assets; and often they have access as well. One access path is from the diagnostic port of substation monitoring and control equipment (see Chapter 7). It is often the case that the manufacturer of a substation device has the ability to establish a link with the device for the purposes of performing diagnostics via telephone and modem (either via the Internet or by calling the device using the public switched telephone network). An unscrupulous employee of the manufacturer may use this link to cause damage or gather confidential information. Additionally, an open link can be accessed by an unscrupulous hacker to obtain unauthorized access to a system. This has occurred frequently in other industries.

Another pathway for employees of the utility or of equipment suppliers to illicitly access computer-based substation equipment is via the communication paths into the substation. Ensuring the security of these communication paths is the subject of Section 17.5.

A third threat is from the general public. The potential intruder may be a hacker who is simply browsing and probing for weak links or who possibly wants to demonstrate their prowess at penetrating corporate defenses. Or the threat may originate from an individual who has some grievance against the utility or against society in general and is motivated to cause some damage. The utility should not underestimate the motivation of an individual outsider or amount of time that they can dedicate to investigating vulnerabilities in the utility's defenses.

A fourth threat may be posed by criminals who attempt to extort money (by threatening to do damage) or to gain access to confidential corporate records, such as those maintained in the customer database, for sale or use.

The fifth, and arguably the most serious threat, is from terrorists or hostile foreign powers. These antagonists have the resources to mount a serious attack. Moreover, they can be quite knowledgeable, since the computer-based systems that outfit a substation are sold worldwide with minimal export restrictions, and documentation and operational training is provided to the purchaser. The danger from an organized hostile power is multiplied by the likelihood that an attack, if mounted, would occur in many places simultaneously and would presumably be coupled with other cyber, physical, or biological attacks aimed at crippling the response capabilities.

However, the most likely "cyber threat" will be the unintentional threat. This is where substation control or diagnostic systems are impacted because of inappropriate testing or procedures. Even if laptop computers, TCP/IP communications, or both are used in substation environments, it does not mean that traditional IT testing such as scanning of networks should be performed without due care and caution.

17.4 Substation Automation System Security Challenges

Traditional IT business systems have been the object of a wide variety of cyber attacks as documented by CERT/CC, Computer Security Institute, and other computer security tracking organizations [17]. These attacks include an exploitation of programming errors in operating systems and application software, guessing or cracking user passwords, taking advantage of system installations that leave extraneous services and open ports open to attack, and improperly configured firewalls that do not exclude unauthorized communications. In addition to manifesting these common vulnerabilities, the control and diagnostic systems in substations have a number of special cyber vulnerabilities.

This section will not attempt to discuss the manifold vulnerabilities of conventional computer systems, which are well documented in other sources. Instead, this section describes some of the characteristics of substation control and diagnostic systems that give rise to special security challenges and vulnerabilities. Section 17.5 covers how the user can reduce the threats to cyber security and some of the characteristics of substation systems that make it difficult to apply conventional protective measures.

17.4.1 Slow Processors with Stringent Real-Time Constraints

One way to strengthen the confidentiality and integrity of messages transmitted across insecure channels is to encrypt and authenticate the communications (see Section 17.5.1.2.2.2 for more details). The two primary types of bulk symmetric encryption are block encryption, where encryption is completed after a block has been filled, and streaming encryption, which is performed as the data are transmitted. Encryption generally is too resource intensive for most legacy-based IEDs (which still form a large portion of the deployed market) and many existing SA systems. Many substation communication channels do not have sufficient bandwidth for the transmission of longer block-encrypted messages. Furthermore, vendor testing has demonstrated that utilizing existing encryption technology will significantly slow down processing and inhibit timing functions on many legacy devices. The RTUs and IEDs in substations in some cases use earlier microprocessor technology. They have limited memory and often have to meet stringent time constraints on their communications. It is often not feasible to require that these RTUs or IEDs enhance communication security by encrypting the data messages because their microprocessors do not have the processing capability to support the additional computational burden. IEEE C4 (IEEE-1689—Trial Use Standard for Retrofit Cyber Security of Serial SCADA Links and IED Remote Access) was formed to address the cyber security of IED access using bump-in-the-wire encryption technologies for serial legacy devices. The standard this group specified provides the possibility of some security where otherwise it might have not been possible. It, however, does not currently meet the requirements specified in NISTIR 7268 [2]. More analysis is needed by cryptographers to determine the true security level of this standard and to further explore the possibility of adopting it to use NIST approved methods exclusively while meeting its difficult resource and network bandwidth constraints.

17.4.2 Legacy Real-Time Operating Systems That Preclude Security

Another security risk is posed by the design of the real-time operating systems that are embedded within many legacy IEDs. At the present time, the suppliers of these embedded operating systems are not in a position to meet modern security requirements because of a limited market or a loss of support because of the age of the supplied platform. Generally, their legacy software systems have been designed to operate in an environment poor in computing resources but where there is a need for deterministic response to events. Such systems are configured to prioritize the execution of control system tasks primarily and communications secondarily, but with limited to no capabilities to implement security functions for encrypted/authenticated communications, access control, and other services.

17.4.3 Insecure Communication Media

The data messages that substation IEDs exchange with the outside world are often transmitted over media that are potentially open to eavesdropping or active intrusion. Dial-in lines are common; IEDs will accept phone calls from anyone who knows or discovers their phone number. Many IEDs are IP enabled; that is, they can be addressed by computers connected to the Internet (if they are on a routed network that has a direct/indirect connection to the Internet).

In addition, much of the data traffic to and from a substation goes over wireless networks (see Chapter 15). Intruders with the proper equipment can record and interpret data exchanges and can insert their own messages to control power system devices for wireless networks that do not implement cryptographic authentication at a minimum (which is the case for many legacy-based deployments). Other data traffic goes over leased lines, passing through telephone company switching centers where they are subject to monitoring or interference. In this latter case, the security of substation operations can be no better than the security of the switching center of the telephone company.

Furthermore, the electronic equipment at substations frequently employs remote desktop applications (such as Windows™ Terminal Server/Client, X-Terminal, pcAnywhere, and Exceed) that are specifically designed to allow users at remote locations to interact with the equipment as if they were present in the substation and directly at the local keyboard. There are numerous vulnerabilities associated with versions of these remote access programs.

Substations exist that are not configured with basic firewalls to help safeguard the systems from intrusion, and IDSs are not yet widely available for substation environments to alert the system operator when cyber intrusions of SA equipment occur (see Section 17.5.2).

17.4.4 Open Protocols

The communication protocols most frequently used in substations are well known. For communications among IEDs, Modbus, Modbus-Plus, and DNP3 are the most frequently used protocols. These protocols are well documented and used worldwide. Many protocols have been used for communications between the substation and the utility's control center. In the past, protocols were often vendor specific and proprietary, but in recent years, the majority of implementations have been with IEC 60870-5 (in Europe) and DNP3 (in North America), and, to a much more limited extent, IEC 60870-6 TASE.2 (also called "ICCP"). These protocols are all nonproprietary, well documented, and available to the general public. Security was not a factor when these protocols were designed, and they contain no features to ensure the

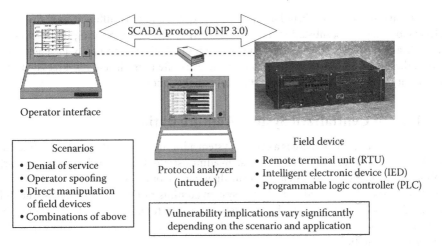

FIGURE 17.1 Bench scale vulnerability demonstrations. (From Dagle, J., Pacific Northwest National Laboratory, Demonstration at *KEMA 2nd Conference*: *Current Status and National Test Bed Planning Workshop*, Denver, CO, April 7–9, 2003, unpublished.)

privacy or authenticity of the data transmitted. However, IEC TC57 Working Group 15 and the DNP Users Group [18] are beginning to address the security concerns of ICCP and DNP3, respectively. It should be noted that the newer substation integration protocol, IEC-61850, is being actively worked on and revised in its various parts to comprehensively address security through encryption and authentication along with supporting key management. An initial attempt at addressing cyber security for IEC-61850, IEC-60870-5, and IEC 60870-6 can be found in IEC-62351. While it was an initial start for the industry, this standard is known to have issues that must be addressed in subsequent revisions [19].

Moreover, devices called "RTU test sets" are commercially available. An RTU test set is typically a portable device with a communication port that interfaces with an RTU or IED. The test set has a user interface that interprets the messages sent to and from an RTU or an IED and that allows the user to define and issue commands to the substation device. Tabletop demonstrations have shown that an intruder can patch into the communication channel to a substation and use a test set to operate devices at the substation (Figure 17.1). Depending on how the protocol has been implemented in the SCADA system, it is possible for an intruder to operate a device using a test set without the SCADA system recognizing the intrusion.

17.4.5 Lack of Authentication

Many communication protocols in current use do not provide a secure means for systems that exchange data to authenticate each other's identity. If an intruder gains access to a communication line to a controllable device, they can execute a control as if they were an authorized user. An intruder could also mimic a data source and substitute invalid data. In most cases, the program receiving the data performs very little effective data-validity checking to detect this kind of interference.

17.4.6 Organizational Issues

Utilities have often zealously guarded their operational systems from perceived interference from corporate IT staff. Yet it is the corporate IT staff that often is most aware of the cyber threats to computer systems and most knowledgeable about the ways to protect these systems. Such knowledge is less frequently present among the staff responsible for SA systems.

Often, the maintenance of substation equipment is divided among different staff: for example, relay technicians for relay IEDs, substation technicians for transformer-monitoring IEDs, and communication technicians for RTUs. There is often no single individual with authority for ensuring the cyber security of these various systems. As a corollary, there are often insufficient resources dedicated to providing security in many organizations, particularly smaller ones.

17.4.7 Lack of Centralized System Administration

Unlike the IT domain, where there is a central system administrator to designate and track authorized users, SA system users are often their own system administrators and as such have the authority to perform all security functions. This often results in allowing personnel who have no reason for access to SA systems to have such access. Additionally, the system administration function allows what is known as "root access." A user with root access has access to all critical functions including assigning passwords, assigning log-in IDs, configuring the system, and adding or deleting software. This can lead to significant cyber vulnerabilities.

17.4.8 Large Numbers of Remote Devices

A typical utility has from several dozen to several hundred substations at geographically dispersed locations, and each automated substation typically has many IEDs. Therefore, there is a high cost to implementing any solution that requires upgrading, reprogramming, or replacing the IEDs.

17.4.9 Substation Diagnostic Systems

RTU and IED suppliers are beginning to address cyber security deficiencies. Substation diagnostic systems such as transformer monitoring and capacitor bank monitoring also utilize remote access and are often integrated into the SA data network. However, many substation diagnostic system suppliers have not begun to address cyber security deficiencies and can be a "backdoor" into the SCADA network.

17.5 Measures to Enhance SA Cyber Security

The principles for enhancing the cyber security of control and diagnostic systems at substations are the same as those for other corporate computer systems: (1) prevent cyber intrusion where you can; (2) detect intrusion where it could not be prevented; (3) respond and recover from an intrusion after it was detected; and (4) improve the preventive measures on the basis of experience.

17.5.1 Protecting Substation Systems against Cyber Intrusion

There are two avenues of potential cyber intrusion to the computer-based equipment in a substation: those originating from the users on the general corporate network and those originating outside. These are treated in separate sections in the following.

17.5.1.1 Cyber Intrusion from inside the Corporate Network

To the extent that substation control and monitoring systems are connected to a utility's corporate wide area network, a large potential threat to these systems is derived from unauthorized users on the corporate network or any network that has general business IT systems. Consequently, the first step in securing substation assets should be to ensure that the corporate network is made as secure as possible and has sufficient points of control and isolation from the SA system network. That is all network traffic coming from the corporate network into the SA system network should be untrusted, monitored, and extensively filtered through the substation network ESP using firewalls and IPSs. The important measures are well known. They include the following:

- Implementing AAA-based services to support role-based access controls (RBAC) and cryptographically strong levels of authentication
- Removing all default user IDs and default passwords on installed systems
- Ensuring that all accounts have strong passwords
- Closing unneeded ports and disabling unneeded services
- Installing security patches from software suppliers in a timely manner
- Removing all sample scripts in browsers
- Implementing firewalls with appropriate rules to exclude all unneeded and/or unauthorized traffic
- Using NAC and endpoint security management solutions to ensure that unauthorized, compromised, or insecurely configured PC hosts are not connecting to the network
- Implementing IPSs, and logging as many events as feasible and providing automated protection against well-known attacks
- Implementing SEMs and aggregating IPS and other system logs into a single and centralized repository for further analysis and investigating all suspicious activity
- Conducting regular audits of all security process-oriented controls as well as technical controls to ensure that policies are being followed and that technologies that are in place to enforce them are doing so correctly
- Conducting regular vulnerability scans (using many of the available tools) to ensure no IT assets are attached to the network with known vulnerabilities
- Conducting regular security penetration tests with suitably experienced or trained and certified professionals to ensure security systems and procedures are keeping pace with the latest cyber offensive technologies and methods
- Regularly testing incident response (IR) for your IT and SA systems, that is, simulating successful attacks that trigger security event alarms and going through the process of isolating attack entry points, closing them down, containing the attack, restoring systems, and collecting forensic data where possible (currently doing this with ICS/SA systems has limited applicability)

The details of these measures, and further measures to protect the corporate network, are the subject of much active discussion elsewhere and will not be covered in this book.

Even though measures have been taken to enhance the cyber security of the corporate network, cyber intrusions may still occur. Therefore, additional measures should be taken to further protect substation systems from successful penetrations onto the corporate network. These measures will also help protect the substation stations from malevolent activity from employees who have access to the corporate network.

17.5.1.1.1 Password Policies

1. The most important measure is one of the simplest, that is, to ensure that all default passwords have been removed from all substation systems and that there are no accounts without any password. (This may not be possible, however, if the equipment supplier has "burned-in" the default password into the system firmware or if the system will not accept passwords.)
2. Password complexity settings should be implemented to ensure that users' passwords are not easily guessable. However, as is well known, passwords that are difficult to guess are also difficult to remember. It defeats the entire sense of having passwords if the users post their passwords on the terminal of the system being protected. Users should be given instruction in ways to generate "difficult" passwords that they can remember without difficulty. Using phrases (instead of single words is common approach) and having substitutions of certain letters for numbers as well as

special characters is a common mnemonic technique that users can be trained to use. In order to provide defense against modern attacks, it is generally recommended that passwords be at least 10 characters, make use of upper and lower case letters, have numbers, and special characters ($, *, !, etc.). Under this paradigm, the difference in protection against password cracking methods is exponential for each character added. Anything smaller than 10 characters in length is generally believed to be trivially breakable with commodity hardware using graphics processor units (GPUs) as cracking engines, which can currently guess in the order of magnitude of billions of passwords per second.

3. Procedures should be in place to immediately terminate a password as soon as its owner leaves employment or changes their job assignment. Ideally, if using RBAC and AAA (see Sections 17.5.1.1.2 and 17.5.1.1.3), centrally revoking/altering a user account is the optimal way to address the issue, as shared and device-specific passwords should not be used.

17.5.1.1.2 RBAC

Different sets of privileges should be established for different classes of users. For example, some users should be allowed only to view historical substation data. Other users may be permitted to view real-time data. Operators will be given control privileges, and relay engineers will be given the authority to change relay settings.

17.5.1.1.3 AAA Server

Centralized authentication, authorization, and accounting of user accounts and roles provide for much more efficient access control and user management. If a user changes role or leaves, all accesses to SA system applications and devices can be revoked from a single point and it should happen with instant assurance. Similarly, permissions on an account can be adjusted and instantly reflected across the entire enterprise as well as all SA applications and devices. This is particularly important when dealing with dismissals of employees that might result in sabotage attempts, or if improper access patterns for a particular account are detected (through SEM monitored and analyzed logs) that indicates its compromise and an instant lock down is required. In cases where the authentication and authorization system are unique and local to each device in the SA system cut across many substations, it practically becomes impossible to provide such monitoring and response, exposing SA systems to considerable risks. To facilitate reliability, many devices can be configured with multiple AAA servers to switch over to other ones in case of failure; in the case of wide area communications being lost, one can also have a locally synchronized server (the credential database copied from a control center AAA server) in the substation local area network (LAN).

17.5.1.1.4 Multifactor Authentications

A utility might consider requiring a stricter measure of authentication of the user before permitting access to a substation system. For example, the utility may consider requiring that a user desiring access to a system present a smart card for authentication or instituting some form of biometric identification of the user (such as a personal fingerprint reader). In the case of the latter, one must take care to make sure the biometric data is being protected using strong cryptographic methods as it is false to assume such systems cannot be spoofed if the individual is not present to provide their unique biometric feature (e.g., finger print, eye scan, and hand geometry). We only need to inject the data that represents the feature into the system. The cost of purchasing the hardware to implement these protective measures is not high, but the administrative costs may make such measures impractical. As is often the case with issues of security, the utility must weigh the costs of the measure against the value of the asset being protected and the perceived risk of damage.

17.5.1.2 Cyber Intrusion from outside the Corporate Network

The possibility of intrusions into the substation by outsiders gaining direct access to substation devices through unprotected communication channels raises new challenges to the cyber security of substation systems. There are two main communication remote paths into the substation that are the potential target for eavesdropping or intrusion: the SCADA or other WAN communication lines and dial-up lines to IEDs.

17.5.1.2.1 SCADA Communication Lines

Classically, the SCADA communication line is the communication link between the utility's control center and the RTU at the substation, increasingly there maybe direct communication to various IEDs using modern networking technologies such as Ethernet/IP routers together with media converters, however, what will be discussed equally applies to modern and traditional communication lines. The SCADA line carries real-time data from substation devices to the utility dispatchers at the control center and control messages from the dispatchers back to the substation. (For substations equipped for SA, a data concentrator/gateway or an SA host processor will play the role of the RTU in sending substation data to the control center and in responding to the dispatcher's control commands.)

A variety of media are used to connect the substation RTU with the control center: power line, leased lines, microwave, multiple address radio, satellite-based communications, fiber-optic cable, etc. The topic is discussed in detail in Chapter 15. It is quite common for communications from the control center to substation to use different media along different segments of the path.

Per the C1 Working Group paper [20], there are a large variety of communication routes for access of devices in substations. The physical media can be point to point (telephone lines), microwave, and higher bandwidth transport (T1, SONET [Synchronous Optical NETwork], or Ethernet).

- POTS (plain old telephone service) dial-up via phone line is the most common medium used to access relays remotely. Modems are required to interface the phone line with the IEDs. Line switchers typically allow one phone line to be switched and used for relay access, meter access, phone conversations, etc.
- Leased lines are typically used for SCADA connection. They are dedicated lines that are connected 24h a day, 7 days a week. They allow constant data acquisition and control capability of substation equipment.
- Wireless communication (cellular phones) is a technology that is useful in the substation environment. Additional wireless communications using IEEE-802 or equivalent networks are increasingly being used to expand the control connectivity boundaries in the substation environment. These networks typically are connected to one of the other communication media described previously, expanding the communication depth and breadth.
- 900 MHz radio is another medium used by utilities. These radios can either be licensed or unlicensed depending on the frequency selected. The unlicensed installations have a lower installed cost, but there is no protection from interference by other users.
- Microwave is a high-frequency radio signal that is transmitted through the atmosphere. Transmitted signals require a direct line of site path and accurate antenna alignment. The Federal Communications Commission (FCC Parts 21, and 94) controls operation and frequency allocations.
- In digital microwave systems, the data modems required in an analog system are replaced by digital channel banks. These channel banks can be combined to form a multiplexed system. The channel banks convert analog voice and data inputs into a digital format using pulse code modulation (PCM). The digital channel bank combines 24 voice channels into a standard 1.544 Mbps DS-1 signal. The DS-1 level is further multiplexed into DS-3 before being transmitted over the radio link.
- Many substations are served by T1, SONET, or Ethernet access equipment to provide a communication path to the substation device. T1 is a term for a digital carrier facility used to transmit a DS-1 formatted digital signal at 1.544 Mbps. T1 was developed by AT&T in 1957 and implemented in the early 1960s to support long-haul PCM voice transmission. The primary innovation of T1

was to introduce "digitized" voice and to create a network fully capable of digitally representing what was, up until then, a fully analog telephone system. T1 is used for a wide variety of voice and data applications. They are embedded in the network distribution architecture as a convenient means of reducing cable pair counts by carrying 24 voice channels in one four-wire circuit. T1 multiplexers are used to provide DS0 (64 kbps) access to higher order transport multiplexers such as SONET. SONET is the American National Standards Institute (ANSI) standard for synchronous data transmission on optical media. Some of the most common SONET applications include transport for all voice services, Internet access, frame relay access, ATM transport, cellular/PCS cell site transport, inter-office trunking, private backbone networks, metropolitan area networks, and more. SONET operates today as the backbone for most, if not all, interoffice trunking as well as trans-national and trans-continental communications.

- IP communications (Ethernet) are growing as a substation access technology. The transport is often over a SONET layer, but Ethernet LANs are also used. The communication network can be privately owned by the utility, or leased from a carrier. A LAN can have its own dedicated communication links or exist as a virtual local area network (VLAN) where the transport layer is shared with other, unrelated traffic. The LAN or VLAN may interconnect with a WAN that carries corporate traffic or is a public transportation network. A demonstration was performed by the Idaho National Laboratory to demonstrate the vulnerability of SCADA networks utilizing IP communications [21]. The demonstration illustrated that breakers could be remotely manipulated and operator displays changed by inserting appropriate scripts in buffer overflow compromises without being detected or blocked by IT security technologies including multiple firewalls. However, it is not clear how well configured these security technologies where, or if they are representative of the last possible solutions and methods.

Some of these media do not use suitable encryption and authentication, thus are vulnerable to eavesdropping or active intrusion. At least one case has been reported in which an intruder used radio technology to commandeer SCADA communications and sabotage the system (in this case, a waste water treatment facility) [22]. The cyber vulnerability of insecure legacy 900 MHZ spread spectrum radio communications utilized in substation communications was demonstrated by Pacific Northwest National Laboratory [23]. Refer to Section 17.5 for a discussion of measures to protect SCADA/SA systems.

17.5.1.2.2 Dial-Up Lines to IEDs

The other path to substation control and monitoring devices is via dial-up lines directly to IEDs. As discussed in Chapter 7, IEDs are devices that intrinsically support two-way communications. IEDs are frequently configured, so a user can dial up the IED. Once the user has logged on to the IED, he may use the connection to do the following:

- Acquiring data that the IED has stored
- Changing the parameters of the IED (e.g., the settings of a protective relay)
- Performing diagnostics on the IED
- Controlling the power system device connected to the IED (e.g., operate a circuit breaker)

These dial-up lines can offer a simple path for a knowledgeable intruder into the substation. There are three lines of defense that a utility can take: (1) strengthen the authentication of the user; (2) encrypt communications with the IED; or (3) eliminate the dial-up lines.

17.5.1.2.2.1 Authentication
Strengthening the authentication of the user authentication refers to the process of ensuring that the prospective user of the IED is the person they claim to be.

As the very first step, the utility should ensure that the default passwords originally supplied with the IEDs are changed and that a set of strong passwords are implemented, if possible.

A simple second step would be to confirm that the telephone call comes from a recognized source. For this purpose, it is not sufficient to get the user ID of the caller and confirm that it is on a preapproved list.

Hackers are often familiar with telephone technology, and the caller ID can be changed or disguised. A more secure approach would be for all dial-in calls to be received by a dial-back device at the substation (also known as a callback device). The device receives the incoming call, requires that the caller enter a user ID and password, searches an internal list for the telephone number that the call should be made from, terminates the incoming call, and dials back the caller at the phone number found in the list. In essence, the incoming call is replaced by an outgoing call.

It should be noted that the use of dial-back is not foolproof, however. According to one source [23]

> There are several ways an intruder can defeat the protection offered by a dial-back modem. For example, if the same modem and line are used for returning the call to the user, the intruder may be able to maintain control of the line while fooling the modem into acting as thought the user had hung up after the original call. The modem would then place the return call, but the intruder's equipment would be mimicking the operation of the telephone system and the return call would be connected to the intruder's modem. Alternatively, the intruder could modify the telephone switch setup to direct the return call to the intruder's telephone number regardless of the pre-arranged number stored in the modem.

The report recommends that the utility consider the use of a separate line for the callback, to defend against this threat. The telephone switch must also be carefully protected since the security of the substation depends on the integrity of the telephone switch.

17.5.1.2.2.2 Encrypting Communications A second approach to enhancing the security of communications to IEDs would be to encrypt and authenticate the messages (using strong methods as per NISTIR 7628 [2] Volume 1: Cryptography and Key Management System requirements) between the user and the IED using built-in application level encryption, bump-in-the-wire encryption (transparent encryption of serial or network link layer data using directly attached external devices, one for the user, and the other for the IED), or VPN technologies as discussed earlier in the chapter. What could be used would depend on the performance capabilities of the system (memory, CPU, bandwidth/latency) as well as the nature of the communication medium involved. What follows is a simplified explanation and many technicalities and details are left out, the reader should go through a true applied text [24] on cryptography to sufficiently understand what is being presented in this section. Encryption helps ensure that only users/systems in possession of an authorized key would be able to communicate commands and data to the IED and/or change IED configuration parameters. There are two types of encryption, symmetric and asymmetric. Symmetric encryption is based on what are known as block and stream ciphers (or algorithms, cipher is a more specialized term) in which each communicating party share the same "secret" key with each other. Asymmetric encryption is based on what is commonly communicated in the literature as public key cryptography. In this form of encryption/cryptography, each party has what is known as a key pair that is unique to them and consists of their public and private key. As the name implies, they freely share their public key with each party they wish to securely communicate with while keeping their private key protected and only under their use and control. Then messages that are encrypted with a public key can only be decrypted by the associated private key and hence the message only is read by the intended recipient and makes it impossible for intercepting attackers to decrypt and view the message. Public key cryptography also provides message integrity (also known as authentication) by using digital signatures. In this case, a special fixed size code (known as a message digest, computed using a cryptographic hash algorithm) is computed on a message (regardless of its size) and encrypted with the private key of a communicating entity, and only the matching public key can decrypt it. This allows any communicating party to verify that a message is not only authentic but specifically came from the user/system that controls the matching private key. This gives the added benefit of nonrepudiation. That is when such a message is sent the party that composed it cannot deny its existence and that it was not them that sent the message, as only they can sign it. This has the same notion of physical signatures that we use every day on legally binding documents and transactions. Nonrepudiation can be important from a forensic perspective if malicious insiders are intent on sending damaging commands and configurations to SA devices for the purposes of sabotage. Hence, digital signatures can often act as a deterrent to such behavior and make users more concerned with how they manage and protect their credentials when they know system actions can explicitly and with cryptographic certainly be attributed to them. Many jurisdictions legally recognize digital signatures and attribution in court and consider it the same as signing your name to a legal contract; this further makes this method advantageous as a deterrent to carelessness and misbehavior by internal personnel. In practice, asymmetric cryptography and associated computations are very CPU and memory intensive (by orders of magnitude) as compared to symmetric methods; hence, in real cryptographic protocols (compositions and use of underlying cryptography algorithms, also known as primitives in the field within this context) both symmetric and asymmetric cryptography is used for encryption and message authentications. Typically, public key cryptography is used to negotiate and/or encrypt generated session-based symmetric keys established (unique keys for each communication session) between two parties, and then block/stream ciphers are used to encrypt each message in a communication session with the session-based key or an individual message if it is a file, for example. An example of a strong NIST approved symmetric ciphers includes the Advance Encryption Standard (AES), and for asymmetric ciphers, elliptic curve cryptography (ECC) and Rivest, Shamir and Adleman (RSA) are other examples. Please see the cryptography and key management system requirements of NISTIR 7628 [2] to further understand the correct use and modes of such cryptography as applied to many classes of SA systems and devices.

In alternative for systems that are performance constrained (in CPU, need hard-real-time messaging latencies, or to maintain high bandwidths) and only require message integrity, one can consider using message authentication codes (MACs) exclusively (without making use of encryption for message privacy) based on keyed hashes (require less computation than total message encryption). Good examples of keyed hash algorithms are HMAC-SHA-256 and AES-GMAC as referenced in NISTIR 7628 [2] cryptography requirements. In a simplified explanation you can consider a keyed MAC method as a function that takes any size of message (e.g., 8 kB communication packet) and turns it into another message of a smaller and fixed size (e.g., 32 bytes) known as the digest, that is, unique for each message and key combination. This MAC is then appended/prepended (in a message header) to the message and sent to the recipient who can verify its authenticity by symmetrically computing the same operation (using the same key each communicating device/party share with each other) to check they get the same MAC as sent with the message. To further ensure the security of each message, it should also be mentioned that sound cryptography protocols make use of a timestamp or a unique nonrepeating random value per message (known as a nonce) as part of computing the MAC, hence not allowing attackers to listen for and intercept certain messages (e.g., SCADA commands and IED configuration changes) only to replay them at a later time. The nonce and/or time stamp ensure the message is unique in time and point of use. Computing a secure hash is much less computationally intensive than encrypting the whole message.

In IED design, the two paramount factors are performance and cost. The high computational requirements of processors to implement some encryption schemes make encryption impractical for the low-performance microprocessors currently used in many IEDs, so external solutions (as discussed earlier in this section) with their own cost and performance considerations maybe the only solution for certain classes of communications. With current VPN and other network link encryption technologies available for substation networks, most, if not all, forms of engineering and configuration access can and should be secured. Many classes of SCADA communications can also be secured in this manner (with a level of consideration and analysis as discussed earlier in this chapter). However, newer forms of SA communications based on rapid peer-to-peer messages as found in IEC-61850 GOOSE communications are unlikely to currently have cost-effective and reliable solutions that can operate in substation environments, but this is likely to change in the next few years as the necessary R&D, product development, and subsequent engineering studies happens in industry to support newer methods and enhanced hardware, firmware, and software. In addition, the SA systems standards community has not yet agreed upon a unified technical approach to encryption (including authentication) and supported key management for the myriad of needed communication protocols, but working groups are being formed in many relevant standard development organizations to address this challenge. In many cases, tunneling SCADA and other SA system protocols over encrypted IP communications with already established and secure standards in the field maybe possible, but for many classes of messaging further development and engineering study is required into more tailored security management models and tools for the SA technical domain. Consequently, it would take a special effort on the part of a utility to encrypt messages to and from IEDs because of the operational issues involved in key management and their associated systems. That is, it is easy to encrypt once the capability becomes present, but the most important factor in encryption is how keys are managed in a coherent system that involves everything from key creation, to reissuing (in cases of compromise) and destruction. The NISTIR 7628 [2] cryptography and key management section outlines significant issues and design considerations in this area that have application to SA systems.

Nevertheless, there are active developments along several fronts that may cause this situation to change. Higher performance microprocessors are being manufactured at ever-lower cost, reducing the cost and performance penalties of encryption. In addition, several groups are making progress in defining encryption standards for the communication protocols used in substations, including IEC Technical Committee 57, Working Groups 7 and 15, and the DNP Users Group. The IEC 61850 protocol is based on international standard communication profiles.

17.5.1.2.2.3 Eliminating the Dial-Up Lines Where possible, another approach to securing the communications to IEDs would be to eliminate dial-up lines into the substation entirely. This is dependent on the manufacturers not embedding wired or wireless modems into the IEDs. This approach is being followed by several utilities that place a high value on cyber security.

Under this approach, all communications to the IEDs originate from within the secure network and are transmitted through and mediated by the data concentrator or substation host processor at the substation. The data concentrator or substation processor forwards the message to the appropriate IED and routes the response back to the original caller. No communications to the substation are permitted that originate outside the secure utility network. Communications to the substation IEDs would be even more secure if, as suggested earlier, fiber-optic lines were used for substation communications.

The security of this approach is dependent, of course, on the success of the utility in preserving the security of its internal network. That issue is beyond the scope of this chapter.

17.5.2 Detecting Cyber Intrusion

One of the axioms of cyber security is that although it is extremely important to try to prevent intrusions into one's systems and databases, it is essential that intrusions be detected if they do occur. An intruder who gains control of a substation computer can modify the computer code or insert a new program. The new software can be programmed to quietly gather data (possibly including the log-on passwords of legitimate users) and send the data to the intruder at a later time. It can be programmed to operate power system devices at some future time or upon the recognition of a future event. It can set up a mechanism (sometimes called a "backdoor") that will allow the intruder to easily gain access at a future time.

If no obvious damage was done at the time of the intrusion, it can be very difficult to detect that the software has been modified. For example, if the goal of the intrusion was to gain unauthorized access to utility data, the fact that another party is reading confidential data may never be noticed. Even when the intrusion does result in damage (e.g., intentionally opening a circuit breaker on a critical circuit), it may not be at all obvious that the false operation was due to a security breach rather than some other failure (e.g., a voltage transient, a relay failure, or a software bug).

For these reasons, it is important to strive to detect intrusions when they occur. To this end, a number of IT security system manufacturers have developed IDSs that have subsequently evolved into IPS as they can take protective measure upon the detection of an attack pattern such as blocking the origin of an attack, but these terms can be used interchangeably to mean the same thing in industry. These systems are designed to recognize intrusions based on a variety of factors, including primarily (1) communications attempted from unauthorized or unusual addresses and (2) an unusual pattern of activity. They generate logs of suspicious event that can be centrally analyzed and processed by SEMs or manually monitored.

Unfortunately, there is no easy definition of what kinds of activity should be classified as unusual and investigated further. To make the situation more difficult, hackers have learned to disguise their network probes, so they do not arouse suspicion. In addition, it should be recognized that there is as much a danger of having too many events flagged as suspicious as having too few. Users will soon learn to ignore the output of an IPS that announces too many spurious events. There are outside organizations, however, that offer the service of studying the output of IPSs and reporting the relevant results to the owner to better help them identify and respond to real attacks. They will also help the system owner to tune the parameters of the IPS and to incorporate stronger protective features in the network to be safeguarded. One can also use SEM systems to provide a centralized and more automated analysis of logs that are more intelligent and effective than trying to do this activity manually.

It is also important to understand that IPS technologies are currently weighted toward "blacklist"-oriented security management, that is, scanning network traffic and applying rule sets that identify known malicious patterns that would be representative of an attack on an SA system. There are two problems with this:

1. There is not enough of a knowledge base to truly understand what generally constitutes SA system attack patterns with sufficient precision and reliability. This is because there is just too much variance in operations and very little understanding of many legacy components found in SA systems. However and most importantly, there have not been a sufficiently large number of understood (or at least reported) attacks to learn from as compared to IT systems.
2. The lack of knowledge to allow for generalization means rates of high false positives with considerable risks of denying legitimate network traffic at a critical time. This is especially true during a critical protection and control action, that if prevented may cause permanent damage to major electrical assets (e.g., high-voltage transformers).

For these reasons, it is more likely that SA systems will make use of detection only in IPSs and rely on human operator response (effectively running them in an IDS mode) for quite some time until the technology can be sufficiently evolved and proven in the field along with associated engineering studies and standards focused on ICS/SA system IPSs. This will carry significant risks as a cyber attack (especially of an automated kind) often spreads and can impact systems faster than a human can respond. However, given the current reliability risks, this is a cyber risk that must be accepted. Lastly, even given these current ICS/SA IPS limitations, we recommend increased deployment of IPS technologies for the following reasons:

1. They are field programmable devices, which mean they can be improved and tuned over time with increasingly better rule sets. The only way to increase IPS performance for ICS/SA systems is to deploy them and learn through in the field experience. Just running them silently and collecting logs can be a very informative analysis and learning activity that is highly recommended. It can sometimes alert system operators to threats that were previously unknown.
2. They can run passively (with no impact on the SA system they are protecting) with security event reporting and alarming tuned to create far less noise with the balance of events going to SEM systems for deeper analysis by security analysts (appropriately cross-trained in ICS/SA systems) to determine if SA system operator intervention/awareness is needed.

3. For specific and known attacks (e.g., Stuxnet [4]), very focused rule sets can be deployed that ensure high detection rates while giving immediate protection.
4. With time and sufficient study and analysis of an SA system one can create a "whitelist" security management paradigm for an IPS. That is having rule sets that codify the correct and accepted behavior of the SA system and provide security event alarming (or possible prevention with sufficient time and proving of the approach and technology) on exception to that accepted behavior. This is a different and easier-to-solve problem for SA systems than looking for malicious attack patterns when not enough is currently known to generalize them. However, because of the relative well-known and deterministic behavior a well-designed and commissioned SA system provides, it should be reasonable to discern exceptional/unauthorized: commands, configuration change attempts, network services/messages, etc., for many devices given their expected state.

Also more research is necessary to investigate what would constitute unusual activity in a SCADA/ICS/SA environment from an SEM analysis perspective. In general, many legacy class SA and other control systems do not have logging functions to identify who is attempting to obtain access to these systems. Efforts are ongoing in the commercial arena and with the National Laboratories to develop these capabilities for SCADA/ICS/SA systems. In summary, the art of automatically detecting/preventing intrusions into substation control and diagnostic systems is early but is starting to see promise with some deployed systems and increased R&D. More effort and resources need to be expended to further develop IPS capabilities for SCADA/ICS/SA systems, there is significant return on security investment to be realized if this is done as it will create a far more resilient SA system overall. Until more and dependable automatic detection tools are developed, system owners will need to have continued focus on (1) preventing intrusions from occurring and (2) recovering from them when they occur, as way of balancing their risks.

17.5.3 Responding to Cyber Intrusion

The "three R's" of the response to cyber intrusion are recording, reporting, and restoring.

Theoretically, it would be desirable to record all data communications into and out of all substation devices. In that manner, if an intruder successfully attacks the system, the recordings could be used to determine what technique the intruder used in order to modify the system and close that particular vulnerability. Secondly, the recording would be invaluable in trying to identify the intruder. In addition, if the recording is made in a way that is demonstrably inalterable, then it may be admissible as evidence in court if the intruder is apprehended.

However, due to the high frequency of SCADA communications, the low cost of substation communication equipment, and the fact that the substations are distant from corporate security staff, it may be impractical to record all communications.

In practice, although theoretically desirable, system owners will probably defer any attempts to record substation data communications until (1) storage media are developed that are fast, voluminous, and inexpensive or (2) SCADA-oriented IPSs are developed, which can filter out the nonsuspicious usual traffic and record only the deviant patterns.

But even if the communication sequence responsible for an intrusion is neither detected nor recorded when it occurs, nevertheless it is essential that procedures be developed for the restoration of service after a cyber attack.

It is extremely important that the utility maintain backups of the software of all programmable substation units and documentation regarding the standard parameters and settings of all IEDs. These backups and documentation should be maintained in a secure storage, not normally accessible to the staffs who work at the substation. It would appear advisable that these backups be kept in a location other than the substation itself to lower the amount of damage that could be done by a malicious insider.

After the utility concludes that a particular programmable device has been compromised (indeed, if it just suspects a successful intrusion), the software should be reloaded from the secure backup. If the settings on an IED had been illicitly changed, the original settings must be restored. Unless the nature of the breach of security is known and can be repaired, the utility should seriously consider taking the device off-line or otherwise making it inaccessible to prevent a future exploitation of the same vulnerability.

17.6 Devising a Security Program

In order to put the recommendations of this chapter into practice, a utility should establish and implement an auditable security program developed specifically for SCADA and SA applications. There are regulatory mandates, NERC CIP [1], as well as business reasons for establishing a control system security program. For federal entities such as TVA, BPA, WAPA, and the Bureau of Reclamation, federal law requires compliance to NIST Special Publication (SP) 800-53, *Recommended Security Controls for Federal Information Systems and Organizations*. Also the latest publication that is more SA system focused and has considerably more fidelity and detail to help form a security program for this technical domain is NISTIR 7628 [2]. The security program should consist of policies, procedures, testing, and compliance with a senior executive role having responsibility and accountability for implementing and maintaining the security program. The security program must be developed specifically to control systems as traditional IT security policies, procedures, and testing have led to impacts on control systems. Additionally, SCADA systems have failed IT security compliance audits as control systems have different operational requirements.

The program for securing the utility's computer systems should also be correlated with a program for ensuring the physical security of the utility's assets. (Some of the components of a plan for cyber security will be very similar to the analogous plan for physical security.) (See Chapter 15.) The utility should consider the following issues when devising the cyber security program.

Assets

- What are the assets of the substation that the policy seeks to protect from critical infrastructure and business perspectives? (As a corollary, what assets are not protected by the policy?)
- What level of protection should be given to each asset (device, control system, communication system, and database)?
- What must a user do or have to gain access to each asset?

Threats

- What are the threats to the security of the substation that the policy seeks to address? (Also, what threats are not addressed?)
- What is the damage that can result from each of the threats?
- What measures should be taken to protect against each threat? (Several alternatives may be considered.)
- What should be done to test the protective measures that have been taken? (Should an outside organization be employed to probe for weaknesses?)
- Who will monitor the changing nature of cyber threats and update the security policy accordingly?

Threat detection

- What measures will be taken to detect intrusion? (Should an outside party be employed to analyze intrusion records?)
- What should be done if an intrusion is suspected?
- Whom should an intrusion be reported to?
- What records should be kept and for how long?

Incident response

- What immediate steps should be taken in response to each type of incident?
- What role will law enforcement play?
- How will the incident be reported to regulatory agencies, reliability councils, or cyber-incident recording centers?
- What improvements must be made (to policy, to documentation, to training, etc.) as a result of lessons learned?

Training and documentation

- What are the training programs for security? (General and control system–specific security awareness, access procedures, and restoration procedures.)
- What are the plans for practicing restoration, and how often will they be applied?
- What are the plans for supporting manual operation if control systems suffer long-term damage?
- Who will issue the documentation for the restoration procedures, and where will the documents be kept?

Administration

- Who has the ultimate responsibility for cyber security at the utility?
- What are the responsibilities of each relevant job category?
- What are the potential consequences for staff of a violation of policy?
- How will compliance with the policy be monitored? (Should an outside organization be used?)
- How will the security policy be revised?

Software management

- Controlling and installing software updates
- Installing security patches
- Maintaining backups
- Password policy and password maintenance

17.7 Future Measures

It should be clear from previous discussions that, at the time of publishing, technologies are still maturing mature for ensuring the cyber security for substation control and diagnostic systems but there have been significant evolutions and recent industry developments discussed in Section 17.1 are creating strong market incentives for innovation. It will, however, take some time for SA systems to totally adopt and deploy newer and tailored security technologies because of reliability concerns and the need for further standards development and engineering study. To a certain extent, a utility will be forced to make do with various mitigating technologies and various compensating procedures until next generation SA equipment begin to be deployed that offer an upgrade path to more advanced and embedded security methods. It is not practical to eliminate all security risks or to close all security vulnerabilities. The utility must evaluate what assets have the highest impact to power system reliability and deserve the greatest effort at protecting.

Strenuous efforts are being taken in several areas, however, to improve the defenses against cyber threats. It will be worthwhile for the utility to monitor these developments and update their security policies to take advantage of technological advances. (As was the case with the arms race, it, however, must be anticipated that adversaries will develop new attack strategies as current vulnerabilities are closed; it is important that the utility monitor and respond to the changing nature of the threat as well.)

This chapter describes some current developments that will make it easier to provide for the cyber security of substation systems and indicates where further work is needed.

17.7.1 Authentication and Encryption

The various standards groups who are responsible for defining the protocols used in substation communications are actively working on defining the standards for authentication and encryption. Concurrently, IEDs are being manufactured with faster microprocessors and more memory, making it feasible to implement encryption in embedded processors. Furthermore, the channel capacity of communication lines to substations is growing, making the performance penalty for encryption less significant. As a result of these trends, it is currently feasible to encrypt selective communications and it will soon be feasible to more widely encrypt communications between control centers and substations. In addition, if there is demand for the function from the user community and sufficient standards are developed, it should soon be possible to implement encryption of many high-speed communications among IEDs within a substation at acceptable cost and reliability.

17.7.2 Secure Real-Time Operating Systems

An ancillary, but important, function is to develop real-time operating systems with more built-in security services that can enable better defensive techniques against a variety of threats, vulnerabilities, and associated exploitation. Of particular importance will be the ability to support firmware integrity—both at boot time and during upgrade/patch. In other words, the ability to ensure only trusted and authentic firmware is loaded or booted from a device using cryptographic-based digital signatures. Secondly having built-in support for AAA and RBAC models as well as built-in network security functions.

17.7.3 Functional Cyber Security Testing and Certification

Also important are emerging network testing and resiliency standards that provide certification that ICS/SA systems can withstand sufficient attacks on their networking interfaces but continue to reliably function. These tests typically focus on creating network storm conditions and malformed packets and sending them to the target device to see if it can continue to operate correctly and to its designed purpose while under attack [25].

17.7.4 Test Beds

It is difficult for a utility, unaided, to discover all the security vulnerabilities in the various systems installed at a substation. In recognition of the need for a more concerted effort, the U.S. government is dedicating resources to investigate the security vulnerabilities of elements of the critical infrastructure. Several national laboratories have taken on the task of establishing the National SCADA Test Bed where the SCADA systems in common use can be studied, their vulnerabilities discovered, and remedies implemented. These test beds have expanded to include control and diagnostic systems in common use in substations.

17.7.5 Incident Reporting Sites

For several years, the CERT Coordination Center (CERT/CC), operated by Carnegie Mellon University [17], has served as a storehouse for reports of security incidents. CERT declares, "Our work involves handling computer security incidents and vulnerabilities, publishing security alerts, researching long-term changes in networked systems, and developing information and training to help you improve security at your site."

The CERT website (http://www.cert.org/) has a form that allows the manager of a computer system or network to report a security incident. CERT also publishes an advisory of cyber security problems, which is emailed to a very large number of destinations. At the current time, the incidents maintained on the CERT database are almost entirely traditional computer problems; problems with SCADA and SA systems have not been identified in their advisories. However that has changed with U.S. ICS-CERT that has focus in this area and has already provided advisories and recommended mitigations to published vulnerabilities in the ICS/SA system domain (http://www.us-cert.gov/control_systems/ics-cert/). It is encouraged that all industry stakeholders make use of the U.S. ICS-CERT incident reporting service to report security incidents, and thereby inform others of common vulnerabilities. They provide a method of making sure that disclosures are responsible and because of the nature of critical infrastructure that what is disclosed and its details can be communicated on a need to know basis with the appropriate asset-owning organizations.

17.7.6 Intrusion Prevention/Detection Systems and Firewalls

It is important to be able to reliably recognize an intrusion into substation computer systems and networks.

Currently, almost all IPSs currently focus on traditional computer networks and have not been adapted for the special circumstances characterizing the systems found at substations. Work is currently in progress to use neural networks or other machine learning methods and heuristics to be able to model and define a normal state of activity and to recognize an intrusion by the change in the patterns, or in exception to norms of operation, also known as anomaly-based detection. This can also be seen as a whitelist security management model (what are the activities that are authorized?) compared to a blacklist model (what are the activities that are unauthorized?). SA systems because of their rigorous configuration and commissioning process do offer a greater opportunity for whitelist-oriented IPSs to execute quite successfully as the blacklist model currently offers limited cyber protections even in many IT systems whose requirements they are classically designed to meet. Currently, these efforts have not yet been proven to be widely effective. It is hoped that these developments will prove successful, and IPSs that are more intelligent and appropriate for utility application will be available. Lastly, DPI IPSs are commonly available for IT systems; they need to become more widely available for substation environments and come with built-in rule sets for doing DPI and analysis for SCADA/SA system protocols to prevent attacks that target actual power and control system elements, as that is where the highest risks are and not in IT-oriented threats alone. There are indications that industry will be engaging in more R&D and product development in this area and we look forward to the results.

17.7.6.1 Secure Recovery

If a computer-based system has been compromised, the process of restoring the system to its pristine state is lengthy, labor intensive, and error prone. It is especially difficult when the software has been modified since it was first installed. The utility must answer the difficult question about when the system was compromised, and whether the software that is being restored perhaps contains the infected code.

Developments that would allow a quick and reliable restoration of uninfected system software would be of great value to the operators of substation control and diagnostic systems. Authenticated and signed files can aid in this development.

17.7.7 Developed and Emerging Technical Standards and Guidelines

1. IEC Technical Committee 57 Working Groups 7 and 15 are addressing security in IEC 60870-6 TASE.2 ("ICCP"), IEC-61850, and IEC 60870-5 (IEC-based DNP3) through IEC-62351 (which as discussed earlier will require further revision).
2. The IEEE Power Engineering Society (PES) is developing several standards that will affect substation cyber security. Currently, it has published IEEE-1689 (Trial Use Standard for Retrofit Cyber Security of Serial SCADA Links and IED Remote Access) and IEEE 1686 Substations IED Cyber Security (which has become dated with latest industry requirements and needs revision).

3. ISA has established a standards committee for process controls cyber security—ISA SP99. These standards will be applicable to substation applications. ISA has issued three standards [26–28] and is working on several others.
4. The DNP committee is addressing security of DNP3 specifically providing authentication measures that make use of symmetric and asymmetric methods. They have recently been working with the NIST Smart Grid Interoperability Panel (SGIP) Cyber Security Working Group (CSWG) to further ensure their methods meet NISTIR 7628 guidelines.
5. Cigré JWG D2/B3/C2-01 has issued CIGRE Technical Brochure, Security for information systems and intranets in electric power systems.
6. CIGRE has formed a new working group on security, D2.22, Treatment of Information Security for Electric Power Utilities (EPUs).
7. OpenSG Security (formally known as UtiliSec) (http://osgug.ucaiug.org/utilisec/default.aspx) is an organization that has started as a utility-based user group to write procurement-oriented standards to address various emerging Smart Grid systems. The first task (not directly related to SA systems) this group engaged in was developing security profiles for advanced metering infrastructure systems (AMI). Currently, this organization is developing profiles for other Smart Grid applications and systems, of note is their recent draft security profile for Wide Area Situational Awareness (i.e., Synchrophasor Systems) that is currently under comment and review that has some bearing on SA systems.

References

1. North American Reliability Corporation, Reliability Standards, Critical Infrastructure Protection (CIP), July 28, 2011. Available from: http://www.nerc.com/page.php?cid = 2|20
2. Smart Grid Interoperability Panel—Cyber Security Working Group, National Institute of Standards, U.S. Department of Commerce, NISTIR 7628 Guidelines for Smart Grid Cyber Security, Vols 1–3, August 2010. Available from: http://csrc.nist.gov/publications/PubsNISTIRs.html
3. ICS-CERT, Advisory, ICSA-10-201-01—USB Malware Targeting Siemens Control Software, July 20, 2010. Available from: http://www.us-cert.gov/control_systems/pdf/ICSA-10-201-01.pdf
4. N. Falliere, L. O. Murchu, and E. Chien, Symantec Security Response, W32.Stuxnet Dossie, Version 1.4, February 2011. Available from: http://www.symantec.com/content/en/us/enterprise/media/security_response/whitepapers/w32_stuxnet_dossier.pdf
5. 18 CFR 39.1, Title 18, Chapter I, Subchapter B, Part 39. Conservation of Power and Water Resources, July 28, 2011. Available from: http://ecfr.gpoaccess.gov/cgi/t/text/text-idx?c=ecfr&sid=617eee924b2d959b3e1f18d1eb087cfc&rgn=div8&view=text&node=18:1.0.1.2.27.0.23.1&idno=18
6. North American Reliability Corporation, Smart Grid Task Force, Reliability Considerations from the Integration of Smart Grid, December 2010. Available from: http://www.nerc.com/files/SGTF_Report_Final_posted.pdf
7. Github, Stuxnet Code Decompilation, July 28, 2011. Available from: https://github.com/Laurelai/decompile-dump/tree/master/output
8. J. Weiss, *Protecting Industrial Control Systems from Electronic Threats*, Momentum Press, New York, May 2010, ISBN: 987-60650-197-9.
9. C. Rigney et al., The Internet Society, RFC 2865: Remote Authentication Dial In User Service (RADIUS), June 2000. Available from: http://www.ietf.org/rfc/rfc2865.txt
10. M. Wahl et al., The Internet Society, RFC 2251: Lightweight Directory Access Protocol (v3), December 1997. Available from: http://www.ietf.org/rfc/rfc2251.txt
11. IEEE Standard C37.1-1994, *IEEE Standard Definition, Specification and Analysis of Systems Used for Supervisory Control, Data Acquisition, and Automatic Control*.
12. ANSI/IEEE Standard 100-1984, *IEEE Standard Dictionary of Electrical and Electronics Terms*, 3rd Edn., p. 695.

13. ANSI/IEEE Standard 100-1984, *IEEE Standard Dictionary of Electrical and Electronics Terms*, 3rd Edn., p. 769.
14. ANSI/IEEE Standard 100-1984, *IEEE Standard Dictionary of Electrical and Electronics Terms*, 3rd Edn., p. 811.
15. S. Kent and K. Seo, The Internet Society, RFC4301: Security Architecture for the Internet Protocol, December 2005. Available from: http://tools.ietf.org/html/rfc4301
16. T. Dierks and E. Rescorla, The Internet Society, RFC5246: The Transport Layer Security (TLS) Protocol Version 1.2, August 2008. Available from: http://www.ietf.org/rfc/rfc5246.txt
17. Carnegie-Mellon Software Engineering Institute, CERT Coordination Center, July 28, 2011. Available from: http://www.cert.org/
18. DNP Users Group, DNP Protocol Primer, July 28, 2011. Available from: http://www.dnp.org/files/dnp3_primer.pdf
19. D. Thanos, Prepared statement of Daniel Thanos, chief cyber security architect of General Electric Digital Energy, before the Federal Energy Regulatory Commission, *Technical Conference on Smart Grid Interoperability Standards*, Washington, DC, January 31, 2011. Available from: http://www.ferc.gov/EventCalendar/Files/20110131084125-Thanos,%20GE%20Digital%20Energy.pdf
20. IEEE C1 Working Group members, Cyber security issues for protective relays, June 2007.
21. Idaho National Laboratory, Demonstration at *KEMA 4th Conference: Advancements in Technology and Business*, Idaho Falls, ID, August 16–18, 2004 (unpublished).
22. Green, G., Hacker jailed for sewage sabotage, *The Courier-Mail*, Brisbane, Queensland, Australia, November 1, 2001.
23. Stan Klein Associates: Draft Unpublished EPRI Report. *Information Security Guidelines for Transmission and Distribution Systems*, October 2000.
24. N. Ferguson, B. Schneier, and T. Kohno, *Cryptography Engineering*, John Wiley & Sons, New York, March 15, 2010. ISBN: 9780470474242.
25. Achilles Certified Communications for Embedded Devices, July 28, 2011. Available from: http://www.wurldtech.com/achilles-certification/achilles-certified-communications/program-summary.aspx
26. ANSI/ISA-TR99.00.01-2004, *Security Technologies for Manufacturing and Control Systems*.
27. ANSI/ISA-TR99.00.02-2004, *Integrating Electronic Security into the Manufacturing and Control Systems Environment*.
28. ANSI/ISA-99.02.01-2009, *Security for Industrial Automation and Control Systems: Establishing an Industrial Automation and Control Systems Security Program*.

18
Gas-Insulated Transmission Line

18.1	Introduction	**18**-1
18.2	History	**18**-2
18.3	System Design	**18**-4
	Technical Data • Standard Units • Laying Methods	
18.4	Development and Prototypes	**18**-10
	Gas Mixture • Type Tests • Long-Duration Tests	
18.5	Advantages of GIL	**18**-24
	Safety and Gas Handling • Magnetic Fields	
18.6	Application of Second-Generation GIL	**18**-26
	Tunnel-Laid GIL • Directly Buried GIL	
18.7	Quality Control and Diagnostic Tools	**18**-30
18.8	Corrosion Protection	**18**-31
	Passive Corrosion Protection • Active Corrosion Protection	
18.9	Voltage Stress Coming from the Electric Power Net	**18**-33
	Overvoltage Stresses • Maximum Stresses by Lightning Strokes • Modes of Operation • Application of External and Integrated Surge Arresters • Results of Calculations • Insulation Coordination	
18.10	Future Needs of High-Power Interconnections	**18**-35
	Metropolitan Areas • Use of Traffic Tunnels	
18.11	To Solve Bottlenecks in the Transmission Net	**18**-38
	Introduction • Transmission Net Requirements • Technical Solutions • Integrating Renewable Energy • Cost of Transmission Losses	
18.12	Conclusion	**18**-42
References		**18**-42

Hermann Koch
Siemens AG

18.1 Introduction

The gas-insulated transmission line (GIL) is a system for the transmission of electricity at high power ratings over long distances. In cases where overhead lines (OHLs) are not possible, the GIL is a viable technical solution to bring the power transmitted by an OHL underground without a reduction of power transmission capacity.

As a gas-insulated system, the GIL has the advantage of electrical behavior similar to that of an OHL, which is important to the operation of the complete network. Because of the large cross section of the conductor, the GIL has low electrical losses compared with other transmission systems (OHLs and cables).

This reduces the operating and transmission costs, and it contributes to reduction of global warming because less power needs to be generated.

Safety of personnel in the vicinity of a GIL is very high because the solid metallic enclosure provides reliable protection. Even in the rare case of an internal failure, the metallic enclosure is strong enough to withstand damage. This allows the use of GILs in street and railway tunnels and under bridges with public traffic. No flammable materials are used to build a GIL. The use of GILs in traffic tunnels makes the tunnels more economical and can solve some environmental problems. If GIL is added to a traffic tunnel, the cost can be shared between the electric power supply company and the owner of the traffic part (train and vehicles). The environmental advantage is that no additional OHL needs to be built parallel to the tunnel. Because of the low capacitive load of the GIL, long lengths of 100 km and more can be built.

Where OHLs are not suitable due to environmental factors or where they would spoil a particular landscape, the GIL is a viable alternative because it is invisible and does not disturb the landscape. The GIL consists of three single-phase encapsulated aluminum tubes that can be directly buried in the ground or laid in a tunnel. The outer aluminum enclosure is at ground potential. The interior, the annular space between the conductor pipe and the enclosure, is filled with a mixture of gas, mainly N_2 (80%) with some SF_6 (20%) to provide electrical insulation. A reverse current, more than 99% of the conductor current value, is induced in the enclosure. Because of this reverse current, the outer magnetic field is very low.

GIL combines reliability with high transmission capacity, low losses, and low emission of magnetic fields. Because it is laid in the ground, GIL also satisfies the requirements for power transmission lines without any visual impact on the environment or the landscape. Of course, the system can also be used to supply power to meet the high energy demands of conurbations and their surroundings. The directly buried GIL combines the advantage of underground laying with a transmission capacity equivalent to that of an overhead power line [1–3].

The changes in the electric power industry coming from deregulation and the separation of power generation, transmission, distribution, and power trade have a very strong influence to the load flow in the electric power net. With new generation units, for example, gas fired turbines or dispersed generation such as wind power generation and photovoltaic, the existing electrical net is used differently to the days when it was planned and installed in the first place. In consequence, this load flow changes are leading to congestions in the electric net and to so-called bottlenecks.

New high-power transmission lines are needed in the future, and the GIL plays an important role when OHL solutions are not possible. When underground solutions are required because of public interest to not have aboveground installations, then the GIL can solve the transmission problem for high power ratings (2000 MVA and more) and long distances (30 km and more) [4,5].

18.2 History

The GIL was invented in 1974 to connect the electrical generator of a hydro pump storage plant in Schluchsee, Germany [3]. Figure 18.1 shows the tunnel in the mountain with the 400 kV OHL. The GIL went into service in 1975 and has remained in service without interruption since then, delivering peak energy into the southwestern 420 kV network in Germany. With 700 m of system length running through a tunnel in the mountain, this GIL is still the longest application at this voltage level in the world. Today, at high-voltage levels ranging from 135 to 550 kV, a total of more than 100 km of GILs have been installed worldwide in a variety of applications, for example, inside high-voltage substations or power plants or in areas with severe environmental conditions.

Typical applications of GIL today include links within power plants to connect high-voltage transformers with high-voltage switchgear, links within cavern power plants to connect high-voltage transformers in the cavern with OHLs on the outside, links to connect gas-insulated switchgear (GIS) with OHLs, and service as a bus duct within GIS. The applications are carried out under a wide

FIGURE 18.1 GIL (420 kV, 2500 A) in Schluchsee, Germany. (Courtesy of Siemens AG, Erlangen, Germany.)

range of climate conditions, from low-temperature applications in Canada, to the high ambient temperatures of Saudi Arabia or Singapore, to the severe conditions in Europe or in South Africa. The GIL transmission system is independent of environmental conditions because the high-voltage system is completely sealed inside a metallic enclosure.

The GIL technology has proved its technical reliability in more than 2500 km/year of operation without a major failure. This high system reliability is due to the simplicity of the transmission system, where only aluminum pipes for conductor and enclosure are used, and the insulating medium is a gas that resists aging.

In the early days, the high cost of GIL restricted its use to special applications. However, with the second-generation GIL, a total cost reduction of 50% has made the GIL economical enough for application over long distances. The breakthrough in cost reduction is achieved by using highly standardized GIL units combined with the efficiencies of automated orbital-welding machines and modern pipeline laying methods. This considerably reduces the time required to lay the GIL, and angle units can be avoided by using the elastic bending of the aluminum pipes to follow the contours of the landscape or the tunnel. This breakthrough in cost and the use of N_2–SF_6 gas mixtures have made possible what is now called second-generation GIL, and it is a very interesting transmission system for high-power transmission over long distances, especially if high power ratings are needed.

The second-generation GIL was first built for EOS (energie ouest suisse) at the PALEXPO exhibition area, close to the Geneva Airport in Switzerland. Since January 2001, this GIL has been in operation as part of the OHL connecting France with Switzerland. The success of this project has demonstrated that the new laying techniques are suitable for building very long GIL transmission links of 100 km or more within an acceptable time schedule [7].

The next step of the second-generation GIL was to the directly buried GIL at the Kelsterbach project in Germany, which was installed in 2010. A 1 km long double system of 400 kV rated voltage was laid close to the new Frankfurt Airport extension. The 80% N_2 and 20% SF_6 gas mixture, the high transmission capacity of 1800 MVA at 2500 A continuous current rating, and the elastic bending radius to follow the bended trench contour are some unique features of this project, the worldwide first at 400 kV voltage rating and a system length of 1 km [6].

Applications of the second-generation GIL laid in a tunnel as done at the PALEXPO project or directly buried as in the Kelsterbach project are seen today as possible solutions for the upcoming change of the transmission network to integrate regenerative energy generation.

18.3 System Design

18.3.1 Technical Data

The main technical data of the GIL for 420 and 550 kV transmission networks are shown in Table 18.1. For 550 kV applications, the SF_6 content or the diameter of the enclosure pipe might be increased.

The rated values shown in Table 18.1 are chosen to match the requirements of the high-voltage transmission grid of OHLs. The power transmission capacity of the GIL is 2000 MVA whether tunnel laid or directly buried. This allows the GIL to continue with the maximum power of 2000 MVA of an OHL and bring it underground without any reduction in power transmission [6,7]. The values are in accordance with the relevant IEC standard for GILs, IEC 62271-204 [8].

18.3.2 Standard Units

Figure 18.2 shows a straight unit combined with an angle unit. The straight unit consists of a single-phase enclosure made of aluminum alloy. In the enclosure (1), the inner conductor (2) is fixed by a conical insulator (4) and lies on support insulators (5). The thermal expansion of the conductor toward the enclosure is adjusted by the sliding contact system (3a, 3b). One straight unit has a length up to 120 m made by single-pipe sections welded together by orbital-welding machines. If a directional change exceeds what the elastic bending allows, then an angle element (shown in Figure 18.2) is added by orbital welding with the straight unit. The angle element covers angles from 4° to 90°.

TABLE 18.1 Technical Data for 420 and 550 kV GIL Transmission Networks

Type	Value
Nominal voltage (kV)	420/550
Nominal current (A)	3150/4000
Lightning impulse voltage (kV)	1425/1600
Switching impulse voltage (kV)	1050/1200
Power frequency voltage (kV)	630/750
Rated short-time current (kA/3 s)	63
Rated gas pressure (bar)	7
Insulating gas mixture	80% N_2, 20% SF_6

Source: Courtesy of Siemens AG, Erlangen, Germany.

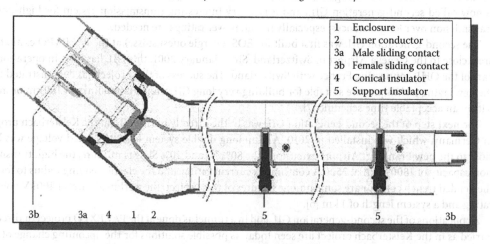

1	Enclosure
2	Inner conductor
3a	Male sliding contact
3b	Female sliding contact
4	Conical insulator
5	Support insulator

FIGURE 18.2 Straight construction unit with an angle element. (Courtesy of Siemens AG, Erlangen, Germany.)

Under normal conditions of the landscape, no angle units are needed because the elastic bending, with a bending radius of 400 m, is sufficient to follow the contour.

At distances of 1200–1500 m, disconnecting units are placed in underground shafts. Disconnecting units are used to separate gas compartments and to connect high-voltage testing equipment for the commissioning of the GIL. The compensator unit is used to accommodate the thermal expansion of the enclosure in sections that are not buried in the earth. A compensator is a type of metallic enclosure, a mechanical soft section, which allows movement related to the thermal expansion of the enclosure. It compensates the length of thermal expansion of the enclosure section. Thus, compensators are used in tunnel-laid GILs as well as in the shafts of directly buried GILs.

The enclosure of the directly buried GIL is coated in the factory with a multilayer polymer sheath as a passive protection against corrosion. After completion of the orbital weld, a final covering for corrosion protection is applied on site to the joint area.

Because the GIL is an electrically closed system, no lightning impulse voltage can strike the GIL directly. Therefore, it is possible to reduce the lightning impulse voltage level by using surge arresters at the end of the GIL. The integrated surge-arrester concept allows reduction of high-frequency overvoltages by connecting the surge arresters to the GIL in the gas compartment [9].

For monitoring and control of the GIL, secondary equipment is installed to measure gas pressure and temperature. These are the same elements that are used in GIS. For commissioning, partial-discharge (PD) measurements are obtained using the sensitive very high frequency (VHF) measuring method.

An electrical measurement system to detect arc location is implemented at the ends of the GIL. Electrical signals are measured and, in the very unlikely case of an internal fault, the position can be calculated by the arc location system with an accuracy of 25 m.

The third component is the compensator, installed at the enclosure. In the tunnel-laid version or in an underground shaft, the enclosure of the GIL is not fixed, so it will expand in response to thermal heat-up during operation. The thermal expansion of the enclosure is compensated by the compensation unit. If the GIL is directly buried in the soil, the compensation unit is not needed because of the weight of the soil and the friction of the surface of the GIL enclosure.

The fourth and last basic module used is the disconnecting unit, which is used every 1.2–1.5 km to separate the GIL in gas compartments. The disconnecting unit is also used to carry out sectional high-voltage commissioning testing [22].

An assembly of all these elements as a typical setup is shown in Figure 18.3, which illustrates a section of a GIL between two shafts (1). The underground shafts house the disconnecting and compensator units (2). The distance between the shafts is between 1200 and 1500 m and represents one single gas compartment.

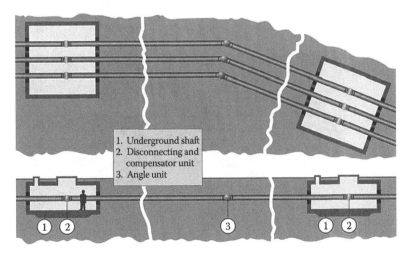

FIGURE 18.3 Directly buried GIL system components. (Courtesy of Siemens AG, Erlangen, Germany.)

A directly buried angle unit (3) is shown as an example in the middle of the figure. Each angle unit also has a fix point, where the conductor is fixed toward the enclosure.

18.3.3 Laying Methods

The GIL can be laid aboveground on structures, in a tunnel, or directly buried into the soil like an oil or gas pipeline. The overall cost for the directly buried version of the GIL is, in most cases, the least expensive version of GIL laying. For this laying method, sufficient space is required to provide accessibility for working on site. Consequently, directly buried laying will generally be used in open landscape crossing the countryside, similar to OHLs, but invisible.

18.3.3.1 Directly Buried

The most economical and fastest method of laying cross country is the directly buried GIL. Similar to pipeline laying, the GIL is continuously laid within an open trench. A nearby preassembly site reduces the cost of transporting GIL units to the site. With the elastic bending of the metallic enclosure, the GIL can flexibly adapt to the contours of the landscape. In the soil, the GIL is continuously anchored, so that no additional compensation elements are needed [10,11].

The laying procedure for a directly buried GIL is shown in Figure 18.4 [16]. The left side of the figure shows a digging machine opening the trench, which will have a depth of about 1.2–2 m. The building shown close to the trench is the prefabrication area, where GIL units of up to 120 m in length are preassembled and prepared for laying. The GIL units are transported by cranes close to the trench and then laid into the trench. The connection to the already laid section is done within a clean housing tent in the trench. The clean housing tent is then moved to the next joint and the trench is backfilled. Figure 18.5 shows the moment of laying the GIL into the trench. Figure 18.6 shows the bended tube and backfilling of the trench.

18.3.3.2 Aboveground Installation

Aboveground GIL installations are usually installed on steel structures at heights of 1–5 m aboveground. The enclosures are supported in distances of 20–40 m. This is because of the rigid metal enclosure. Because of the mechanical layout of the GIL, it is also suitable to use existing bridges to cross, for example, a river.

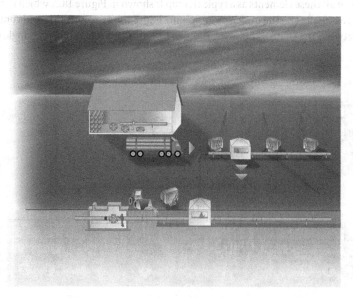

FIGURE 18.4 Laying procedure for a directly buried GIL. (Courtesy of Siemens AG, Erlangen, Germany.)

Gas-Insulated Transmission Line

FIGURE 18.5 Laying the GIL into the soil. (Courtesy of Siemens AG, Erlangen, Germany.)

FIGURE 18.6 Bended tube and backfilling. (Courtesy of Siemens AG, Erlangen, Germany.)

The aboveground installations are typical for installations within substations to connect, for example, the bay of a GIS with an OHL, where larger distances between the phases of the three-phase system are used, or to connect the GIS directly with the step-down transformer. The GIL is often chosen if very high reliability is needed, for example, in nuclear power stations.

Another reason for GIL applications in substation power is that the aboveground installations are used for the transmission of very high electrical power ratings. The strongest GIL has been installed in Canada at the Kensington Nuclear Power Station in a substation with GIS where single sections of the GIL bus bar system can carry currents of 8,000 A and can withstand short circuit currents of 100,000 A.

Aboveground GIL installations inside substations are widely used in conjunction with GIS. Usually, the substations are fenced and, therefore, not accessible to the public. If this is not the case, laid tunnel or directly buried GIL will be chosen for safety reasons. Accessibility of GIL to the public is generally avoided so as not to allow manipulations on the GIL (e.g., drilling a hole into the enclosure), which can be dangerous because of the high-voltage potential inside.

18.3.3.3 Tunnel Laid

If there is not enough space available to bury a GIL, laying the GIL into a tunnel will be the most appropriate method. This tunnel-laying method is used in cities or metropolitan areas as well as when crossing a river or interconnecting islands. Because of the high degree of safety that GIL offers, it is possible to run a GIL through existing or newly built street or railway tunnels, for example, in the mountains.

Modern tunneling techniques have been developed during the past few years with improvements in drilling speed and accuracy. So-called microtunnels, with a diameter of about 3 m, are economical solutions in cases when directly buried GIL is not possible, for example, in urban areas, in mountain crossings, or in connecting islands under the sea. Such microtunnels are usually the shortest connection between two points and, therefore, reduce the cost of transmission systems. After commissioning, the system is easily accessible. Figure 18.7 shows a view into a GIL tunnel at the Institut Prüffeld für elektrische Hochleistungstechnik (IPH) test field in Berlin. This tunnel of 3 m in diameter can accommodate two

FIGURE 18.7 View into the tunnel. (Courtesy of Siemens AG, Erlangen, Germany.)

Gas-Insulated Transmission Line

FIGURE 18.8 Tunnel-laid GIL for voltages up to 550 kV. (Courtesy of Siemens AG, Erlangen, Germany.)

systems of GIL for rated voltages of up to 420/550 kV and with rated currents of 3150 A. This translates to a power transmission capacity of 2250 MVA for each system.

Figure 18.8 shows a view into the tunnel at PALEXPO at Geneva Airport in Switzerland with two GIL systems [15]. The tunnel dimensions in this case are 2.4 m wide and 2.6 m high. The transmission capacity of this GIL is also 2250 MVA at 420/550 kV rated voltage with rated currents up to 3150 A.

In both laying methods—directly buried and tunnel laid—the elastic bending of the GIL can be seen in Figures 18.6 and 18.8, respectively. The minimum acceptable bending radius is 400 m.

Figure 18.9 shows the principle for the laying procedure in a tunnel. GIL units of 11–14 m in length are brought into a tunnel by access shafts and then connected to the GIL transmission line in the tunnel. In cases with horizontal accessibility—such as in a traffic tunnel for trains or vehicles—the GIL units can be much longer, 20–30 m by train transportation. This increase in length reduces the assembly work and time and allows major cost reductions. A special working place for mounting and welding is installed at the assembly site [12]. As seen in Figure 18.9, the delivery and supply of prefabricated elements (1) is brought to the shaft or tunnel entrance. After the GIL elements are brought into the shaft to the mounting and welding area (2), the elements are joined by an orbital-welding machine. The GIL section is then brought into the tunnel (3). When a section is ready, a high-voltage test is carried out (4) to validate each section.

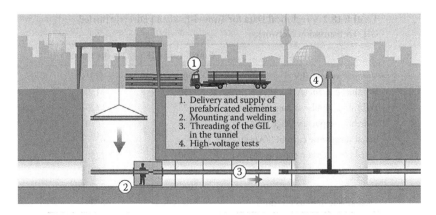

FIGURE 18.9 Laying and testing in a tunnel. (Courtesy of Siemens AG, Erlangen, Germany.)

18.3.3.4 Covered Trench

To lay the GIL in a covered trench is another low-cost solution. The trench is usually a U-shape concrete structure laid underground. The U-shape concrete structure is usually prefabricated and transported on site for laying in the open trench and connecting section by section.

Once the U-shape concrete structure is laid, the GIL can be inserted and fixed to the bottom or to the sidewalls by steel structures. The steel structures can be fixed by bolts to the concrete U-shape structure and will carry the GIL pipes on rolls or slide pads if flexibility is needed for thermal expansion. Or the GIL pipes can be fixed to the steel structure by a steel band to realize a fix point where no thermal expansion is allowed. The GIL pipes are usually laid with expansion joints allowing flexibility to the pipes concerning the thermal expansion and the bending radius.

To close the U-shape concrete, different covers are possible depending on the surrounding and protection. A totally closed cover made of concrete, steel, or aluminum can be chosen to avoid any direct contact to the GIL outer pipe. This is usually the case in areas where public access is given. In other cases, it can be possible to use cattle grid or other open covers. This may be the case if the area is fenced, for example, within a substation or power plant.

18.4 Development and Prototypes

Development of the second-generation GIL was based on the knowledge of gas-insulated technologies and was carried out in type tests and long-duration tests. The type tests proved the design in accordance with IEC 60694, IEC 60517, IEC 61640, and related standards [13,14]. An expected lifetime of 50 years has been simulated in long-term duration tests involving combined stresses of current and high-voltage cycles that were higher than the nominal ratings. At the IPH test laboratory in Berlin, Germany, tests have been carried out on tunnel-laid and directly buried GIL in cooperation with the leading German utilities.

A prototype tunnel-laid GIL of approximately 70 m length has been installed in a concrete tunnel. The jointing technique of a computer-controlled orbital-welding machine was applied under realistic on-site conditions. The prototype assembly procedure has also been successfully proved under realistic on-site conditions.

The directly buried GIL is a further variant of GIL. After successful type tests, the properties of a 100 m long directly buried GIL were examined in a long-duration test with typical accelerated load cycles. The results verified a service life of 50 years. Installation, construction, laying, and commissioning were all carried out under real on-site conditions. The test program represents the first successfully completed long-duration test for GIL using the insulating N_2–SF_6 gas mixture. The technical data for the directly buried and tunnel-laid GIL are summarized in Table 18.2 [19].

TABLE 18.2 Technical Data for Tunnel-Laid and Directly Buried GIL Transmission Networks

	Tunnel-Laid GIL	Directly Buried GIL
Nominal voltage (kV)	420/550	420/550
Nominal current (A)	3150	3150
Lightning impulse voltage (kV)	1425	1425
Switching impulse voltage (kV)	1050	1050
Rated short-time current (kA/3 s)	63	63
Rated transmission capacity (MVA)	2250	2250
Insulating gas mixture	80% N_2 20% SF_6	80% N_2 20% SF_6
Pipe outside dimension (mm)	520	600

Source: Courtesy of Siemens AG, Erlangen, Germany.

The values shown in Table 18.2 are chosen for the application of GIL in a transmission grid with OHLs and cables. Because the GIL is an electrically closed system, meaning the outer enclosure is completely metallic and grounded, no lightning impulse voltage can directly strike the GIL. Therefore, it is possible to reduce the lightning impulse voltage level by using surge arresters at the ends of the GIL. The integrated surge-arrester concept allows the reduction of high-frequency overvoltages by connecting the surge arresters to the GIL in the gas compartment [6,9,15,16].

18.4.1 Gas Mixture

Like natural air, the gas mixture consists mainly of nitrogen (N_2), which is chemically even more inert than SF_6. It is therefore an ideal and inexpensive admixture gas that calls for almost no additional handling work on the gas system [17]. The low percentage (20%) of SF_6 in the N_2–SF_6 gas mixture acquires high dielectric strength due to the physical properties of these two components. Figure 18.10 shows that a gas mixture with an SF_6 content of only 20% has 70% of the pressure-reduced critical field strength of pure SF_6. The curves are defined in Figure 18.10. A moderate pressure increase of 40% is necessary to achieve the same critical field strength of pure SF_6 [27].

N_2–SF_6 gas mixtures are an alternative to pure SF_6 if only dielectric insulation is needed and there is no need for arc-quenching capability, as in circuit breakers or disconnectors. Much published research work has been performed and properties ascertained in small test setups under ideal conditions [18]. The arc-quenching capability of N_2–SF_6 mixtures is inferior to pure SF_6 in approximate proportion to its SF_6 content [19]. N_2–SF_6 mixtures with a higher SF_6 concentration are successfully applied in outdoor SF_6 circuit breakers in arctic regions in order to avoid SF_6 liquefaction, but a reduced breaking capability has to be accepted [29].

In the event of an internal arc, the N_2–SF_6 gas mixture with a high percentage of N_2 (80%) behaves similar to air. The arc burns with a large footpoint area. Footpoint area is the area covered by the footpoint of an internal arc during the arc burning time of typically 500 ms. Consequently, the thermal power flow density into the enclosure at the arc footpoint is much less, which causes minimal material erosion of the enclosure. The result is that the arc will not burn through, and there is no external impact to the surroundings or the environment.

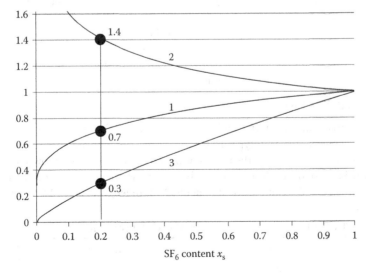

FIGURE 18.10 Normalized ideal intrinsic properties of N_2–SF_6 mixtures. 1. Pressure-reduced critical field, 2. necessary pressure for mixtures of equal critical field strength, and 3. necessary amount of SF_6 for mixtures of equal critical field strength. (Courtesy of Siemens AG, Erlangen, Germany.)

FIGURE 18.11 High-current and internal-arc test setup. (Courtesy of Siemens AG, Erlangen, Germany.)

18.4.2 Type Tests

The type tests were based on the new IEC 62271-204 standard [8]. The test parameters for additional tests to assess GIL lifetime performance were defined with reference to the CIGRE technical brochures TB 176 "Gas Mixtures" [17], TB 218 "GIL" [20] and TB 351 "GIL in long structures" [21], and IEC 62271-204. For the type tests, full-scale test setups were installed, containing all essential design components.

18.4.2.1 Short-Circuit Withstand Tests

The short-circuit withstand tests were carried out on the test setup shown in Figure 18.11. The GIS test setup was assembled using the different GIS units: straight unit, angle unit, compensator unit, and disconnector unit. From left to right in Figure 18.11 are the straight unit, a 90° angle unit, and at the far right a disconnection unit. Table 18.3 lists the parameters for the short-circuit withstand test. The different values for the duration of short-circuit currents are not related to design criteria but, rather, reflect regional market requirements.

After these tests, no visible damage was seen, and the functionality of the GIL prototype was not impaired. The contact resistivity was measured after the test and was well within the range of what was allowed by the IEC 62271-204 standard. Actually, the contact resistivity of the GIL sliding contact after the test was even a little lower than before, indicating the system's very good current-carrying capability.

18.4.2.2 Internal-Arc Test

To check whether arcing due to internal faults causes burn-through of the enclosure, an internal-arcing test was performed on the GIL prototype. Tests were carried out with arc currents of 50 and 63 kA and arc duration times of 0.33 and 0.5 s. The results of the internal-arcing tests showed only little damage, with the wall thickness of the enclosure eroding by only a few micrometers. The pressure rise was very low because of the size of the compartment of about 20 m in length. Figure 18.12, a view into the GIL

TABLE 18.3 Parameters for Short-Circuit Withstand Test

GIL Test Parameter	Tunnel-Laid GIL	Directly Buried GIL
Short-circuit peak current (kA)	185	165
Short-time current (kA)	75	63
Duration of short-circuit current (s)	0.5	3

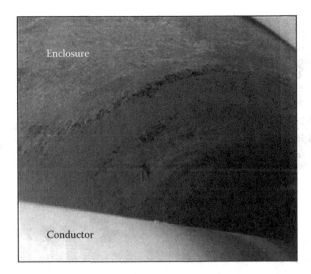

FIGURE 18.12 View into the GIL after an arc fault of 63 kA and 0.5 s. (Courtesy of Siemens AG, Erlangen, Germany.)

after the arc fault test, shows very few distortions. The resistance to arcing damage means that the GIL can use the autoreclosure function, the same as with OHLs.

Results of the internal-arc tests can be summarized as follows:

- No external influence during and after the internal-arc test was noticed.
- No burn-through of the enclosure occurred. Very low material erosion was observed on the enclosure and conductor.
- The pressure rise within the enclosures during arcing was so low that even the rupture discs did not open.
- The arc characteristic is much smoother compared with the characteristics in pure SF_6 (e.g., large arc diameter and lower arc traveling speed).
- Cast-resin insulators were not seriously affected.

All of these results speak in favor of the safe operation of the GIL. Even so, in the very unlikely event of an internal arc, the external environment is not affected. The results of the arc fault test also showed that in the case of a tunnel-laid GIL, there is no danger to the people traveling through the tunnel. This makes the GIL the only high-power transmission system that can be used in public traffic tunnels together with trains and street traffic.

18.4.2.3 Dielectric Tests

Dielectric type tests were carried out on the full-scale test setup in the high-voltage laboratory of Siemens in Berlin (Figure 18.13) and in the IPH test laboratory, also in Berlin. The tunnel-laid and directly buried GIL systems were tested according to the rated voltages and test voltages given in IEC standard 62271-204. The gas pressure was set to 7 bar abs. Test parameters are presented in Table 18.4. The tests were applied with 15 positive and 15 negative impulses, and the power-frequency withstand-test voltage was applied for 1 min. All tests were passed.

18.4.3 Long-Duration Tests

To check the GIL system's suitability for practical use, every effort was made to implement a test setup that came close to real conditions [23]. Therefore, the tunnel-laid GIL was assembled on site and installed in a tunnel made of concrete tubes (total length: 70 m). The directly buried GIL was laid in soil (total length: 100 m). The test parameters are given in Table 18.5.

FIGURE 18.13 Test setup for high-voltage tests. (Courtesy of Siemens AG, Erlangen, Germany.)

TABLE 18.4 Parameters for Dielectric Type Tests at 7 bar Gas Pressure

Test Parameters GIL		
	Directly Buried and Tunnel Laid (kV)	Directly Buried (kV)
Maximum voltage of equipment Um	420	550
AC withstand test, 1 min	630	750
Lightning impulse test	1425	1600
Switching impulse test	1050	1200

Source: Courtesy of Siemens AG, Erlangen, Germany.

The test values are derived from typical applications for directly buried and tunnel-laid GIL. The lower test voltages for the tunnel-laid GIL reflect the fact that such systems are typically used in metropolitan areas, where they are usually connected to cable systems and therefore have lower overvoltages from the net. The higher voltages for the directly buried GIL represent the typical application as part of the OHL net, with higher overvoltages due to lightning. In any case, both applications of test voltages can be used for directly buried and tunnel-laid GIL.

The duration and cycle times of the current and high-voltage sequences were chosen to apply maximum stress to heat up and cool down the GIL system. After a heat cycle of 12 or 24 h, the current was switched off, and the high voltage was applied to the GIL at the moment when the strongest mechanical forces were coming with the cooldown phase of the GIL. The sequences are listed in Table 18.6.

The total time of the long-duration test was 2500 h, which represents a lifetime of 50 years due to the overvoltage (double value) and the mechanical stress. The complete long-duration test is shown in Figure 18.14.

GIL conductor and enclosure temperature, as well as GIL movement due to thermal expansion/contraction, were monitored during load cycles. All tests were performed successfully.

TABLE 18.5 Parameters of the Commissioning Test and the Recommissioning Test after Demonstration of a Repair Process

	Test Parameters GIL, Directly Buried		Test Parameters GIL, Tunnel Laid	
Commissioning	AC withstand test, 1 min with PD monitoring	630 kV	AC withstand test, 10 s	550 kV
	Lightning impulse test	1300 kV	AC withstand test, 1 min with PD monitoring	504 kV
	Switching impulse test	1050 kV	Lightning impulse test	1140 kV
Re-commissioning, after demonstration of repair process	AC withstand test, 1 min with PD monitoring	630 kV	AC withstand test, 10 s	550 kV
	Lightning impulse test	1300 kV	AC withstand test, 1 min with PD monitoring	504 kV
	Switching impulse test	1050 kV	Lightning impulse test	1140 kV
	AC, 48 h with PD monitoring	480 kV		
Final test (tunnel-laid, after 2500 h) (directly buried, after 2880 h)	AC withstand test, 1 min with PD monitoring	630 kV	AC withstand test, 10 s	550 kV
	Lightning impulse test	1300 kV	AC withstand test, 1 min with PD monitoring	504 kV
	Switching impulse test	1050 kV	Lightning impulse test	1140 kV
	AC, 48 h with PD monitoring	480 kV		

Source: Courtesy of Siemens AG, Erlangen, Germany.

TABLE 18.6 Load Cycles and Intermediate Tests of the Long-Duration Test

	Test Parameters GIL, Tunnel Laid		Test Parameters GIL, Directly Buried		
Load cycles	Total duration	2500 h	Load cycles, time parameters change every 480 h	Total duration	2880 h
	Duration of one cycle	12 h		Duration of one cycle	12/24 h
	Number of cycles	210		Number of cycles	120/50
	Heating current, 7 h	3200 A		Heating current, 8 h	4000 A
	High voltage, 5 h	480 kV		High voltage, 4/16 h	480 kV
Intermediate tests, every 480 h	Switching impulse test	1050 kV	Intermediate tests, every 480 h	Lightning impulse test	1140 kV

Source: Courtesy of Siemens AG, Erlangen, Germany.

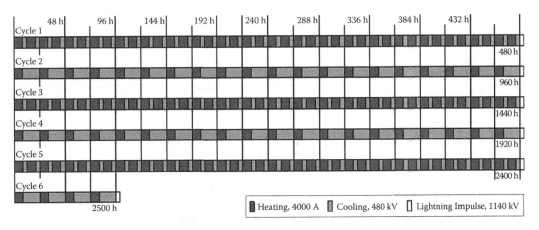

FIGURE 18.14 Long-duration test cycle of the directly buried GIL. For the tunnel-laid GIL, the test sequences of cycle type 1 had been applied for the total duration. (Courtesy of Siemens AG, Erlangen, Germany.)

FIGURE 18.15 Tunnel arrangement of two systems GIL for the long-term test. (Courtesy of Siemens AG, Erlangen, Germany.)

18.4.3.1 Long-Duration Test on a Tunnel-Laid GIL

A 70 m long prototype was assembled and laid in a concrete tunnel of 3 m diameter (Figure 18.15). The arrangement contained all major components of a typical GIL, including supports for the tunnel installation. The tunnel segments are original concrete units that are laid 20–40 m under the street level. The technology of drilling such tunnels has improved during the past years, and a large reduction in costs can be obtained through today's improved measuring and control techniques.

Figure 18.16 shows a top view of the long-duration test setup, which consists of a 50 m straight-construction unit, an angle unit, and another 20 m section after the directional change. The axial compensator took care of the thermal expansion of the enclosure during the load cycles. The disconnecting unit separates the GIL toward the high-voltage connection and the connection to the

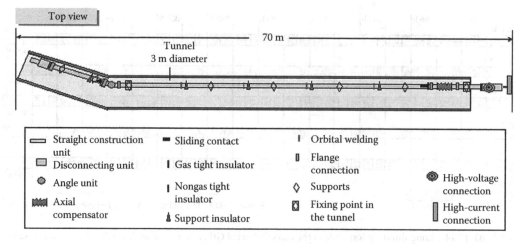

FIGURE 18.16 Arrangement of long-duration test setup. (Courtesy of Siemens AG, Erlangen, Germany.)

FIGURE 18.17 Computer-controlled orbital welding on site. (Courtesy of Siemens AG, Erlangen, Germany.)

high-current source. Sliding contacts inside the GIL compensate for the thermal expansion of the conductor, which slides on support insulators.

The segments of the GIL are welded with an orbital-welding machine, as seen in Figure 18.17. The orbital-welding machine is highly automated and gives a high-quality, reproducible weld. Together with the orbital welding, an automated, ultrasonic measuring system provides 100% quality control of the weld, which guarantees a gastight enclosure with a gas leakage rate of almost zero.

In addition to the aforementioned long-duration test with extremely high mechanical and electrical stresses, the sequence was interrupted after 960 h and a planned repair process—including the substitution of a tube length—was carried out (Figure 18.18). The total process of exchanging a segment of the GIL, including the recommissioning high-voltage testing, was finished in less than 1 week. The results demonstrate that the GIL can be repaired on site and then returned to service without any problems. The repair process requires only simple tools and is easily carried out in a short time.

Mixing of the gas was performed on site using a newly developed computer-controlled gas mixing device. The mixing process is continuous and arrives at a very high accuracy of the chosen gas mixture in the GIL. The gas mixture can be stored in standard high-pressure gas compartments (up to 200 bar) and can be reused after recommissioning.

18.4.3.2 Long-Duration Test on a Directly Buried GIL

The long-duration test for the buried GIL was carried out on a 100 m long test setup. Figure 18.19 shows the site arrangements. The laying procedure was carried out under realistic on-site conditions. Installation of the GIL under these conditions has proved the suitability of this laying procedure. The economy of the tools and procedures developed for this solution has been successfully demonstrated.

The tunnel-laid GIL described here is the first one with a N_2–SF_6 gas mixture to be tested and qualified in a long-duration test. The long-duration test—which involves extremely high stresses over a period of 2500 h (simulating a lifetime of more than 50 years) and a planned interruption to simulate a repair process—was concluded successfully. The results demonstrate once again the excellent performance and high reliability of the GIL [22,23].

FIGURE 18.18 Cutting the enclosure pipe with a saw. (Courtesy of Siemens AG, Erlangen, Germany.)

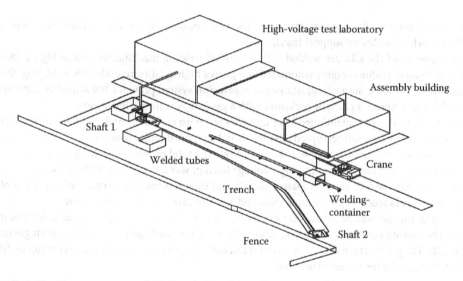

FIGURE 18.19 Site arrangements of the directly buried long-duration test. (Courtesy of Siemens AG, Erlangen, Germany.)

Figure 18.19 shows the IPH high-voltage test laboratory in Berlin with the high-voltage connection to shaft 1. From shaft 1, the trench with the directly buried GIL of 100 m length, including elastic bending and a directly buried angle module, proceeds to shaft 2 at the end. In shaft 2, a ground switch closes the current loop. The current-injection devices are in shaft 1. The shaft structures at the ends of the tunnel-laid GIL accommodate the separating modules and expansion fittings. The secondary equipment with the telecommunications system is also located there.

A crane transports the assembled GIL unit from the nearby assembly building to the welding container situated beside the trench, where the straight GIL segments are joined using an

Gas-Insulated Transmission Line   **18**-19

FIGURE 18.20 Trench of 55-m-long section during laying and view into the trench. (Courtesy of Siemens AG, Erlangen, Germany.)

orbital-welding machine. The final on-site assembly takes place either beside the trench or in the shaft structures. The place of assembly depends on the civil engineering design dictated by local conditions. The installation finishes with the laying of the GIL in its final position. Figure 18.20 shows the trench laying of the GIL with cranes. The trench follows a spherical curve with a bending radius of 400 m, which can be seen in Figure 18.20.

The process of constructing the trench and laying the GIL is quick and cost effective. The thermal expansion of the enclosure is absorbed by the surrounding bedding of coarse material by means of frictional forces. The bedding must also have sufficient thermal conductivity to dissipate the heat losses from the GIL. The temperature at the transition from the enclosure to the ground does not exceed 50°C when 2250 MVA are transmitted continuously by the GIL.

For the purpose of commissioning, comprehensive electrical and mechanical tests are necessary to verify the properties of the directly buried GIL. In addition to verifying the dielectric properties and checking the secondary equipment, the tests listed in Table 18.7 must be performed.

In addition, the typical elements of the secondary equipment of the GIL were employed; thus, PD measurement was carried out during commissioning and on-line during the test. The gas properties, such as temperature and pressure, were monitored on a continuous basis. Arc-location sensors were implemented. Radio sensors measured conductor temperature, gas density at the conductor, and the enclosure temperature in the ground at several points. The mechanical behavior of the GIL was studied by monitoring data from displacement sensors in the shaft structures and along the route. These sensors record the movement of the GIL relative to the ground or to the building.

During the course of the long-term test, the essential physical variables that describe the GIL—and that are used to prove the parameters of the calculations—are recorded. In addition to the

TABLE 18.7 Tests on Commissioning and Recommissioning

Pressure Test	Verification per Pressure Vessel Regulation
Gastightness test	Checking of flange joints
State of gas mixture	Mixture ratio
	Filling pressure
	Dew point
Corrosion protection coating voltage test	10 kV/1 min
Resistance test	Main circuit

Source: Courtesy of Siemens AG, Erlangen, Germany.

electrical stress imposed on the system by voltage and current, the aforementioned temperatures and movement were recorded.

18.4.3.3 Results of the Long-Duration Testing

18.4.3.3.1 Thermal Aspects

The GIL and its surrounding soil is a system of thermally coupled bodies, with inner heat produced by circulation of electrical current in both the conductor and the enclosure. Convection and radiation remove the heat losses from the conductor to the enclosure, whereas heat transfer in the annulus by conduction is negligible. This heat, adding to the losses by Joule effect from the enclosure, dissipates in the soil mainly in the radial direction to the surface of the soil and then flows into the ambient air by convection. The soil parameters were obtained from various literature sources documenting the soil properties in Berlin.

Before performing the unsteady-state study of the thermal behavior of the GIL, a steady-state model was developed taking into account the mechanisms of conduction in a solid body, natural convection in a cylindrical cavity, and radiation and convection in the interface between the soil surface and the air. The thermal system was divided into two parts—the GIL and the surrounding soil—and the physical phenomena occurring in each part was modeled. The finite element method (FEM) method (ANSYS program) was used first to check the accuracy of the developed analytical model and then to carry out the unsteady-state analyses of the thermal behavior of the buried GIL.

18.4.3.3.2 Calculation Model

Calculations were carried out using the finite element method. Heat loss, heat-transfer coefficient, and thermal resistance in the annular gap between the conductor and enclosing tube were calculated using a steady-state method according to the IEC 60287 standard [22,24]. These results were then used as constants in the transient calculation.

Calculations for the GIL cross section at the first location were carried out with the following parameters:

Cover $h = 0.7$ m ($h = 2.6$ m for the second location)
Thermal conductivity, $l = 1.6$ W/m K
Soil temperature, $T_s = 15°C$
Initial values for soil temperature, $T_i = 20°C$

The thermal resistance of the soil was measured at the start of the test at three different places (at the ends of the line and in the center). At each of these points, measurements were taken at two depths between 0.9 and 2.3 m. The average thermal resistance measured varied from 0.46 to 0.80 m K/W, a 70% difference between the extreme measured values. The measurements show a wide scatter from the mean value. The thermal resistance that was used in the calculations was taken as the mean values of the measurements.

The boundary conditions used in the calculations were as follows:

- Interface between soil and air: heat-transfer coefficient 20 W/m² K
- Air temperature is taken as an approximation of the measured air temperature by a sine function
- Temperature of soil: 15°C (20 m away from the GIL)
- Initial temperature of soil: 20°C
- Bisecting line: heat loss 0 W/m² (symmetry conditions)

Calculations were carried out for the following cycles load:
Short cycle

- 8 h, I = 4000 A, loss = 145 W/m
- 4 h, I = 0 A, loss = 0 W/m

Long cycle

- 8 h, I = 4000 A, loss = 145 W/m
- 16 h, I = 0 A, loss = 0 W/m

18.4.3.3.3 Comparison of Calculations and the Test Results

In order to compare the measured temperatures with the calculated temperatures, heating of the GIL during the whole test time was simulated. Comparisons of the measured and the calculated temperatures for a 1 m depth and 16 days (September 1, 1999 to September 16, 1999), with two cycles occurring above, below, and to the side of the enclosing tube. The calculations agree well with the measured values. The maximum temperatures rose slowly during the short cycles and reached 35°C after 8 days. During the second period, the cooling phase was extended from 4 to 16 h, which was why the temperatures in the GIL system fell (Figure 18.21). In this case, the maximum enclosing-tube temperatures were less than 33°C.

The calculations, unlike the measurements, show that the maximum temperature is to be found around the circumference of the underside of the enclosing tube, since the effect of heat transfer by natural convection from the inner conductor to the enclosing tube is not taken into account in the calculations.

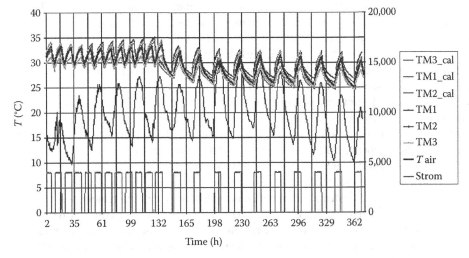

FIGURE 18.21 Comparison of numerical and experimental results of overload current rating of a directly buried GIL long-duration test, short and long cycle. (Courtesy of Siemens AG, Erlangen, Germany.)

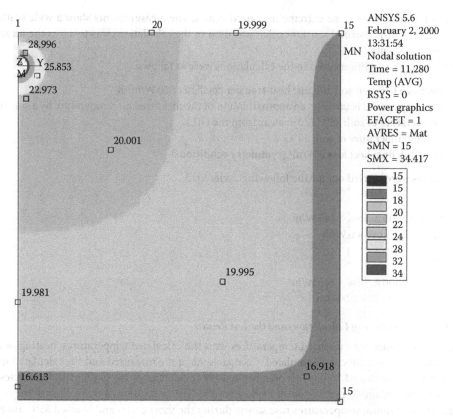

FIGURE 18.22 Temperature distribution at time t = 188 h and at depth 1.2 m, heating phase. (Courtesy of Siemens AG, Erlangen, Germany.)

In the upper and lower parts of the enclosing tube, the different temperatures can be explained by variations in the resistance of the soil.

The diagram in Figure 18.22 shows the temperature distribution of the calculations in the soil at 188 h (7.8 days) during the heating phase. The temperature measured during the test is assumed as a marginal condition for the air temperature. Temperature distributions during the day and the night show a difference only in the higher layer of soil immediately below the ground surface. This can be explained by the heat transmission between air and the surface of the ground due to the lower air temperature during the night. Heat transmission from the GIL is better at night, since the temperatures in the soil are slightly lower. The fluctuation in air temperature between night and day did not have as great an influence on the temperature distribution in the GIL and the soil as those that caused the load variation.

Further simulations of the test were carried out for a depth of 2.9 m during the same period as mentioned earlier (September 1, 1999 to September 16, 1999). A comparison of calculations and measurements showed good agreement between the calculations and the measurements.

The calculations show that the temperatures at the bottom and the top are higher than the temperature at the sides, and the temperature difference is less than 2°C. The temperature at the bottom is slightly higher than the temperature at the top (DT = 0.5°C). In contrast to this, the test showed a considerable temperature difference between the bottom, the top, and the sides, with a high value at the top and a low value at the bottom. In this example, the temperature at the circumference of the pipe is not constant because the not-unsteady effect of the natural convection between the inner conductor and the enclosing tube was included in the calculation.

18.4.3.3.4 Mechanical Aspects

At measurement points in the middle of the right section on the buried GIL, only very minor movements were recorded. The measured values vary between −1.1 and 0 mm. This corresponds to the maximum absolute movement of the long section of pipe near the bend enclosure in the direction of shaft 1, which connected to the longest section (Figure 18.19). The two sections of pipe can be regarded as an adhesion zone.

A measurement point in the shaft at the end of the test section measures the movement of the expansion joint and at the same time corresponds to the change in the pipe. Measurements of the pipe movements are shown in Figure 18.23. The enclosing-tube temperatures at the first cross section at a distance of about 9 m from the shaft vary, on average, between 28°C and 34°C in the case of the short cycles (DT = 6°C) and between 25°C and 33°C in the case of the long cycles (DT = 8°C). During this period, the enclosing tube in the shaft moved from −3.4 to +0.8 mm, which corresponds to an absolute distance of 4.2 mm (see Table 18.8).

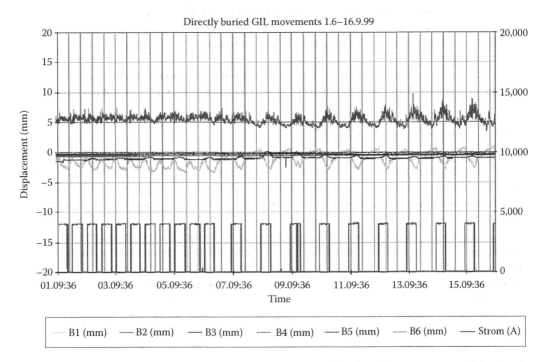

FIGURE 18.23 Mechanical aspects, movements of the enclosing tube during long-term testing of the directly buried GIL. (Courtesy of Siemens AG, Erlangen, Germany.)

TABLE 18.8 GIL Movement in Long-Duration Tests

	Movement (mm)	Absolute Distance (mm)
Pt 2	−0.6/−0.4	0.2
Pt 3	−0.5/−0.3	0.2
Pt 4	−0.1/0	0.1
Pt 5	−1.1/−0.6	0.5

Source: Courtesy of Siemens AG, Erlangen, Germany.

18.5 Advantages of GIL

The GIL is a system for the transmission of electricity at high power ratings over long distances. Current ratings of up to 4000 A per system and distances of several kilometers are possible in tunnel-laid or directly buried GILs. As a gas-insulated system, the GIL has the advantage of electrical behavior similar to that of an OHL, which is important to the operation. Furthermore, the gases do not age, so there is almost no limitation in lifetime, which is a huge cost advantage given the high investment costs of underground power transmission systems.

Because of the large cross section of the conductor, the GIL has the lowest electrical losses of all available transmission systems, including OHLs and cables. This reduces operating costs while reducing the utility's contribution to global warming, since less power needs to be generated.

Personnel safety in the presence of a GIL is very high because the metallic enclosure provides reliable protection. Even in the rare case of an internal failure, the metallic enclosure is strong enough to withstand the stress of failure. The inherent safety of the GIL system, which contains no flammable materials, makes it suitable for use in street or railway tunnels and on bridges. The use of existing tunnels has obvious economic advantages by sharing the costs and can solve some environmental problems because no additional OHL is needed. Because of the low capacitive load of the GIL, long lengths of 100 km and more can be built.

The GIL is a viable and available technical solution to bring the power transmitted by OHLs underground, without reducing power transmission capacity, in cases where OHLs are not possible.

18.5.1 Safety and Gas Handling

The GIL is a gas-filled, high-voltage system. The gases used, SF_6 and N_2, are inert and nontoxic. The 7 bar filling pressure of the GIL is relatively low. The metallic enclosure is solidly grounded and, because of the wall thickness of the outer enclosure, offers a high level of personal safety. The mechanized orbital-welding process ensures that the connections of the GIL segments are gastight for the system's lifetime.

Even in case of an internal failure, which is very unlikely, the metallic encapsulation withstands the internal arc so that no damage is inflicted on the surroundings. In arc fault tests in a laboratory, it was proven that no burn-through occurs with fault currents up to 63 kA, and the increase of internal pressure during an arc fault is very low. Even under an arc fault condition, no insulating gas is released into the atmosphere.

For the gas handling of the N_2–SF_6 gas mixture, devices are available for emptying, separating, storing, and filling the N_2–SF_6 gas mixtures. Figure 18.24 shows the closed circuit of the insulation gas with all devices used for gas handling. The initial filling is done by mixing SF_6 and N_2 in the gas mixing device (5) in the required gas mixture ratio. The initial filling is normally sufficient for the whole lifetime of the GIL because of the system's high gastightness. For emptying the GIL system, the gas is pumped out with a vacuum pump (1), filtered, and then separated (2) into pure SF_6 and a remaining gas mixture of N_2–SF_6. This N_2–SF_6 gas mixture has an SF_6 content of only a few percent (1%–5%), so it can be stored under high pressure up to 200 bar in standard steel bottles (3). Three sets of steel bottles can hold the gas content of a 1 km section for storage. The pure SF_6 is stored (4) in liquid state. To fill or refill the GIL system, a gas mixing device (5) is used, including a continuous gas monitoring system for temperature, humidity, SF_6 percentage, and gas flow. The gas mixing device has input connections for pure N_2 (6), pure SF_6 (4), and gas mixtures containing a low percentage of SF_6. The mixing device adjusts the required N_2–SF_6 gas percentage used in the GIL, for example, 80% N_2.

With these gas-handling devices, a complete cycle of use and reuse of the gas mixture is available. In normal use, the SF_6 and N_2 will not be separated completely because the gas mixture will be reused again. A complete separation into pure SF_6, as used, for example, in GIS, can be done by the SF_6 manufacturers. Thus, the requirements of IEC 60480 [13] and IEC 62271-4 [3] are fulfilled.

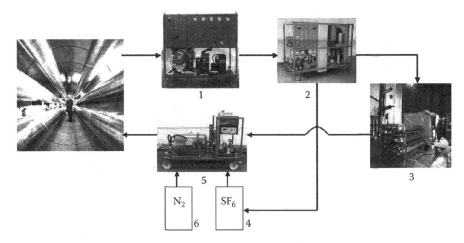

FIGURE 18.24 Gas-handling devices. 1 = Evacuation pump; 2 = Gas demixing device; 3 = Gas mixed storage; 4 = SF_6 containers; 5 = Gas mixing device; and 6 = N_2 containers. (Courtesy of Siemens AG, Erlangen, Germany.)

18.5.2 Magnetic Fields

18.5.2.1 General Remarks

Magnetic fields can disturb electronic equipment. Devices such as computer monitors can be influenced by magnetic-field inductions of $\geq 2\,\mu T$. Furthermore, magnetic fields may also harm biological systems, including human beings, a subject of public discussion. A recommendation of the International Radiation Protection Association (IRPA) states a maximum exposure figure of $100\,\mu T$ for human beings. In Germany, this value has been a legal requirement since 1997 [25].

Several countries have recently reduced this limit for power-frequency magnetic fields. In Europe, Switzerland, and Italy were the first to establish much lower values. In Switzerland, the maximum magnetic induction for the erection of new systems must be below $1\,\mu T$ in buildings, according to NISV [26]. Today some exceptions may be accepted. In Italy, $0.5\,\mu T$ has been proposed for residential areas in some regions, with the goal of allowing a maximum of $0.2\,\mu T$ for the erection of new systems. This trend suggests that, in the future, electrical power transmission systems with low magnetic fields will become increasingly important.

The GIL uses a solid grounded earthing system, so the return current over the enclosure is almost as high as the current of the conductor. Therefore, the resulting magnetic field outside the GIL is very low. The installation at PALEXPO in Geneva demonstrates that GILs can fulfill the high future requirements that must be expected in European legislation.

18.5.2.2 Measurements of the Magnetic Field at PALEXPO, Geneva

The measurements at PALEXPO in Geneva were carried out with both GIL systems under operation with a current of $2 \times 190A$. Based on the measured values, the magnetic induction was calculated for the load of $2 \times 1000\,A$. Inside the tunnel between the two GIL systems, the maximum magnetic induction amounts to $50\,\mu T$.

The magnetic field at right angles to the GIL tunnel is presented in Figure 18.25. The measurements were taken at 1 and 5 m above the tunnel, which is equivalent to the street level and to the floor of the PALEXPO exhibition hall. The magnetic induction on the floor of the fair building is relevant for fulfilling the Swiss regulations for continuous exposure to magnetic fields. The 1 m maximum value amounts to $5.2\,\mu T$ above the center of the tunnel. The maximum induction at 5 m above the tunnel is $0.25\,\mu T$. This result is only 0.25% of the permissible German limit [14] and 25% of the new Swiss limit [26].

It is worth mentioning that cross-bonded high-voltage cable systems need to be laid at a depth of 30 m or more to achieve comparable induction values. There are a range of possibilities for reducing the

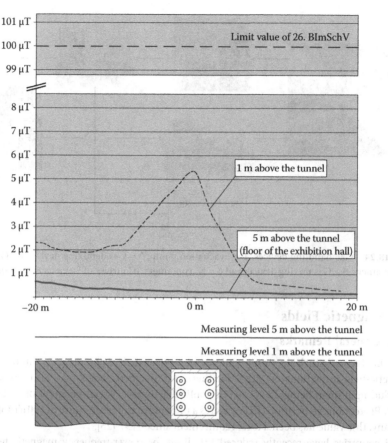

FIGURE 18.25 Measured values of the magnetic induction above the GIL tunnel at PALEXPO, Geneva, at a rated current of 2 × 1000 A. (Courtesy of Siemens AG, Erlangen, Germany.)

magnetic field in cable systems, such as ferromagnetic shielding, compensation wires, or laying in steel tubes. All these measures, however, increase the losses markedly. Table 18.9 provides a comparison of different 400 kV transmission systems.

A comparison of calculations and measurements made at PALEXPO shows that it is not sufficient to focus on the GIL only. The current distribution through the grounding systems around the GIL also has a significant influence. Along the OHL, a current is induced into the ground wire and then conducted through the enclosure of the GIL. The increase in the magnetic field at a distance of 20 m from the GIL (Figure 18.25) is related to induced currents in the ground grid.

The magnetic inductions above the GIL trench are negligible and meet the Swiss requirements under full-load conditions. However, the results show that the magnetic fields induced by the grounding system also need to be considered in the system design. All measured and calculated values of the induction from the GIL are far smaller than those for comparable OHLs and conventional cable systems.

18.6 Application of Second-Generation GIL

18.6.1 Tunnel-Laid GIL

The first application of the second-generation of GIL was implemented between September and December 2000. After only 3 months erection time, the OHL was brought underground into a tunnel. Graphical explanation is given in Figure 18.27. In January 2001, the line was energized again.

TABLE 18.9 Comparison of Different 400 kV Transmission Systems

	VPE-Cable 2XS(FL)2Y[a] 1 × 2500, Cross Bonding	OHL 4 × 240/40 Al/St	GIL[a] 520/180
Rated voltage (kV)	400	400	400
Thermal limit load (MVA)	1080	1790	1790
Overload (60 min)	1.2 times	1.2 times	1.5 times
Reactive power compensation	Needed	Not needed	Not needed
Maximum induction in (µT) at 2 × 1000 MVA at ground level	29	23.5	1.4
Thermal system losses[b] at 1000 MVA (W/m)	71	194	43
External influences (environment, animals)	No	Yes	No
Behavior in case of fire	Fire load with plastic	No additional fire load	No additional fire load
Damage to neighboring phases in the event of failure	Possible	Possible	Not possible
Maintenance	Maintenance free[c]	Needed	Maintenance free[c]

Source: Courtesy of Siemens AG, Erlangen, Germany.

[a] Tunnel laying, natural cooling, level aboveground 2 m.
[b] Conductor temperature 20°C.
[c] Corrosion protection test is required only for direct burial.

FIGURE 18.26 Delivery of transport unit to the preassembly area. (Courtesy of Siemens AG, Erlangen, Germany.)

Figure 18.26 shows the delivery of GIL transport units to the preassembly area. The preassembly tent was placed directly under the OHL and above the shaft connected to the tunnel right under the street. The space was limited because of an airport access road on one side and the highway to France on the other side. Nevertheless, the laying proceeded smoothly. The positive experience from this project shows that even GIL links for long distances can be installed within a reasonable time. The highly automated laying process has proved to guarantee a consistent quality on a very high level over the complete laying process, and the commissioning of the system was carried out without any failures.

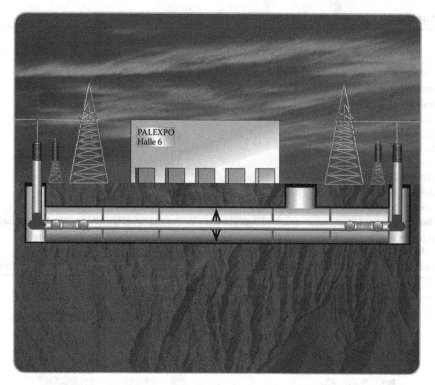

FIGURE 18.27 Principle of the PALEXPO project. (Courtesy of Siemens AG, Erlangen, Germany.)

During erection of the GIL, a preassembly tent was placed directly above the access shaft connected to the tunnel near Pylon 175 (Figure 18.27). The narrow space between an airport access road on one side and the highway A1 to France on the other side necessitated the use of the space directly under the existing OHL for the site work. A total of 162 pieces of straight GIL units, each 14 m long, were preassembled, brought into the tunnel, welded together, and continuously pulled toward the end of the tunnel at Pylon 176. Thanks to advanced site experience, it was possible to double productivity for assembly of the GIL sections from two connections per shift per day to four.

Erection of the PALEXPO GIL started in September 2000 and was completed within 3 months. The double-circuit OHL was brought underground into a square tunnel (Figure 18.8). The elastic bending of the GIL aluminum alloy tubes (minimum bending radius of 400 m) was sufficient to follow the layout of the tunnel route.

The technical data for the PALEXPO GIL are shown in Table 18.10. The power transmission capacity of the tunnel-laid GIL allows the maximum power of an OHL to continue underground without any power transmission reduction. Surge arresters are used at the GIL terminations.

For monitoring and control of the GIL, secondary equipment is installed to measure the gas density. An electrical measurement system is used to detect arc location. Very fast transient electrical signals are measured at the ends of the GIL, and the position of a very unlikely internal fault can be calculated with an accuracy of ±25 m.

Given the importance of this high-voltage line for the power system—and despite the security constraints for the construction site—the operator endeavored to keep the line in service during the whole construction period. Operation of the link was suspended only for a span of 3 weeks during the GIL commissioning tests and connection [32].

TABLE 18.10 Technical Data for the PALEXPO GIL

Type	Design	Project
Nominal voltage	420 kV	300 kV
Nominal current	3150 A/4000 A	2000 A
Lightning impulse voltage	1425 kV	1050 kV
Switching impulse voltage	1050 kV	850 kV
Power frequency voltage	650 kV	460 kV
Rated short time current	63 kA/3 s	50 kA/3 s
Rated gas pressure	7 bar	7 bar
Insulating gas mixture	80% N_2, 20% SF_6	80% N_2, 20% SF_6

Source: Courtesy of Siemens AG, Erlangen, Germany.

18.6.2 Directly Buried GIL

The first application of a 400 kV directly buried GIL was installed at the Frankfurt Airport to connect the Kelsterbach substation with the 400 kV transmission network in parallel to the new airport runway north. The continuous high power rating requirement of 1800 MVA and some reserves for overloading and the very low external magnetic field strength were two main advantages of the GIL. The two systems of the OHL are brought underground with two GIL systems simply by connecting the GIL to the OHL via bushings.

The technical data of the project are show in Table 18.11. It is remarkable that the high power transmission capability is reached by the standard GIL design with 500 mm enclosure pipe diameter and the insulating gas mixture of 80% N_2 and 20% SF_6.

The on-site laying technology is designed to cover long distances in a cross country laying condition by using the elastic banding as shown in Figure 18.28. The single pipe sections have been transported on site and were welded on site to form the total system length of 1 km. A total of 4 km pipe length have been laid in about 6 months. The laying process is very similar to oil or gas pipe line laying as a continuous laying process. The aluminum enclosure pipe is coated with a polyethylene (PE) laid to protect against corrosion in conjunction with an active corrosion protection system. The pipes are laid on sand with a good and constant thermal conductivity. The trench finally has been closed with the material excavated before with a minimum soil cover of 1 m above the GIL.

The GIL is in safe and reliable operation since April 2011 and provides the power connection of the Kelsterbach substation for the city of Frankfurt directly buried in parallel to the new run way north, you might remember next time landing in Frankfurt.

TABLE 18.11 Technical Data 400 kV GIL at Kelsterbach, Germany

Type	Value
Nominal voltage (kV)	400
Nominal current (A)	2500
Lightning impulse voltage (kV)	1425
Switching impulse voltage (kV)	1050
Power frequency voltage (kV)	630
Rated short-time current (kA/3 s)	50
Rated gas pressure (bar)	8 abs.
Insulating gas mixture	80% N_2, 20% SF_6
Power transmission (MVA)	1800

Source: Courtesy of Siemens AG, Erlangen, Germany.

FIGURE 18.28 Two system of 400 kV GIL at the Frankfurt Airport of the Kelsterbach project.

18.7 Quality Control and Diagnostic Tools

In the complex insulation system of a real GIL, the intrinsic dielectric strength of N_2–SF_6 mixtures—presented in Section 18.4.1 by the pressure-reduced critical field—is affected by many factors. The usual surface roughness of metal surfaces is well understood in gas-insulated systems [27]. Metal protrusions or mobile particles are also studied in many cases, and their influence is also well understood [28].

The statistical distribution of breakdown voltage in gas mixtures was found to be similar to that of pure SF_6 of equal dielectric strength [29]. Therefore, the approved conventional test procedures for GIS can be applied to confirm the required withstand levels [27]. For insulation coordination (to make the correct choice of high voltage and best voltages for type testing) of extended transmission lines, the Weibull distribution has to be applied because of the statistical distribution of the flashover voltage levels [30]. Thus, the knowledge and experience of more than 25 years of GIS installations and operation can be fully applied to GIL installations because GIS and GIL show the same statistical behavior for high-voltage flashovers.

With careful assembly and efficient quality control, defects are practically ruled out. Mobile particles are the most common defect, and these are usually eliminated by conditioning procedures during power-frequency high-voltage testing [10]. In a conventional GIS with a complex insulation system, particles are moved by a stepwise increased AC field stress into low-field regions that act as natural particle traps. However, there are no such natural particle traps in the plain insulation system of a GIL. Therefore, it has been equipped with artificial particle traps all along the GIL extension, and these traps have proved to be very efficient.

Modern diagnostics are applied for the detection, localization, and identification of defects. The VHF method proved to be most efficient [10]. Its application is restricted by signal attenuation and the correspondingly limited measuring range of installed sensors. In a GIS, this attenuation is mainly caused by the conical spacers that are usually installed. The maximum distance between sensors should therefore normally not exceed 20 m. In a GIL, an efficient VHF PD measurement can be carried out even with distances between sensors of several hundred meters. This enables use of the ultra-high-frequency method in a GIL, as successfully performed for the first time on site in Geneva [11].

Moisture penetration by diffusion through the enclosure and from the bulk of the insulators into the gas occurs much less frequently in a GIL than in a conventional GIS because of the excellent gastightness and the low amount of solid insulating material. The insulation quality of the insulator surfaces can therefore reliably be preserved by conventional measures to avoid dewy surfaces of reduced dielectric strength.

Altogether, it can be expected that the GIL will give the same or even better long-term performance than a GIS, which demonstrates a long service lifetime with no critical aging even after 30 years of operation [31]. The GIL uses almost the same materials, while the amount of solid insulating material and SF_6 is considerably reduced. Moreover, the requisite quality control can be obtained by means of tests and modern diagnostics. In conclusion, it can be said that the on-site high-voltage quality control and the diagnostic tools used have proved to be very successful. The GIL in Geneva went into service without problem.

18.8 Corrosion Protection

For applications where aluminum pipes are used in air aboveground or in a tunnel, aluminum generates an oxide layer that protects the enclosure from any kind of corrosion. The oxide layer of an aluminum pipe is very thin, only a fraction of a micrometer, but it is very hard and very resistive against a gaseous environment like the atmosphere. In most cases it is not necessary, even in outdoor applications, to protect the aluminum pipes against corrosion with, for example, coloring.

Going underground for directly buried systems, the situation regarding corrosion changes and a corrosion protection is required. Two basic methods are used today: a passive corrosion protection and an active corrosion protection. The passive corrosion protection is an added layer of noncorrosive materials, for example, PE or polypropylene (PP), whereas the active protection system uses voltage protection to direct corrosion from the protective aluminum enclosure toward a loss electrode.

18.8.1 Passive Corrosion Protection

Passive corrosion protection is used widely for all kinds of metallic underground systems that have direct contact with the soil, for example, electric power cables, oil or gas pipelines, and all kinds of other pipes.

There are several different technologies available on the market to add a coating to a pipe as a passive corrosion protection. In all cases, the processes used and the materials are similar. The surface of the metallic aluminum enclosure needs to be degreased, and the oxide layer needs to be removed. Acid fluids or mechanical brushes can be used to accomplish this. In some cases, both of them are used. If fluids are used to prepare the enclosure for the coating process, the pipe is run through a curtain of fluid acid. If mechanical treatment is used, then brushes treat the surface accordingly, sometimes together with a fluid. These processes are the same for steel and aluminum pipes and are run with the same machines. After the surface of the aluminum pipe is prepared, a first layer of a corrosion protection fluid is brought onto the pipe to stop corrosion. This first layer is the active part of the passive corrosion protection and is only a few micrometers thick. On top of this layer, a 3–5 mm layer is added mainly for mechanical protection reasons. For this protective layer, two basic processes are applied: the extruded layer and the tape-wrapped layer methods. The use of these passive corrosion protection methods has a long history and thousands of kilometers of experience as well as years of operating experience.

Figure 18.29 shows a GIL with passive corrosion protection. In the middle of the photo, the blank aluminum enclosure is shown, prepared for the welding process. To the right and left of the blank, the small dark bands are the active corrosion protection layer. Finally, to the far right and left is the white cover for mechanical protection, which is a PE or PP coating 3–5 mm thick.

After the pipe segments are welded together, it is also necessary to protect the welded area. There are various corrosion protection processes available that can be applied on site. Figure 18.30 shows how one such on-site corrosion protection method is applied. In this case, a corrosion protection system based on a shrinking method is used. Other methods involve granulates or tapes and are also widely used in the pipeline industry.

FIGURE 18.29 GIL with passive corrosion protection. (Courtesy of Siemens AG, Erlangen, Germany.)

FIGURE 18.30 On-site corrosion protection of the welding area—shrinking method. (Courtesy of Siemens AG, Erlangen, Germany.)

18.8.2 Active Corrosion Protection

With an active corrosion protection method, the induced current generates a voltage potential of the metallic enclosure toward the soil. If this voltage level is at a potential of around 1 V toward a loss electrode, then the loss electrode corrodes instead of the aluminum enclosure. The active corrosion protection system is a backup to the passive corrosion protection system. It is installed as an additional quality insurance system if the passive corrosion protection fails. Failures in the passive corrosion protection can occur over the lifetime of the system by outer damage through other earthworks or by cracks or voids in the protective material. If necessary, the active corrosion protection system prevents corrosion in the event of cracks and voids in the passive corrosion protection system.

Experience with installed, directly buried pipe systems worldwide shows that, over the decades, some cracks or voids in the passive corrosion protection system can occur, which increases the induced current of the active protection system. The positive effect of the active corrosion protection system is that each passive corrosion protection failure need not be repaired immediately, and a guaranteed lifetime of the passive corrosion protection system of 50 years can be extended by many more years. Repairs of passive corrosion protection systems can be planned and concentrated on troublesome segments. Experience with oil and gas pipelines shows that the lifetime can be extended significantly without opening the pipe.

The active corrosion protection system, also called cathodic corrosion protection, uses an induced current to adjust the protective voltage of approximately 1 V against the lost electrode. To reach this 1 V protective voltage, an induced current of approximately $100\,\mu A$ is needed. The induced current is related to the total of the surface to be protected and increases with the length of the system and the numbers of failures. In practice, several kilometers can be protected with only one DC-voltage source because the current is low.

To obtain cathodic protection, the buried GIL must be a non-earthed system, which means that it must be insulated toward the ground potential. To allow the 1 V protection potential, the ground system is coupled with the GIL through a decoupling element, which could be a battery or a diode. This battery or diode ensures that a protective voltage of about 1 V is applied to the aluminum shield. If an earth fault of the electric system occurs, the failure current is conducted to the ground potential through the diode or battery.

Active corrosion protection can be easily installed along with the electrical transmission system, with no interferences between electrical transmission and the corrosion protection voltage potential. Such electrical transmission systems have been operating for many years with no failures reported. The high reliability observed for pipelines and cables also applies to GIL systems.

18.9 Voltage Stress Coming from the Electric Power Net

18.9.1 Overvoltage Stresses

Two typical GIL applications are represented by the connection of 400 kV OHLs to a GIL with a length of 1 and 10 km. The OHL is protected by two shielding wires along its full length. The height of the last three towers is about 65 m, and the maximum footing resistance for the towers is about 7.5 W.

18.9.2 Maximum Stresses by Lightning Strokes

Based on these configurations of the OHLs and the lengths of their insulator strings, the following maximum stresses by lightning strokes were evaluated and used to calculate the maximum overvoltage stresses on the GIL:

Remote stroke: 2000 and 2100 kV
Nearby direct strokes: 35 and 18 kA
Stroke to towers: 200 kA

18.9.3 Modes of Operation

The basic arrangement allows calculation of lightning strokes on the GIL for the following modes of operation:

Transport

- OHL connected by the GILs of 1 and 10 km length
- Overvoltage stresses caused by lightning strokes to the OHL

Open end

- GILs of 1 and 10 km length connected on one end to the open bay of a substation
- Overvoltage stresses caused by lightning strokes to the OHL

18.9.4 Application of External and Integrated Surge Arresters

To protect the GIL against high lightning and switching overvoltage stresses, external surge arresters located at the last towers of the OHL, as well as encapsulated metal oxide surge arresters immediately connected to the GIL at certain locations (integrated surge arresters), can be applied. For 400 kV systems with an earth-fault factor of £1.4, the special integrated surge arresters have the following characteristic data:

Rated voltage, U_r = 322 kV
Continuous operating voltage, U_c = 255 kV
Residual voltage at 10 kA, U10 kA = 740 kV
External metal oxide surge arresters commonly used in German 400 kV systems—those with U_r = 360 kV, U_c = 288 kV, and U10 kA = 864 kV—were taken into account.

18.9.5 Results of Calculations

For each mode of operation, for both lengths of GIL, and for the different possibilities of surge-arrester application, the maximum overvoltage stresses (depending on the distance from the left-hand-side end of the GIL) have been calculated for all kinds of possible lightning strokes. In all cases, the maximum stresses are caused by nearby direct strokes to line conductors. For the various possibilities of surge-arrester application, the maximum lightning overvoltage stresses are listed in Table 18.12 for the GIL of 1 km length and in Table 18.13 for the GIL of 10 km length.

TABLE 18.12 Maximum Overvoltage Stresses Depending on Mode of Operation and Number of Surge-Arrester Sets for a GIL of 1 km Length

	Number of Surge Arrester Sets at		Maximum Overvoltages in kV Caused by Nearby Direct Strokes to	
Mode of Operation	GIL	Tower L1/R1	N2 (35 kA)	N3 (18 kA)
Transport	2	—	1013	913
	2	2	952	890
Open end	2	—	1066	958
	2	2	989	938

Source: Courtesy of Siemens AG, Erlangen, Germany.
---, >990 kV; —, >904 kV.

TABLE 18.13 Maximum Overvoltage Stresses Depending on Mode of Operation and Number of Surge-Arrester Sets for a GIL of 10 km Length

Mode of Operation	Number of Surge Arrester Sets at		Maximum Overvoltages in kV Caused by Nearby Direct Strokes to	
	GIL	Tower L1/R1	N2 (35 kA)	N3 (18 kA)
Transport	2	—	996	904
	2	2	983	890
	4	—	867	837
	4	2	842	829
Open end	2	—	1048	950
	2	2	1035	940
	4	—	902	883
	4	2	893	877

Source: Courtesy of Siemens AG, Erlangen, Germany.
---, >990 kV; —, >904 kV.

18.9.6 Insulation Coordination

At least up to a length of some tens of kilometers, lightning overvoltage stresses are decisive for the insulation coordination of a GIL, since stresses by switching overvoltages at those lengths will be much lower than on OHLs because of their lower surge impedance (60 W compared with 300 W for an OHL). The insulation coordination of a GIL of up to 10 km length considered here is therefore based on the maximum stresses by lightning overvoltages.

Given the large amount of experience already gained with design tests and on-site tests of huge GIS, the following procedures are proposed for selecting test voltages for on-site tests on GIL sections of up to 1 km length and of type tests on a representative length of GIL.

18.9.6.1 On-Site Tests

On-site tests are designed to verify that the GIL is free of irregularities after laying and assembling. Taking into account the safety factor of $K_s = 1.15$ (according to IEC 60071-2), a withstand voltage of $U_w \geq 1.15$ ULE_{max}—with ULE_{max} = maximum overvoltage stress from the calculations—should be verified by these tests.

18.9.6.2 Type Tests (Design Tests)

On the other hand, the on-site test voltage corresponds to 80% of the required rated lightning-withstand voltage, U_{rw}, when type testing a representative length of GIL. At these tests, single flashovers on self-restoring insulation are permitted according to the applicable IEC standard 62271-204.

18.10 Future Needs of High-Power Interconnections

18.10.1 Metropolitan Areas

Metropolitan areas worldwide are growing in load density, mainly at their centers. Demand for power has grown because of the construction of huge residential and tall office buildings with air conditioning and lots of electronic equipment, leading to increases in electric loads of up to 10% per year in metropolitan areas [32]. The following short historical overview explains how the power supply of metropolitan areas has developed over the last 30 years.

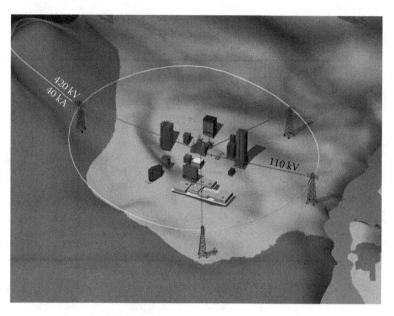

FIGURE 18.31 Power supply of metropolitan areas in 1970. (Courtesy of Siemens AG, Erlangen, Germany.)

Figure 18.31 shows the principle for power supply in a metropolitan area. Power generated in a rural area is connected to a metropolitan area by 420 or 550 kV OHLs with a short-circuit rating of 40 kA. Several substations are placed around the city as overhead towers using a bypass or a ring structure around the metropolitan area, from which 110 kV cables transport electrical energy into the center of the city, where medium-voltage energy is distributed.

In Figure 18.32, the metropolitan area has grown, with more tall buildings in the center. Most cities still have a 420/550 kV ring around the city, but the short-circuit rating has been increased to 50 kA or, in some places, to 63 kA. Note the second connection to the ring from another rural power generation area. More 110 kV cables are connected to the ring to transport the energy to the substations in the city

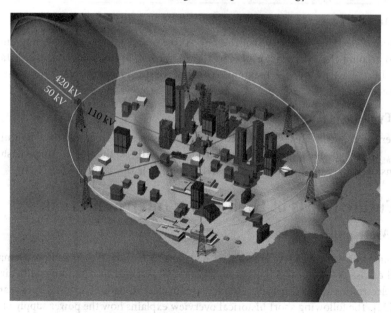

FIGURE 18.32 Power supply of metropolitan areas in 2000. (Courtesy of Siemens AG, Erlangen, Germany.)

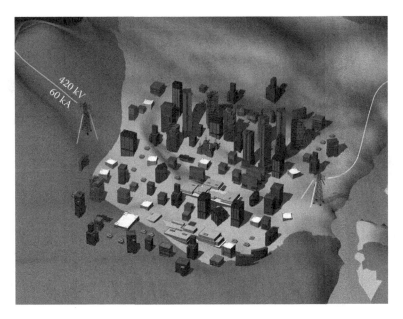

FIGURE 18.33 Power supply of metropolitan areas in 2010. (Courtesy of Siemens AG, Erlangen, Germany.)

for distribution. To increase the power transportation into the center of the city, it is very difficult to build OHLs 1000 kV voltage rating because of space requirement and dielectric problems. Moreover, worldwide experience with very high short-circuit ratings shows that short-circuit rating values cannot go far above 63 kA because of mechanical problems. So the only way to increase the power transportation into the city is to lay 400 kV underground bulk power transmission systems right into the center. In such cases, the GIL offers a good underground solution laid in a tunnel or directly buried.

Figure 18.33 shows the metropolitan area as it may appear in 2010. The same metropolitan area with more buildings has grown further, and a 420/550 kV, 63 kA, double-system GIL was built as a diagonal connection underpassing the total metropolitan area. This GIL could have a length of 30–60 km. The solution illustrated here allows splitting of the short-circuit ratings of the ring into two half rings and to connect directly to the 400 kV high transmission power GIL in the center of the metropolitan area. The underground connection is a tunnel for GIL, as discussed in Chapter 4.

18.10.2 Use of Traffic Tunnels

GILs can safely be routed through tunnels carrying traffic on rails or streets. This new application for electrical transmission systems with solid insulated cables was not possible until today because of the risk of fire or explosion. The GIL has a solid metallic enclosure and does not burn or explode, as explained in Chapter 4. The combinations of GIL and street or railroad tunnels are shown in Figure 18.34. Three examples are given. The first one is a traffic tunnel with cars and a GIL mounted on top of the tunnel; the second is a double railroad tunnel system with a separate GIL tunnel; the third example is a double-track railroad tunnel with a GIL included.

The use of such traffic tunnels with GIL is now under investigation in different parts of the world. In the European Alps, interconnections between Germany, Austria, Switzerland, Italy, and France are now planned to improve the traffic flow and to allow trade of electric energy. In China and Indonesia, interconnections between the mainland and islands or between outlying islands are under investigation. In the near future, GILs will become economically viable and will be widely used as high-power, long-distance transmission lines.

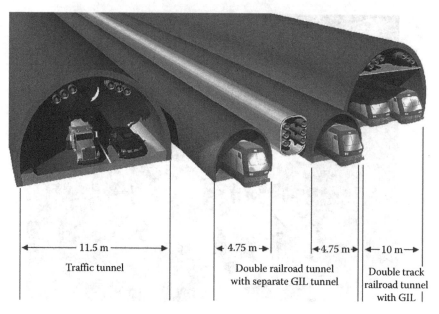

FIGURE 18.34 Different types of traffic tunnels to be used for GIL. (Courtesy of Siemens AG, Erlangen, Germany.)

18.11 To Solve Bottlenecks in the Transmission Net

18.11.1 Introduction

The European Transmission Net has a regional development history. In the beginning of the transmission net, the highest voltages were used to connect larger electrical load areas with large scale power generation in fossil and nuclear power plants. These connecting lines had a regional structure inside national country borders. The main power flow was from the regional power generators to the regional power consumers [33,34].

In a next step of the development of the transmission net, line connections were made to connect regional transmission networks with the neighbor. The main reason for such transmission lines was the need of emergency power supply if the own power generation was not high enough. With interconnecting transmission lines between the regional networks also the reliability was improved by redundancy.

These interconnecting transmission lines were built to connect regional transmission network inside national country borders and sometimes are also cross border connections. For all regional power suppliers, it was the first goal to generate the electricity inside their own network to be sure to have control onto the power generation costs, which was seen as essential for the usually public utility. In most cases, the power transfer on interconnecting transmission lines in and out of one regional power supplier was balanced.

The interconnecting transmission lines were typically not the strongest in power transmission capability because of their function as an emergency backup connection. These solutions were found long before the deregulated market was found, and trading with electrical energy was normal.

Strict contracting with regional monopolistic structures ensured the status of regional power transmission and distribution. This regional transmission and distribution network structure reached its final stage in the late 1980s and in the 1990s but is still in operation today.

The deregulation of the energy market in Europe in the 1990s until today caused big changes in the way how the transmission net will be used and what are the new requirements. This transmission process from a closed and protected regional electrical energy market to an open, competitive one is still ongoing today and in the near future. Changing requirements to the transmission net of tomorrow ask for new solutions to avoid costly and reliability decreasing bottlenecks.

18.11.2 Transmission Net Requirements

The transmission net follows the requirements that are coming from the user and public. First, the user needed an economical (low cost, high reliability) technical solution that was developed and manufactured by the industry in the form of the OHL.

The OHL for transmission networks uses high voltages of 400 kV in Europe or 245/550 kV in other regions in the world. Typical technical values are shown in Table 18.14. With the technical layout of four aluminum wires in a bundle of 240 mm² with a 40 mm² steel core, the transmission rating is for the thermal limit about 1800 MVA. This requires a current of about 2600 A, also a thermal limit value.

The low capacitance of the OHL of 14.2 nF/km allows to build long lines without the need of phase angle compensation. A mean maximum length of OHL without compensation is around 1000 km.

The resistance per kilometer of the OHL is relatively high due to the reason of the limited cross section of the conductors or wires possible with OHL structures. With typical values of 30.4 mΩ/km the losses in the upper part of the transmission capability cumulate to high transmission losses. In Table 18.15, the losses of such a line are shown.

The GIL offers instead a higher transmission capability of 2000 MVA and a thermal current limit of about 3000 A. The lower resistance of 9.4 mΩ/km produces much less thermal transmission losses compared to an OHL as it is shown in Table 18.15 [11–13].

The requirements to the transmission net in the past were solved best with the OHL if erection cost and operation cost were evaluated in an $n - 1$ system environment. That means if the transmission of 2000 MVA was needed between two points, the OHL was built following the $n - 1$ rule with two systems of 2000 MVA transmission capability each. In case of one failing system, the second is able to carry the full current. That means if both systems are in normal operation, the current in each system is half and below to grant the full redundancy, and, in consequence, the transmission losses are relatively low. This changes with the current increase close to the thermal limits. The driver of today for using the assets much more close to the limit are asset management and an open trading market. This will result in higher ratings for power transmission lines.

A second change of the requirements for power transmission is coming from the public and the low or nonacceptance of OHLs in most places in Europe. The reasons are purely aesthetic or related to health concerns because of the electromagnetic fields.

A third driver in Europe that is changing the requirement for power transmission is coming from regenerative energy source like wind, solar, or bio power. The strongest increase today and in the near future is coming from wind energy land installed, and in the near future also from offshore wind parks of large scale units. The consequences to the transmission net are the increase of load in general, the possibility of bottlenecks, and the loss of the $n - 1$ redundancy.

TABLE 18.14 Technical Data

	OHL 4 × 240/40 Al/St	GIL
Thermal load limit (MVA)	1800	2000
Thermal current limit (A)	2600	3000
Resistance per kilometer (mΩ/km)	30.4	9.4
Capacitance per kilometer (nF/km)	14.2	54.4

TABLE 18.15 Transmission Losses

Transmitted current (A)	500	1,000	1,500	2,000	2,500	3,000
Transmitted power (MVA)	350	700	1,000	1,400	1,800	21,000
Losses OHL	20	91	205	364	570	820
Losses GIL	5	28	63	112	176	254

A fourth influence of changing the requirements for the future transmission network in Europe is coming from the open, competitive electricity market. Large scale of electricity flow changes in the transmission net will require reinforcement and new transmission lines to fulfill the needs of the future.

These four influences will come simultaneously with influences to each other and will cumulate in the effects. The weakest point in the transmission net is often called the "bottleneck." This weak point will come up first and will cause transmission net problems. All large area outages in the world of the last years like Italy in 2004 or United States and Canada in 2003 can be technically connected to these bottlenecks of electrical power transmission, which finally causes the blackout.

18.11.3 Technical Solutions

There are several possibilities to solve bottlenecks in the transmission network [35,36]. Bottlenecks are typically local overloads of the transmission net that finally can cause a large area outage.

The influences of one bottleneck or congestion in the transmission net can reach large areas if the load on the single lines is high and an $n - 1$ criterion is not fulfilled anymore. One trip of one branch then can end in regional blackout.

Increasing the transmission capability, for example, with an underground GIL may improve this complex transmission net situation. This should be part of net studies to bring understanding of overload situations in the net. Various cases need to be studied and simulated by computer software.

The first and mostly done in the past is the reinforcement of OHLs by upgrading the voltage, for example, from 245 to 420 kV, or the current by adding more wires to increase the current rating to the thermal limit at 2600 A. In this case, a bundle of four wires is needed. Besides upgrading, new OHLs could also be built to solve the bottleneck problem.

The second principle solution is to control the power flow by use of electronic equipment in the transmission net using FACTS and HVDC equipment. FACTS stands for Flexible AC Transmission System, which is able to control the power flow on a transmission line via electronic valves (thyristors). HVDC stands for High Voltage DC and is using also electronic valves for power flow control with a DC transmission line between the two HVDC converter stations at its ends. This electronic control can prevent outages in cases when the power flow can be rerouted without creating new overload sections, bottlenecks, and other locations.

A third way to solve the bottleneck problem is to go underground at the regional overload sections to reinforce the transmission capability. This underground solution offers the possibility to use the way of rights of OHL and to reduce the time for getting the permission by authorities. It is clear that the cost of underground lines is higher.

There are three technical principle solutions: solid insulated cables, gas-insulated lines, and superconductive cables.

Solid insulated cables are used for underground power transmission since the very beginning of the installation of the transmission network, mostly used in cities or other applications where OHLs cannot be used. The use of solid insulated cables is limited in length and current rating, even if these values have been increased in the last years.

Gas-insulated lines are used for more than 30 years worldwide, many projects offering a very high-power transmission capability like the OHL, and are practically not limited in length.

The superconductive cable is still in a stage of industrial implementation with a few projects, mainly in the United States. It will need some more time before a wide use will come. A bottleneck is usually in a limited region, limited to some kilometers of length. An existing OHL is in operation and under normal conditions close to the thermal limits of the transmission capability.

The way of right is given and limited to the line built. In Europe, in most cases, it is not possible to reinforce the line by upgrading to higher voltages without a new commissioning process with the authorities.

This process may last very long, before the first work can be started. In some cases, the public opinion might be so strong that an OHL cannot be built or reinforced.

The underground solutions for solid insulated cables and GIL have much higher investment cost but show advantages in operation during the lifetime. Transmission losses are lower, maintenance and repair expenses are lower, reliability is higher, public acceptance might be higher, commissioning time might be shorter, and the impact to the public, for example, reduced value of the property, is lower.

All the single points do have different impacts and values in the single region where the bottleneck problem is to solve. How to evaluate the single points is depending on the individual situation. But some calculations can be made.

18.11.4 Integrating Renewable Energy

The next future will bring a strong increase of renewable energy generation worldwide. Driven by the CO_2 limitations and the latest experiences made with the catastrophic situation at Fukushima in Japan with the problems of the nuclear power reactors, the use of all sorts of regenerative energy will develop quickly to high amounts of electric power usually far away from the load centers. And the existing transmission network can offer the required transmission capacities many from remote areas to the cities and industrial regions. In consequence, new transmission lines are required to solve this requirements. Most of these new transmission lines will be built as OHLs, but in some areas, it might be necessary to go underground and here the GIL offer a good and reliable technical solution.

Most of the regenerative energy will come from wind farms and photovoltaic or solar thermal power generation. These locations are in deserts, mountains, or even at sea, in any case away from the electric power consumers of today. This long distance transmission requirement first has been solved in china by connecting large resource of hydro or coal power over distances of 1000–2000 km to the dense populated areas along the coast line, e.g. in shanghai region. High-voltage and ultra-high-voltage technologies in AC and DC have been developed and are in use with 1100 kV AC and 800 kV DC today. These technologies can also be used to interconnect the regenerative energy resource to balance the fluctuation energy generation over a long distance. The technology is available including AC GIL with the highest voltage level of 800 kV AC installed today also in China. 1100 kV AC for GIL is not a problem, but only an upgrading. For DC GIL, new technology will be needed, which needs some years of development. First, DC GIL experiences are made in Japan since 1998 at the Kii Chanel project.

18.11.5 Cost of Transmission Losses

Transmission losses can be calculated and evaluated with the cost per kWh of the nondelivered energy that cannot be sold to the customer. Taking the losses of Table 18.15 for the rated current of 2000 A in comparison of GIL and OHL, and a 0.10 €/kWh, the values are given in Table 18.16.

TABLE 18.16 Cost Comparison of GIL and OHLs

	OHL	GIL
Transmission power (MVA)	2,100	2,100
Losses per kilometer (kW/km)	820	254
Losses per 50 km (MW)	41	12,7
Cost of losses (0,1 €/0.7 * 8600 h * losses) (€)	24,682,000	7,645,400
Savings in losses using GIL (€)	13,036,600	
Investment cost for GIL (1 system) (€)	175,000,000	

18.12 Conclusion

The impact of blackouts can reach very large areas and affect millions of people, as we have seen during the last years.

One reason for these blackouts is coming from the fact that the existing transmission network is higher loaded than decades ago, creating bottlenecks and then, in consequence, the blackouts.

The public situation is that OHLs are not welcome and the net situation is that power flow directions are changing with new power generation plants and distributed generation. The GIL is one technical solution to overcome bottlenecks in the transmission net.

The applications at PALEXPO and Kelsterbach have shown that the GIL offers reliable solution and the capability to transmit high power ratings also over long distances.

References

1. Koch, H. and Schuette, A., Gas-insulated transmission lines for high power transmission over long distances, Paper presented at *EPSR Conference*, Hong Kong, China, December 1997.
2. Koch, H., Underground gas-insulated cables show promise, *Modern Power Systems*, 17(5), 21–24, May 1997.
3. Baer, G., Diessner, A., and Luxa, G., 420 kV SF6-insulated tubular bus for the Wehr pumped-storage plant, electric tests, *IEEE Transactions on Power Apparatus and Systems*, 95, 1976, p. 2.
4. Koch, H., *To Solve Bottle-Necks in the European Transmission Net*, IASTED, Benalmádena, Spain, June 2005.
5. Koch, H., *AC Bulk Power Systems in Metropolitan Areas Application*, IEEE/PES T&D Asia Pacific, Dalian, China, August 2005.
6. Henningsen, C.G., Kaul, G., Koch, H., Schuette, A., and Plath, R., Electrical and mechanical longtime behaviour of gas-insulated transmission lines, *CIGRE Session*, Paris, France, 2000, p. 3.
7. Koch, H. et al., N2/SF6 gas-insulated line of a new GIL generation in service, *CIGRE Session*, Paris, France, 2002, p. 3.
8. International Electrotechnical Commission, *Rigid High-Voltage, Gas-Insulated Transmission Lines for Rated Voltages of 72.5 kV and Above*, IEC 61640, IEC, Geneva, Switzerland, 1998–2007.
9. Völcker, O. and Koch, H., Insulation co-ordination for gas-insulated transmission lines (GIL), *IEEE Transactions on Power Delivery*, PE-102 PRD (07-2000), 16(1), 2001.
10. Schichler, U., Gorablenkow, J., and Diessner, A., UHF PD detection in GIS substations during on-site testing, Paper presented at *8th International Conference on Dielectric Materials, Measurements and Application*, Edinburgh, U.K., 2000, pp. 139–144.
11. Schöffner, G., Boeck, W., Graf, R., and Diessner, A., Attenuation of UHF signals in GIL, *12th International Symposium on High Voltage Engineering*, Bangalore, India, 2001, pp. 453–456.
12. Siemens, *Power Engineering Guide*, 5th edn., 2008, pp. 24–25. www.siemens.com/energy
13. International Electrotechnical Commission, *Guide to the Checking of Sulphur Hexafluoride (SF6) Taken from Electrical Equipment*, IEC 60480, IEC, Geneva, Switzerland 1974–2001.
14. International Electrotechnical Commission, *High-Voltage Switchgear and Controlgear—Use and Handling of Sulphur Hexafluoride (SF6) in High-Voltage Switchgear and Controlgear*, IEC 62271-4, IEC, Geneva, Switzerland, 2011.
15. Alter, J., Ammann, M., Boeck, W., Degen, W., Diessner, A., Koch, H., Renaud, F., and Pöhler, S., N2/SF6 gas-insulated line of a new GIL generation in service, *CIGRE Session*, Paris, France, 2002, p. 9.
16. Piputvat, V., Rochanapithyakorn, W., Pöhler, S., Schoeffner, G., Hillers, T., and Koch, H., 550 kV gas-insulated transmission line for high power rating in Thailand, *CIGRE Session*, Paris, France, 2004, p. 6.
17. CIGRE Working Group 23.02, *Task Force 01, Guide for SF6 Mixtures*, Brochure 163, CIGRE, Paris, France, 2000.

18. Christophorou, L.G. and van Brunt, L.R., SF6–N2 mixtures, *IEEE Transactions on Dielectrics and Electrical Insulation*, 2, 952–1003, 1995.
19. Knobloch, H., The comparison of arc-extinguishing capability of sulphur hexafluoride (SF6) with alternative gases in high-voltage circuit breakers, paper presented at *8th International Symposium on Gaseous Dielectrics*, Virginia Beach, VA, 1998, p. 6.
20. CIGRE Technical Brochure, Gas Insulated Transmission Line (GIL), Brochure 218, CIGRE, Paris, France, 2003.
21. CIGRE Technical Brochure, Application of GIL in long structures', Brochure 351, CIGRE, Paris, France, 2009.
22. Diessner, A., Koch, H., Kynast, E., and Schuette, A., Progress in high voltage testing of gas-insulated transmission lines, paper presented at *ISH Conference*, Montreal, Quebec, Canada, 1997, p. 5.
23. Koch, H. and Schuette, A., Gas-insulated transmission lines (GIL)—Type tests and prequalification, *Jicable*, Versailles, France, 1999, p. 13.
24. International Electrotechnical Commission, *Calculation of the Continuous Current Rating of Cables (100% Load Factor), IEC 60287*, IEC, Geneva, Switzerland, 1982.
25. Germany: Sechsundzwanzigste Verordnung zur Durchführung des Bundesimmissionsschutzgesetzes (Verordnung über elektromagnetische Felder—26. BImSchV), December 16, 1996.
26. Switzerland: Verordnung über den Schutz vor Nicht Ionisierender Strahlung (NISV), December 23, 1999.
27. Diessner, A., Finkel, M., Grund, A., and Kynast, E., Dielectric properties of N2/SF6 mixtures for use in GIS or GIL, Paper presented at *11th ISH*, London, Vol. 3.67, 1999, p. 11.
28. Graf, R. and Boeck, W., Defect sensibility of N2–SF6 gas mixtures with equal dielectric strength, CEIDP Annual Report, Vol. 1, CEIDP, Victoria, Canada, 2000, pp. 422–425.
29. Graf, R. and Boeck, W., Statistical breakdown behaviour of N2–SF6 gas mixtures under LI stress, Paper 3.96.S20 presented at *11th International Symposium of High Voltage Engineering*, London, U.K., 1999, p. 11.
30. Finkel, M., Boeck, W., Jänicke, L.-R., and Kynast, E., Experimental studies on the statistical breakdown characteristic of SF6, CEIDP Annual Report, Vol. 1, CEIDP, Victoria, Canada, 2000, pp. 405–408.
31. CIGRE Task Force 15.03.07 of Working Group 15.03, Long term performance of SF6 insulated systems, Report 15–301, CIGRE, Paris, France, 2002.
32. Hillers, T. and Koch, H., Gas insulated transmission lines (GIL): A solution for the power supply of metropolitan areas, Paper presented at *CEPSI Conference*, Pattaya, Thailand, 1998, p. 28.
33. Kahnt, R., Entwicklung der Hochspannungstechnik 100 Jahre Drehstromübertragung, Elektrizitätswirtschaft 90 (1991), H. 11, S. 558–576.
34. "FACTS Overview", IEEE and Cigré, Catalog Nr. 95 TP 108.
35. Müller, H.-C., Haubrich, H.-J., and Schwartz, J., Technical limits of interconnected systems, Cigré Report 37–301, Paris, France, Session 1992.
36. Koch, H., Influences of bottle-necks in the transmission network, *7th International Power Engineering Conference*, Singapore, 2005.

19
Substation Asset Management

H. Lee Willis
Quanta Technology

Richard E. Brown
Quanta Technology

19.1	Business-Driven Approach	19-2
	Asset Management Framework • Coordinated Cross-Functional Decision Making	
19.2	Important Functional Elements of Asset Management	19-6
	Portfolio Management of Projects and Resources • Evaluation of Projects Based on Multi-KPI Contributions • Prioritization Based on Total KPI Contribution • Probabilistic Risk Management • Managing an Aging Infrastructure to a Sustainable Point	
19.3	Asset Management in an Electric Utility	19-14
	Shift from Standards-Driven to Business-Based Management • Pareto Curves and the Efficient Frontier • Use of Risk-Based Asset Management Methods	
19.4	Asset Management Project and Process Example	19-19
	Substation Example • Implementing Project Evaluation and Portfolio Selection • "Half-Measure" Approaches Fail to Deliver Good Results • Limiting "Must-Do" Projects	
19.5	Changes in Philosophy and Approach	19-26
	Asset Management Does Not Lower Standards • Asset Management Is Not Necessarily Trying to Reduce Spending • Changes in Perspective and Culture Specific to Substation Personnel	
19.6	Substation Asset Management	19-28
	Utilization and Life Cycle Management • Condition-Based Maintenance • "Big Bad Outages" and Setting Priorities for Substations • Advanced Substation Technology • Different *Philosophy* with Regard to Equipment Stewardship	
19.7	Summary	19-36
References		19-37

Asset management is a way of making spending decisions throughout an organization that aligns all asset-level spending with high-level business objectives. Generally, this means that spending decisions are made with the explicit objective of maximizing business performance while proactively managing risks, budgets, resources, and key performance indicators (KPIs). By its very nature, asset management applies to all aspects of a business. For a vertically integrated electric utility, true asset management must span generation, transmission, substations, distribution, and all company-owned retail and wholesale customer systems. Therefore, one cannot talk about "substation asset management" per se. Instead, one must talk about how substations are affected by and fit into an overall asset management utility organization. However, there are several concepts of asset management that can be applied specifically to substations and groups of substations. For the people who operate, design, engineer, plan, or manage substations, asset management means a shift from past perspectives and practices about the role of the substation, how budgets are allocated to substations, and what is expected from both the substations and the people who manage and operate them.

19.1 Business-Driven Approach

Asset management is a management paradigm that seeks to maximize an organization's business performance in a rigorous and data-driven manner with regards to multiple considerations like profitability, risk tolerance, cash flow, regulatory relations, reliability, environment, customer satisfaction, employee satisfaction, and safety. In other words, strategic business goals directly drive engineering and operating decisions (Figure 19.1). As used within the power industry, the term "asset management" has slight variations in interpretation. However, invariably it means a strategy, decision-making, and prioritization system that includes a closer integration of capital and O&M planning and budgeting that was traditionally the case, with the goal of managing the life cycle of equipment while considering system perspectives, throughout aimed at achieving an *overall* "lifetime optimum" business case for the acquisition, use, maintenance, and disposal of equipment and facilities (Center for Petroleum Asset Risk Management).

Conceptually and strategically, asset management is a *business-driven paradigm*, but functionally it is a fact-based, *data-driven process* that requires more data than utilities traditionally used in planning and operations, applied in a more comprehensive and rigorous decision-making structure than they are traditionally used (Brown and Humphrey, 2005). This is the fundamental reason that, when done well, asset management yields improved performance when compared to traditional approaches (Morton, 1999). The power industry is gradually moving to asset management for several reasons. First, asset management methods provide executives and financial managers with more and better information about how much budget is enough and on exactly where and how spending should be directed. It makes the T&D systems' needs more visible to financial decision makers, including the role of substation investments. In addition, when applied comprehensively, asset management can simultaneously increase profitability, reduce risk, and improve customer satisfaction.

Finally, asset management is being adopted now because it *can now be done*. It is not a new concept, having been used for decades in industries that could afford the data and information costs to apply it within their venue (Humphrey, 2003). But it is only in the last decade that power systems and business technologies have permitted it to be economically applied to a business that operates a network of many very geographically dispersed but heavily interacting equipments and units and resources. Most modern automation and control technologies, enterprise information systems, and computerized engineering analysis methods can support the more data-driven, customized decision-making approach required to apply asset management. Decades ago, when equipment could not be monitored routinely, when supervisory control and data acquisition (SCADA) system measurements were not archived and analyzed by comprehensive advanced algorithms, when photocopiers and faxes did not permit wide

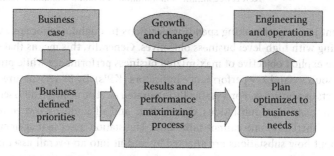

FIGURE 19.1 Asset management is a business-driven approach, in which multiple priorities, which may conflict or interact, are balanced and leveraged to create the maximum possible performance.

dissemination of information throughout an organization, and where different departments and functions within a utility could not share information at near-real-time on an enterprise level, asset management was simply not a viable management paradigm (Philipson and Willis, 2005).

19.1.1 Asset Management Framework

Asset management incorporates all of the following into a comprehensive analysis, planning, and execution framework: business-driven, multiple objectives, and risk management. A description of these three aspects is now provided.

19.1.1.1 Business Driven

Asset management makes business performance, rather than engineering and operations measures, the "standard" by which success is measured. All decisions about what the company owns, how it uses and takes care of (or neglects) what it owns, how it invests in new assets it wants to own, and how it operates those assets and uses its human and other resources are made on the basis of if and how they contribute to the overall business performance. From an asset management perspective, substations are part of an investment portfolio that are (1) competing for resources and (2) are expected to contribute cost-effectively to corporate goals.

19.1.1.2 Multiple Objectives

"Business performance" does not mean a focus only on financial profit (even purely financial businesses must consider both profit and risk exposure). The need to consider multiple objectives is especially true for regulated utilities, where it is the job of regulators to determine acceptable levels of profit. The job of the utility is to provide adequate service to all customers within its service territory for the lowest possible rates. Profitability will always be a major goal, but there are other considerations that are quite important such as well-managed risk, high customer satisfaction, high reliability, low environmental impact, employee safety, public safety, full compliance with all laws and regulations, being a good corporate citizen, and so forth. Table 19.1 shows the KPIs used by one utility, including the metric measurement used for each and the targets desired. It is representative of the comprehensive nature of KPIs used in the asset management approach, but not necessarily typical of most utilities. Goals and needs vary widely. Most of the KPIs are self-explanatory, but several are not. Here, "DivBSpl" is the largest

TABLE 19.1 KPIs Used by One Asset Management Utility

KPI	Meaning	Metric	Target
ALEL	Average loss of expected equipment life	Years/year	<2%
CAPEX	Capital spent in the year	Dollars	<$135 M
OPEX	O&M spend in the year	Dollars	<$340 M
Total $	Sum of CAPEX and OPEX	Dollars	<$450 M
DivBSpl	Divisional budget split	Max Δ	<10%
SAIFI	Ann. avg. customer inter. events	Events per customer	<1.3
SAIDI	Ann. avg. customer inter. duration	Minutes per customer	<95
$CEMI_3$	Customers with more than three inter.	% of customers	<3%
BBOM	"Big Bad Outage" contribution to SAIDI	% of SAIDI	<4%
PSAFE	Public Safety	Nonemployee accidents	0
ESAFE	Employee Safety	OSHA events per million hours	<1
LGL	Compliance with legal, codes, etc.	No. of outstanding violations	0
PWOM	PW of future O&M spending	Dollars	As low as possible
PWCR	PW of future capital spending	Dollars	As low as possible
SAEFA	System average equivalent failure age	Years/year	At strategic target

deviation among the company's many operating divisions from the ratio of budget spent in the division to revenues collected in the division annually. DivBSpl is a measure of the "fairness" of spending. BBOM (big bad outage measure) is a KPI aimed at reducing the likelihood of "outages that get into the newspapers" and are discussed in more detail later in this chapter.

All but two of the KPIs are *targets* (there is no compelling reason to spend money to go beyond it). The remaining two KPIs are *objectives* (areas where the utility desires to drive performance as far as possible). Overall, this set of KPIs can therefore be interpreted to mean

> Spend no more than $450 million in the year, distributed so that CAPEX is no greater than $135 million and OPEX no more than $340 million, with spending by division no more than 10% out of proportion to corresponding revenues. Achieve a SAIFI less than 1.3, $CEMI_3$ less than 3%, a SAIDI below 95 min with less than 4% of that caused by widespread (big bad) outages, have no public safety events, no more than 1 OSHA reportable accident per 100,000 h of company labor, and no deviations from legal and regulatory requirements. Subject to achieving those targets, minimize the present worth of future capital and O&M spending.

Financial asset management is about balancing risk versus return. Infrastructure asset management is more complicated and must balance performance, cost, and risk. Sometimes the need for considering risk is obvious, as when considering whether to spend money on a critical substation to reduce the risk of a catastrophic failure. A utility can spend a lot of money and have low risk, spend no money and have high risk, or do something in between.

Most utilities are not perfectly efficient when it comes to performance, cost, and risk. In this situation, an asset management utility seeks to *finesse* rather than balance competing targets and objectives. It may be possible through synergy, leverage, optimization, and efficiency gains to improve all KPIs simultaneously but not to the initial target levels that have been set. In this case, a utility may find that it makes more sense to focus on certain KPIs before others. A utility may also find that it could slightly lower a few KPI targets so that other KPIs can substantially improve. In the end, a utility should have a set of KPI targets that can all be reached. Much of the asset management process involves the finessing of these targets into an acceptable spending plan.

The utility which used Table 19.1 effectively told its asset management planning process:

- Spend up to $450 million next year. CAPEX and OPEX can vary a bit, but keep each one, and the spending allocation to divisions, within set limits.
- Satisfy all of the KPI targets, but there is no need for improvement beyond the target. It is OK to further improve any KPI in order to get other things done that need to be done.
- To the extent possible within the two previous requirements, minimize the present worth future CAPEX plus OPEX.

When asset management portfolio planning is done well, it will either (1) determine ways to achieve these goals or (2) show utility planners why this set of targets cannot be achieved.

19.1.1.3 Risk Management

Risk involves uncertain knowledge about the future. For targets, risk is associated with the probability of not achieving the target. For objectives, risk is associated with the uncertainty surrounding the expected outcome of the objective. Uncertainty stems from unpredictable or uncontrollable external and internal factors.* Asset management seeks to quantify and manage risk through explicit, fact-based consideration of the uncertainties and application of investment and decision principles that minimize

---

* An external factor is something "outside" of the utility, anything from future tax rates (set by politicians, not utility managers) to the weather, which no one can control or forecast. Internal factors of concern to the risk management process are variance from budget and completion time (not all projects are brought in on time and on budget), etc.

risk impact and balance it against expected gain. Approaches to risk management vary but must always involve the quantification of risk and the weighing cost of mitigation to its impact on expected performance and expected cost.

Common approaches to risk management are the use of probabilistic techniques, multiscenario analyses, or a mixture of both. In a probabilistic approach, while one does not know the eventual outcome of a particular factor (e.g., weather), one can characterize the uncertainty with a probability distribution that is known. In a scenario approach, one studies the implications of some large event or shift in conditions with respect to the entire plan.

Traditionally, electric utilities have been very risk averse. The move to risk-based approaches is motivated by a growing recognition that often this risk aversion is not in alignment with overall corporate objectives. Instead, risk-based asset management seeks to balance risk (the likelihood of bad outcomes) against the cost of avoiding them. Consider the experience of a one large investor owned utility (IOU) in the central United States. The capacity requirements and design standards at this utility required a minimum ratio of substation capacity to projected peak demand for substation equipment. A risk assessment showed that these criteria, resulting in about 4% of capital substation spending, would only be useful less than once every 40 years. A risk perspective revealed two important points about this expenditure:

1. *This spending was effective.* It did assure that the bad outcomes it was meant to prevent (outages due to insufficient substation capacity) would be avoided. However, not spending it did not necessarily mean the outcomes would always happen: operating measures and emergency reactions in many cases might avoid them.
2. *There were more effective ways to spend the money.* The utility could buy superior *overall performance* for its customers and stockholders with 4% of the substation capital budget by spending it elsewhere: as on items like automation and monitoring, which would provide improvement more often than every 40 years and improve more than just customer reliability.

Thinking in terms of risk management is often a difficult adjustment for organizations that have institutionalized a dislike for risk, as have many traditional electric utilities. However, risk management is fundamental to all businesses subject to substantial uncertainty, and sound asset management is impossible without risk management. A utility using a risk-based approach might set the KPI targets shown in Table 19.1 and then set *probabilistic targets* for their achievement: "Probability of budget exceeding limits will not exceed 5%, nor that of failure to meet any one target exceed 10% or all targets simultaneously, 20%." Probabilistic analysis then looks at the probability distribution of uncertain and uncontrollable events and factors, as well as stochastic factors such as how much project cost varies from forecast, etc., and how all influence the KPIs, to determine if a spending plan meets these probabilistic criteria. Proper numerical methods can optimize the problem, finding the best spending plan to meet the KPI targets within the desired level of confidence.

19.1.2 Coordinated Cross-Functional Decision Making

Asset management calls for all policies and decisions about the use of existing assets, investment of new capital, and application of O&M resources to be coordinated so that they maximize their *joint* contribution to achieving goals and managing risk. This is related to the pursuit of synergy and leverage between projects when attempting to achieve all targets at once. Synergy and leverage can only be managed through the close coordination of investments, resource allocations, and policies. At many utilities, asset management starts with an initiative to prioritize only large capital projects using some sort of cost-to-benefit ratio ranking method. This is a good first step. Large capital projects are an important budget and performance category, and asset management principles can be easily and effectively applied. Further, there are typically good data. Last, the cost analysis for each large and important project can be separately justified. However, true asset management links all spending decisions into a coherent whole. It does not consider only all large capital projects. Rather, asset management looks at small capital projects, large capital projects, inspection projects, maintenance projects, and operational policies together, with full consideration of their positive and negative interactions.

For an electric utility, this means that capital spending, maintenance spending, and operational policies must all be coordinated and managed through the asset management process. Asset management means substantially more communication and cooperation among departments and across the disparate functions scattered throughout the utility through a common asset management process. The hope is that all departments and functions mutually work toward a broader corporate goals. For substations, this means that a substation asset management process must include substation planning, substation design, substation equipment standards, substation operating guidelines, and the inspection and maintenance of all substation equipment.

19.2 Important Functional Elements of Asset Management

Many functions within an asset management process already exist within the existing utility framework. However, asset management has several important aspects that are distinct from traditional utility planning and engineering. This includes portfolio management, multi-KPI assessment, multi-KPI prioritization, and probabilistic risk management.

19.2.1 Portfolio Management of Projects and Resources

The term *portfolio* has long been used within the investment community to refer to the set of investments that a person or institution has made. In utility asset management, the portfolio is best thought of as the set of investments and resource commitments that the utility makes in order to get the business results it wants (Figure 19.2). A utility obtains its KPI targets by "buying" its portfolio of assets.

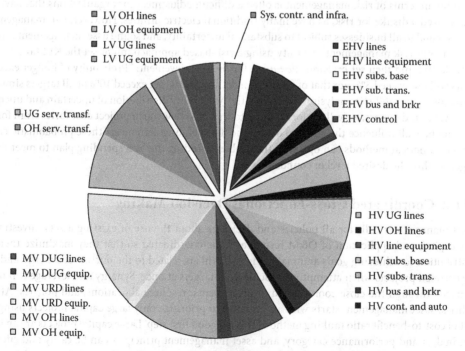

FIGURE 19.2 The "portfolio" is the distribution of "investment" by the utility in order to obtain the results it wants. This pie chart shows the asset value of the T&D system owned by an invested-owner utility in the southeastern United States.

A point often missed about portfolio selection is that the portfolio is selected as an integrated whole. All the projects within a portfolio are selected simultaneously or in a way that effectively accomplishes the same result. This is a large departure from past practices. In the traditional utility, all projects were *approved* simultaneously or at least within the same process in the same budget year. However, each project was typically created to address a particular "problem," and this was done in relative isolation from other projects related to other problems. In contrast, asset management links not only the decision about whether to approve a particular project to the approval of other projects but also the decision about what to do in each instance to what is decided to be done in every other instance. The integration of project identification and selection is much tighter than in traditional utility planning and management. All projects are effectively part of a single, broader decision.

Consider an old substation transformer that is showing signs of deterioration. Options to address this issue include the following: do nothing, do weekly visual inspections, install online condition monitoring equipment, do minor maintenance, do a major overhaul, transfer load away, replace the unit, and many other possibilities. The traditional approach would examine each option and select the "best." Asset management does not prejudge the answer. Instead, all options are provided to the portfolio optimization process and the choice is made in full coordination with all other projects and with all of their associated options.

19.2.2 Evaluation of Projects Based on Multi-KPI Contributions

Table 19.1, as discussed previously, highlights the multi-KPI nature of asset management's focus. Functionally, this means that in the evaluation, planning, and approval stages, the merits of any particular project option are assessed on how it contributes to all business goals, *and* on whether improvement in each of those KPI areas is needed (i.e., a particular project may provide a good deal of improved reliability, but none may be needed). Not every project contributes to every KPI, and some projects will actually have negative impacts on certain KPIs. Many projects will incrementally contribute to several KPIs, not just one, even if they were created to address one particular KPI. Asset management explicitly recognizes this by evaluating projects based on their value overall.

19.2.3 Prioritization Based on Total KPI Contribution

A key concept of asset management is that projects and options the utility *could* select for approval are assessed based on their total KPI contributions and those with the highest overall contribution are selected. This is often called "bang for the buck" prioritization because projects are evaluated on the "results" (bang) they provide compared to their cost (bucks). At its simplest, asset management assembles a single-valued "score" for each project based on the KPIs and a scoring or conversion formula. For example, if the KPIs were as listed in Table 19.1, this formula would be something like that shown in the following:

$$\text{Score for a project} = \frac{\text{Project value}}{\text{Project cost}} \qquad (19.1)$$

Project value

$$= A \cdot \Delta SAIDI + B \cdot \Delta SAIDI + C \cdot \Delta CEMI_3 + D \cdot \Delta BBOM + E \cdot \Delta PSAFE + F \cdot \Delta ESAFE + G \cdot \Delta LGL \qquad (19.2)$$

$$\text{Project cost} = PWCR + PWOM \qquad (19.3)$$

where
    A → G are scalar weighting factors proportional to the importance of each KPI
    PWC is the present worth of the project's capital cost
    PWOM is the present worth of O&M

Projects are then ranked on the basis of their score and selected "from the top" moving down the list until either budget or resource constraints are reached, or all KPI targets are achieved. Once a particular target is reached, no additional value should be placed on further improvement of that KPI. Therefore, value scores for additional projects should be recomputed with a zero weight associated with any KPI that has reached its target. The list can then be "reranked" in a dynamic reevaluation manner each time another KPI is achieved, and the selection process continues. The portfolio (group of selected projects) should meet all KPI targets for the minimum possible present-worth cost, regardless of the weighing factors that are used. One should note that the "value" score and the value weighting factors used are only a means to an end and not of material importance or even particularly meaningful. They are used only to enable a simple solution methodology to work.

If a solution that does not violate any constraints cannot be found, the asset manager must relax one or more constraints, generate a new portfolio, and repeat this process until an acceptable solution is found. Methodologies that use a ranking approach, even those that use dynamic reevaluation and reranking (as described previously), are simple to implement but are generally not able to find optimal solutions, for several reasons. First, the solution will generally be dependent on the weights chosen for the ranking formula, although this is generally a small concern as long as all KPI targets are reached. But a weighting-factor ranking method that finds a solution may not find the *best* solution.

More important, ranking methods have difficulty handling practical constraints (e.g., you can select project X only if you have previously selected project Y or project Z; you cannot do both M and N in the same year; you must do one but only one of U, V, or W). Last, it is nearly impossible to properly consider the interdependencies of project benefits with simple methods (e.g., project A will deliver more value if project B is also done; project C will deliver less value if project D is also done). In practice, the authors have found it is impossible to accommodate more than a handful of such issues with customized reranking methods.

For complicated project portfolio selection problems, the use of rigorous optimization algorithms is generally preferred. These algorithms can automatically address any shift in values as KPIs are achieved one by one, can handle any number of practical constraints, and can simultaneously select projects while considering interdependencies. However, the concept of ranking based on total contribution to business goals is central to asset management and is useful as an intuitive guide in understanding its goals and methods.

While tools and techniques vary from utility to utility, all asset management approaches decide which projects or options to select for the portfolio (i.e., which to approve for funding) in a simultaneous manner where all are effectively in competition with one another for funding. In general, projects selected for funding (i.e., to be included in the portfolio) contribute broadly to many KPIs for which improvement or performance gain is needed. Few approved elements of a plan will be "single criterion" projects that achieve or contribute to only one aspect of the business goal set.

An example from an actual asset management plan for a large IOU in the central United States is shown in Table 19.2. In the interests of space, only a few KPI columns and a few project rows in a very big table are shown. The actual plan involved 18 KPIs and over 430 different potential strategic project options, 108 of which were selected for the portfolio. The approval optimization determined the best set of projects meeting the specified KPI targets and ranks projects based on a benefit-to-cost score. The entire set of 108 projects from which this list is taken is the optimum portfolio to achieve the KPI goals the utility has set.

19.2.4 Probabilistic Risk Management

Asset management methods in the financial industry are based on sophisticated risk quantification based on probabilistic techniques. The goal is to identify the maximum expected return that can be achieved for each level of risk (defined as the standard deviation of expected return). This topic is vast, but the point is that there are many mature probabilistic techniques proven to work well. Many of these also work when applied to utility asset management and substation asset management.

TABLE 19.2 Ranked List of Projects from a Utility Asset Management Plan

#	Project Name	Project Option	CAPEX	OPEX	SAIFI	SAIDI	NPV	Utility/$	Cumulative Cost
1	Dist. sub. transformers >10 MV	−10%$ I&M., cond. based	$15	−$523	317	57,060	$2458	870.9	($520)
2	Dist. sub. transformers <10 MV	+10%$ I&M, cond. based	$18	$484	8,004	960,526	$3667	837.0	($33)
3	MV OH lines	+15%$ perf. based sch. 3	$22	$674	32,193	2,961,756	$4779	813.7	$644
4	EHV breakers >115 kV	Repl. 50 bad@ 10/year	$2110	$780	11,035	1,986,300	$6070	713.7	$1783
5	MV breakers <34 kV	+10%$ I&M., cond. based		$284	4,002	80,044	$1716	549.0	$2067
6	3-Ph Pad. Serv. transformer	Replace units rated <S	$1637	$281	350	63,015	$1608	513.2	$2626
7	Dist. sub. transformers <10 MV	Monitoring@ remote sites	$1180	$24	4,803	864,474	−$757	487.2	$2851
8	EHV sub. transformers	Full 8-gas moni. and alarms	$8967	$56	4,911	206,262	−$283	487.2	$4431
9	HV breakers (35 and 69 kV)	−10%$ I&M., cond. based		−$345	−1,252	−11,780	$2255	399.9	$4086
10	OH serv. trans TLM	TLM repl. prog. ongoing	$234	$71	154,819	270,000	$20	389.2	$4197
11	HV circuits (35 and 69 kV)	Fault red. focus program	$84	$196	32,018	1,344,756	$900	371.2	$4407
12	MV circuits (<69 kV)	Fault red. focus program	$84	$996	40,022	3,682,018	$7740	342.2	$5418
13	MV UG lines circuits	−20%$ I&M., cond. based		$180	−3,200	−576,000	$1012	323.7	$5598
14	EHV breakers >115 kV	−20%$ I&M., cond. based		$134	−2,087	−1,789	$484	270.6	$5732
15	EHV OH lines	0%$ I&M., cond. based	$84		1,252	22,536	$680	256.8	$5746

Probabilistic risk management falls into two categories: cash and noncash. Cash flow uncertainty related to a project is reflected in the present value calculations for the project. Projects with cash flow risk that is similar to the overall cash flow risk of the company should use the weighted average cost of capital (WACC) for the discount factor. Projects with cash flow risk that is higher to the overall cash flow risk of the company should use a higher discount factor.

Noncash risks generally relate to nonfinancial KPIs. Consider a utility with a SAIDI target of 110 min, but SAIDI will vary from year to year depending on unpredictable factors such as weather and the stochastic nature of equipment failures. It may not be enough to target SAIDI at 110 min on average. Should the 110 minute target be 105 minutes instead? Should it be achieved with 80% likelihood (4 years out of 5), or 90% likelihood, or 99% likelihood? A deterministic asset management methodology that works with numerical KPI targets essentially assumes that they are expected values. As such, it is about 50% likely that expected values will not be achieved in a particular year due to random chance. A probabilistic method is needed to determine how often the utility will achieve the target. A probabilistic methodology can determine a statistical confidence level for all noncash KPI targets (e.g., achieve a SAIDI of 110 min with

90% confidence). A probabilistic portfolio optimization requires probability distribution functions for key inputs (such as expected project benefit) and use of Monte Carlo techniques that are able to compute the probability distribution functions for key outputs (e.g., KPIs).

The impact of each project on SAIDI is described as a probability distribution, typically a triangular distribution consisting of a minimum benefit, a most likely benefit, and a maximum benefit. A Monte Carlo simulation will use random numbers to determine the benefit that each project has on SAIDI in a particular year, with the sum of the benefits of the five projects equal to the total SAIDI benefit. This process can then be repeated many times. The percentage of times that SAIDI is lower than the target is equal to the confidence of achieving the target. Assume that the Monte Carlo simulation is performed 1000 times, and the SAIDI target is met in 850 of these simulations. The probability of meeting the SAIDI target is 850/1000 = 85%. If the SAIDI confidence level is set higher than 85%, the portfolio is unacceptable. If the SAIDI confidence level is set lower than 85%, the portfolio is acceptable.

19.2.5 Managing an Aging Infrastructure to a Sustainable Point

When a system is composed of all-new equipment, failure rates are low and operating, maintenance, and repair costs are quite predictable, and equipment and system performance is generally satisfactory. As the equipment ages—remains in service—the condition of materials and components deteriorates from wear and tear and the effects of time, heating, and operating stress. Breakdowns and failures occur more frequently, and maintenance, repair, and unexpected service costs are higher and less predictable, and system performance may not longer be as satisfactory as it once was. Business and operating costs are less predictable than they were due to the higher failure rate and the higher standard deviation of failure and repair costs that usually develops in an older system.

Any set of equipment that is kept in operation and maintained to a prescribed set of operating, repair, and replacement policies will eventually reach a *sustainable point*, where all these factors remain stable thereafter: their expected values will not change from year to year (Figure 19.3). If the equipment operating policy is "run to failure," the average age of equipment in service will initially increase by roughly 1 year of age in each year of operation. But eventually, as equipment ages and conditions deteriorate, failure rates will rise: older units will fail and be removed. The newer replacements will have a lower

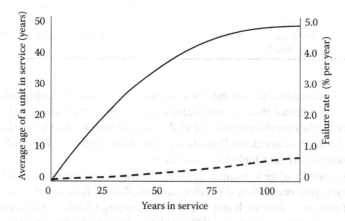

FIGURE 19.3 The concept of the sustainable point. In this example, a group of power transformers—perhaps 1000—are installed in the same year, and average age that year is zero. As they stay in service, their average age increases. Initially, few fail each year and their average age increases by nearly 1 year per year. But over time, more fail and are replaced, so average age gradually decreases. Here, average age gradually increases until a sustainable point average age of 50.0 years is reached, but this takes far longer than 50 years, as can be seen. Similarly, the average failure rate for the entire set, and all other statistics for it, will reach a stable value at the sustainable point, too.

failure rate. Older units continue to age and fail, and those that do not fail this year will develop even higher failure rates as they remain in service into the following year, etc. Given enough time, a type of balance is reached—the average age remains constant at the sustainable age from year to year thereafter, as will be expected failure rate, reliability levels, and operating costs. As a simple example, the reader can imagine a situation where failure rate reaches 2.0% at age 50. When average age reaches 50 years, 2% of units would fail, which would mean that going into the following year, those who had not failed (98%) would be 51 years of age and 2% of age zero, for an average of 50 years.*

If the operating policy were different, "replace at 40 years" rather than "run to failure," a different sustainable age, level of reliability, and expected operating cost would result. Failure rates would be lower than if all units were left in service until they failed. Replacement costs are higher and very likely business costs more predictable. But again, given time, the equipment set and the system would converge on a sustainable point—a different combination of expected average age, failure rate, and operating costs. Any set of similar equipment such as power transformers, wooden poles, steel lattice towers, etc., that is put into service under a constant set of operating policies has a sustainable point, a stable average age/failure-rate/operating-cost combination toward which it moves, asymptotically, incrementally, and generally very slowly, every year. This sustainable point is a function of the equipment type, design and its age and condition *and* the utility's operating policies.

Ultimately, equipment in service will age and these issues will become important to the utility. For any particular set of transformers, breakers, towers, poles, capacitors, etc., statistical analysis of historical operating data can determine the sustainable point's expected average age, distribution of ages of the equipment set about that average age (it is almost never Gaussian), operating costs, etc., for any utility system. Managing the aging infrastructure—by which one means the condition deterioration caused by continuing service and *the operating issues and business costs it causes*—it is the very essence of asset management. More details on these concepts can be found in Willis and Schrieber (2012).

19.2.5.1 Useful Management Concept

In almost all cases, the sustainable point for a utility infrastructure, even one with average ages of 50 years or more, is two or more decades into the future and represents a combination of age, failure rate, and operating cost volatility all much greater than at present. Average age might now be 52 years, and failure rate 0.4%, but the sustainable point might represent an average age of 58 years and a failure rate of 0.6%. Average age will increase slightly each year until it reaches 58, and failure rate will eventually increase by 50%. These changes will take much longer than 6 years (the difference between the 58 year sustainable age and the current age)—probably two decades or so. That future sustainable point may—in fact usually does—represent a situation that is completely untenable to the utility, a combination of failure rates, reliability problems, and operating costs that is unacceptable in every way. The trend will be a very gradual but steady worsening each year. The important points about the sustainable point and the "inexorable" trend toward it are as follows:

- Both can be estimated through analysis of historical data: *they are predictable.*
- Both are functions of equipment type and design and the operating policies for loading, maintenance, repair, replacement, etc. Change equipment or policies or both and the sustainable values and the trend change: *they are manageable.*
- Among all possible sets of equipment types and operating policies, there is one that leads to a sustainable point/trend combination that is optimum for the utility: *they are optimizable.*

---

* This example serves to illustrate the point quickly and simply but has characteristics that seldom exist in the real world and that readers must realize almost never exist. Very rarely are the failure rate (2% in this case) and the age (50 years) reciprocal, so that 1/age = failure rate. Failure rates are nonlinear, the distribution of ages of an equipment set uneven, and a host of other issues are involved and interact with one another, to the point that the situation is quite complex. Generally, sustainable failure rate $\ll$ 1/sustainable age.

Very rarely is the optimum business solution to quickly change equipment and operating policies to a new, acceptable sustainable point. Where cost is included in the analysis, "acceptable" sustainable points just do not exist: they simply cost more than the utility can currently afford. Conceptually, it may be crystal clear that, eventually, the utility will have to either accept an "unacceptable" sustainable combination of failure, reliability, and costs that it finds very unpalatable or select one it prefers and change operating policies, equipment types, and spending to comply.

But from a practical standpoint, the fact that the sustainable point is far in the future, that the trend of worsening performance and cost is gradual, and that the time value of money affects spending decisions, means that the optimum business solution usually involves the following:

Mitigating the symptoms. A margin against the worsening reliability can be "bought back" through various improvements in operations or the use of automation or other technologies that has a lower present worth cost. Higher service and breakdown costs can often be countered by technology for monitoring and condition tracking, and better processes and practices for service.

Gradually adjust spending. A utility may determine that the least unpalatable sustainable point may involve operating costs that are 20% higher than today's. A 20% jump in spending is not feasible, but adjusting cost gradually, by about 1% a year over the next 15 years, is much more realistic.

19.2.5.1.1 Value of a Master Lifetime Management Plan

An asset management framework that recognizes that aging and its effects are inevitable is useful even when the utility determines it cannot afford to fully implement the plans it recommends. The analysis will provide the utility with a forecast of how the situation will change and what it faces in the future. It will provide information that can help it do the best it can. Knowledge of the optimum solution, even if unaffordable, will help it do as well as possible. Perhaps most important, all of this will provide a foundation from which the utility can demonstrate to regulators and all stakeholders that eventually spending will have to increase.

19.2.5.1.2 Practical Issues of Lifetime Management

For almost all power equipment, expected lifetimes are decades long. Deterioration rates are quite slow, and often their effects subtle and hard to measure. For this reason, a lot of historical data are typically required to built a base from which analysis can determine trends and sustainable points. An asset tracking database is useful for monitoring condition and triggering proactive service and usually more than pays for itself in reduced costs.

Age is used as a proxy for condition. Age is almost universally used as a proxy for condition and an overall measure of condition-related issues on the system. In some cases, such as for poles or service transformers, age is the only practical condition-related metric available and is used directly in analysis and management of the assets. But even for equipment that is managed using comprehensive condition definitions and cost, age is still tracked and reported as a useful overall statistic. A utility may track and make service and operating decisions for its power transformers based on IEEE condition codes for core integrity and gassing. But even so, it will track and internally report the average age and age-related statistics for its power transformers, and overall those numbers are generally used most as general indicators of long-term trends.

Long-term data are needed for modeling trends and planning. Asset registries that collect, track, and archive data on equipment condition, loading, events, and service and alert operators to issues regarding equipment are useful and generally justify their cost several times over on the basis of reduced total cost. These systems need very comprehensive data of a type that has often been collected for only a few recent years. By contrast, analysis of long-term trends for detailed determination of expected lifetimes, aging trends, etc., usually requires much longer historical records: 50 years or more of data are desirable. This is accommodated most often by the development of "sustainable point models" based on basic equipment and system data going back many decades.

Reliability should be converted to a cost in order to provide an effective framework for lifetime management. Most often, this is done by determining the avoided cost of making up for any deterioration of reliability. A utility might determine that continued deterioration of its medium voltage breakers indicates their contribution to SAIDI will increase by three-quarters of a minute per year over the next decade. The avoided cost of "buying" SAIDI improvements amounting to 45 s/year through the most cost-effect means defines a portion of the cost that this trend represents to the utility. In many cases, this "reliability impact" cost is two to three times greater than any other costs associated with the worsening condition and its effects.

Failure and breakdown. For major equipment, it is best to track failure (cannot be or not worth being repaired: must be replaced) and breakdown rates (can or should be repaired and returned to service).

Basic metrics for aging and its effects should be put into place and tracked. Their purpose is both to raise awareness of aging and its effects and issues overall and to provide the means to measure trends and consider mitigation plans. Average age of equipment in various categories is an obvious measure but often not the most meaningful or useful. Table 19.1 showed metrics for reliability often used by utilities and identifies their basis.

Average age. It is sometimes computed on a simple equipment count basis ("the average 138/12.47 kV transformer in our system is 47 years old") but often weighted by capacity or other considerations ("the average MVA of 138 kV power transformer capacity is 44 years old").

Age distribution. It is typically tracked and displayed (Figure 19.4) in order to provide more details.

Aging rate. It is the annual increase in the average age of the asset set.

Average loss of expected life (ALEL). It is a measure of how fast equipment is "being used up." Tracking this provides a helpful indicator of how closely operation is actually following the optimum business case. For example, optimum lifetime for a particular class of transformer from an overall business case

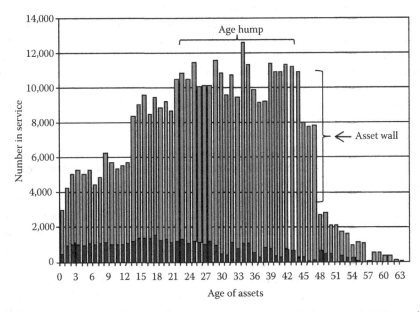

FIGURE 19.4 Distribution of ages for a set of equipment in a power system. Average unit is 32 years but an "age hump"—large number of units of one age category, means that average and statistics may not be reliability indicators of trends. Here, the "asset wall"—sharp drop-off in age distribution at age 48—is due to when units were installed, not because units fail in large numbers at age 48.

might target a lifetime of 63 years meaning average "use of life" ought to be 1.59% (1/63rd per year). Extremely hot summer temperatures in a year and imperfect balancing of loads may result in a rate of 1.8% for the year, which both identifies overall performance trends and provides a measure that can be used in making analysis and decisions about where and how to make improvements.

System equivalent failure age. It is the age at which the failure rate curve for the equipment equals the current average failure rate for equipment. It is perhaps the most meaningful measures of aging and its impacts. Due to a host of factors, including the nonlinearity of failure rate as a function of age, the equivalent failure for a system is often greater than the average age. ("While our average medium voltage breaker is 38 years old, the current failure rate is equal to that of a 46 year old unit because the small number of units over 55 years of age have much higher failure rates.")

System average equivalent failure aging rate. It is the annual change in system-wide equivalent failure age. This is a good indicator of the stability of operating issues and strongly recommended as a metric to use as a leading indicator of trouble. While average age can increase no more than 1 year per year, equivalent failure age can and sometimes does increase at a faster rate.

System average equivalent breakdown age and rate. These are breakdown-related metrics similar to those covered previously.

19.3 Asset Management in an Electric Utility

19.3.1 Shift from Standards-Driven to Business-Based Management

Traditionally, utility planners recommended, and management generally approved, projects and resource commitments based on minimizing the cost of satisfying planning criteria, engineering standards, and operating guidelines (Seevers, 1995). Typically, this was carried out on a department level (e.g., transmission, substation, distribution) with a "standards-driven approach," in that a set of guidelines determined when and how the utility would commit spending and resources (Figure 19.4). For example, a utility might have an equipment loading standard stating that a substation transformer should not be loaded past 80% of its nameplate rating under normal (weather adjusted, no contingencies) peak demand conditions. If load growth forecasts predicted that the 80% loading limit would be exceeded in the future, capital planners would determine what and how and when to change the system. This might include the addition of transformer capacity, so the loading limit would no longer be violated while spending the least amount of money possible. Similarly, the company's O&M guidelines would call for medium voltage breakers to be inspected and serviced at specified intervals regardless of the condition of a breaker and regardless of the importance of a breaker to overall system performance.

Such standards and guidelines were developed, fine-tuned, and proven over time. For the most part, they led to the desired results. The lights stayed on. Equipment usually lasted a long time, and there was little drama with respect to operation of the utility system. But the traditional engineering and operating guidelines (standards) were, in effect, "one size fits all." They assure that performance would be satisfactory, but they did not always spend money and use resources and existing assets optimally.

Occasionally, they called for projects that had very high marginal costs for very low marginal benefit. Sometimes they passed up opportunities where a small additional expenditure could provide a large business benefit outside of the mainstream context of the standards. Such situations were exceptions to these "one size fits all" rules, but there were enough exceptions that a more case-by-case approach like asset management can improve cost-effectiveness and overall results by noticeable amounts.

While good planners generally did minimize the cost of each project within the traditional standards-driven framework (e.g., planning all substation upgrades so that the loading guideline would no longer be violated), each project was considered in relative isolation to other projects being studied, other decisions being made, and how those decisions and plans might interact with other goals the organization had. As a result, the traditional paradigm *did not* necessarily minimize the overall cost of achieving the ultimate result (e.g., satisfactory service and equipment lifetime). Projects and expenditures were not selected based on being "put together" in an overall optimum plan. Then, too, only capacity-based KPIs were considered, leaving other KPIs to be dictated implicitly by design standards. Asset management goes beyond the traditional paradigm by considering all KPIs explicitly and considering all projects simultaneously in the same way. It provides improved business performance *and* improved cost-effectiveness because it seeks to achieve satisfactory overall performance at the lowest possible overall cost by looking at the ultimate goal, not at subsidiary goals such as loading standards, and by considering how projects, programs, and policies "fit together" into a coordinated portfolio.

Finally, the standards-driven process did not fully exploit opportunities for leverage, nor drive synergy among different functions and departments. Certainly, good managers and planners at traditional utilities were aware of and tried to create synergy and cross-functional leverage where they could. But the process, and the institutionalized framework within which they worked, typically did not encourage this type of behavior. This is especially true of departments and budgets. It was often perceived that a bigger the budget means a more important department. Asset management, ideally, is quite the opposite. Importance of departments is measured not by budget but by contribution to corporate success. Asset management is a different way of making spending decisions and requires a dramatic change in corporate culture. This culture change is perhaps the most difficult aspect of implementing an asset management strategy. For substations, this culture change will primarily be a shift away from and engineering and "equipment stewardship" mindset toward a business and risk management mindset.

19.3.2 Pareto Curves and the Efficient Frontier

Optimization methods cannot, by themselves, determine the recommended portfolio or the recommended plan. What they can do is analyze thousands of possible portfolios and compare results. A simple way to do this is to plot the present value of CAPEX and OPEX for a each portfolio against the weighted KPI improvement, as shown in Figure 19.5. Each circle, whether shaded or unshaded, represents one possible portfolio consisting of a set of projects and programs prioritized to implement over a 5 year period. Each circle is plotted based on its "bang" (KPI improvement) versus its cost ("bucks").

Most of the project portfolios (asset plans) in this diagram are not optimal in the sense that other portfolios exist that are both less costly and result in more KPI improvement. For example, portfolio "A" costs the same amount as portfolio "B" but provides noticeable less "bang." Clearly, "B" and "C" are better choices than "A." Since there are no projects that are unambiguously better than "B" or "C," they are referred to as *efficient*. Their KPI improvement level cannot be achieved without higher cost, and cost reduction cannot be achieved for either without sacrificing some of their KPI improvement.

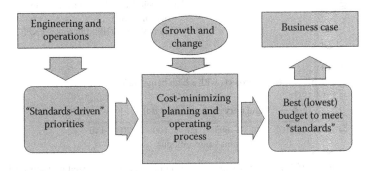

FIGURE 19.5 Many of the priorities and most decisions to spend money in a traditional electric utility were driven by "standards" on what and how "things should be done." When followed, standards typically resulted in low levels of risk and operational stability. Compare this approach to Figure 19.1.

The *efficient frontier* is the set of all efficient solutions and can be thought of as the upper boundary of this set of portfolios. The efficient frontier represents the set of choices the utility has with regard to how much performance it wants to buy. It describes how much performance can be purchased for various levels of spending and is a cornerstone of asset management in all industries (Markowitz, 1952). Conversely, it describes how much a utility must spend to achieve various levels of performance improvement. This efficient frontier is also known as a Pareto curve, since it identifies a set of optimal solutions based on the same efficiency concepts used by the Italian economist Vilfredo Pareto when he examined issues related to social welfare (Pareto, 1906).

Other graphs similar to the efficient frontier are often useful. Consider a portfolio of projects where inside the portfolio projects are ranked based on the ratio of total KPI contribution to the present value of cost. For budgeting reasons, a utility may want to plot KPI improvement versus budget impact rather present worth. Projects are still ranked based on present worth, but the plot is based on cash requirements. Another useful approach is to plot the improvement of each KPI separately. Projects are still ranked based on total KPI contribution, but results show improvement for each KPI versus various levels of spending.

Figure 19.6 shows an actual Pareto curve from the asset management study for a large investor-owned electric utility in the central United States. The curve was formed by a constrained optimization method that started with "zero budget" and then determined optimum portfolios (sets of approved projects) for every 5 year budget level up to $1,500,000,000 in even increments of $1 million. The optimization worked on multiple KPIs, but Figure 19.6 shows only SAIDI result. Characteristics worth noting here are as follows:

1. *Must-do region.* Starting at the origin and up to about $122 million, SAIDI improves only by about a minute for each $20 million spent. But beyond $122 million, the slope of the curve increases dramatically. From there to $250 million, SAIDI increases an average of 1 min for each $2.5 million spent. The initial low slope and poor performance in buying SAIDI is due to a number of "must-do" projects. These projects must be funded for $122 million but do not contribute greatly to SAIDI. In the optimization process, "must-do" projects are approved before any other spending. But once these are funded, the algorithm can search for the most efficient ways to buy SAIDI, and slope increases dramatically as it begins spending beyond $122 million.

   It is rare to see a "must-do" slope depicted in theoretical descriptions of optimization and asset management. However, "must-do" projects are a reality in any practical utility situation

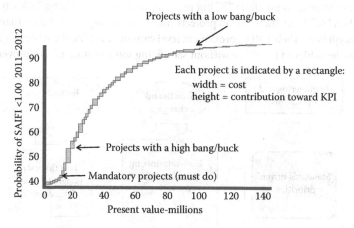

FIGURE 19.6 Potential portfolios (asset use and spending plans) can be plotted by their cost and performance results as shown here and explained in the text. Only the best portfolios lie on the efficient frontier, also called the Pareto curve, which denotes the maximum performance possible at every budget level. Optimization used in asset management finds only the portfolios on the efficient frontier and draws the curve for the utility planners.

Substation Asset Management

and are often quite an important part of the corporation's focus and its potential KPI performance achievement. Accommodating them well is a key success factor in any practical asset management approach.

2. *Decreasing marginal return.* After the "must-do" projects are approved, the Pareto curve is concave. This means that each subsequent dollar spent results in slightly less KPI improvement. Early spending has relatively high benefit to cost. Later spending has lower benefit to cost. This characteristic indicates a correct analysis and an optimum "performance buying process." The optimization process starts with the most effective (highest bang for the buck) projects and saves the rest for later. For example, in Table 19.2, the first project bought has an effectiveness (bang/buck) of 870 per dollar, the next has 837 per dollar, the next has 813 per dollar, the next 713 per dollar, and so forth.

3. *Bumps and small shifts in slope.* Figure 19.6 is not a smooth curve like the one shown in Figure 19.5. It has several points where, despite the general condition of concavity discussed previously, the slope increases briefly as spending is further increased. Any discontinuities and deviations from a strictly decreasing slope are due to the facts that (1) real projects are being optimized and (2) constraints are being applied. Deviations from the gradual smooth decrease in slope like that occurring at $760 million in Figure 19.6 are due to "big projects," as described in the following paragraph.

4. *Portfolios on the curve are not cumulative as a function of cost.* In other words, the portfolio (set of projects) on the curve that costs $800 million (point H) does not necessarily consist of all the projects included in the lesser cost portfolio (G, $700 million) with an additional $100 million of projects included. In fact, in this particular case, H contains only $630 million of overlap with G. With $800 million to spend rather than $700 million, the optimization took a different approach, which the discontinuity indicates. When the optimization reached $760 million, it "dumped" $70 million in other projects it had bought up to that point, to spend $71 million on a system-wide substation automation system that it could not previously afford, while still meeting all other requirements and constraints it had to meet, at any lesser budget. This $71 million project provides much better SAIDI impact than the $70 million in "dumped" projects did, hence the nearly straight upward slope at that point. Curve behavior like this is common in real, practical utility planning situations.

19.3.3 Use of Risk-Based Asset Management Methods

Asset management principles are most effective when applied with an explicit consideration of risk management, as discussed earlier. The purpose of asset management is to maximize business performance. While this can be interpreted to mean "maximize" the expected of performance, it can also be interpreted to mean "minimize" the likelihood of failure (Brown and Spare, 2004, 2005). A mature asset management process will explicitly consider all risk factors in a balanced manner. Risk management is particularly important for substations since utilities are often concerned more about avoiding large, but rare event rather than just the contribution of substations to broader customer-level KPIs such as SAIDI. Risk exists whether the planning method a utility uses explicitly considers it or not.

Deterministic asset management methods do not evaluate, display, or allow planners to deal with uncertainties and the risk they create, but the risk is still there. As an example, the left side of Figure 19.7 shows the probability of SAIDI corresponding to an investment plan that was deterministically optimized to an expected SAIDI of 100 min. The authors calculated the probability distribution shown based on analysis of the probabilities for factors such as weather, load growth, on-time completion, budget compliance of large projects, large equipment failures, etc. The average expected SAIDI is 100 min, but as shown the utility can expect it to exceed 115 min in about 3 out of every 10 years

An alternate plan was developed using a risk-based methodology that did not optimize based on average SAIDI. Rather, the objective was to minimize likelihood of SAIDI exceeding 115 min (the utility's regulatory definition of "poor performance" in this KPI). The risk-based plan spends the same budget but directs spending in a way that will minimize performance-based rate penalties in "bad luck" years

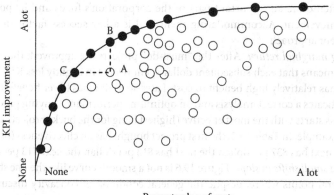

FIGURE 19.7 Actual Pareto Curve for an electric delivery and retail utility in the central United States shows characteristics common to actual, practical applications. See text for details.

(Brown and Burke, 2000). The resulting expected SAIDI is 106 min, a 6% increase over the first plan. However, the likelihood of SAIDI exceeding 115 min is less than 8%, a nearly fourfold reduction in the likelihood of "poor performance" when compared to the first plan.

Risk-based asset management methods typically look at sources of uncertainty in the utility's performance that cannot be precisely predicted or controlled (Table 19.3). Independent stochastic events such as equipment failures are analyzed probabilistically and their impact minimized in the portfolio optimization. Major stochastic events are analyzed on a scenario basis with the goal of developing plans that are robust in a variety of alternative scenarios (Willis, 2004).

Risk-based asset management methods perform analysis of projects and portfolios on a probabilistic basis, providing indications of the probability of success rather than deterministic evaluations (Figures 19.6 and 19.7). Figure 19.8 shows information developed by a utility in the western United States. A $400 million budget gives an expected SAIDI of 90 min. However, to be confident that it will achieve a SAIDI of 90 min or better 9 years out of 10, it would have to target an expected SAIDI of 80 min and spend

TABLE 19.3 Factors Typically Analyzed as Risk Sources in Utility Asset Management

	External Factors	Internal Factors
On a probabilistic basis	Weather	On-time completion of projects
	Load growth	On-budget completion of projects
	The economy	
	Price of power/fuel/labor/resources	
	Cost of money	
On a scenario basis	Regulatory relationship	Labor union–utility relationship
	Change in regional employment base	

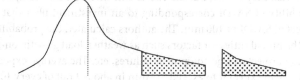

FIGURE 19.8 Left, probability of SAIDI for a plan optimized to a deterministic target of 100 min. There is a 28% likelihood of exceeding 115 min in any year. Right, distribution of a plan optimized to minimize the likelihood of SAIDI exceeding 115 min. The probabilistic plan leads to a higher probability of success in avoiding unacceptable results even though its mean expected SAIDI is greater.

$520 million (dotted line). The diagram shown is for a plan developed deterministically. A probabilistic optimization eventually developed a plan that spent $470 million and had the same probability of SAIDI exceeding 90 min. Information derived and presented in this manner is used to support decisions on both spending and forecasts, that is, executive management uses risk-based methods both to study what targets they can achieve and the likelihood of their success (i.e., set realistic, achievable goals) and to manage to those targets on a strategic and operational basis.

19.4 Asset Management Project and Process Example

The following example highlights the difference between a traditional utility approach and asset management. Suppose Big State Electric's Eastside distribution substation has two 40 MVA transformers, two low-side buses (each fed by one of the two transformers), and a total of eight feeder circuit breakers (four per bus). These transformers, buses, and breakers have been cared for according to well-established guidelines. All equipment is visually inspected annually. Breakers are completely serviced every 6 years or so many operations, whichever comes first. The substation currently serves a 57 MVA peak demand balanced fairly well between the two transformers' bus combinations and among the eight circuits.

19.4.1 Substation Example

The utility's guidelines call for no more than 75% loading of transformers in any two-transformer substation at normal peak demand, which means a limit on the substation load (when balanced between the two units) of 60 MVA. This guideline is meant to limit the stress on substation power transformers and thus ensure a long service life. Other guidelines call for no more than four breakers per low side bus (so that a bus outage does not cause too many circuit outages) and no more than 8 MVA loading per feeder (based on the present size of feeder getaway cable). All of these guidelines have their justification in analytic technical evaluations done in the past and have been proven over time. These guidelines seem to work well when they are followed.

Suppose that peak demand at this substation is expected to rise to 68 MVA in the next 4 years, continuing over the next decade to 78 MVA. Under the traditional paradigm, utility planners would evaluate options and select the best (least-cost) one so that the substation did not violate this or any other guideline as this load growth occurred, in a process diagrammed in Figure 19.9. The options studied would include transferring load to neighboring, less-loaded stations, as well as more capital-intensive options

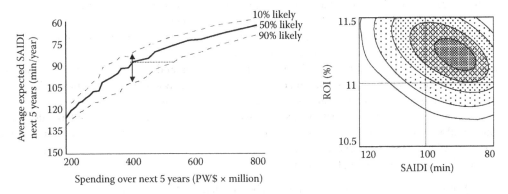

FIGURE 19.9 Information developed by a utility in the western United States during studies for its asset management strategy, which was targeted at 90 min SAIDI and an 11.2% return of investment. The left graph shows a Pareto curve of spending versus SAIDI, including 10%, 50%, and 90% confidence levels. The right graph shows the probability of SAIDI and ROI outcomes for the selected plan. It is seen that SAIDI and ROI are negatively correlated, mainly due to the probabilistic effects of weather: hot summers are good from a profit standpoint (high revenues) but place additional stress on system and equipment (harmful from a SAIDI standpoint).

involving addition and replacement of equipment. The planners would pick the option with the lowest total cost, usually meaning the option with the lowest net-present value of capital and future O&M costs.

In this example, the least-cost options that are selected include (1) the transfer of what will eventually be 6 MVA to circuits out of neighboring substations, (2) the replacement of the two existing transformers with 50 MVA units (thus serving the 72 MVA of remaining projected peak within the utility's standard), and (3) upgrading the two buses and all eight feeder getaways so that the circuits can each carry 8 MVA (i.e., so all other aspects are within other standards). To justify these projects, the planners have provided documentation that among all options examined, this is the one with the lowest cost that satisfies all the relevant guidelines.

Asset management would decide what to do in this case by looking at a *wider set* of considerations, with a focus on the ultimate business outcomes the utility desires and without absolute adherence to the substation design guidelines. The decision about if and what should be done at Eastside would be made in conjunction with evaluation of other needs and opportunities, and with it in "competition" with possible projects to be done elsewhere in the system. While this represents a significant departure from the traditional approach, it recognizes that the utility is not really interested in adhering to design guidelines per se. Rather, the utility is interested in good customer service quality, satisfactory and sustainable equipment condition, and a low spending level *overall*. Traditionally, design guidelines were indirect means of achieving those overall goals.

Some engineers and planners will be concerned that asset management has the potential to lower the utility's standards. Although this may seem true from a narrow perspective, it is important to realize what issues traditional approach *did not* consider. It did not take into account the condition of the substation's equipment. Perhaps much of Eastside substation's equipment is old and worn, certain to require expensive maintenance and expected to provide marginal reliability in the future, and likely to fail without warning and need replacement in the foreseeable future. Or perhaps this substation is new, in very good condition, and of a particularly robust design that can take a lot more stressful service, particularly if selected equipment is inspected and serviced more frequently in proportion to the greater wear and tear that those higher loadings create.

Furthermore, the tradition process did not consider the impact that substation outages would have on KPIs and their interaction with other resources and assets. Perhaps the feeder network around Eastwood substation is old, worn, and prone to frequent failures. Because of operational restrictions in the area served by the substation, it may be difficult and time consuming to restore customers after an outage. As a result, customers served by this substation see service reliability somewhat poorer than system average, regardless of any substation-related issues. With this broader perspective in mind, a new substation along with feeder upgrades might solve the loading violation and provides a needed reliability boost for the region. Or perhaps the opposite is true. The circuits out of this station, and their tie points to circuits from neighboring stations, could be automated so they can detect problems during emergencies and switch load very quickly. In this case, reliability of the substation and adherence to the strict standard would be less important because substation outages could be quickly and effectively mitigated.

Although substation guidelines are still useful as guidelines, asset management recognizes that broader considerations should affect the decision about what needs to be done at the Eastside substation, as well as how Big State Electric intends to spend money and assign priorities elsewhere throughout its system. When all this and perhaps other factors are taken together, and depending on the specific situation for Eastside substation, the utility might decide to

1. Do nothing. The "zero base" option should be part of any and all planning.*
2. Do exactly what would be done in the traditional case: upgrade the substation and feeder getaways, at considerable capital expense.

---

* Analysis of it will document why "doing nothing" is not acceptable and provide a baseline to identify the benefits, and thus determine a completely valid bang for the buck, for the other options.

Substation Asset Management

3. Do nothing to upgrade the substation or its feeder system but only increase the frequency and comprehensiveness of inspection and service. This is in recognition of increased equipment loading and the potential increase in failure probability.
4. Do nothing to upgrade the transformers, buses, or existing breakers and feeders but add two additional 8 MVA breaker/circuits (one per bus). This is in violation of the traditional four-per-bus limit but allows the 72 MVA peak demand to be served within the 8 MVA limit. This plan would also increase the frequency and comprehensiveness of transformer and bus inspections.
5. Do nothing to upgrade the substation or its feeder capacity but automate and upgrade selected feeder tie points in the substation area so dispatchers can monitor loading and switch remotely during emergencies. This plan may also increase the frequency of inspections.
6. Build a new substation and feeder network nearby.
7. Do something else entirely.

All of these options, and any others evaluated, should be assessed on a multi-KPI basis within the context of a broader project portfolio. Whether a particular option "solves" the projected loading violation is only important to the extent that these loading levels, in conjunction with other system characteristics, will impact the utility's KPIs. In addition, an asset management approach should apply a risk assessment to this decision. This would typically begin with an evaluation of the load forecast. It might be high, so there really will not be a loading violation. It might be low, so the situation will be worse than forecast. The risk assessment should (as much as possible) identify the probability of extreme outcomes and the cost to address these outcomes. This allows the utility to make an informed decision about how much money to spend to mitigate specific sources of risk.

19.4.2 Implementing Project Evaluation and Portfolio Selection

Many utilities making the transition to asset management see project evaluation and portfolio selection as its most important element. However, the changes in how and why people work together for project evaluation and portfolio selection, and in learning to think "asset management," are just as or more important. These *cultural changes* results will improve the way that engineers and planners throughout the utility make business decisions. This will improve the quality of spending focus in itself, even without a sophisticated project evaluation, and portfolio selection, and risk management process.

Figure 19.10 shows portion of the decision-making process as it applies to Eastside substation under an asset management approach, drawn in a manner as similar as possible to the traditional steps shown in Figure 19.9. The reader must keep in mind that the final selection in this case is not done in "isolation" as it is in Figure 19.9 but instead through portfolio prioritization, using optimization, as depicted in Figure 19.11. There, the decision about what to do at Eastwood substation and the decision about how to best address all other needs and opportunities throughout the utility system are made simultaneously. This includes a comprehensive consideration of all costs and business impacts needed to achieve the aggregate KPI targets set (e.g., Table 19.1). The decisions about what will be done at this example substation, and in all those other situations, are made in a way where each potential project competes against others based on benefits, costs, and uncertainties (Figure 19.10).

When addressing an issue or problem, traditional planning and asset management both strive to generate a number of options to address that issue or problem. In traditional planning, the "best" (lowest total cost) of these options is typically recommended for implementation. Asset management is completely different in this regard. The same several project options may be identified, but all options are retained during the optimization process. Some options may be low cost and low benefit, while others may be high cost and high benefit. Without looking at all other spending decisions simultaneously, it is not obvious which option is superior from the broader KPI perspective of *how it fits within the entire portfolio of selected projects*. In asset management, all asset-related spending decisions must be made at the same time and at the same place and within the same process.

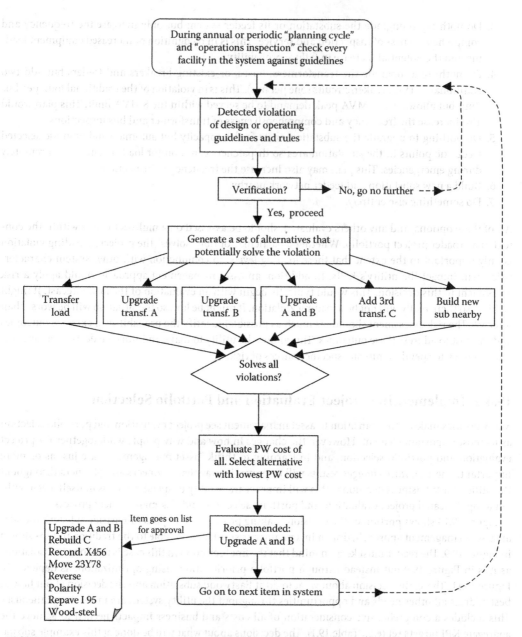

FIGURE 19.10 Traditional process of "spending" by a utility is initiated by either planning studies of system capability against forecast or condition assessment of equipment. Expenditure is triggered by detection of a possible guideline (standards) violation. This initiates something like the process shown, a search for a project that will fix the deviation (and cause no more) at the least possible cost.

19.4.2.1 Constraints Are a Key to Practical Success

For the asset management process shown in Figure 19.11, the identification and proper use of the constraints is a key factor and warrants further discussion. The thoughtful use of constraints is largely what determines how practical an asset management portfolio plan is, and proper use greatly increases overall value of the plan. Even if constraints are properly identified, ranking algorithms tend to have difficulty in finding optimal solutions subject to these constraints. This is the reason that mature asset

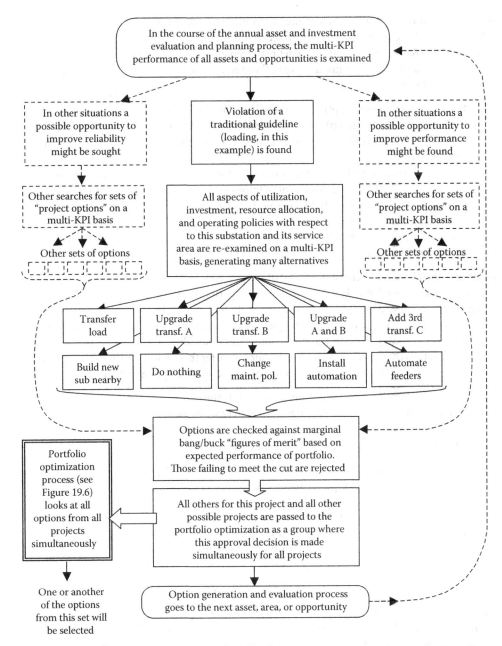

FIGURE 19.11 Like the traditional utility process, asset management periodically looks at the entire system for deficiencies that need correcting, as well as for opportunities to invest well or improve performance. Options, not selected recommendations, are passed to the portfolio optimization.

management processes will increasingly turn toward rigorous optimization algorithms instead of ranking algorithms in the portfolio selection step.

In any practical utility asset management process, there are *many* constraints, not just a few dozen. As a rough rule of thumb, there will be more constraints than projects but not more than the total number of project options. For example, one utility that has 582 projects with 1480 options has more than 900 constraints in the portfolio planning setup. Most of these constraints are of the "common sense" nature, generated automatically by the optimization algorithm, so that it will find only reasonable solutions.

An example is a constraint for each project than has more than one option requiring that only one option be selected (e.g., only one of the 10 project options shown in Figure 19.10 for Eastwood substation can be selected). Beyond this, the utility asset management planners can enter additional constraints to represent goals they know they must achieve in the execution of their plan or limits they know they face. Examples might be the following:

- Only one of the project X options can be selected if prior decisions elected to do project Y option 2 or 3 (e.g., the utility can only build new feeders out of Eastwood substation if it has first picked an option that builds the new bus needed to connect them into the system and supply them with power).
- Capital construction projects in Area 7 must be at least $250,000. (The utility promised local politicians it would make a certain amount of improvement during its franchise renewal hearings.)
- No more than $12.4 million in breaker projects can be done. (There are only so many qualified breaker crew hours available.)
- At least $7.8 million in breaker projects must be done. (Internal breaker crews will be fully utilized on breaker work.)

Setting up constraints represents a good deal of the practical work in asset management planning, and the ability to use them effectively will increase with experience and study.

Overall, the process depicted in Figures 19.9 and 19.10 might decide to do absolutely nothing at Eastwood substation, despite the traditional guidelines. In rare cases it might decide to spend more, or to spend in a different way, than the traditional paradigm would dictate. The important point is that asset management selects the project and expected performance for this project based on the needs and priorities of the entire portfolio. This project is *only one of a set* of selected options which in combination would

- Stay within the corporate budget limits and constraints.
- Achieve all the KPI performance targets. In each case, the target would be met but not necessarily exceeded.
- Minimize the present worth of CAPEX and OPEX.

In selecting the recommended portfolio, optimization seeks to find project options that cumulatively achieve all KPI targets while being cost-effective over time. In some cases, a utility's initial "optimization run" will spend up to its budget limits without achieving all KPI targets. In this case, there is no viable solution and either budgets must increase, KPI targets must be reduced, or more cost-effective project options must be identified. Understanding the interrelationship of KPIs to budgets, KPIs to themselves, KPIs to available spending, and project options to all of these is critical when adjusting setup. This is where diagnostics and graphics aids (item 5 in Figure 19.11) are invaluable.

19.4.2.2 Conditional Approval in Risk-Based Planning

Figure 19.11 shows the optimized portfolio (item 6) as including two classes of approved projects, those approved outright and those "conditionally approved." This second class includes several subcategories of projects or programs that might be approved, or needed, depending on if and how any number of probabilistic outcomes occur, in order to *reduce risk*. For example, if the utility experiences a very hot summer, revenues will probably exceed expectations for an "average year," but expected stress on substation and other major equipment will also be higher than average (see right side of Figure 19.7). The conditional project set would contain one or more projects to mitigate the negative outcomes of this hotter-than-average weather (e.g., perhaps a project to inspect equipment thought to be in jeopardy during a hot summer, the money coming from the additional revenues, to be approved only upon seeing that the summer in fact turns out to be hot). Similarly, the conditional project set would contain one or more projects that effectively say "If load growth is above expectation, we would spend this additional money on connecting the greater-than-expected number of customers." Such projects permit the risk-based portfolio plan to have the necessary probabilistic elements to control risk.

19.4.3 "Half-Measure" Approaches Fail to Deliver Good Results

A number of utilities have attempted to apply asset management, risk-based, or otherwise, using a "half-measure" approach to portfolio selection that is between the traditional and full portfolio selection approach outlined earlier. In this method, instead of passing all the project options (the product of Figure 19.10) as input to the project data set (item 3 for the portfolio optimization in Figure 19.11), they select and pass only the one option that would be traditionally selected for each planning or operating study (i.e., the product of Figure 19.9 in the case of the example) to the prioritization/optimization. Utilities that take this approach do so because it simplifies and reduces the amount of data that must be prepared and input into the portfolio optimization, greatly reducing the work required. In addition, it permits a simpler algorithm to be used (simple spreadsheets in most cases). This approach assures that only projects that adhere to the traditional cultural "comfort level" are considered for inclusion in the project portfolio. Almost always, only projects meeting traditional standards, or something like them, make it *into* the process. This renders the move to asset management much easier to accept throughout the organization because the "answer" looks pretty much like business as usual.

The problem with this approach is that the resulting portfolio is essentially what would have been planned in the traditional approach. With this "half-measure" approach, performance and bang for the buck increases only slightly, if at all, when compared to the traditional paradigm. If the process concludes that KPI targets cannot be achieved, management will likely conclude that all possible trade-offs have been considered when this is definitely not the case. *In practice, asset management is only effective if it has the ability to spend a little less on one project and shift the savings to other projects.* In the authors' experience, good results come only from asset management approaches that apply all of the following in the portfolio selection process:

1. The use of multiple KPIs covering all major interests of the utility, with only quantitative targets
2. The use of rigorous optimization algorithm (as opposed to simple ranking)
3. The use of constraints to define limits and requirements for execution
4. The use of project options that provide a range of results and costs for each potential project

An asset management process that applies these concepts will likely recommend a portfolio of projects that is significantly different than those that would have come out of the traditional approach. Since this can create a certain amount of discomfort for traditionalists, there is always a need for strong executive championing to anticipate and manage culture change, at least in the early stages of the transition. Executive involvement is warranted because asset management can make a significant positive impact on the utility's performance for customers, stockholders, and employees alike.

19.4.4 Limiting "Must-Do" Projects

Any successful practical optimization process will need to recognize a project designated as "must do," which means that regardless of project KPI contributions or cost, it will be selected and approved. From time to time every utility has projects that are truly "must do," and so this feature is needed in any practical decision-making process. Furthermore, there are reasons why planners want to have this feature available just because of the study capability it provides.* But for asset management to be successful, "must-do" projects cannot represent a large percentage of spending. The utility must make every effort to minimize project options identified as "must do" to the portfolio optimization process. Failure to address this particular point has been the downfall of many asset management initiatives, particularly those using a half-measure approach. Many utilities taking up asset management for the first time end up labeling far too many projects "must do." This is the case when a hybrid or half-measure approach is used as described in Section 19.4.3. It is not uncommon for utilities taking that approach to end up with upward of 70% of their total capital budget committed to "must-do" projects.

---

* A project can temporarily be designated as "must do" to force the optimization to select it—thus showing the planner what other projects and decisions it would or would not make given that this project will be done.

What happens is that with regard to a particular need (for change to the system, as when an area of new growth needs to be connected to the system), *something must be done* and since only one project is to be submitted into the portfolio optimization to address this need, and that project seems reasonable (particularly from the traditional perspective), it is clearly "must do." Consider the following example. There is a vacant field that will be developed to serve 300 homes being built in what will be "Valley Oaks" subdivision. Traditionally, the utility would look to its design standards and submit a project calling for the extension of a distribution feeder and construction of laterals and service drops. If the utility was not using project options and had entered only this one option for the Valley Oaks project, very likely it would label that project as "must do" because the new customers *have to be* connected. Likewise, something similar would be done in many other situations. The result is that the portfolio optimization is presented with a plethora of "must-do projects" which both "use up" much of the available budget and also give it far less flexibility to pick and choose so it can achieve synergy and optimization. "Asset management" carried out this way produces little if any improvement in performance.

In an option-based approach (Figures 19.9 and 19.10), planners would instead enter a set of options for this project along with a *must-perform constraint*. The options might include the feeder extension project described earlier, as well as others. One might be to build a new substation near the subdivision, another to use different construction standards for the feeder, another to extend other circuits from other stations, yet another to use distributed generation, and still another to do a combination of these things, and so forth. These options would provide a range of KIP impact and price variations to the portfolio optimization, allowing it to mix and match its plans to maximum bang for the buck. But none of these would be a "must-do" project. Each would have a KPI or contribution tag identifying it as "serves new Valley Oaks customers" and the optimization would be given a constraint that one project option meeting this requirement must be included in the portfolio.[*] As with the use of options in general, this provides more flexibility to the portfolio selection algorithm and greatly improves the likelihood of performance improvement and/or cost reduction. There is no doubt that this approach represents more work, in identifying and inputting options to the optimization, but most of this can and should be automated (to both reduce labor and improve consistency): that is what computers and modern technology are all about.

19.5 Changes in Philosophy and Approach

To engineers, operators, and managers in an electric T&D utility, asset management means their company will undergo a philosophical shift in its concept of why the power system exists and "what it is there to do." Under asset management, the system exists to achieve the company's *business* goals, rather than deliver power to customers. Over time, this leads to subtle but significant differences in attitudes and values throughout the organization. It will also mean some differences in how people work.

First, nearly everyone will have to work "together" to a greater extent than in the past. Comprehensive and balanced integration of all decisions requires that managers, planners, operators, and executives communicate more, share more information, cooperate, and "compromise" more than in the past. For some, there will be only a small change in their work practices and responsibilities. For others, their entire function and purpose will be transformed. Regardless, information systems will become more central, both in their ability to make data available to decision makers and in their ability to facilitate communication and coordinate work processes. Sophisticated information systems in themselves do nothing without changes in the mindset of employees. However, asset management, by nature, is a data-driven process that requires supporting information systems to be successful.

Second, more of a business-case approach permeates all decisions and processes. Even traditionally "pure engineering" venues like substation equipment specifications may have an element of business-case evaluation, with all decisions having to be written in a business-case manner before gaining approval.

---

[*] Another way to do this is to enter the options, each of which connects to the 300 customers somehow, as a group and require the program to select one of them.

Third, there will be a company-wide "standardization" of the way project documentation and justification is required for approval and performance tracking. In the traditional paradigm, it was relatively easy for planners to show that their plans satisfied their goals by adhering to the company standards. (e.g., loading is within the 83% limit. Maintenance was performed within guideline periods.) Often, this was done through department-specific analysis and documentation. In an asset management organization, everyone documents a more diverse set of business attributes, as well demonstrate that their proposal optimizes, and not just satisfies, company requirements. Furthermore, an asset management utility puts much more emphasis on tracking results against targets. In the past, it was straightforward for the traditional utility to check that it was, in fact, keeping substation transformer loading within guidelines once the money was spent to upgrade the substation. In contrast, the asset management utility focuses on a more difficult set of measures: "Did we get the maintenance cost reduction, customer service quality improvement, and capital cost containment we expected?"

Asset management represents a big change from the traditional utility framework for managing, budgeting, planning, prioritizing, and operating. However, if implemented correctly, asset management will lead to improved system performance, reduced costs, and better managed risk.

To be successful, asset management must be applied across the entire utility, covering capital, inspection and maintenance, and operating budgets and priorities in all departments and for all functions. For utility personnel involved in the management and operation of electrical substations, asset management means that the performance expectations their company has and the decisions they make will become more business based and that over time they will have to work within a wider range of considerations and communicate and cooperate with a broader base of coworkers in the utility.

19.5.1 Asset Management Does Not Lower Standards

Like any organizational change, resistance is expected when shifting away from the traditional approach to one utilizing asset management. Many people will insist that the organization is lowering its standards because asset management considers "cheap alternatives" and might not decide to do as much as the traditional approach would have in many cases. This is a narrow and misleading view. In reality, the utility is not lowering its standards, but it is *changing* them to be more directly linked to business objectives. From an executive perspective, this represents a *raising* of standards.

Regardless, culture resistance to the change is to be expected. Some of this resistance is due to the natural resistance to change in general. However, there is an additional driver. Some of the work required for asset management is more difficult and requires additional skills. Planners and engineers may feel like technical experts within a traditional approach but may initially feel insecure about their abilities within the context of a business-driven approach (many engineer may also be genuinely uninterested in business issues).

If change management is effective and asset management principles are implemented well, a utility will be assured to two results:

1. *Planning and decision making will be more difficult and costly.* In the past, hard and fast rules dictated what was done "Do this. Period. Then move on." Now, the utility must gather and use a wide range of information to make multi-KPI decisions that are often not as black-and-white as in the past. In addition, planners must generate and consider within this broader context a wider range of options and alternatives. Finally, the utility must track this expected multi-KPI performance against its expectations to validate and improve assumptions and models. All of this takes additional data, analysis, and modern information systems to make it all work.
2. *Business performance will improve.* Why? Because, asset management can always pick "the traditional solution" when it is best from the standpoint of buying performance (but not necessarily just adherence to arbitrary standards). But it selects different alternatives whenever they provide more business performance or less risk exposure per dollar spent.

Management needs to be honest with employees throughout the organization about the added skills and effort required by asset management. Management must also reinforce the message that the end (better performance on all fronts) more than justifies the added cost and organizational disruption.

19.5.2 Asset Management Is Not Necessarily Trying to Reduce Spending

Asset management methods can and have been used by utilities striving to reduce costs (Brown and Marshall, 2000). In such cases, the utility could not afford to do everything that it had done in the past and looked to asset management to prioritize spending and "minimize the pain" of budget cutbacks. As a result, asset management has gained a reputation in some parts of the industry as a cost-cutting measure.

There is no doubt that asset management can help a financially challenged company determine how to spend a limited budget in the best possible manner. There are cases where early success in reducing spending levels without unacceptable impacts on KPIs leads to a focus on further cost cutting. ("Since that did not hurt so much let us try again until it does.") But asset management can just as easily be used by utilities looking to increase spending (it will spend the additional budget as effectively, from a business perspective, as possible) or just to more effectively allocate existing levels of spending.

Overall, asset management often leads to better recognition of the consequences and benefits of *not* spending than the traditional approach provides. It can also lead to spending increases in the area of risk mitigation and for projects that do nothing themselves but create synergy among other projects and programs. In addition, there are supply-side effects in the long term. And finally, because the cost of improving performance is reduced, the laws of supply and demand dictate that more spending will sometimes be desired, especially in areas that are shown to be highly cost-effective: when "bang for the buck" *ratio* is improved, as often happens with asset management, the "bangs" cost less and as a result utilities often decide to buy more and improve customer and system performance even further.

19.5.3 Changes in Perspective and Culture Specific to Substation Personnel

An asset management approach will bring about a further change in perspective and organizational culture for those engineers, managers, and operations personnel most closely associated with substations: a broadening of their role within the organization to one of enterprise-level monitoring and information management. Traditionally, substations have been regarded as important and allocated some degree of priority in attention within a utility purely because of their role in power delivery: they are key "way stations" in the transportation and control of power transmission and distribution. Certain this role will be no less important in the future.

However, as Section 19.6 explores, a combination of asset management priorities and modern technological advances is almost certain to lead to utilities using substations as "data hubs" for corporate enterprise information technology (EIT) systems. In the future, substations will be regarded not just as key hubs in the power delivery chain, but as the cornerstones of a distributed data gathering system the utility maintains to monitor its assets and performance and share throughout company-wide data warehousing systems where it can be used in a wide variety of applications, many of which will be outside the substation, or even power delivery, part of the corporation. This second, IT, function for substations will become, in many ways and to many people in other parts of the company, as important as the original function of power delivery. Substation personnel will have to work in an environment where this second function of the area of their responsibility is as important, in many regards and to many people, as the power delivery role with which they have long been familiar.

19.6 Substation Asset Management

By its nature, asset management spans all elements of a utility since it is interested in overall business performance. However, the concepts of asset management can, to a certain extent, be applied effectively to substations in an approach commonly called "substation asset management." Basically, this approach will look at all spending decisions, look at their impact on substation-related KPIs, and look at areas of risk that are specifically related to substation performance. Substation asset management will typically mean more flexible decision making, much more data collection and usage, and much more rigor in the decision-making process. Overall, most utility personnel associated with substations will like the results of the change.

This section looks at what utilities can expect with regard to the substations when they use an asset management approach to strategy and decision making. Example results are from a United States utility and represent a typical situation. However, many utilities differ from the norm in some way: service area geography, climate, customer demographics, regulatory environment, employee base and internal skill sets, system design and age profile, power pool requirements, or business goals. Therefore, the example results are difficult to generalize since specific results for specific utilities can vary widely.

19.6.1 Utilization and Life Cycle Management

Asset management generally spends slightly more on the care of major power equipment and switching facilities than was traditional in the industry but then uses that equipment (loads it, operates it) more heavily than traditional practice. Asset management process will try to *finesse* (balance one against the other while trying to create and leverage synergy) O&M, remaining lifetime, and utilization rates to maximize the *business* case for ownership of the equipment. There are certain obvious trade-offs. Cut inspection and maintenance too much and

- Performance will suffer due to breakdowns.
- Equipment lifetime will decrease due to outright failures.
- Operation and cost will become less predictable.

Either way the value derived from the assets will drop. Conversely, if the utility spends too much on inspection and maintenance, the payback is diminished since there are other areas where that money can be spent to buy more KPI performance.

Similarly, high utilization policies throughout the utility will "use equipment fully" and therefore increase the business value obtained from it today, but these policies could seriously erode remaining lifetime. On the other hand, low loading or usage will result in longer equipment lifetime but will mean that the equipment provides less business value every day it operates. Finally, business needs of the corporation will dictate how these trade-offs are viewed. In every case, asset management (at least if done well) will attempt to optimize these trade-offs from a slightly broader perspective than the traditional paradigm. It will look at them and their secondary and unintended consequences, in aggregate, as they affect overall business performance.*

Usually, a multi-KPI perspective will look at these trade-offs and recommend slightly more spending on preventive substation inspection and maintenance than the utility did traditionally. However, the recommended expenditure may not be for traditional types of "O&M." For example, instead of increasing the frequency of maintenance, an asset management process may recommend the capital addition of online condition monitoring equipment for critical pieces of equipment.

The reason asset management often leads to increased spending on substations is that substations influence a broad range of KPIs. First, substations involve a good deal of very expensive equipment. For this reason alone, there is a business case for taking good care of substations. But beyond this, substations effect most of the KPIs listed in Table 19.1, including all of the reliability-related metrics and the "big bad outage" measure. The only KPI listed in Table 19.1 in which substations are not as involved

---

* For example, high loading limits not only will reduce equipment lifetimes but also they are almost certain to drive up annual unexpected maintenance needs in certain equipment categories, too. Thus, high loading policies and increased maintenance costs go hand in hand. On the other hand, higher loading and stress on equipment increases the marginal value seen from scheduled inspection and maintenance. Good asset management planning processes take such interactions into account.

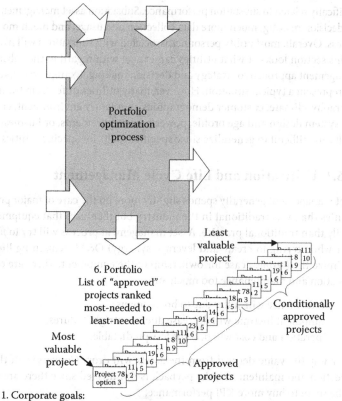

6. Portfolio
List of "approved" projects ranked most-needed to least-needed

Least valuable project

Conditionally approved projects

Most valuable project

Approved projects

1. Corporate goals:
KPIs and their targets

2. Future scenario
Assumptions about future: customer and load forecast, weather, changes in regulations or laws, conditions, etc. might be probabilistic in nature to drive a risk-based analysis

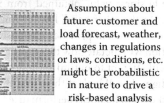

3. Possible projects,
Programs and policy commitments submitted for spending approval (many more project option sets than shown here, perhaps hundreds)

5. Analytic aids
Graphics, statistical diagnostics and data tables to help analyze options and results

4. Constraints and limits
Practical as well as desired limits, requirements, or characteristics for project execution and results

as some other part of the T&D system is public safety, which is mostly concerned with distribution. As a result, substations and their equipment tend to receive slightly higher priority from a multi-KPI approach than under the traditional utility paradigm, where spending priorities were typically based mostly, if not exclusively, on equipment-care considerations alone.

While asset management tends to take good care of the substation equipment, this business-driven approach will also want to "get its money's worth" by loading that equipment to high levels. A pure return-on investment business analysis of loss-of-life as a function of loading often will conclude that the "optimum" utility use of a transformer is to load it to where its expected lifetime is only 25 years (Willis, 2004). Results vary from utility to utility and case to case, but usually the optimum loading sought by asset management will be higher than what the utility had been using traditionally. It is worth nothing that some utilities dramatically increased loading levels during financially troubled times in the late 1990s. Asset management may actually reduce these "recommended" loading levels.

In addition to normal loading levels, asset management will tend to increase acceptable levels of loading during emergency conditions, a result of risk balancing (Figure 19.12). A business-case study of probabilities, costs, and expected outcomes often leads to a decision not to spend money in many cases where a traditional utility would have bought capacity or facilities to mitigate exceptionally high loadings during a major equipment outage. Asset management will often reduce this type of spending with the following thinking, "the additional capacity may never be needed, I will not purchase it and accept the fact that I will have to load equipment to extraordinarily high levels if certain rare situations happen to occur."

19.6.2 Condition-Based Maintenance

The use of asset management over a period of several years invariably moves a utility closer to some condition-based, reliability-centered, on a performance-oriented maintenance approach (van Schaik et al., 2001). Ultimately, management to a sustainable point of system condition and performance will be achieved, either explicitly or implicitly. Regardless, the evolution of the organization's focus leads to more emphasis on inspection and use of inspection data. "Bang for the buck" is enhanced whenever the utility can direct maintenance activities where they are truly needed and avoid or defer maintenance activities on equipment for which it is not needed (Ostergaard and Jensen, 2001). For example, suppose Big State Electric owns 1600 medium voltage breakers. Among them there may be 150 that are much more in need of maintenance than average, because they are in poor condition or because they are in especially critical positions, or both. Similarly, there may be 150 that are less in need of attention than average. Asset management will spend much more on inspection and maintenance for the former group of breakers than the latter.

Big State Electric may need to maintain and service far more than just 150 breakers in any year. But regardless of the number it decides it should service, it will probably not see much value in doing service on the 150 that *do not* need attention. Therefore, it should find a way to exclude them from scheduled maintenance activities. Conversely, it should make certain it does include those 150 most in need of attention. Taking both these steps will significantly improve the cost-effectiveness of its maintenance program.

FIGURE 19.12 Asset management's portfolio optimization process evaluates all the options for all the projects in order to select the portfolio (set of project options) that achieves all performance targets and maximizes objectives. It has two strategic inputs: the corporations goals expressed as KPI targets and values (1) and a scenario description giving trends and conditions and assumptions about the future (2). "Tactical" inputs include (3) the options for all projects (from Figure 19.5) and a set of constraints (4) which set practical limits ("You can do project A or B but not both.") and preferences. Outputs include graphs, diagnostics, and statistics to aid planners in understanding their options, the portfolio, and why items were selected, and the overall performance result (5), and the portfolio (6), the set of project options recommended for implementation (many more options were not selected and are not shown). Conditionally approved project set is explained in the text.

Inspection and the subsequent good use of inspection data are the keys to effective condition-based maintenance (Butera, 2000). In order to focus maintenance well, the utility needs to know about the condition of those 1600 breakers. There is more than just inspection. It is *the use of inspection results for tracking and targeting that is needed*. Therefore, inspection programs (institutionalized processes that include inspection, retention of inspection records, and use of that archived data for condition assessment) are among the highest priorities for substation asset management.

19.6.3 "Big Bad Outages" and Setting Priorities for Substations

Several large investor–owned utilities in the United States use a KPI factor in their planning that is a measure of the likelihood or severity of expected "big bad outages." These are typically defined as unexpected outages that are less severe than a regional blackout but more widespread than typical outages and is likely to make headlines in the local newspapers. One of these utilities even refers to its measure as BBOM—"big bad outage measure." Specifically, it defines a "big bad outage" as any nonstorm or nonblackout-related event that takes 30,000 customers or more out of service for 4h or longer. Its BBOM is the expected customer-minutes per year due to this class of outage and is computed for various plans or policies using probabilistic reliability assessment methods (Brown, 2001).

Most utilities consider a widespread interruption to have a higher cost to the utility per affected person than a more limited interruption. Thus, the avoided cost of a minute of BBO customer interruption is deemed higher than that of less widespread outages. Additionally, big bad outages garner media and public attention, which is bad for the utility's public image and raises the cost of public, municipal, and regulatory interaction. More limited outages, even if far more frequent, do not "get in the newspapers." The use of a BBOM KPI and target focuses decision making on reducing these types of outages, and not just on driving down broad measures of reliability performance such as SAIDI and SAIFI.

The use of a BBOM pushes spending priorities toward substations, particularly higher voltage and large capacity substations. Few, if any, elements of the distribution system, and many elements of smaller electrical substations, have an impact on a measure requiring a minimum of 30,000 customers or on any similar measure. Therefore, use of a BBOM will focus a certain amount of spending large substations and transmission facilities and their control systems, as shown in Table 19.4.

In many cases, it costs very little for a utility to add a big bad outage focus to its reliability program. In the case illustrated here, a hypothetical case run by the authors using data from the utility in Table 19.1 (which did use BBOM) with and without the measure, the overall difference in spending required to address BBOM is only $1.6 million. The reason is that BBOM's major effect is to shift $7 million (2.3%) in spending away from distribution and service level spending, where it worked down SAIFI and SAIFI, to the high-voltage and very-high-voltage levels, where it helps work down SAIDI, SAIFI, and BBOM. Much of this money goes for enhanced substation investment and maintenance. The additional $1.6 million in spending is required because the shifted spending, focused on BBO, is slightly less effective at reducing SAIDI and SAIFI, so a bit more must be spent to cover that small shortfall.

TABLE 19.4 Spending Allocation across the System—Millions of Dollars

System Level	Attain All Goals but BBOM	Attain All Goals Including BBOM
EHV (345–230 kV)	$63	$64.3
HV (161–69 kV)	$81.3	$88.7
Distribution (>69 kV)	$163.5	$160.9
Service transformers and secondary	$65.3	$61.1
New customer connections	$75	$74.8
Total spending	$448.2	$449.8

19.6.4 Advanced Substation Technology

One characteristic of many asset management portfolio plans is that a good deal of capital is devoted to improved control and monitoring systems rather than to the purchase of raw equipment capacity. Current trends toward wider use of automation, particularly in substations, are not driven predominately by an industry shift to an asset management perspective. A good deal of advanced technology monitoring, control, and information systems would be adopted regardless of the decision-making methods being used by utilities. Among other improvements, online condition monitoring can reduce failure rates and level of damage sustained from failures of major components like transformers (McCullough, 2005; Timperley, 2005).

But beyond those benefits, a decision-making framework seeking broad, multi-KPI impacts can often derive great value from certain types of substation automation and systems. First and foremost, condition monitoring systems can help lower breakdown and failure rates and extend equipment lifetime. In conjunction with automated switching, condition monitoring can also shorten customer service interruptions caused by equipment failure or storms. Thus, substation automation improves performance in a number of KPI areas.

Additional value can be derived if the substation automation system is linked into corporate-wide enterprise systems, allowing decision makers throughout the company to use data gathered in the substations during the asset planning process. For these reasons, many utilities making the transition to an asset management approach see themselves also moving more quickly than they had expected into substation automation.

19.6.5 Different *Philosophy* with Regard to Equipment Stewardship

19.6.5.1 Change in Perspective and Culture

While asset management's approach to valuing and funding maintenance will be welcome, many long-time utility personnel are often uncomfortable with the bigger context of the changes required to implement it, particularly in its more comprehensive forms. Traditionally, many utilities and many utility personnel viewed their company and themselves as "equipment stewards." As equipment stewards, their job was to take "good care" of system equipment by keeping failures and breakdowns to a minimum. From their perspective, failures of equipment are really failures of the utility and its personnel. In the words of one utility Director of Substation Maintenance, "If all the substation equipment that was here on the day I first started thirty five years ago is not here and in good condition on the day I retire, then in some sense I have failed to do my job."

Asset management takes a very different perspective. Equipment is acquired for business reasons. It is a business tool, to be utilized well within the context of overall business performance, and taken care of only in the sense that that care improves the business's prospects of achieving its goals. In many cases, those business needs as interpreted and applied within the asset management approach will dictate heavier loadings and reduced service budgets as compared to traditional approaches, to the extent that equipment lifetime will be shorter than in the traditional standards-driven approach. However, value received from the equipment will be higher, and if the asset management is being done correctly, more than compensate for that shorter lifetime. Particularly, whenever a sustainable-point approach is being used with respect to managing any and all aging infrastructure issues, the asset management approach is explicitly managing equipment lifetime to quite literally "use it up" at a rate that is best for the business.

This perspective, which ultimately treats equipment as almost "disposable" in some manner, represents a major departure from the traditional equipment stewardship culture that still prevails in some utility Operations Departments. The move to an asset management perspective will make many equipment experts and managers uncomfortable and resistant to the change. Utility management that addresses these personnel concerns through communication and training will find the transition both smoother and less stressful for all concerned.

19.6.5.2 Organization of All Capital Investment "Causes" into One Process

Asset management, done well, also represents a major culture change for utility planners and decision makers. The traditional standards-driven utility paradigm led to the type of organization shown in Figure 19.13. Here, separate groups of planners and engineers focus on the three major factors that lead to capital spending: growth, reliability, and aging issues. Each year, each group concentrated on its particular concern, identifying its problem areas with respect to that and developing its list of projects to correct any deficiencies found. These requested lists of projects were then submitted separately to management and resolved into one approved list through application of a type of multiattribute evaluation.

A major benefit that the multiattribute, portfolio-oriented, Pareto-optimization asset management approach has in comparison to traditional utility planning is that it can evaluate projects on the basis of a wide range of considerations, and in the presence of several different goals and needs that ultimately must be traded off, one to the other. Reorganization of the utility's capital investment planning into one group, rather than three, permits a better utilization of the planning advantage asset management provides. In Figure 19.14, the three planning groups from Figure 19.13 are combined. All candidate projects are evaluated for their impact in all three areas: on if and how much they help meet customer and demand growth needs, with respect to if and how they make improvements in customer reliability of service, and for if and how they affect aging, condition deterioration, and issues affecting those factors. A single list of requested projects, prioritized against the companies' multiattribute targets and needs (e.g., something like Table 19.1), is developed (Figure 19.15).

This approach can provide noticeably better results in terms of overall budget performance and total planning labor required by the utility. The primary reason is that very few projects ever affect only one of the three planning focal areas. A project to add a new low side bus and two new feeders out of a substation might be developed and designed to address a need to run those new circuits to a part of the system where there is new customer growth. But those feeders will also provide new tie points for existing feeders and thus improve reliability in another part of the system. Similarly, most projects that address equipment condition and deterioration also improve reliability, at least in small measures. And reliability augmentation projects often increase system capability. Planners focusing only on growth might see increased capability as something they can use to serve a growing load.

These overlaps and the issues they create are to a certain extent thrashed out in the traditional (Figure 19.13) structure through communication and give and take by the three groups of planners.

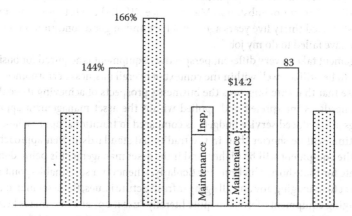

FIGURE 19.13 Comparison of distribution substation transformer ownership and operating policies at one utility before and after its use of multi-KPI asset management. Unshaded blocks show practice fairly typical of the U.S. industry in the mid-1980s and early 1990s. Afterward (shaded), much more was expected of the equipment but more was spent on preventive maintenance and service. Despite the more intense care given to the equipment, the expected service lifetime of the units decreased to an optimum from a business standpoint. Breakdown of inspection and maintenance bars shows the amount of spending devoted to each. See text for details.

Substation Asset Management

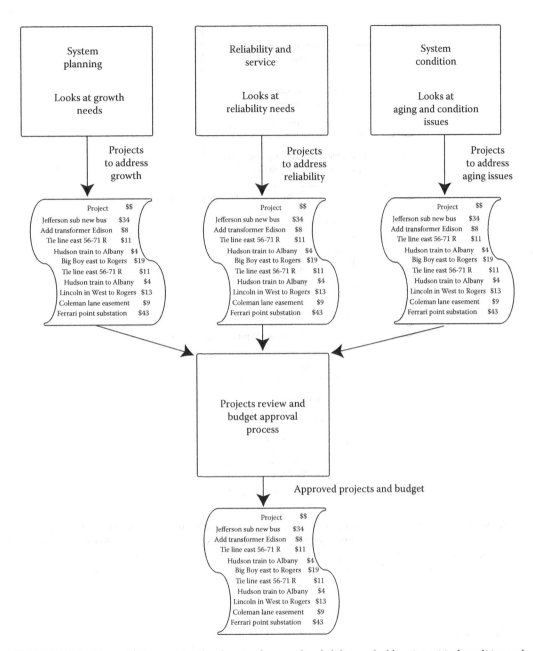

FIGURE 19.14 Many utilities organize the planning for growth, reliability, and addressing critical condition and aging issues, into separate groups, in order to concentrate skills specific to each area in a group that focus on that need. Competing lists of projects are submitted through a process of comparison, resolution, and approval, leading to the utility's overall budget.

The single-planning group approach (Figure 19.14) does this from the beginning of the planning process and leads to a thought process that not only views all projects as potentially leading to improvements in all three areas but also tries to achieve synergies so as many as possible do, each to the greatest extent possible. Asset management works best and provides its benefits when all projects that are being considered are evaluated on a consistent basis and considered as one group in one decision-making process.

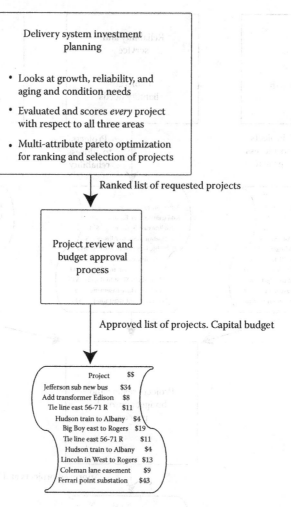

FIGURE 19.15 Comprehensive asset management approach aggregates these separate groups into a group and process that evaluates all projects from the perspective of all three needs and for their contribution to all KPIs and corporate goals. Multiattribute Pareto optimization is used to determine the best set of projects to achieve various levels of overall corporate performance. Management picks the achievement of various KPI and corporate goals it wants, which defines the selection from the list and finalizes the ultimate budget.

19.7 Summary

Asset management is a data-driven process that strives to link all asset-related spending decisions to their impact on corporate objectives. The result is a more way of making decisions that is more rigorous and cuts across traditional functions and budgets. For substations, asset management means a move away from standard-based decisions based on equipment-level considerations and toward multi-KPI-based decisions based on system-level considerations. When a utility pursues substation asset management, the following are likely to occur:

- Increased focus on data collection
- Increased focus on risk management
- Increased ability to trade-off capital spending with maintenance spending
- Increased focus on developing options for projects

- Increased ability to quantify the impact of projects on KPIs
- Ability to prioritize projects and project options
- Increased loading of equipment
- Less focus on periodic maintenance and more focus on condition-based and reliability-centered maintenance
- Increased usage of data collected from substation automation systems

Since asset management concerns itself with overall corporate objectives, asset management activities relating to substations must ultimately be considered within a broader context. Regardless, many of the concepts and techniques of asset management can be directly applied to substations so that performance can be better managed, spending can be more efficient, and risks can be better managed.

References

Brown, R. E., *Electric Power Distribution Reliability*, Marcel Dekker, New York, 2001.

Brown, R. E. and J. J. Burke, Managing the risk of performance based rates, *IEEE Transactions on Power Systems*, 15, 2, May 2000, 893–898.

Brown, R. E. and B. G. Humphrey, Asset management for transmission and distribution, *IEEE Power and Energy Magazine*, 3, 3, May/June 2005, 39–45.

Brown, R. E. and M. M. Marshall, Budget constrained planning to optimize power system reliability, *IEEE Transactions on Power Systems*, 15, 2, May 2000, 887–892.

Brown, R. E. and J. H. Spare, Asset management, risk, and distribution system planning, *IEEE Power Systems Conference and Exhibition*, New York, October 2004, pp. 1681–1686.

Brown, R. E. and J. H. Spare, Asset management and financial risk, *DistribuTECH Conference and Exhibition*, San Diego, CA, January 2005, pp. 311–316.

Butera, R., Asset management for the distribution pole plant—Closing the performance gap between traditional maintenance and asset management, *IEEE Power Engineering Society Summer Meeting*, Seattle, WA, July 2000, pp. 561–565.

Center for Petroleum Asset Risk Management, Section 3: A brief background of real asst risk management in petroleum E&P, 6 pp., 1999. www.cpge.utexas.edu/cparm/cparm_prosp_v4_03.pdf

Humphrey, B., Asset management, in theory and practice, *Platts Energy Business and Technology*, 3, March 2003, 50–53, VOL III.

Markowitz, H., Portfolio selection, *Journal of Finance*, 7, 1, 1952, 77–91. (This paper covered work for which Harry Markowitz was eventually awarded the Nobel Price in Economics.)

McCullough, J. J., GSU transformer health monitors combining EMI detection with acoustic emissions measurement, *Proceedings of the 38th Annual Frontiers of Power Conference*, Oklahoma State University, Stillwater, OK, October 24–25, 2005, pp. 87–92.

Morton, K., Asset management in the electricity supply industry, *Power Engineering Journal*, 13, 5, October 1999, 233–240.

Ostergaard, J. and A. N. Jensen, Can we delay the replacement of this component? An asset management approach to the question, *CIRED 16th International Conference and Exhibition*, Paris, June 2001, pp. 405–408.

Pareto, V., *Manuale di economia politica,* Centre d'Etudes Interdisciplinaires, Université de Lausanne, 1906.

Philipson, L. and H. L. Willis, *Understanding Electric Utilities and De-Regulation*, 2nd Edn., CRC Taylor & Francis, Boca Raton, FL, 2005.

Seevers, O. C., *Management of Transmission and Distribution Systems*, Fairmont Press, Lilburn, GA, 1995.

Timperley, J. E., EMI diagnostics detects transformer defects, *Proceedings of the 2005 Doble Conference*, Boston, MA, April 2005, pp. 317–322.

Van Schaik, N. et al., Condition based maintenance on MV cable circuits as part of asset management; philosophy, diagnostic methods, experiences, results and the future, *CIRED 16th International Conference and Exhibition*, Paris, June 2001, pp. 873–878.

Willis, H. L., *Power Distribution Planning Reference Book*, 2nd Edn., Marcel Dekker, New York, 2004

Willis, H. L. and R. R. Schrieber, *Aging Power Delivery Infrastructures*, 2nd Edn., Marcel Dekker, New York, 2012 (fourth quarter, 2012).

20
Station Commissioning and Project Closeout

Jim Burke (retired)
Baltimore Gas & Electric Company

Rick Clarke
Baltimore Gas & Electric Company

20.1 Commissioning 20-1
  Testing • Coordination • Site Issues • Notification
20.2 Project Closeout 20-4
  Final Walk-Through or Inspection • Punch List • "As-Built" Information • Invoices • Closure of Outstanding Permits, Sureties, or Bonds • Archive Records • Develop Unit Costs • Closeout Project Accounting • Notify Stakeholders • Development of Lessons Learned

Once the construction is complete, its time to determine if all the systems work as specified, connect to the system, straighten out any site issues, and close out the project and accounting. This chapter will take us through the various issues that should be addressed in order to finally complete the project.

20.1 Commissioning

Final tests of the completed substation work in a partially energized environment are required in order to determine the acceptability and conformance to customer requirements under conditions as close as possible to normal operating conditions. Coordination should be achieved with other entities in order to successfully connect to the electrical system and all outstanding site issues have to be finalized in order to provide the operating staff with a functional product. Additionally, there are a number of public and utility organizations that should be made aware that the facility is ready for operation.

20.1.1 Testing

- Power supplies
- Relays
- Protection schemes
- Communications
- Grounding
- Fire protection
- Major equipment
- Security systems

Although testing had been performed on individual items during the construction phase, functional testing should be performed on all subsystems in order to ensure proper function. Verify that all the factory acceptance and site acceptance tests have been satisfactory and then proceed to the process of checking the functionality of the entire package. It is possible that vendors and consultants will have to be involved at this stage in order that warrantees and specifications might be accommodated.

All AC or DC systems providing power for subsystems or major equipment should be checked. This includes any batteries, transformers, generators, switches, breakers, panels, chargers, hydrogen sensors, and associated fans or ventilation. Off-site power may also be involved if the substation is required for black start of the system or if the station is a new generation switchyard.

Verification is required that all relay devices, instrument transformers, transducers, meters, and IEDs, located at both the major equipment and control house, provide the intended control and monitoring functions as well as provide the proper inputs to the protection, automation, and communication schemes. As stated in an earlier chapter, the key to a commissioning test plan is to make sure that every input and output that are mapped in the system is tested and verified. The systems that operate, monitor, and protect the substation equipment should be verified to function according to the functional diagram. The interface with SCADA and other communication systems should also be tested as well as the operator interface. In addition to the proper communications between components with the substation, the communications with the energy control center and other utility elements should also be tested. This may also involve communications with other utilities or interconnection entities such as a non-utility owned power plant that seeks connection to the electric system. It may also be necessary to include systems to monitor or advise customers and suppliers.

It is necessary to determine the integrity of the station ground grid prior to connection to the system. The grid resistance must be measured prior to connection to the rest of the system in order to verify that the grid will provide the proper operation of the electrical equipment and the protective relays as well as the personnel safety margins that were intended.

The various fire or smoke detectors located at the equipment or within buildings or other enclosures should be functionally tested. Their interface to control, communication, and alarm systems should be checked along with any pumps, valves, and spray systems. Alarm systems should be found to be functional along with the interface to the necessary utility entities and the fire department. If direct alarming of the fire department is not provided, the proper notification scheme needs to be verified. Since this scheme has probably been negotiated with the fire marshal, several entities may be involved.

Major items of electrical equipment with their own controls, and monitoring should be successfully operated. Power transformers need to be properly charged prior to applying load. Breaker and switch operation needs to be verified both locally and remotely. It is not uncommon to involve manufacturer, vendor, or consultant personnel in this verification.

The integrity of any walls, fences, locks, or any other personnel barriers should be checked along with the function of any intrusion detection systems such as motion sensors, video cameras, and door alarms. This is also the time to verify the functionality of the notification process for corporate security and the police department.

20.1.2 Coordination

- Generation
- Interconnection
- Distribution
- Public sector

In the case of power plant switchyards, it is necessary to coordinate testing of the interface between the substation and the plant as well as the timetable for final energization. This may involve off-site power if black start of the plant is involved. If there is separate ownership of these facilities, the coordination will also involve legal and contractual issues. Bulk power substations and power plant switchyards must not only integrate into local utility systems, but must also properly interface with regional interconnection entities.

Communications systems for monitoring and control should be tested not only for proper function, but also verified that they provide the features necessary to meet the contractual obligations of the interconnection. Area supply substations must properly interface with energy control center, but also with distribution automation schemes. Also, the availability of incoming and outgoing feeders must be coordinated so as to meet the agreed service dates.

The police and fire departments will be involved in the testing of any security and fire detection systems. Public works departments will be involved in issues associated with water mains, drains, and traffic. It may also be necessary to advise the general public of activity that may impact the neighborhood.

20.1.3 Site Issues

- Permits
- Roads
- Aesthetics
- Landscaping
- Drainage
- Storm-water management

The preponderance of permits are usually associated with the construction of the facility, but there may also be permits necessary for occupancy or operation. Now is the time to make sure that the requirements have been met for these permits. Also, the procurement of any special use permits should be verified, for example, Federal Aviation Administration clearance for any high structures or communication towers. Paving for driveways, roads, turnarounds, and any other vehicle access needs to be completed. This may also include deceleration or merge lanes associated with public roads along with any required curbing. Any stone covered access or parking area should be final dressed. Final touches need to be applied and final inspection undertaken of any features of the installation that serve special aesthetic purposes in order to obtain community acceptance. Decorative walls, special fences, fence inserts, custom coloring, or any other treatments need to be finalized. The final landscaping arrangement needs to be checked against the approved landscaping plan to ensure compliance. Due to drought or seasonal requirements, it may be prudent to delay some plantings until conditions are optimal. Should this be the case, it may be necessary to advise the appropriate public agencies that plantings will be delayed. In any case, care should be taken to comply with any warranty requirements.

Cleanup should be conducted on all drainage systems, including removal of all silt fences and installation of stone cover at the outfall of trenches. Should any grading have been necessary for drainage, the final stone layer must be installed or turf repaired. Check valves need to be tested and protective facilities such as fences around storm-water management ponds need to be verified. The function of any oil–water separator systems also needs to be tested. Should direct connection to public storm drains be involved, these connections should be checked for proper function.

20.1.4 Notification

Once the facility has been made available for service, various elements within the utility organization need to be notified. Besides the obvious notification of the operating departments, planning organizations, corporate security, general services, and legal staff need to be advised. In addition, the accounting group needs to ensure that the facility is now included in the rate base.

Public safety organizations, such as the police and fire departments, need to be advised of the operation of new infrastructure. In addition, legal notification of the local political district and several state agencies may be required along with federal entities, such as the Federal Aviation Administration, Corp of Engineers, etc. Regional interconnection entities may need notification along with special customers, for example, a power generator.

20.2 Project Closeout

20.2.1 Final Walk-Through or Inspection

20.2.1.1 Owners or Customers

A final walk-through or inspection of the completed substation project is undertaken as a beneficial measure that allows the substation owner or customer the opportunity to view the finished product first hand. On internal utility substation projects, the typical owner or customer of the project is the area of the company that possesses both the authority and ability to operate the station. On most internal utility substation projects, the system operating area is broadly familiar with the project intent and deliverables. Yet the final walk-through or inspection can provide the opportunity for these operating personnel to unquestionably verify their full understanding of the project objective or perhaps it can provide the opportunity for a learning experience when new technology was implemented to achieve a familiar deliverable. On substation projects that are pursued by the utility to meet an explicit external customer need and where final ownership of the substation falls outside the boundaries of the utility, the final walk-through or inspection can take on a broader customer satisfaction dynamic. In these cases, the owner or customer may not be thoroughly familiar with the substation business; therefore, the final walk-through serves as a key opportunity for the owner or customer to begin their education and training process.

20.2.1.2 Contractors

It is quite necessary to include in the final walk through and inspection of all the contactor disciplines that were involved in the project. Since these were the entities that directly executed the project deliverables, they are partly accountable with respect to ensuring that the project deliverables were provided as engineered. Due to the contractor's role and responsibility on the project, they would be a primary contributor during any question and answer session with the owner or customer that may ensue. The walk-through activity is also beneficial to the contractors from the standpoint that it can provide internal learning and training opportunities for their additional staff that may not have been directly involved in the project.

20.2.1.3 Vendors

Equipment and material vendors are also necessary participants to include in the walk-through and inspection activity. Their on-site participation allows them to see their various products in service firsthand. This visual observation opportunity provides several unique benefits for the equipment and material vendors. It offers the ability to verify that what they are providing indeed meets their customer's expectation and it perhaps provides for the forum to learn of improvement opportunities from the contractors or stakeholders that are also on site. These learning opportunities, if implemented, cannot only provide better products for that same utility on future projects, but perhaps the improved products can also be beneficial to the vendor's additional customer base. Finally, along with the contractors, the vendors can also serve as key contributors during any question and answer session where they can provide immediate and comprehensive feedback on their products.

20.2.2 Punch List

20.2.2.1 Development and Ownership Establishment of Specific Items

A key purpose of the need to conduct a final walk-through or inspection of the completed project, with all pertinent members of the project team, is to develop a punch list of project items that require full closure. The punch list is primarily a compilation of construction related issues that, although typically have no bearing on the ability to energize the deliverable of the substation project, require additional attention in order to bring all elements of the project to a thoroughly safe and acceptable closure. Typically, the project manager, responsible engineer, or the construction manager leads the punch list development exercise. The punch list items can involve all engineering and construction disciplines and can range from nominal issues to significant project elements that must be addressed immediately in order to eliminate their possibility to negatively impact the project deliverable at a future date. Punch lists routinely include such items as site erosion issues, insignificant equipment problems, minor material corrosion issues, as well as various unsafe conditions that require immediate and full closure. However, the final punch list can be comprised of any and all project issues that are either collectively agreed upon, by all involved in the punch list development exercise, as being worthy of inclusion or issues seemingly insignificant in nature that are deemed worthy of inclusion by the punch list development leader.

Full completion of the punch list is customary prior to the primary project stake holder, project sponsor, or customer accepting formal ownership of the completed project. Prior to accepting ownership, all elements of the project must be completed in their entirety in a fully functional, operationally sound, and quality manner. The punch list and the corresponding ability to verify the completion of its contents are the necessary control mechanisms that are put into place in order to protect the project stake holder, project sponsor, or customer from accepting ownership of an incomplete project. An additional control mechanism to ensure the timely completion of the punch list items is the practice of identifying firm ownership of each punch list item. In leading the punch list development exercise, the project manager, responsible engineer, or the construction manager has the responsibility of soliciting and identifying a specific owner who is singularly accountable for ensuring the acceptable completion of a particular punch list item. This ownership establishment practice is commonplace and allows for the ability to expand the responsibility of the completion effort throughout the team membership, thus increasing the success rate of punch list item completion as well as improving the timeliness of completion.

20.2.2.2 Ensure That Each Item Is Properly Completed

As mentioned, the effort to identify and verify firm ownership of each punch list item is an important practice that seeks to establish accountability for ensuring specific item completion. This ownership identification approach allows for the establishment of working relationships between the punch list item owner and the project manager, responsible engineer, or the construction manager. This approach streamlines the completion accountability verification process in that direct lines of communications can be established and clear performance expectations can be set. The need for a follow-up walk-through or inspection of the project site, to ensure the completion of all punch list items, is a function of the complexity of the overall punch list in addition to being contingent upon the successful performance of the communication links. Although the need to revisit the site to verify the completion effort firsthand will vary on a project by project basis, overall the ownership identification approach dramatically increases the completion performance while significantly improving the ability to indirectly ensure completion fulfillment. All of which is necessary prior to the primary project stake holder, project sponsor, or customer accepting formal ownership of the completed project.

20.2.3 "As-Built" Information

20.2.3.1 Construction Drawings

Although every effort is typically put forth to produce perfectly engineered construction drawings, site conditions, situational unknowns, and incorrect original record documentation issues are routinely encountered that require the construction process to deviate from what was specified on the guiding construction documents. These deviations range from slight in nature, which require no approval to implement, to recommended changes whereby implementation is only pursued upon the approval of the project manager or responsible engineer or both. Construction document changes can be encountered throughout the construction life cycle of a project within all involved engineering disciplines. Regardless of the magnitude of the change implemented, it is necessary to capture and record the change via what is known as the as-built process. The as-built process is usually a manual effort of documenting the construction changes and deviations from the original plan onto a hard copy of the construction drawings. The as-built process is typically initiated and completed by the construction forces involved in the project. The manual effort consists of using a red or other conspicuously colored writing instrument to record the acceptable changes onto the original construction drawings.

Upon completion of the as-built drawings by the construction forces, the as-built package is forwarded to the responsible engineering discipline for their use to permanently transfer the as-built information onto the original construction documents. This information transfer practice is necessary for appropriate legacy record keeping purposes. This will help to ensure that any future use of these record documents will accurately reflect the field conditions expected to be encountered. The transfer of the as-built information onto the permanent record drawings should be completed as soon as possible following the completion of the project. This timely completion effort will help to eliminate future drawing confusion issues that may perhaps surface if the lingering as-built information were to be inadvertently overlooked or mistakenly discarded. It is customary for the original engineering personnel who created the construction documents to be involved in the as-built transfer process. This back-end involvement offers the engineering personnel an opportunity to perhaps learn from their own mistakes and misjudgments or it can serve as a reminder that a seemingly perfectly engineered product can sometimes encounter unforgiving field conditions that warrant an acceptable deviation from the desired product.

20.2.3.2 Equipment Manuals and Operations Instructions

In addition to the as-built construction documentation, it is necessary to ensure that other forms of important project information are disseminated to the appropriate project stakeholders. Both equipment manuals supplied by the equipment vendor as well as any substation operational instructions fall into this category of vital information. This information may not be utilized during the construction or commissioning phase of the project, but rather it becomes a necessary tool for future equipment maintenance needs or during times when the station requires an operational change or modification that deviates from the station's normal mode of operation.

The equipment manuals are typically provided by the equipment manufacturers and, upon their dissemination, are usually stored in a central office environment where they are readily accessible to the appropriate equipment maintenance personnel. These manuals become very useful in that they provide the necessary technical guidance during trouble shooting events as well as during future maintenance cycles. Advancing and evolving equipment technology along with routine workplace attrition issues can present challenges with trying to maintain a fully educated and proficient equipment staff. The equipment manuals provide for a safeguard to ensure that all recommended maintenance practices are thoroughly followed and at the same time they offer a comprehensive education on the equipment being serviced. The substation's operational instructions, typically created during the engineering phase by appropriate members of the engineering team, are routinely stored on site within the energized substation. On location at the substation, the instructions are readily accessible by operational stakeholders to provide the oversight necessary to guide one through a safe operational change or modification exercise.

20.2.4 Invoices

20.2.4.1 Resolve Outstanding Issues or Conflicts

As the substation project enters the closeout phase, the processing and payment of service and product invoices typically represents the primary final project activity that may require firm oversight. Invoice issues or conflicts can involve any of the consulting or contracting service entities involved in the project as well as involve any of the vendors involved that provided equipment and material.

Effective project management techniques, if implemented throughout the life cycle of the project, generally result in a minimal number of adverse invoice issues or conflicts to resolve. Yet, there are times when a detailed invoice analysis is required to ensure that a service or product invoice is accurately reflecting the proper and fair charges that are required to be rendered by the utility. If the project is executed according to its original plan, and thorough scope of work plans were developed and broadly communicated, the invoicing process is ordinarily administered in a successful manner as expected. On projects, with complex deliverables, where scope of work changes were routinely encountered and perhaps project site nuances caused a deviation from the original work plan, the resulting invoice process can become somewhat complicated, especially in the absence of implementing effective project management techniques.

Since many of the same engineering and design consultants, contractors, and vendors will provide a duplication of the services and products on future substation projects, in the spirit of team unity and in the effort to retain successful working relationships with these entities, it is in the best interest of all parties to effectively and fairly resolve the invoice issues. It is not uncommon for many utilities and selected service or product providers to establish alliance relationships. These alliance relationships essentially provide the means for the utility to retain the continued ability to procure services and products at the most economical cost possible. These alliance relationships are equally attractive to the service and product providers in that they are routinely and continually contracted for substation projects in a non-competitive bid manner. These mutually beneficial business partnerships lend themselves well to establishing effective invoice validating processes that are rooted in a give and take approach that serves as a win–win outcome for all vested parties.

20.2.4.2 Complete All Payments

The contractor, consultant, or equipment vendor invoicing process is not complete until the utility renders full payment to those entities. Rendering full payment is important to help ensure that the final project cost accurately reflects the thoroughly relevant and factual cost to complete the project. This cost knowledge is essential for not only providing an accurate picture of the estimate's performance, but it is also vital historical, financial information to capture that will serve as a useful reference when estimating future similar projects. A final activity involves the effort to verify, with the utility accounting personnel, that the invoice payments have been officially surrendered to the invoicing company. This verification activity should be pursued prior to formally closing the project's charge accounts, thus to ensure that the invoice payments will be captured in their appropriate project charge account numbers.

20.2.4.3 Submit Invoices for Reimbursable Items or Services

Typically, substation projects are initiated by utility companies to meet an exclusive internal need of that utility. Therefore, the utility remains the primary customer on routine substation projects and services by others, outside of the utility employment arena, are rendered to the utility rather than by the utility. This would translate into the fact that, generally, the utility company would not be submitting an invoice for a reimbursable item or service for their routine projects. Yet there are times when rare projects are pursued by the local utility to institute a change or modification to an existing substation that is required to ensure the operational integrity of the interconnection grid between itself and a neighboring utility or perhaps a merchant independent power provider (IPP). These types of projects are initiated by either the IPP or neighboring utility with oversight by the regional grid interconnection entity, otherwise known as an independent system operator (ISO), to address transmission system and substation operational impacts between all involved entities. Since the local utility is not the originator of this type of project, and would not otherwise engage in the work, the cost of the work performed locally by the utility is categorized as a reimbursable expense to be rendered by the IPP or neighboring utility. In these cases, the local utility would submit an invoice for full reimbursement following the completion of all work and after capturing all associated project costs. The ISO is usually the recipient of the invoice and manages the reimbursement payment process.

20.2.5 Closure of Outstanding Permits, Sureties, or Bonds

20.2.5.1 Permits or Sureties Required
- Grading
- Storm-water management
- Landscaping

Utility substation projects that fall into the categories of either entirely new substation installations or existing substations, whose scope of work entails an extensive modification or significant expansion effort, typically require some type of local governing agency permit to be secured. The need for a certain permit depends directly upon the nature and scale of the proposed substation work. Each governing agency interprets a project's nature and scale differently; therefore, exact permit mandates and requirement thresholds can vary between governing jurisdictions. The types of substation permits that are usually required range from building permits for foundation work to more extensive types of permits such as those for site grading activities and storm-water management implementations.

When certain permits are necessary, it is customary for the governing agency to require that the permit requestor also secures a surety to be linked directly with a certain permit. In the case of a substation project, the surety is a control mechanism that endeavors to ensure that the utility complies with the permit requirements and performs the substation's project scope of work to the full satisfaction of the governing agency. Having ownership of the surety, the governing agency would invoke their fiduciary authority on the utility to require them to indemnify the governing agency in the event of a permit non-compliance classification. Also, by requiring the surety, the governing agency is protecting itself from any financial loss in the event that it must assume ownership of various substation construction activities to make certain that project deliverables result in sound engineered products from both a general public and environmental protection perspective. Generally, for substation projects, the permits that require a surety to be obtained are site grading permits, storm-water management permits, as well as landscaping permits.

When a surety is required, the specific financial instrument utilized can vary dramatically between governing jurisdictions. The financial instruments that are typically authorized range from bonds, certified checks, letters of credit, to letters of guarantee. Each jurisdiction has the independent authority to determine which instrument is necessary for the specific permit purpose. Although there are several unique financial instruments that can be utilized, they all equally grant the governing agency firm fiduciary authority to render the utility financially responsible for a permit violation. In each case, the financial instrument is submitted to the governing agency, along with the permit application, where the governing agency retains ownership of the surety while the substation work is pursued.

20.2.5.2 Permit Closure Process or Final Governing Agency Inspections

Although all secured permits require some type of proper closure process, the permits that have associated sureties necessitate a more stringent procedure to bring those permits to full closure. These types of permits are appropriately viewed as covering project elements that require strict governing agency oversight to ensure that project deliverables adhere to acceptable engineering practices. In bringing those permits to proper closure, the agencies have established a firm policy that requires a final on-site inspection to be performed by an agency representative for the purpose of reviewing first hand the completed project element. This final governing agency inspection is routinely initiated by the permit applicant and represents the first step in the effort to secure the release of the submitted surety.

Final governing agency inspections are usually required for site grading permits, storm-water management permits, as well as landscaping permits. It is not uncommon for the actual completed field construction work, pursued under these permits, to deviate from the design product. Sometimes the deviation from the design is appreciable. For example, it is virtually impossible to perform the site grading or construct the storm-water management facility to perfectly match the engineered design.

Various unknown conditions and other site nuances typically arise during the construction phase and contribute to the finished product being different from what was originally engineered. In these cases, another control measure invoked by the governing agency is the requirement to create and submit as-built documentation of the completed site grading and storm-water management facility. In receiving this as-built documentation, the governing agency endeavors to prove that, although the final actual site product deviates from the engineered design, it will still adequately perform as engineered. If the design performances of the actual conditions still prove to be acceptable, then the permit's surety is rendered unnecessary and the surety release process proceeds to its final administrative phase. Eventually the surety, regardless of the financial instrument utilized, is returned to the utility for their archive record purposes.

20.2.6 Archive Records

In an effort to perpetually retain key project documentation, a hard copy archive file of the project should be created upon its completion. The archive file exists for the primary purpose of offering both an engineering and financial history of the completed project. The archive file serves as a repository of vital and sometimes esoteric project knowledge that must be retained for future awareness needs. This historical information can be useful or even necessary at a future date for perhaps gleaning lessons learned for a similar project or possibly utilized for a research analysis of the archived project.

Key data to be stored in the archive file should be limited to items of information that cannot be readily reproduced via another storage mechanism or information that is perhaps already being stored elsewhere, as a normal practice, in the office environment. The archive file data should typically include documentation that centers on key customer communications and agreements, original zoning, permitting or surety information, as well as the final analysis of estimate and schedule performance. In addition, the archive file should contain any lessons learned documentation and any project nuances that are deemed unique and valuable engineering experiences that warrant legacy capture.

The archive file creator is typically the project manager or the responsible engineer or both. In being appropriate stewards of the project file documentation, it is important that the archive file creator possesses the level of experience necessary to effectively differentiate between project documentation that requires archiving and project documentation that can be readily discarded. The historical archive file should not be merely a full collection of the project working file, but rather a conscientious recovery effort of vital documentation only. All project archive files should be stored in hard copy format in a central and accessible area of the office environment. However, in establishing the archive area, some thought should be given to the ability to secure the files when considering their significance from a homeland security perspective. Electronic historical archive file efforts should be discouraged due to continuing technological advances that may render current recovery efforts obsolete.

20.2.7 Develop Unit Costs

In support of the never ending pursuit of improving project estimate performance, it is customary to develop unit costs, derived from actual project labor hours consumed as well as actual project cost data. The computation of unit costs allows for the establishment of detailed estimate building blocks that can be used to better develop accurate estimates for future projects with similar activities. This building block approach enables elements of a complex project scope to be fragmented into quintessential project activities in order to allow basic cost and labor hour components to be developed. Future estimates based on actual derived unit costs enable the estimate to be built utilizing these discrete building blocks, thus usually leading to significant improvements in estimating accuracy. In addition, the unit costs can also serve as achieved performance measurements that can be established as successful benchmarks or target performance goals on future similar projects.

Unit costs can be developed for each of the engineering disciplines associated with a project as well as for each of the construction trades that are involved. Examples of unit costs that can be derived are as follows: labor hours per pound of steel erected, labor hours per yards of concrete poured, labor hours per length of underground duct bank constructed, labor hours per length of cable installed, and labor hours per construction drawing developed. Essentially, unit costs can be derived for almost any singular project activity whereby it is possible to clearly differentiate the labor hours and expenses that were consumed to complete that specific activity. These unit costs become the essential estimate building blocks for future projects that possess similar scope activities, thus contributing immensely to the continued effort to improve overall project estimate performance. As project estimates continue to be built with these unit cost building blocks, repeating the process of computing unit costs at the completion of the project serves as a calibration tool to verify the accuracy of the base data, further improving the accuracy of the future estimating activity.

20.2.8 Closeout Project Accounting

20.2.8.1 Cancel Charge Numbers

A timely and key activity to pursue shortly following the completion of the project involves the complete closure of the project's dedicated accounting charge numbers. The timely closure of the charge numbers is important in order to avoid the inadvertent or inappropriate charging of the project's account numbers, for unrelated work, that would directly result in the distortion of the project's estimate performance. The closure of the charge numbers can be pursued in a segmented fashion. This approach allows certain charge numbers to be closed immediately or shortly after project completion, while other numbers remain active while they wait final charges from project activities or invoices that may linger. This segmented closure strategy is a successful project estimate control practice routinely followed by the project manager or the responsible engineer as an added measure that limits the availability of active charge accounts. This strategy directly endeavors to minimize the inadvertent or inappropriate overcharging of a specific account number, thus generally improving the project's estimate performance.

20.2.8.2 Verify Invoices Have Been Paid

Typically, second party invoices for contractor services, purchased equipment, and material represent the category of outstanding financial responsibilities that remain following the completion of the project deliverable. The effort to properly process the payment and full accounting of this category of invoices can often linger for a few months following the project's completion. The project's charge numbers should only be closed upon the verification that all appropriate project financial responsibilities have been fully and completely satisfied.

20.2.9 Notify Stakeholders

20.2.9.1 Project Completed

A formal project completion announcement should be appropriately and thoroughly disseminated shortly following the accepted completion of the project. The announcement should be widely circulated to all levels of the project team as well as to all project stakeholders or project sponsors. Prior to the announcement, stakeholder or sponsor acceptance of the completed deliverable should be verified to avoid any false or inaccurate claims of completion. The thorough and fully disseminated successful project completion announcement provides all interested parties with the knowledge that the project has been brought to a successful closure. Equally, the completion announcement provides various team members with the completion awareness that may perhaps initiate their own associated project activity closure processes. The project completion announcement should include commentary that centers on schedule performance, estimate performance, as well as scope and quality performance. Thus, the announcement serves as a direct performance feedback mechanism to all project team members and project stakeholders.

20.2.9.2 Charge Numbers No Longer Valid

At the appropriate time, a follow-up announcement should be widely disseminated that declares that the project charge numbers have been closed and are no longer valid. The announcement serves as a final reminder that the project is considered 100% complete and can no longer accept any financial responsibility for the project. The announcement is typically offered as a courtesy to provide the full awareness that the project charge numbers have been rendered invalid, which serves as a project control mechanism to eliminate the inadvertent or inappropriate charging of the project's account numbers for unrelated work.

20.2.10 Development of Lessons Learned

Although many projects can seemingly be categorized as routine pursuits or perhaps basic in nature, most projects offer possibilities to glean some type of lessons learned that can be successfully leveraged on the next project opportunity that possesses either a similar scope or duplicate activities. The quantity and quality of the lessons learned and developed are not necessarily a function of the complexity of the project. Although, typically the greater the project complexity, the more the opportunity exists to develop quality lessons learned, all projects can usually offer learning experiences worthy of noting in order to be implemented on future projects. In general, many project lessons learned are identified during the final walk-through or inspection phase of the completed project and speak uniquely to the construction phase. However, the full project life cycle should be thoroughly examined to discover these improvement opportunities. The derived and implemented lessons learned generally endeavor to improve upon both the construction practices and safety performance of a project; however, all project phases and disciplines involved can benefit directly from this lessons-learned identification task. Effective project management techniques call for the lessons-learned to be identified during the planning phase of a future project, thus typically leading to appreciable improvements in the project's estimate performance, quality of the project deliverable, and safety performance as the project moves through the construction execution phase.

The process of developing or identifying lessons-learned opportunities can be accomplished in a variety of forums. These forums can range from personal notations based on individualized experiences, informal conversations amongst limited project team members, or sometimes the process can consist of engaging in a formal meeting setting. The formal meeting setting option is usually held for projects of a very complex or unique nature, which required the contribution of several engineering disciplines and involved a wide range of construction trades. The meeting would be attended by all pertinent project-activity owners and is customarily facilitated by the project manager or the responsible engineer who follows a structured agenda. Essentially, a lessons-learned item can be anything that either an individual derives or a project team identifies that is considered a worthwhile implementation on future projects to continually improve towards achieving higher levels of project success.

21
Energy Storage

21.1	Why Storage?..	21-1
21.2	Wholesale Energy Applications	21-3
21.3	Transmission and Distribution Applications..........................	21-4
21.4	Voltage Support...	21-4
21.5	Transmission Time/Load Shifting..	21-4
21.6	Stability-Related Applications	21-5
21.7	Transmission/Distribution Deferral..	21-6
21.8	Radial System Reliability Improvement..................................	21-7
21.9	Community Energy Storage...	21-7
21.10	Transmission and Distribution Capital Deferral....................	21-9
21.11	Storage and Distributed Solar...	21-10
21.12	Why Solar-Storage Applications?..	21-10
21.13	Financial Impacts of Benefits..	21-11
21.14	Energy Storage Technologies...	21-11
21.15	Lead-Acid Batteries	21-12
21.16	Nickel-Cadmium and Nickel-Metal Hydride Batteries..........	21-12
21.17	Lithium-Ion Batteries...	21-14
21.18	Sodium-Sulfur Batteries...	21-14
21.19	Zinc-Bromine Batteries..	21-15
21.20	Vanadium Redox Batteries..	21-16
21.21	Flywheels..	21-18
21.22	Technologies in Development...	21-19
21.23	Comparing Technologies	21-20

Ralph Masiello
KEMA, Inc.

21.1 Why Storage?

Storage is used in just about every industry that operates today. In all aspects of commerce, storage is used throughout the value chain to balance supply and demand and to buffer the upstream supply from the fluctuations in downstream demand. This is true of commodities like wheat and corn, of fuels like oil and gas, and of consumer products. However, this basic component of everyday processes in every industry does not exist in the electricity industry. There is no warehouse, silo, or tank for electricity.

From the outset, the electricity industry has been operating under a just-in-time delivery system, where all that is produced is delivered. As such, in order to maintain operations, the focus has been on balancing of the generation and load while ensuring reliable delivery of electricity. Without the ability to store electricity, this has been an acceptable mode of operation.

As more systems look to adopt renewable generation into the electricity system, it becomes more difficult to balance. Before the advent of renewables, variability was evident only in load; now generation also becomes more variable, making the system more unpredictable.

As our electricity grids become more dynamic, efficient, and real time in their decision processes, more tools and applications will be needed to provide the flexibility required for this state. For utilities and operators of the grid, storage offers the potential for a device that can meet this requirement. Its addition to the system can add the flexibility required for the system to adjust to the future requirements by acting as a bridge, buffer, and reliability component.

The first power system—built by Thomas Edison in New York—used lead acid batteries to store DC power and provide that balancing/buffering effect. But until recently, the dominant forms of storage on the grid have been large installations such as pumped hydroelectric and pilots of compressed air energy storage; with lead acid remaining in a "backup" role providing added reliability for critical installations. Today, however, we see new storage technologies being piloted to meet new needs.

Key characteristics of the new storage technologies that make them interesting for utility applications include

- *Fast response*: New battery technologies possess fast response capabilities, essentially providing square wave outputs over the ramping of traditional alternatives.
- *Multiple cycles*: Technologies are providing multi-thousand range cycles, allowing systems to be utilized for long periods of time and in applications that require daily and sub-hourly use.
- *Durations*: Some advanced storage technologies are providing economical durations in the 4–8 h range, introducing a real potential for electricity production shifting with renewables.
- *Transportability*: Technologies are mobile or have the ability to be placed anywhere there is a power system interconnection, opening a range of possibilities on where to apply the technologies.
- *Scalability*: Storage technologies have the ability to be scaled. The characteristics of the battery, whether aggregated in sizes of 10 kW or 10 MW, are essentially the same in each case.
- *Technology fit to application*: There are numerous storage technologies available today with more under development. This means that the user/developer can choose the technology that best fits the application.

For utilities and merchant operators, the three criteria test shows that storage has the potential to be applied to the following areas:

- *Ancillary services*: Frequency regulation, spinning reserves, supplemental reserves, real-time dispatch, and potential products such as synthetic governor and inertial frequency response.
- *Transmission and distribution*: Deferral, congestion relief, voltage control and stability, reliability improvements, substation upgrade deferral, reduced costs from transmission and distribution losses, and component life extension.
- *Renewables*: Integration of the technologies to the grid for renewable applications of all sizes, with particular focus on large wind and solar systems. Applications can range from short-term balancing and ramping support to time shifting of production across the diurnal cycle.
- *Demand response*: Storage can serve as a demand response tool or a key enabling technology to allow other devices, such as small commercial and residential renewables, to serve as demand response tools, providing the ability to defer the need for increased generation.
- *End user*: Storage can fulfill a traditional backup role with improved life span, maintenance, and environmental performance compared to traditional storage. It also can be used in conjunction with distributed renewable production for time-shifting energy to match end-user needs.

This list is by no means exhaustive. However, the ability of storage to perform the application roles listed earlier is not limited to utility-scale devices. Electricity storage is unique in the ease with which the technologies can be scaled. Whether the device is packaged as a 15 kW application or a 15 MW application, the characteristics of the device are the same. The advantages of these characteristics are no more apparent than with electric vehicle batteries and community energy storage (CES) applications.

The magnitude of the benefits that electricity storage can provide to the U.S. energy industry has been assessed by the Sandia National Laboratory and the summary of that assessment is shown in Exhibit 21.1.

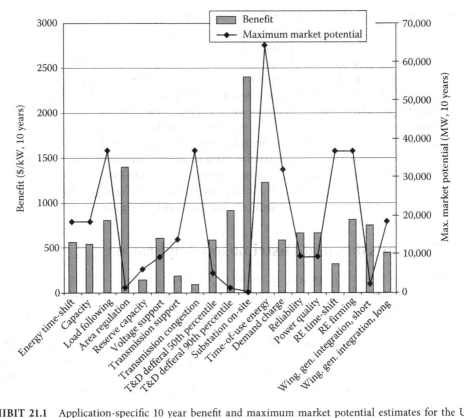

EXHIBIT 21.1 Application-specific 10 year benefit and maximum market potential estimates for the United States. (Reproduced from the SANDIA Report, *Energy Storage for the Electricity Grid: Benefits and Market Potential Assessment Guide*, SAND 2010-0815, Unlimited Release, SNL Report, February 2010.)

21.2 Wholesale Energy Applications

Wholesale applications are not themselves directly relevant to substation engineering. However, the impact that wholesale applications can have on the substation (and the feeders connected to it) very much is an issue, as storage facilities greater than a few megawatts in size will always be interconnected at a substation or power plant.

When a storage facility is colocated at a generation facility for provision of ancillary services, time arbitrage, or renewable firming, it has little additional impact on the grid or the substation—indeed usually it would act to reduce variability and is probably a positive factor in terms of the substation itself. For instance, a battery can be used to reduce voltage fluctuations caused by wind generators and to reduce harmonics.

When the storage facility is located at a transmission substation near a load center or which is a distribution substation serving distribution feeders, but is primarily intended to provide wholesale services, then each application has to be examined for its impact on the station and adjacent grid. While it is impossible to develop a general formula for this, we can describe some of the applications and their impacts:

- *Regulation services*: Storage (batteries, flywheels especially) is attractive for regulation service because it can respond rapidly to control signals. It would not be atypical to see a unit providing regulation swing from full charge to discharge almost instantaneously within periods less than an hour. If the facility is of a size that is significant in terms of the connected load of the station, then this rapid change in real power can cause issues of power factor and voltage. There are not as yet interconnection standards well developed for this application, but clearly the inverter associated with the storage facility must be controlled so as to avoid causing undue voltage fluctuation and problems for other voltage control equipment at the station.

- *Time arbitrage of energy*: If the storage system is discharging when demand is high, then presumably this will be a beneficial effect as the station is itself likely to be serving peak loads at that time.
- *Congestion relief*: The storage facility would be discharging at a period when the station "node" was experiencing high congestion prices—meaning high localized load and limited transmission access to low cost generation. As with time arbitrage, this should be a positive effect.
- *Renewable firming*: If the storage facility is firming "downstream" or distributed photovoltaic on the connected feeders, then it will discharge when its production falls off due to cloudiness or the like—this can assist in managing the voltage issues that photo voltaic (PV) variability can cause on the feeder (and storage on the feeder itself is even better). If the facility is providing wind firming services to remote wind farms, then the same issues as with regulation services apply.

21.3 Transmission and Distribution Applications

The benefits on the transmission system are a result of cost reductions due to the time and location shifting of energy for congestion relief. In addition, fast-acting storage devices can provide reliability benefits via enhanced stability and fast response to outages and also provide incremental voltage support, once the storage device and its power electronics are in place.

Between the transmission and distribution systems, at the sub-transmission or distribution substation, storage can be a vehicle for the deferral of capital expenditures for power transformer upgrades to meet peak load conditions. In remote areas served by radial subtransmission, storage can also be a vehicle for reliability improvement as a way to ride through uncleared faults on the transmission line(s) that serve the substation.

21.4 Voltage Support

On both transmission and distribution systems, a storage facility and its power electronics can provide similar benefits to capacitors and static VAR compensators. However, storage systems have the added ability to provide some real power as well as reactive power. In general, capacitors and static VAR compensators are less costly for the same functionality than are storage devices at today's costs. However, once a storage device is installed, it can be configured to provide the functionality of a static VAR compensator at little incremental cost. It also has greater flexibility, as it can provide both real and reactive power when necessary and thus provide power factor compensation over a wide range.

21.5 Transmission Time/Load Shifting

As with generation area applications, a significant benefit that storage can bring to the transmission system is shifting the energy commodity in time and location. A storage system at a wind farm can help avoid curtailment due to transmission congestion—when the wind farm production in excess of local load exceeds the transmission path's available capacity. This allows the wind farm to shift its power deliveries to a time frame when such transmission capacity is available, because either the local load has increased or the total local production is reduced. A similar transmission benefit can also be realized from time/load shifting by locating the storage at the load side of the congested transmission path.

In this case, the transmission path does not have the capacity to carry the lower cost or greener power to the load center at certain peak times. This congestion is currently alleviated through expensive, local generation and/or demand response at the load side to match demand to available local production plus transmitted power.

If the remote production can be transmitted to the load center off-peak but it exceeds local demand, then it may make sense to locate the storage at the load side. In this case, the energy has been shifted from off-peak at the production resource to on-peak at the load center a day later by the addition of storage near the load. The benefits of this application are twofold:

- The congestion price at the load center can be reduced if enough storage is available; then expensive local resources may not be needed. This reduces the hourly price for all consumers in the load center, as in a market environment the nodal price is applied to all parties, reflecting the congestion costs.
- It allows the regional transmission owner(s) to defer costly transmission capacity expansion that may be hard to justify if the congestion only occurs during limited number of hours a year.

As a further argument for this application, it is rarely the case that a single transmission path is the source of congestion due to line loading at maximum capacity. It is more often the case that the power system's security constrained unit commitment solution has limited the total transfers. This means that, after the first transmission contingency, the lines will not be overloaded beyond their short-term ratings and local resources can be brought into service before the line sag is too great. Local storage is a very attractive way to mitigate this issue, because the storage only has to have duration equal to the time it takes to bring local resources to bear. It will then allow the full path capacity to be used.

So far this application has not been developed in transmission and distribution systems. The principal reason is that the storage device in this application would be a regulated asset. In such a case, the utility would have to gain regulatory approval for the investment and the regional transmission operator would have to validate the congestion savings. This is unlikely unless the use of storage for transmission contingency relief becomes an accepted planning concept with approved standards and methodologies. A second reason is that the storage asset would be taking physical possession of the energy commodity at one time (and presumably one price) and delivering it at a second time and price. This is, to some extent, a merchant function that would not be allowed for a regulated transmission asset today. In other words, our current planning and regulatory paradigm locks in the profits of the production resource in the congested zone. (As a further complication, in many markets these resources may well be reliability-must-run resources that provide local ancillary services—which storage can also support—on a cost-recovery basis, however expensive they may be.)

21.6 Stability-Related Applications

A variation on this application is when the contingency limits are not set by post-contingency thermal limits or when the local resources can respond, but rather by transient stability limits. Stability limits occur in regions where large production resources are very remote from the load (so that the transmission path has a high reactance or X parameter), and in the milliseconds immediately after a fault, the generators accelerate ahead of the load until the fault is cleared at which time they decelerate.

A full exposition of this phenomenon is beyond the scope of this book, but the end result is that in these cases it is not unusual to limit the transmission path to levels below the thermal capacity due to stability limits. This ensures that the generators do not accelerate so far that they go over the top of the sinusoidal power-angle curves and are unable to decelerate back to a stable position when the fault is cleared. When the system actually encounters an unstable fault, protective relays will operate, ultimately causing a loss of load. Hence, the system is operated so as to avoid such situations. The cost of this is usually quite high as the remote resources tend to be lower cost. The lower limits mean that the low cost production is limited, and customer loads are supplied with power from expensive gas units.

As described earlier, the contingency congestion relief storage application provides not only a congestion cost/benefit but can also defer transmission capital expansion until the need for the capacity of a whole new line, conductor upgrade, or line voltage increase is justified. In the real world, where gaining approval for transmission expansion is more time consuming than construction, another way of stating this is that the storage can provide a bridge over increased transmission needs until the lines can be sited, approved, and built.

21.7 Transmission/Distribution Deferral

Between transmission and distribution, there are storage applications in the substation that can improve reliability and defer capital expansion. One example for a substation application is to support power transformer contingency operations during peak load periods that have grown in excess of the N-1 contingency capacity of the station.

For a simple example of this application, consider the case where the station has two power transformers connecting the sub-transmission side (say 68 or 115 kV) to the distribution side (say 12 kV), as shown in Exhibit 21.2. The power transformers are normally sized, so that if one is out of service for scheduled maintenance or because it has failed, the other transformer can carry the full load of all the distribution circuits (by closing the normally open bus tie breaker on the low side). When the peak load exceeds the rating of the remaining power transformer, the utility must either upgrade the transformer or add a third transformer if the substation configuration allows. In the meantime, a load curtailment scheme must be implemented (usually opening a feeder breaker and shedding load).

The problem presents both a temporary risk of decreased reliability due to outages at peak and a capital expansion plan that is unattractive because the incremental increase in capacity may put the total capacity well over the peak load and far in excess of the average load. Also, oversized transformers will incur increased losses due to circulating power, so there is a small but significant operating power loss cost—borne by the consumer via the market or the rate structure. However, a storage device of more modest size on the low-side bus can be used for contingency/outage relief of the peak load until it is feasible and economic to upgrade the power transformers. An additional benefit of this application is that the storage would be a temporary asset that would be there for a few years and could then be moved to another station that needs such a solution.

This potential application, however, suffers from the same barriers today as the transmission congestion relief application:

- The planning and operational methodologies are not established.
- The regulatory process for approval is nonexistent.
- The storage device again crosses the boundary between transmission- and distribution-regulated functionality and merchant functionality because it is potentially shifting off-peak energy to on-peak delivery.

As this typical substation application is likely to be in the range of 1–10 MW, this would not require storage systems of extreme physical size. And as previously noted these could be portable or semiportable in nature.

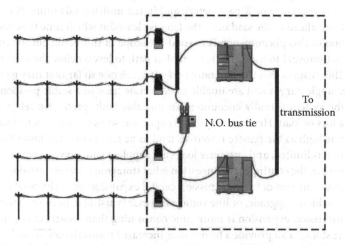

EXHIBIT 21.2 Example of storage sited at the substation. (Courtesy of KEMA, Inc., Chalfont, PA.)

21.8 Radial System Reliability Improvement

A variation on the aforementioned substation application occurs when the station is served by a radial sub-transmission line, as is typical in rural and remote service areas. In these cases, the system average interruption duration index can suffer because a single line outage results in a loss of service to all the feeders and customers served from that station. In such locations, it is usually not possible to cut over the feeder to an alternative source, due to the long distances and low load densities. Such situations often lead customers to install diesel or gasoline-powered backup generation at significant cost to ensure reliable electricity supply.

A storage device that is large enough to carry the entire station load for several hours will greatly improve reliability. Only those outages that require numerous hours to resolve when significant physical damage occurs would exceed the storage duration. Typical outages that require a few hours to fix should not result in any loss of service. One pilot project to test this application is underway in AEP's West Texas service territory today.

21.9 Community Energy Storage

CES straddles the transmission and distribution domain as well as the customer-side domain of applications. Users of electricity want reliable electric service at a reasonable price with as little environmental and aesthetic impact as possible. Reliable electric service is a challenge for electric utilities because so much has to go right over long distances for electricity to be available when customers want it. The generators have to work and adjust for increased demand immediately. Transmission lines and distribution lines have to be fault free over many miles. One tree limb or animal in the wrong place and the flow of electricity is stopped. Even underground systems are subject to people or contractors accidentally digging into cables.

Providing electricity at a reasonable price has been made far more difficult due to the fact that the difference between peak load and average load is substantial. If electricity demand could somehow be flattened, the size of much of the electric utility infrastructure could be less than half of what it is today to meet the same load. The practical difference is that far less money would be spent on wire and transformers.

CES is a small distributed energy storage unit connected to secondary transformers serving a few houses or small commercial loads. As the name implies, the primary benefit of the storage device is to benefit a local community by enhancing reliability, reducing required capital investment by flattening peak loads, and compensating for the variability of distributed renewable resources such as solar on roof tops. Typical sizes for single-phase community storage devices in a residential neighborhood would match the size of distribution transformers—10, 25, 50 kVA, etc. For larger three-phase applications, the devices could be installed in a bank just like distribution transformers. The batteries will be able to carry their full demand capacity for 1–3 h or longer, depending on the application. A typical specification is as shown in Exhibit 21.3.

A CES device has several advantages over other forms of storage. CES technology is similar to the technology being developed for plug-in electric and hybrid electric vehicles. As CES devices

Key Parameter	Value
Power (active and reactive)	25 kVA
Energy	50 kWh
Voltage	120/240 V

Source: Courtesy of KEMA, Inc., Chalfont, PA.

EXHIBIT 21.3 Typical CES specification.

and batteries for vehicles will be produced by several manufacturers in relatively high numbers, prices will likely be reduced faster than other storage technologies being used in various utility applications.

Inherently, the closer the storage device can be to the load, the more it can reduce losses on the utility system. In addition to loss reduction on the distribution primary and station transformer, CES devices can reduce loss on distribution transformers. In addition, the CES device can more effectively mitigate the variability of solar energy at individual homes and businesses.

The solution is enhanced by the potential offered by smart grid systems. Distributed CES devices can be controlled by a smart grid to act locally or for the greater benefit of the utility as necessary. Forty CES devices in a residential subdivision could act like a single 1 MW device by utilizing appropriate controls. On the other hand, in an outage the device could be used to maintain service to customers served by a single distribution transformer. With communications to homes, end customers could be given information to let them know they are being fed by a battery with limited available energy. The customers could then reduce load as necessary to extend the time that the battery alone could provide service.

Although there are no major technical barriers to CES, some significant details need to be developed and tested in pilot programs. An open standard has been created by AEP for CES as a starting point. Even when the devices are deployed without a smart grid, limited communications with the utility and the end customer will be highly desirable.

Standardization of community energy devices will help to reduce cost and speed the rate of implementation, including standards for

- Acceptance test plans (electrical, physical, and environmental interfaces)
- Handling momentary overload (inrush) for motor starting
- Overhead (pole-mounted) version
- Communications with the utility (CES hub) and the end customer (home area network interface)
- Functions of the battery based on communications from the utility or the customer
- Communication and security
- Mitigation or avoidance of power quality issues

CES can provide multiple benefits from the same installation, depending on the circumstances and the changing needs of the utility system or end customer. Certainly, a CES close to the customer can provide continuing service even when the utility system is not able to provide electricity. With a fully charged battery, the CES device should be able to provide full service for 2 or 3 h or much longer if the customers reduce usage. A CES can also improve power quality for end customers. Momentary interruptions often irritate customers as digital equipment needs to be reset after momentary interruptions. A CES device can provide the ability to ride through most momentary interruptions caused by disturbances on the utility.

One difficult issue for electric utilities is power factor correction. Normally, capacitor banks are used on distribution systems to improve efficiency and reduce the load on wires and equipment. Many of the capacitors are switched at appropriate times to try to change the power factor correction according to actual load. The capacitor banks often require maintenance and repairs. A CES device at the customer location can correct power factor with greater accuracy and without the maintenance issues of a capacitor bank. The benefits of power factor correction on the distribution system also help both the transmission and generation aspects of utility operations. High photovoltaic penetration on a feeder causes aggravated voltage and power factor fluctuation as the PV output varies with cloudiness. Storage devices can be used to help manage this avoiding excessive cycling of capacitors.

Using a smart grid, a group of CES devices can be aggregated to look like one large device. Once aggregated, the CES devices could be used for load leveling based on substation and grid needs. The total generation required at peak could also be reduced. In addition, the CES devices could be used to provide ancillary services through further aggregation at the grid level.

CES near end customers can solve multiple problems at the same time and as a result look attractive to utilities, end customers, and regulators. Imagine the widespread use of CES by a utility with the following benefits:

- Reliability is improved to customers, with only 10% of them with worse reliability.
- By targeting areas with the need for significant capacity upgrade, a high percentage of the cost of the batteries can be paid for by capital deferral of transmission and distribution systems.
- A large generation project that has been deferred pays for 100% of the cost of the CES devices.
- As solar energy becomes prevalent on roof tops, the CES devices provide the voltage regulation required making additional distribution upgrades unnecessary.
- Customer-owned solar can be teamed with the smart grid and CES devices to make the solar energy a dispatchable resource at peak load.

In short, CES could substantially change the way capital is deployed in the electric utility industry and the energy options available to end customers.

21.10 Transmission and Distribution Capital Deferral

In general, no one is proposing the use of storage as a way to defer new transmission that is required to alleviate congestion or to bring new renewable production to load centers. However, the siting of new transmission can be difficult and the cost allocation for it equally so, whether allocated to the new resources or to the load in a socialized way.

In any case, storage cost points currently appear to be too high to make capital deferral a viable, generic application for transmission. If 3000 MW of remote wind generation are constructed with an expected average capacity factor of 30% on a daily basis, and that amount of new transmission capacity is required to bring the wind power to market, we can make simple calculations. Storage of 6–8 h duration would typically be required to levelize the capacity factor adequately. If an estimate of the cost of storage of $1000/kWh is used, this implies that 8 h × 2000 MW × $1000/kWh of storage is required or about $16 billion. This would compare with the incremental cost of building 3000 MW of transmission capacity instead of 1000 MW, approximately. This could mean using 345 or 400 kV instead of 500 or 765 kV, for instance, or could mean deferring a circuit but probably not a set of towers. The physical distance to be traversed is a critical parameter in any such costing. If new transmission costs $2 million/mile, 1000 miles of transmission would cost two billion dollars. In any of those cases, the savings are much less than $16 billion. Storage would have to drop an order of magnitude or more in order for this mathematics to work. Assuming that in 5–10 years these kinds of cost reductions do occur, storage then becomes a viable alternative for capital deferral, at least worthy of consideration.

Note that this problem as stated is very different than the problem of using storage to levelize wind production and avoid curtailment where a load center nodal price sets the price of the curtailment.

However, other less generic cases where storage might allow deferral or avoidance of transmission capital do exist. When reliability is the key driver instead of renewable energy economics, much smaller amounts of storage can be used as a solution to remote isolated loads. When a 6 MW battery is required to provide reliability to a small remote town served by one 69 kV line, the trade-off is between $60 million of storage versus $50 million (at $500,000/mile for 100 miles)—that is, the numbers are in the same ballpark.

In other examples, when a 1 MW storage device can avoid the replacement of two power transformers or the addition of a third at a distribution station in order to meet peak load for a few hours a year, the savings are also in the ballpark. When construction costs are factored in, the savings are probably attractive if the upgrade can be deferred 5 or more years.

In general, the transmission deferral application still needs to be evaluated case by case and is generally difficult to "pencil out," as recently noted by the conclusions of New York State Energy Research and Development Authority (NYSERDA):

> The market is believed to be necessary due to the difficulty in siting transmission lines, and then once sited, the cost of building the transmission lines. Storage can be utilized to defer the need for the additional lines.
>
> This market has not been demonstrated and is in a conceptual stage. To date, it is the duration that would be a driving factor and general acceptance of a storage system to serve the role.
>
> Though this would most likely only be used at limited times during critical peak periods, the application is believed to require a duration of one to two hours for the individual device. The complete solution may require four to six hours for it to be considered an adequate solution for utilities and transmission operators. The risks are mainly focused on eventual acceptance of the device being used in this specific application, and the duration of the device.

For the expected storage value of $1200/kW, assuming the amount of deferral for the U.S. market to be one-thirtieth of the peak lead, a quick calculation of the cumulative value over a 5 year period assuming the U.S. DOE Energy Information Administration's average peak load for the United States being 800,000,000 kW, the market for storage deferral is $3.2 billion/year.

21.11 Storage and Distributed Solar

Another use for storage is in combination with distributed generation devices, or what is now really considered to be solar applications. Many groups, from utilities and state agencies to solar manufacturers, have been examining this application. The concept is summarized in Section 21.2. However, the barriers to deployment have focused as much on cost and paybacks for the application as the technical issues. The reason for this is the revenue that can actually be gained from the application is limited in relation to the cost of adding the system. As with most ideas with emerging technologies, the concepts are valid, but the expense of the early development system makes the practical implementation difficult.

21.12 Why Solar-Storage Applications?

For solar-storage applications, there are a number of benefits that are driving development. The first of these applications are to cover the mismatch between the solar peak and a second peak being witnessed in most utility territories. This is illustrated in Exhibit 21.4.

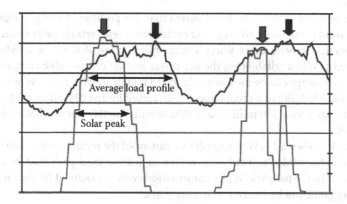

EXHIBIT 21.4 Comparison of system peak versus solar peak output. (Courtesy of KEMA, Inc., Chalfont, PA.)

In Exhibit 21.4, the average system load profile is shown in the dark curve while the average solar generation curve is shown in the light gray curve. Where an average load profile extends into late afternoons and evenings, the solar system does not have the ability to cover the full peak. Solar provides relief during the mid-afternoon, but not during the late afternoon and early evening peaks. As tariffs are often based on peak demand charges, solar would not have an impact on peak demand charges. This limits the benefits of solar for an owner. For such a case, storage can play a roll by extending the peak. The system can simply charge during the day and discharge during late afternoons to capture the benefit of peak reduction.

The second benefit that storage can offer is by allowing solar to be a resource that can be dispatched for demand response programs. In Exhibit 21.4, the second solar peak shows an intermittent output of the solar system. This intermittency means that utilities and solar system owners are unable to rely on the generation for dispatchable power. In addition, owners of distributed solar are not able to be included in demand response programs. If storage was combined with the system, the solar system would be able to provide an output, whether from the solar panel or the storage system when called upon by the utility. This essentially provides a guaranteed response for "green demand response" programs.

A third benefit is simply accrued to the end user itself. Currently, most small solar systems are not allowed to act as backup power systems. The reason is that interconnection standards typically require the device to drop off line when the power grid is down. When combined with the proper inverter systems, storage can provide the power necessary to ensure the system is able to run off-grid and then fully synchronize back to the grid at the moment grid power is restored.

21.13 Financial Impacts of Benefits

Though these benefits are easily understood, the relatively low paybacks offered in comparison to the price of the application place an initial hurdle on the solar-storage system. Solar applications tend to be high-cost devices even prior to any additional equipment, averaging around $6000–$8000/kW for installation (absent any tax incentives or grants). Adding storage to the application would add another, approximately, $1200–$1500 to that cost. In the first two cases listed earlier, the payback that an owner would receive by (1) participating in demand response programs or (2) by extending the peak-period output of the device is small, year on year. Even when additional benefits are added into the equation, such as emission savings, increased reliability with backup power system, or entering real-time energy tariffs with the capabilities provided by the storage system, the economics are still difficult to make a compelling case.

Though the economics are still difficult, mapping the financial paybacks against the cost of an emerging technology ignores the potential prices reductions of new electricity storage technologies that may be entering the markets. Today, there are enough sound reasons to pursue the applications and storage pricing trends are heading in the same direction as solar trends.

Is there an alternative answer as well? One answer may be the CES system. Though currently planned to be on the utility side, having the community storage device work in concert with the end-user solar system may be a means to overcome the cost barriers of placing the installation burden on the end-use customers. If the CES device is allowed to interact with the solar system, the system would be justified through the utility application. The end user could simply be paid for having their solar system participate in demand response, and society would then be able to capture the benefits of dispatchable renewable devices.

Determining how these benefits accrue becomes even more difficult when the storage resource is capable of bridging multiple roles that normally fall in different categories or merchant or regulated sectors. Implicit in the aforementioned categorizations are some examples and possible treatments.

21.14 Energy Storage Technologies

Storage is unique in that no single technology fits all applications. Some applications may require fast response while others may want duration as a key attribute. Examples are seen in the fuel cell world, where solid oxide, molten carbonate, and phosphoric acid provide different approaches to one solution.

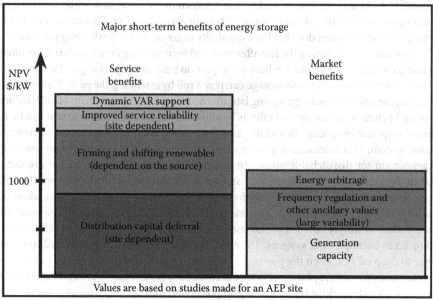

EXHIBIT 21.5 Calculated service and market benefits for 1 MW, 7 h discharge substation battery. (Courtesy of AEP/A. Nourai and C. Schafer.)

As storage becomes increasingly adopted by the electricity industry, it can be expected that advancements in specific technologies and introductions of new technologies will occur—all focused on providing a cheaper, more elegant solution.

21.15 Lead-Acid Batteries

Lead-acid batteries are one of the most developed battery technologies. Electricity is stored in the form of a chemical reaction based on lead, using proven technology. The life expectancy of these batteries is long, and although the batteries have high operating and maintenance costs and low energy density, they are used in a wide range of applications, such as transport, uninterrupted power supply, and grid-connected electricity storage (Exhibit 21.5).

The properties (lifetime and energy) of the batteries are highly dependent on the depth of discharge of the system. If the system is only partly discharged, lifetimes are remarkably longer than when the system is fully discharged and continuously operated. To have a satisfactory lifetime, these lead-acid systems are often over-dimensioned in comparison with the required specifications.

Lead-acid systems, such as that shown in Exhibit 21.6, can be operated at powers in the range of 0.1–40 MW and identical energy ranges. The amount of energy (kilowatt hours) that a lead-acid battery can deliver is not fixed and depends on its rate of discharge. The response time of this technology depends on the applied power electronics; however, response times of less than a second are possible.

21.16 Nickel-Cadmium and Nickel-Metal Hydride Batteries

One of the oldest types of batteries is the nickel-cadmium (NiCd) battery. As cadmium is toxic, in the 1980s and 1990s, NiCd batteries were replaced by nickel-metal hydride (NiMH) batteries. Since the introduction of the NiMH technology, NiCd batteries are only used for stationary applications. NiMH batteries are used for mobile applications and in areas where toxicity and environmental

EXHIBIT 21.6 Lead-acid storage system at So Cal Edison—10 MW, 40 MWh site. (Courtesy of Electric Power Research Institute, Palo Alto, CA.)

concerns are important. As the energy density of NiMH is higher, it requires shorter charging times in comparison with NiCd.

Like lead-acid batteries, the energy rating and lifetime are dependent on the depth of discharge. Because of the required over-dimensioning of the system to meet lifetime requirements, the costs for these systems can be relatively high. Response time is dependent on the applied power electronics. Nevertheless, this technology can and is being used for fast-response applications like frequency control, spinning reserve, and voltage control.

NiCd and NiMH can be operated in a power range from 0.5 to 30 MW and 0.005 to 7 MWh. A NiMH system is shown in Exhibit 21.7.

EXHIBIT 21.7 Nickel-metal hydride battery system. (Courtesy of Pennwell, Tulsa, OK.)

21.17 Lithium-Ion Batteries

Lithium-ion batteries have been best known for their use in mobile applications, such as cell phones, laptops, personal electronics, as well as for military electronics. However, in recent years this technology has developed considerably and expanded its portfolio of applications. Today lithium-ion batteries are available for applications in electric vehicles as well as for ancillary services in grid-connected storage.

With this technology, lithium-ions are exchanged between the electrodes during charging and discharging. This is also known as the rocking-chair principle.

Historically, the development of lithium-ion batteries has focused on increasing the performance in the area of energy density, power density, and cycle life in comparison with lead-acid and NiMH batteries. However, recently, the development focus has shifted more toward safety aspects as lithium is a highly reactive material. The high energy density of these batteries results in greater safety risks for large-scale systems. These risks should always be taken into account when developing a large-scale lithium-ion system.

This technology is now evolving into utility-scale applications. Recently, AES implemented two 1 MW lithium-ion systems that can store 15 min of energy for ancillary services, making it one of the largest lithium-ion systems in operation. Because of its characteristics, the technology is very well suited for fast-response applications.

21.18 Sodium-Sulfur Batteries

Sodium-sulfur (NaS) batteries have been demonstrated at over 190 sites in Japan for daily peak shaving. Today, the first NaS systems are in operation in the United States and Europe for grid-connected applications. With NaS batteries, electricity is stored through a chemical reaction, which operates at 300°C or above. At lower temperatures, the chemicals become solid and reactions cannot occur.

These systems are currently produced in Japan. Each module is 1 MW and about 7 MWh. By combining modules, larger systems can be realized. This system is a fast-response system with good lifetime expectancy. Exhibit 21.8 shows an NaS battery system and Exhibit 21.9 shows an NaS battery schematic. Exhibit 21.10 shows reactions for NaS batteries.

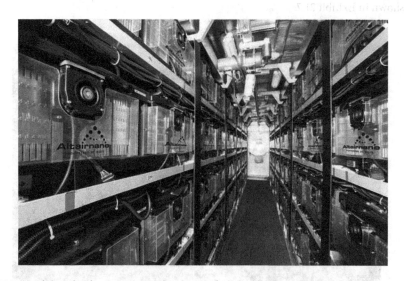

EXHIBIT 21.8 Mobile utility battery storage. (Courtesy of Altairnano, Reno, NV.)

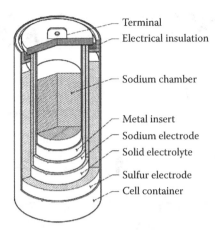

EXHIBIT 21.9 Sodium-sulfur battery schematic. (Courtesy of NASA John Glenn Research Center, Cleveland, OH.)

EXHIBIT 21.10 Sodium-sulfur battery system. (Courtesy of Energy Storage Association, Washington, DC.)

21.19 Zinc-Bromine Batteries

Zinc-bromine systems are attractive for use in utility energy storage and electric vehicle applications. In this type of storage system, electrolytes are stored in separate tanks. Zinc is deposited on the electrodes during charging, and during discharging the zinc reacts with bromine.

The response time of this technology is not good enough for fast-response applications; however, it is a promising technology for balancing power generation and consumption. The operational power range is 0.25–2 MW and up to 5 MWh.

This system has not yet been commercialized. A demonstration project in the United States is currently being conducted. A zinc-bromine battery system is shown in Exhibit 21.11.

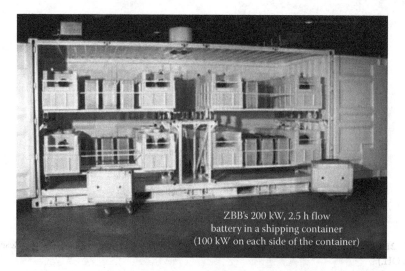

EXHIBIT 21.11 Zinc-bromine battery system. (Courtesy of ZBB Energy Corporation, Menomonee Falls, WI.)

21.20 Vanadium Redox Batteries

Redox flow storage technologies use tanks with electrolyte to store the energy. The electrolyte is charged and discharged with the use of a proton exchange membrane. The main advantage of this system is that power and energy properties can be tuned separately to the required specifications as the energy tanks can be freely sized. Flow systems are developed especially for stationary applications.

Several different redox flow systems are being developed. The vanadium redox technology is the most mature of all flow system technologies. In an all-vanadium system, vanadium is used as electrolyte to store electricity. Energy capacity ranges from 0.5 to 6 MWh and the power ranges

EXHIBIT 21.12 Vanadium redox battery system. (Courtesy of Cellstrom, Brunn am Gebirge, Austria.)

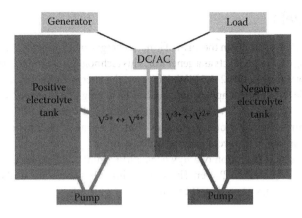

EXHIBIT 21.13 Diagram of a vanadium flow battery. (Courtesy of Wikipedia, free public domain.)

Technology	Reactions
Lead-acid batteries	Positive: $PbO_2 + 3H^+ + HSO_4^- + 2e^- \leftrightarrow PbSO_4 + 2H_2O$
	Negative: $Pb + HSO_4^- \leftrightarrow PbSO_4 + H^+ + 2e^-$
	Overall: $PbO_2 + Pb + 2H_2SO_4 \leftrightarrow 2PbSO_4 + 2H_2O$
Nickel-cadmium/nickel-metal hydride batteries	Positive: $NiO(OH) + H_2O + e^- \leftrightarrow Ni(OH)_2 + OH^-$
	Negative: $Cd + 2OH^- \leftrightarrow Cd(OH)_2 + 2e^-$
	Overall: $2NiO(OH) + Cd + 2H_2O \leftrightarrow 2Ni(OH)_2 + Cd(OH)_2$
Lithium-ion batteries	Cathode half reaction: $LiCoO_2 \leftrightarrow Li_{1-x}CoO_2 + xLi^+ + xe^-$
	Anode half reaction: $xLi^+ + xe^- + 6C \leftrightarrow Li_xC_6$
Sodium-sulfur (NaS) batteries	Positive: $2Na^- \leftrightarrow 2Na^+ + 2e^-$
	Negative: $4S + 2e^- \leftrightarrow S_4^-$
	Overall: $2Na + 4S \leftrightarrow Na_2S_4$
Zinc-bromine batteries	Positive: $2Br^- \leftrightarrow Br_2(aq) + 2e^-$
	Negative: $Zn^{2+} + 2e^- \leftrightarrow Zn$
	Overall: $ZnBr_2(aq) \leftrightarrow Zn + Br(aq)$
Vanadium redox batteries	Charge: $V^{3+} + e^- \leftrightarrow V^{2+}$
	$V^{4+} \leftrightarrow V^{5+} + e^-$
	Discharge: $V^{2+} \leftrightarrow V^{3+} + e^-$
	$V^{5+} + e^- \leftrightarrow V^{4+}$

Source: Courtesy of KEMA, Inc., Chalfont, PA.

EXHIBIT 21.14 Reactions by battery technology.

from 0.5 to 4 MW. Depending on the applied power electronics, the response varies from fast to slow response.

The vanadium technology was demonstrated successfully in several projects. Systems are commercially available. Exhibit 21.12 shows a vanadium redox battery system coupled with a solar panel. Exhibit 21.13 is a diagram of a vanadium flow battery. Exhibit 21.14 details reactions for the battery technologies presented in this chapter.

21.21 Flywheels

With a flywheel, electricity is stored in the form of kinetic energy/rotating mass. During charge, the flywheel acts as a motor; during discharge, it acts as a generator. This technology is extremely suitable for fast-response applications in a power range of 0.02–1.5 MW. However, the energy rating of this technology, maximum 0.25 MWh, makes it less suitable for applications such as balancing power generation and demand.

Today, flywheels are commercially in use for grid-connected storage at different sites in the United States. The most common application is frequency control. Large-scale flywheel storage systems, up to 20 MW, are mostly composed of several smaller subsystems.

Exhibit 21.15 shows a cross section of a flywheel, and Exhibit 21.16 shows a flywheel system concept that has been proposed by Beacon Power. The system shows how multiple units are linked together in order to form large capacities or megawatt capabilities.

EXHIBIT 21.15 Cross section of a flywheel. (Courtesy of Beacon Power, Tyngsboro, MA.)

EXHIBIT 21.16 Flywheel system. (Courtesy of Beacon Power, Tyngsboro, MA.)

21.22 Technologies in Development

A number of established players have been manufacturing batteries for the military and aerospace markets for many years. These firms will most likely try to extend their offerings into the electricity industry. Following are some examples of the technologies that are being advanced:

- *Zinc-metal air*: Typically, these technologies are cheap and easy to produce. However, in the past, they have been limited because of the inability to recharge the devices. Companies are exploring this area to try to implement innovative solutions to solve the charge–recharge issues. If successful, the device will be able to provide an alternative to some of the current systems.
- *Liquid-metal battery*: Researchers at MIT are developing a liquid-metal battery. This battery consists of three layers of liquids. The goal of the effort is to create a low-cost option for high-energy batteries. The first prototypes of the devices utilized molten salts and it is hoped that the device will reach a commercialized level in 5 years. Though the devices operate at high temperatures, it is projected to be a low-cost alternative battery technology.
- *Electrochemical capacitors*: They store electrical energy in the two series capacitors in the electric double layer that is formed between each of the electrodes and the electrolyte ions. The distance over which the charge separation occurs is just a few angstroms. The capacitance and energy density of these devices is thousands of times larger than electrolytic capacitors. The electrodes are often made with porous carbon material. The electrolyte is either aqueous or organic. The aqueous capacitors have a lower energy density due to a lower cell voltage but are less expensive and work in a wider temperature range. The asymmetrical capacitors that use metal for one of the electrodes have a significantly larger energy density than the symmetric ones and have lower leakage current. Compared to lead-acid batteries or advanced battery technologies, electrochemical capacitors have lower energy density but they can be cycled tens of thousands of times and are much more powerful than batteries (fast charge and discharge capability) (Exhibit 21.17).
- *Ultracapacitors*: Emerging battery technologies offer the potential of power and energy. However, even the emerging battery technologies seem to be able to focus on only one area. Batteries that are capable of high power tend to not provide the duration or energy. However, for very high power applications, advanced battery technologies may not be the best choice of technologies. Ultracapacitors are offering the promise of very high power applications, and advancements are now dramatically reducing the price of the technologies. Devices that are currently under

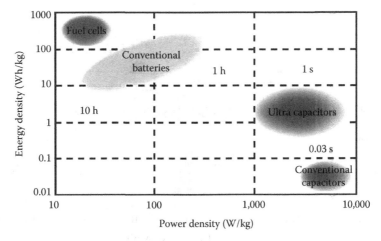

EXHIBIT 21.17 Energy density/power density for various energy storage devices. (Courtesy of S. Zurek, based on data provided by Maxwell Technologies, San Diego, CA.)

development at researcher and development facilities such as MIT are utilizing nanotechnologies to increase the power capability and reduce the cost. These devices already have the ability to operate in temperature ranges above that of most battery systems as well as offering an extreme cycle life. However, as the cost comes down, the applications may no longer need to be limited to utility applications but may also be extended to commercial and industrial as well.
- *Thermal storage*: Like pumped hydro systems, not all storage needs to be an advanced technology. Thermal storage is a proven technology that is being advanced in ICE as well as heat systems. ICE Energy is advancing the concept of using its ICE Bear system to reduce air-conditioning load, while companies such as Steffes Corporation are offering thermal electric devices to offset heating loads.
- *Fuel cells*: Fuel cells use the oxidation of hydrogen to form water as the electrochemical reaction producing electricity. As such, the fuel or energy stored is hydrogen gas under pressure. When combined with electrolysis to form hydrogen from water, fuel cells can function as an electricity storage device.

21.23 Comparing Technologies

The importance of labeling the efforts and understanding key storage characteristics becomes even more important as comparisons to technologies are attempted and made from data that is currently available. The following charts are a valid attempt to put boundaries on how devices and technologies are expected to perform.

The first chart in Exhibit 21.18 examines specific system ratings of various storage technologies and Exhibit 21.19 examines the cost of the unit per power and per energy. Though both of these charts provide a great deal of assistance when comparing technologies, accuracy of the information that is provided is unclear due to the fact that most technologies are developing rapidly and changes are constantly occurring. In addition, as is seen in Exhibit 21.18, there is a great deal of overlap in the technology ratings and characteristics.

Each of these charts reinforces the need for testing of the technologies that are being proposed. To answer the question of which is the best storage device, that answer is simply the storage device that best fits the specific needs of the application and, if a driver, at the lowest cost.

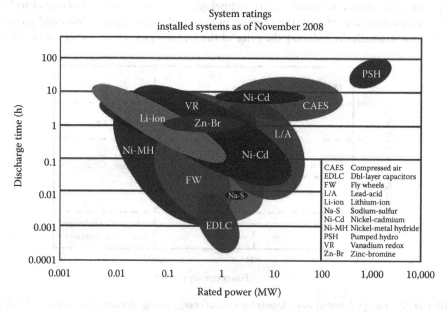

EXHIBIT 21.18 System ratings for storage technologies. (Courtesy of Energy Storage Association, Washington, DC.)

Energy Storage

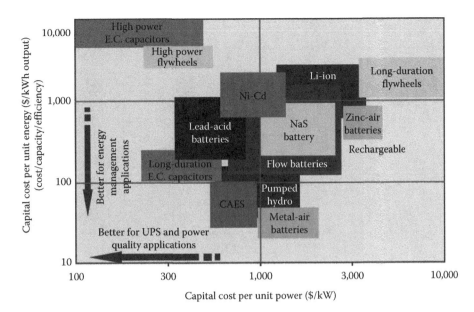

EXHIBIT 21.19 Capital costs for storage technologies. (Courtesy of Energy Storage Association, Washington, DC.)

To help differentiate each of the technologies, typical testing protocols are used that allow devices to be independently verified or to confirm that the specific technology is meeting its performance specifications. However, depending on the stage of development, performance testing has different types of tests as the product develops.

When a manufacturer develops a technology, whether it is an advanced compressed air energy storage, flywheel, or advanced battery technology, the initial characteristics of the device need to be quantified. However, as can be seen in Exhibit 21.9, the characteristics have a range that the technology is capable of meeting. In addition, the final commercialized device is going to be used in a specific application—whether frequency regulation, ramping, diurnal shifting—that will tell us how the storage device really needs to do. Hence, for a manufacturer that is developing a storage device, the process that is typically followed is to (1) create a base technology, (2) confirm the device can be safely operated and scaled up to the appropriate level, and (3) apply that final version to the intended application. This process is typically covered in performance testing, safety testing, application testing, and lifecycle testing.

22
Role of Substations in Smart Grids

Stuart Borlase
Siemens Energy, Inc.

Marco C. Janssen
UTInnovation

Michael Pesin
Seattle City Light

Bartosz Wojszczyk
GE Energy–Digital Energy

22.1 What Is Smart Grid, Why Now? 22-1
    Smart Grid or Smarter Grid? • Smart Grid Drivers • Benefits, More Than a Business Case • Technology Framework
22.2 Transformation of the Grid 22-7
    Engineering and Design • Information Infrastructure • Operation and Maintenance • Enterprise Integration • Testing and Commissioning
22.3 Substation Technology Advances 22-10
22.4 Platform for Smart Feeder Applications 22-14
22.5 IEC 61850 in Smart Substations 22-15
    Paradigm Shift in Substation Design • Interoperability and IEC 61850 • Impact of IEC 61850 in the Substation • Station Bus–Based Architecture • Station and Process Bus Architecture • Summary
22.6 Smart Grid, Where Do We Go from Here? 22-29
References 22-30

22.1 What Is Smart Grid, Why Now?

22.1.1 Smart Grid or Smarter Grid?

The Smart Grid has seen major hype over the past few years. Many consider Smart Grid a cliché in the utility industry. What was initially given the name of "intelligent grid" (maybe that implied too much) and is currently interpreted a hundred and one different ways, Smart Grid has gained worldwide recognition, and the "smart" catchphrase seems to have carried over into other industries—for good reasons or not. Almost everything is dubbed "smart" in the electric system, even down to the "smart bolt" on a transmission tower. The many interpretations of "Smart Grid" depend on the perspective of those talking about Smart Grid—the utility, vendors, consultants, academics, and consumers. One thing is clear that Smart Grid is a definite change in the way people are thinking about the generation, delivery, and use of electric energy. Thus, the "Smart Grid" has become an essential component to addressing the energy demand, security, and environmental challenges we face.

    While the grid is a marvel in engineering design and may be one of mankind's greatest achievements, it has yet to be transformed into a modern grid, a sustainable grid, a truly Smart Grid that takes advantage of proven, cleaner, cost-effective technologies that are available today or under development. Whether you believe in the Smart Grid or not, most of the utility industry agrees that the Smart Grid efforts are certainly placing a new emphasis on improved and new technologies for the utility grid as well as consumers. The Smart Grid effort has still a long way to go; it will continually evolve, driven by the economies and regulations across the world; and who knows the final state of the effort, but it looks like it has sufficient momentum to bring significant advances in technology and changes in the industry. There will be winners and there will be losers, not just in technologies and vendors, but also in the utilities and their Smart Grid deployments. Consumers will also be among the winners and losers, as has been seen recently by some challenges with consumer acceptance of Smart Grid.

Some could argue that the electric grid is really not that "dumb." The electric grid is operated by complex software programs and automation routines and protected by microprocessor-based relays. This may be true in some parts of the world, but electric infrastructure evolution has been slow, affected by economic factors, demographics, regulation, and many other factors. What seems to be similar across the world are the challenges of operating the electric system and delivering power reliably and cost-effectively to consumers. People compare the electric infrastructure to the telecommunications industry—"Graham Bell would not recognize cell phones if he were alive today, but Thomas Edison would still recognize the electric utility system if he were alive today." So there are many reasons for this difference—the massive electric infrastructure cannot be replaced as quickly and cost-effectively as the telecommunications infrastructure, plus developments in processing and IT devices have far exceeded those of the electric industry. Also, the environment driving the need for faster communications has changed a lot faster for the communications industry than the electric industry. Plus one could argue that comparing electricity as a commodity to consumer-driven communications, at least in the personal communications industry, is not the same—consumers are more willing to pay for additional services on a cell phone than electric service. So maybe that is where the Smart Grid should be heading—focus on the consumer, but not just to try reducing consumer electricity bills, but also to provide other services, or, something more rudimentary, focus on supplying more reliable service to customers or, in some countries, simply making electricity a commodity accessible to more people. However, in the case of the electric industry, utilities also have to drive Smart Grid in realizing capital and operating benefits that can be achieved with more advanced technologies. So while we may not have such a "dumb" grid, our thinking should be more along the lines of a "smarter grid"—not really about doing things a lot differently than the way they are done today, rather about doing them smarter—sharing communication infrastructures, filling in product gaps, and leveraging existing technologies to a greater extent while driving a higher level of integration to realize the synergies across enterprise integration. Smart Grid initiatives will certainly help drive the development of more advanced technologies, perhaps faster than the pace we have seen over the past few decades. The technologies will be more pervasive and will include a much-needed platform for enterprise-wide solutions that deliver far-reaching benefits for both utilities and their end customers.

So, what constitutes a Smart Grid? Numerous utilities are making claims to be first with a Smart Grid. While these claims may be true in the sense of the name, most Smart Grid stories today are focused on responding to metering regulations and some kind of pilot as a showcase under the Smart Grid banner. A Smart Grid is not an off-the-shelf product, nor something you install and turn on the next day; it is an integrated solution of technologies driving incremental benefits in capital expenditures, operation and maintenance expenses, and customer and societal benefits. It is the integration of electrical and communication infrastructures and the incorporation of process automation and information technologies with our existing electrical network. Smart Grid represents a complete change in the way utilities, politicians, customers, and other industry participants think about electricity delivery and its related services. This new thought process will likely lead to different Smart Grid technologies and solutions, but will benefit from greater integration of utility engineering and operations and new business models.

The solutions that provide those benefits are vast, and Smart Grid benefits may not be achievable by all utilities (and consumers) depending on their needs and operational strategies. Global, regional, and national economic recovery and growth will serve as the cornerstones of increasing investments in development of smart electric power infrastructure and increased reliance on integrated communication and information technology. Drivers will include national and state government issuance of clear policy directives and incentives concerning energy futures and development of smart infrastructure. When the stimulus funding and country initiatives have come and gone, the next 3–5 years will indicate how much today's fervor and hype will influence the decades to follow in investment, innovation, and achievable benefits of a "smarter grid."

The concept of Smart Grid has many definitions and interpretations dependent on the specific country, region, and industry stakeholder's drivers and desirable outcomes and benefits. A preferred view of Smart Grid may not be what it is, but what it does and how it benefits utilities, consumers, the environment, and the economy. The Smart Grids European Technology Platform (which is comprised of European stakeholders, including the research community) defines "a Smart Grid [as] an electricity network that can intelligently integrate the actions of all users connected to it—generators, consumers and those that do both, in order to efficiently deliver sustainable, economic and secure electricity supply."[1] In North America, two dominant definitions for Smart Grids are proposed by the Department of Energy (DOE) and the Electric Power Research Institute (EPRI).

U.S. DOE: Grid 2030 envisions a fully automated power delivery network that monitors and controls every customer and node, ensuring two-way flow of information and electricity between the power plant and the appliance, and all points in between.[2]

EPRI: The term "Smart Grid" refers to a modernization of the electricity delivery system so it monitors, protects, and automatically optimizes the operation of its interconnected elements—from the central and distributed generator through the high-voltage network and distribution system, to industrial users and building automation systems, to energy storage installations and to end-use consumers and their thermostats, electric vehicles, appliances, and other household devices.[3]

Beyond a specific, stakeholder-driven definition, Smart Grids should refer to the entire power grid from generation, through transmission and distribution infrastructure all the way down to a wide array of electricity consumers. Smart Grid is essentially modernizing the twentieth century grid for a twenty-first century society. A well thought-out Smart Grid initiative builds on the existing infrastructure, provides a greater level of integration at the enterprise level, and has a long-term focus. It is not a one-time solution, but a change in how utilities look at a set of technologies that can enable both strategic and operational processes. Smart Grid is the means to leverage benefits across applications and remove the barrier of silos of organizational thinking. Smart grid pilot projects driven by regulatory pressure that focus on the impact of new meters on consumers will evolve to technology-rich, system-wide Smart Grid deployments that demonstrate well-proven and quantified benefits. A key component to effectively enable the full-value of Smart Grid realization is technology with the functionalities and capabilities to achieve cohesive end-to-end integrated, scalable, and interoperable solutions.

22.1.2 Smart Grid Drivers

The Smart Grid is essential to enable a future that is prosperous and sustainable. All stakeholders must be aligned around a common vision to fully modernizing today's grid. Throughout the twentieth century, the electric power delivery infrastructure has served many countries well to provide adequate, affordable energy to homes; businesses; and factories. Once a state-of-the-art system, the electricity grid brought a level of prosperity unmatched by any other technology in the world. But a twenty-first-century economy cannot be built on a twentieth-century electric grid. There is an urgent need for major improvements in the world's power delivery system and in the technology areas needed to make these improvements possible.

A number of converging factors will drive the energy industry to modernize the electric grid. These factors can be categorized into five major groups as follows:

Policy and Legislative Drivers

- Electric market rules that create comparability and monetize benefits
- Electricity pricing and access to enable Smart Grid options

- State regulations to allow Smart Grid deferral of capital and operating costs
- Compatible Federal and state policies to enable full integration of Smart Grid benefits

Economic competitiveness

- Creating new businesses, new business models, adding "green" jobs
- Technology regionalization
- Alleviate the challenge of a drain of technical resources in an aging workforce

Energy reliability and security

- Improve reliability through decreased outage duration and frequency
- Reduce labor costs, such as manual meter reading and field maintenance, etc.
- Reduce nonlabor costs, such as the use of field service vehicles, insurance, damage, etc.
- Reduce T&D system delivery losses through improved system planning and asset management
- Protect revenues with improved billing accuracy and prevention and detection of theft and fraud
- Provide new sources of revenue with consumer programs, such as energy management
- Defer capital expenditures as a result of increased grid efficiencies and reduced generation requirements
- Fulfill national security objectives
- Improve wholesale market efficiency

Customer empowerment

- Respond to consumer demand for sustainable energy resources
- Respond to customers' increasing demand for uninterruptible power
- Empower customers so that they have more control over their own energy usage with minimal compromise in their lifestyle
- Facilitate performance-based rate behavior

Environmental sustainability

- Response to governmental mandates
- Support the addition of renewable and distributed generation to the grid

Many of these drivers are country and region specific and differ according to unique governmental, economic, societal, and technical characteristics. For developed countries, issues such as grid loss reduction, system performance and asset utilization improvement, integration of renewable energy sources, active demand response, and energy efficiency are the main reasons in adopting Smart Grid. Many developed countries experience system reliability degradation resulting from aging grid infrastructure. Inadequate access to "strong" transmission and distribution grid infrastructure limits the potential benefits of integration of renewable energy generation.

22.1.3 Benefits, More Than a Business Case

Smart Grid delivery should not be based only on enabling solutions, but integrated solutions that address business and operating concerns and deliver meaningful, measurable, and sustainable benefits to the utility, the consumer, the economy, and the environment (Figure 22.1).

Various components come into play when considering the impact of Smart Grid technologies. Utilities and customers can benefit in several ways. Rate increases are inevitable, but Smart Grids can offer the prospect of increased utility earnings, together with reduced rate increases (plus improved quality of service). Viewing Smart Grid programs in the context of, for example, a "green" program for customer choice or a cost reduction program to moderate customer rate increases can help define utility drivers and shape the Smart Grid roadmap. A Smart Grid program should have a robust business case where numerous groups in the utility have discussed and agreed upon the expected benefits and costs of

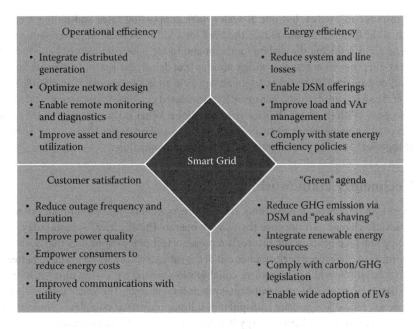

FIGURE 22.1 Smart Grid benefits.

Smart Grid candidate technologies and a realistic implementation plan. In some cases, the benefits are modestly incremental, but a Smart Grid plan should minimize the lag in realized benefits that typically occur after a step change in technology. A Smart Grid deployment is also intended to allow smoother and lower cost migrations to new technologies and avoid the need to incur "forklift" costs. A good Smart Grid plan should move away from the "pilot" mentality and depend on wisely implemented field trials or "phased deployments" that provide the much-needed feedback of cost, benefit, and customer acceptance that can be used to update and verify the business case.

Of utmost importance in implementing Smart Grid solutions are the tangible, quantifiable, and meaningful results:

- Improving the utility's power reliability, operational performance, and overall productivity
- Delivering increases in energy efficiencies and decreases in carbon emissions
- Empowering consumers to manage their energy usage and save money without compromising their lifestyle
- Optimizing renewable energy integration and enabling broader penetration

Improving grid reliability and operational efficiency is possible using more intelligence in the delivery network to monitor power flow in real time and improve voltage control to optimize delivery efficiency and eliminate waste and oversupply. This will reduce overall energy consumption and related emissions while conserving finite resources and lowering the overall cost of electricity. Software applications including smart appliances, home automation systems, etc., that manage load and demand distribution help to empower consumers to manage their energy usage and save money without compromising their lifestyle—encouraging consumers to become smart consumers in smart homes, by giving them access to time-of-use rates and real-time pricing signals that will help them to save on electricity bills and cut their power usage during peak hours. This also helps to improve overall system delivery efficiency and reduce the number of power plants and transmission lines that will need to be built.

Smart Grids will enable broader deployment and optimal inclusion of cleaner, greener energy technologies into the grid from localized and distributed resources, including rooftop solar and combined heat and power plants, and distributed generation, thereby reducing dependence on coal and foreign oil and promoting a sustainable energy future. Electric and PHEV integration will bring another distributed resource to market, but one at scale—with supporting rates and billing mechanisms that can help flatten the load profile and reduce the need for additional peaking power plants and transmission lines potentially reducing the carbon footprint and fostering energy security and independence.

The Smart Grid provides enterprise-wide solutions that deliver far-reaching benefits for both utilities and their end customers. Utilities that adopt Smart Grid technologies can reap significant benefits in reduced capital and operating costs, improved PQ, increased customer satisfaction, and a positive environmental impact. With these capabilities arise questions: What is the potential of the Smart Grid? Is there one set of technologies that can enable both strategic and operational processes? How do the technologies fit together? How do you leverage benefits across applications?

22.1.4 Technology Framework

The Smart Grid is a framework for solutions. It is both revolutionary and evolutionary in nature because it can significantly change and improve the way we operate the electrical system today, while providing for ongoing enhancements in the future. It represents technology solutions that optimize the value chain, allowing us to drive more performance out of the infrastructure we have and to better plan for the infrastructure we will be adding. It requires collaboration among a growing number of interested and invested parties, in order to achieve significant, system-level change. The Smart Grid will embrace more renewable energy, public and private transport, buildings, industrial complexes, and houses; increase grid efficiency; and transfer real-time energy information directly to the consumer—empowering them to make smarter energy choices.

From a high-level system perspective, the Smart Grid can be considered to contain the following major components:

- Smart sensing and metering technologies that provide faster and more accurate response for consumer options such as remote monitoring, time-of-use pricing, and demand-side management (DSM)
- An integrated, standard-based, two-way communication infrastructure that provides an open architecture for real-time information and control to every end point on the grid
- Advanced control methods that monitor critical components, enabling rapid diagnosis, and precise responses appropriate to any event in a "self-healing" manner
- A software systems' architecture with improved interfaces, decision support, analytics, and advanced visualization that enhances human decision making, effectively transforming grid operators and managers into knowledge workers

Interoperability between the different Smart Grid components is paramount. A framework can be used that defines the components at three levels, the electricity infrastructure level, the smart infrastructure level and the Smart Grid solutions level (Figure 22.2). At each of these levels, different applications exist that need to interoperate among themselves (horizontally) and with the levels above or below (vertically).

The Smart Grid will provide a scalable, integrated architecture that delivers not only increased reliability and capital and O&M savings to the utility, but also cost savings and value-added services to customers (AMR/AMI/ADI). A well-designed Smart Grid implementation can benefit from more than just advanced metering infrastructure (AMI). Numerous technologies now touted under the Smart Grid banner are currently implemented to various degrees in utilities. The Smart Grid initiative uses these building blocks to drive toward a more integrated and long-term infrastructure than is intended to realize incremental benefits in operational efficiency and data integration while leveraging open standards. For example, building on the benefits of an AMI with extensive communication coverage across the distribution system helps to improve outage management and enables integrated volt/VAr control (IVVC). In addition, a high-bandwidth communication network provides opportunities for enhanced customer service solutions, such as Internet access, through a home area network (HAN), and a more

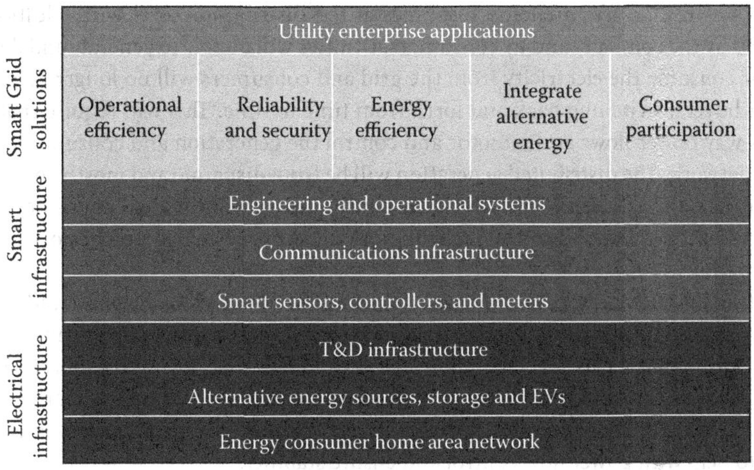

FIGURE 22.2 Smart Grid technology framework.

attractive return on investment. New Smart Grid–driven technologies, such as advanced analytics and visualization, will continue to offer incremental benefits and strengthen a renewed interest in the consumer interface, AMI, demand side management (DSM), and other customer-centric technologies, such as plug-in hybrid electric vehicles (PHEVs).

The development of new capabilities and enabling technologies will be critical to fulfilling the grand promise of the Smart Grid. Smart Grid investments should be directed toward holistic grid solutions that will differentiate utility Smart Grid initiatives. Smart Grid, however, is more than simply a new technology. Smart Grid will have a significant impact on a utility's processes. Perhaps more importantly, it is also about the new information made available by these technologies, and the new customer–utility relationships that will emerge. Enabling technologies such as smart devices, communications and information infrastructures, and operational software are instrumental in the development and delivery of Smart Grid solutions. Each utility customer will begin the Smart Grid journey based upon past actions and investments, present needs, and future expectations.

22.2 Transformation of the Grid

Current transmission and distribution grids were not designed with Smart Grid in mind. They were designed for the cheap, rapid electrification of the country or region. The requirements of Smart Grid are quite different and, therefore, the reengineering of the current grid is imminent. This engineering work will take many forms including enhancements and extensions to the existing grid, inspection and maintenance activities, preparation for distributed generation and storage, and the development and deployment of an extensive two-way communication system.

The "heavy metal" electric delivery system of transmission lines, distribution feeders, switches, breakers, and transformers will remain the core of the utility transmission and distribution infrastructure. Many refer to this as the "dumb" part of the grid. While some changes in the inherent design of these components can be made, for example, the use of amorphous metal in transformers to reduce losses, the "smarts" in the T&D system are related to advances in the monitoring, control, and protection of the "dumb" equipment. Substations therefore play an essential role as the operational interface to the T&D equipment in the field. Advances in technology over the years and the introduction of microprocessor-based monitoring, control, protection, and data acquisition devices have made a marked improvement in the operation and maintenance of the transmission and distribution network. However, changes in the way the T&D system is utilized and operated in a smarter grid will create significant challenges.

There will be an increase in energy resources at the distribution level with additional generation sources placed by the company and by consumers. Utilities will be able to generate and deliver electricity to the grid or consume the electricity from the grid and consumers will no longer be pure consumers, but sellers or buyers, switching back and forth from time to time. This will require that the grid operate with two-way power flows and monitor and control the generation and consumption points on the distribution network. The distributed generation will be from disparate and mostly intermittent sources and subject to great uncertainty. The electricity consumption of individual consumers is also of great uncertainty when they respond to the real-time pricing and rewarding policies of power utilities for economic benefits.

From the transmission perspective, increased amounts of power exchanges and trading will add more stress to the grid and bring the challenge of reducing grid congestion while ensuring grid stability and security and optimizing the use of transmission assets and low-cost generation sources. In order to keep generation, transmission, and consumption in balance, the grids must become more flexible and more effectively controlled. The transmission system will require more advanced technologies such as FACTS and HVDC to help with power flow control and ensure stability.

Substations in a Smart Grid will move beyond basic protection and traditional automation schemes to bring complexity around distributed functional and communication architectures, more advanced local analytics, and data management. There will be a migration of intelligence from the traditional centralized functions and decisions at the energy management and distribution management system level to the substations to enhance reliability, security, and responsiveness of the T&D system. The enterprise system applications will become more advanced in being able to coordinate the distributed intelligence in the substations and feeders in the field to ensure control area and system-wide coordination and efficiency.

Some envisaged smart substation needs and transformations are described in the following.

22.2.1 Engineering and Design

Future substation designs will be driven by current and new well-developed technologies and standards, as well as some new methodologies that are totally different from the existing philosophy. Design requirements will be aimed at either cost reductions while maintaining the same technical performance or performance improvements while assuring no or minimal cost increase. Based on these considerations, smart substation design may take the form of (1) retrofitting existing substations with a major replacement of the legacy equipment while maintaining minimal disruption to the continuity of the services; (2) deploying brand new substation designs using latest off-the-shelf technologies; or (3) greenfield substation design that takes energy market participation, profit optimization, and system operation risk reduction into combined consideration.

Designing the next generation substations will require not only an excellent understanding of primary and secondary equipment in the substation but also the role of the substation in the grid, the region, and for the customers connected to it. Signals for monitoring and control will migrate from analog to digital and the availability of new types of sensors, such as nonconventional current and voltage instrument transformers, will require shifting the engineering and design process from a T&D network focus to also include the substation information and communication architecture. This will require a better understanding of communication networks, data storage, and data exchange needs in the substation. As with other communication networks used in other process or time critical industries, redundancy, security, and bandwidth are an essential part of the design process. Smart substations will require protocols specific to the needs of electric utilities while ensuring interconnectivity and interoperability of the protection, monitoring, control, and data acquisition devices. One approach to overcoming these challenges is to modify the engineering and design documentation process so that it includes detailed communication schematics and logic charts depicting this virtualized circuitry and data communication pathways.

22.2.2 Information Infrastructure

Advances in processing technology have been a major enabler of smarter substations with the cost-effective digitization of protection, monitoring, and control devices in the substation. Digitization of substation devices has also enabled the increase in control and automation functionality and, with it, the proliferation of real-time operational and nonoperational data available in the substation. The availability of the large amounts of data has driven the need for higher speed communications within the substation as well as between the substation and feeder devices and upstream from the substation to supervisory control and data acquisition (SCADA) systems and other enterprise applications, such as outage management and asset management. The key is to filter and process these data so that meaningful information from the T&D system can be made available on a timely basis to appropriate users of the data, such as operations, planning, asset maintenance, and other utility enterprise applications.

Central to the Smart Grid concept is design and deployment of a two-way communication system linking the central office to the substations, intelligent network devices, and ultimately to the customer meter. This communication system is of paramount importance and serves as the nervous system of the Smart Grid. This communication system will use a variety of technologies ranging from wireless, radio frequency (RF), and broadband over power line (BPL), most likely all within the same utility. The management of this communication network will be new and challenging to many utilities and will require new engineering and asset management applications. Enhanced security will be required for field communications, application interfaces, and user access. An advanced EMS and DMS will need to include data security servers to ensure secure communications with field devices and secure data exchange with other applications. The use of IP-based communication protocols will allow utilities to take advantage of commercially available and open-standard solutions for securing network and interface communications.

IEC 61850 will greatly improve the way we communicate between devices. For the first time vendors and utilities have agreed upon an international communication standard. This will allow an unprecedented level of interoperability between devices of multiple vendors in a seamless fashion. IEC 61850 supports both client/server communications as well as peer-to-peer communications. The IEC process bus will allow for communication to the next generation of smart sensors. The self-description feature of IEC 61850 will greatly reduce configuration costs and the interoperable engineering process will allow for the reuse of solutions across multiple platforms. Also, because of a single standard for all device training, engineering and commissioning costs can be greatly reduced.

22.2.3 Operation and Maintenance

The challenge of operations and maintenance in advanced substations with smart devices is usually one of acceptance by personnel. This is a critical part of the change management process. Increased amounts of data from smart substations will increase the amount of information available to system operators to improve control of the T&D network and respond to system events. Advanced data integration and automation applications in the substation will be able to provide a faster response to changing network conditions and events and therefore reduce the burden on system operators, especially during multiple or major system events. For example, after a fault on a distribution feeder further, instead of presenting the system operator with a lockout alarm, accompanied by associated low volts, fault passage indications, battery alarms, and so on, leaving it up to the operator to drill down, diagnose, and work out a restoration strategy will instead notify the operator that a fault has occurred and analysis and restoration is in progress in that area. The system will then analyze the scope of the fault using the information available; tracing the current network model; identifying current relevant safety documents, operational restrictions, and sensitive customers; and locating the fault using location data from the field. The master system automatically runs load-flow studies identifying current loading, available capacities, and possible weaknesses, using this information to develop a restoration strategy. The system then attempts an isolation of the fault and maximum restoration of customers with safe load transfers, potentially involving multilevel feeder reconfiguration to prevent cascading overloads to adjacent circuits. Once the reconfiguration is complete, the system can alert the operator to the outcome and even automatically dispatch the most appropriate crew to the identified faulted section.

22.2.4 Enterprise Integration

Enterprise integration is an essential component of the Smart Grid architecture. To increase the value of an integrated Smart Grid solution, the smart substation will need to interface and share data with numerous other applications. For example, building on the benefits of an AMI with extensive communication coverage across the distribution system and obtaining operational data from the customer point of delivery (such as voltage, power factor, loss of supply, etc.) helps to improve outage management and IVVC implementation locally at the substation level. More data available from substations will also allow more accurate modeling and real-time analysis of the distribution system and will enable optimization algorithms to run, reducing peak load and deferring investment in transmission and distribution assets. By collecting and analyzing nonoperational data, such as key asset performance information, sophisticated computer-based models can be used to assess current performance and predict possible failures of substation equipment. This process combined with other operational systems, such as mobile workforce management, will significantly change the maintenance regime for the T&D system.

22.2.5 Testing and Commissioning

The challenge of commissioning a next generation substation is that traditional test procedures cannot adequately test the virtual circuitry. The best way to overcome this challenge is to use a system test methodology, where functions are tested end to end as part of the virtual system. This allows performance and behavior of the control system to be objectively measured and validated. Significant changes will also be seen in the area of substation integration and automation database management and the reduction of configuration costs. There is currently a work underway to harmonize the EPRI CIM model and Enterprise Service Bus IEC 61968 standards with the substation IEC 61850 protocol standards. Bringing these standards together will greatly reduce the costs of configuring and maintaining a master station through plug and play compatibility and database self-description.

22.3 Substation Technology Advances

The early generations of SCADA systems typically employed one remote terminal unit (RTU) at every substation. With this architecture, all cables from the field equipment had to be terminated at the RTU. RTUs have typically offered limited expansion capacities. For analog inputs, the RTU required the use of transducers to convert higher level voltages and currents from CT and potential transformer (PT) outputs into the milliamp and volt level. Most RTUs had a single communication port and were only capable of communicating with one master station. The communication between an RTU and its master station was typically achieved via proprietary bit-oriented communication protocols. As technology advanced, RTUs became smaller and more flexible. This allowed for a distributed architecture approach, with one smaller RTU for one or several pieces of substation equipment. This resulted in lower installation costs with reduced cabling requirements. This architecture also offered better expansion capabilities (just add more small RTUs). In addition, the new generation of RTUs was capable of accepting higher level AC analog inputs. This eliminated the need for intermediate transducers and allowed direct wiring of CTs and PTs into the RTU. This also enabled RTUs to have additional functionality, such as digital fault recording (DFR) and power quality (PQ) monitoring.

There were also advances in communication capabilities, with additional ports available to communicate with intelligent electronic devices (IEDs). However, the most significant improvement was the introduction of an open communication protocol. The older SCADA systems used proprietary protocols to communicate between the master station and the RTUs. Availability of an open and standard (for the most part) utility communication protocol allowed utilities to choose vendor independent equipment for the SCADA systems. The de facto standard protocol for electric utilities SCADA systems in North America became DNP3.0. Another open communication protocol used by utilities is MODBUS. The MODBUS protocol came from the industrial manufacturing environment. The latest communication standard adopted by utilities is IEC 61850. IEC 61850 supports a very powerful and flexible network-based, object-oriented communication implementation that allows for flexible and expandable multivendor solutions and standardized engineering and design.

Another technology that aided SCADA systems was network data communications. The SCADA architecture based on serial communication protocols put certain limitations on system capabilities. With a serial SCADA protocol architecture,

- There is a static master/slave data path that limits the device connectivity.
- Serial SCADA protocols do not allow multiple protocols on a single channel.
- There are issues with exchanging new sources of data, such as oscillography files, PQ data, etc.
- Configuration management has to be done via a dedicated "maintenance port".

The network-based architecture offers a number of advantages:

- *Significant improvement in speed and connectivity*: An Ethernet based local area network (LAN) greatly increases the available communication bandwidth. The network layer protocol provides a direct link to devices from anywhere on the network.
- *Availability of logical channels*: Network protocols support multiple logical channels across multiple devices.
- *Ability to use new sources of data*: Each IED can provide another protocol port number for file or auxiliary data transfer without disturbing other processes (e.g., SCADA) and without additional hardware.
- *Improved configuration management*: Configuration and maintenance can be done over the network from a central location.

The network-based architecture in many cases also offers a better response time, ability to access important data, and reduced configuration and system management time. Take, for example, SCADA systems that have been around for many years. These were simple remote monitoring and control systems exchanging data over low-speed communication links, mostly hardwired. In recent years, with the proliferation of microprocessor-based IED, it became possible to have information extracted directly from these IEDs either by an RTU or by other substation control system (SCS) components. This is achieved by using the IED communication capabilities, allowing it to communicate with the RTU, data concentrator, or directly with the master station. As more IEDs were installed at the substations, it became possible to integrate some of the protection, control, and data acquisition functionality. A lot of the information previously extracted by the RTUs now became available from the IEDs. However, it may not be practical to have the master station communicate directly with the numerous IEDs in all the substations. To enable this data flow, a new breed of devices called substation servers is utilized. A substation server communicates with all the IEDs at the substation, collects all information from the IEDs, and then communicates back to the central master station. Because the IEDs at the substation use many different communication protocols, the substation server has to be capable of communicating via these protocols, as well as the master station's communication protocol. A substation server allows the SCADA system to access data from most substation IEDs, which were only accessible locally before.

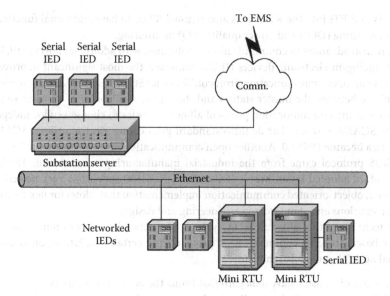

FIGURE 22.3 Server-based substation control system architecture.

With the substation server–based SCADA architecture (Figure 22.3), all IEDs (including RTUs) are polled by the substation server. The IEDs and RTUs with network connections are polled over the substation LAN. The IEDs with only serial connection capabilities are polled serially via the substation server's serial RS232 or RS485 ports (integrated or distributed). In addition to making additional IED data available, the substation server significantly improves overall SCADA system communication performance. With the substation server-based architecture, the master station has to communicate directly with only the substation server instead of multiple RTUs and IEDs at the substation. Also, a substation servers' communication capability is typically superior to that of an IED. This, and the reduced number of devices directly connected to the master station, contributes to a significantly improved communication performance in a polled environment.

Modern IEDs, such as protection relays and meters, have a tremendous amount of information. Some of these devices have thousands of data points available. In addition, many IEDs generate file type data such as DFR or PQ files. A typical master station is not designed to process this amount of data and this type of data. However, a lot of this information can be extremely valuable to the different users within the utility, as well as, in some cases, the utility's customers. To take advantage of these data, an extraction mechanism independent from the master station needs to be implemented. Operational data and nonoperational data have independent data collection mechanisms. Therefore, two separate logical data paths should also exist to transfer these data (Figure 22.4). One logical data path connects the substation with the EMS (operational data). A second data path transfers nonoperational data from the substation to a separate engineering data warehouse (EDW) system. With all IEDs connected to the substation data concentrator, and sufficient communication infrastructure in place, it also becomes possible to have a remote maintenance connection to most of the IEDs. This functionality is referred to as either "remote access" or "pass-through." Remote access or pass-through is the ability to have a virtual connection to remote devices via a secure network. This functionality significantly helps with troubleshooting and maintenance of remote equipment. In many cases, it can eliminate the need for technical personnel to drive to a remote location. It also makes real-time information from individual devices at different locations available at the same computer screen that makes the troubleshooting process more efficient.

An advanced substation integration architecture (Figure 22.5) offers increased functionality by taking full advantage of the network-based system architecture, thus allowing more users to access important information from all components connected to the network. However, it also introduces

Role of Substations in Smart Grids   **22**-13

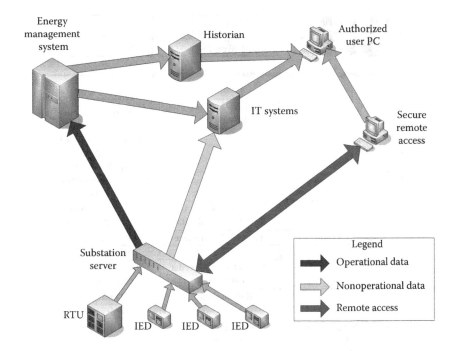

FIGURE 22.4 Substation data flow.

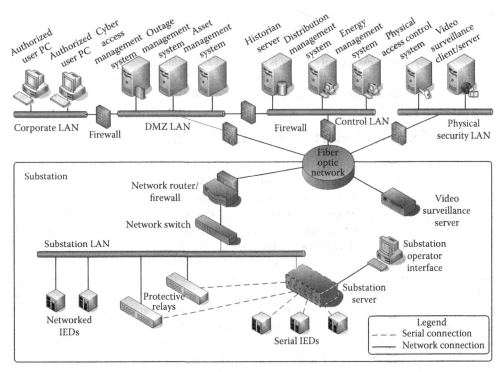

FIGURE 22.5 Utility communication and security architecture with smart substation. (Courtesy of Michael Pesin. Copyright 2012.)

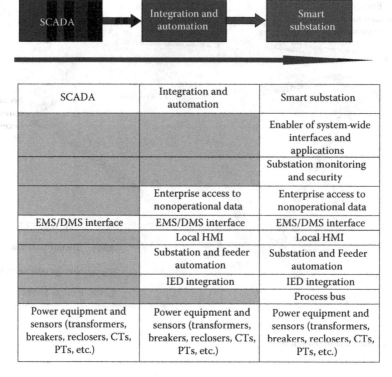

FIGURE 22.6 Substation smart grid migration.

additional security risks into the control system. To mitigate these risks, special care must to be taken when designing the network, with special emphasis on the network security and the implementation of user authentication, authorization, and accounting. It is very important that a substation communication and physical access security policy is developed and enforced.

Figure 22.6 shows the conceptual migration path from basic SCADA functionality, through integration and automation to a full Smart Grid substation solution.

22.4 Platform for Smart Feeder Applications

Distribution networks have not always been the focus of operational effectiveness. As supply constraints continue, however, there will be more focus on the distribution network for cost reduction and capacity relief. Monitoring and control requirements for the distribution system will increase, and the integrated Smart Grid architecture will benefit from data exchange between smarter distribution field devices and enterprise applications. The emergence of widespread distributed generation and consumer demand response programs also introduces considerable impact to the DMS operation.

Monitoring, control, and data acquisition of the electricity network will extend further down to the distribution pole-top transformer and perhaps even to individual customers, either through the substation communication network, by means of a separate feeder communication network, or tied into the AMI. More granular field data will help increase operational efficiency and provide more data for other Smart Grid applications, such as outage management. Higher speed and increased bandwidth communications for data acquisition and control will be needed. Fault detection, isolation, and service restoration (FDIR) on the distribution system will require a higher level of optimization and will need to include optimization for closed-loop, parallel circuit, and radial configurations. Multilevel feeder reconfiguration, multiobjective restoration strategies, and forward-looking network loading validation will be additional features with FDIR. IVVC will include operational and asset improvements, such as identifying failed capacitor banks and tracking capacitor bank, tap changer, and regulator operation to provide sufficient statistics for opportunities to optimize capacitor bank and regulator placement in the network. Regional IVVC objectives may include operational or cost-based optimization.

Dashboard metrics, reporting, and historical data will be essential tools for tracking performance of the distribution network and related Smart Grid initiatives. For example, advanced distribution management will need to measure and report the effectiveness of grid efficiency programs, such as VAr optimization, or the system average interruption duration index (SAIDI), the system average interruption frequency index (SAIFI), and other reliability indices related to delivery optimization Smart Grid technologies. Historical databases will also allow verification of the capability of the Smart Grid optimization and efficiency applications over time, and these databases will allow a more accurate estimation of the change in system conditions expected when the applications are called upon to operate. Alarm analysis, disturbance, event replay, and other PQ metrics will add tremendous value to the utility and improve relationships with customers. Load forecasting and load management data will also help with network planning and optimization of network operations. This will all require advanced smart substation and feeder solutions and a broader perspective on how integration of substation and feeder data and T&D automation can benefit the Smart Grid.

> Realizing the promises and benefits of a smarter grid—from improved reliability, to increased efficiency, to the integration of more renewable power—will require a smarter distribution grid, with advanced computing power and two-way communications that operate at the speed of our 21st Century digital society. The problem—only approximately 10% of the 48,000 distribution substations on today's grid in the U.S. are digitized. Upgrading these substations to meet today's energy challenges will require time, resources and money.[4]

22.5 IEC 61850 in Smart Substations

22.5.1 Paradigm Shift in Substation Design

For many years, the current generation substation designs have been based on that functionality and over time we have developed several typical designs for the primary and secondary systems used in these substations. Examples of such typical schemes for the primary equipment are shown in Figure 22.7 and include the breaker and a half scheme, the double busbar scheme, the single busbar scheme, and the ring bus scheme. These schemes have been described and defined in many documents including Cigré Technical Brochure 069 General guidelines for the design of outdoor AC substations using facts controllers.

For the secondary equipment (protection, control, measurement, and monitoring) typical schemes have also been in use, but here we have seen more development in new concepts and philosophies. Typical concepts for secondary equipment include redundant protection for transmission system using different operating principles and manufacturers and separate systems for control, measurements, monitoring, data acquisition, operation, etc. At distribution, integrated protection and control at feeder level is a common solution. In general, it can be said that the concepts used for the secondary systems have been based on the primary designs and the way the utility wants to control, protect, and monitor these systems.

In general, the existing or conventional substations are designed using standard design procedures for high-voltage switchgear in combination with copper cables for all interfaces between primary and secondary equipment.

Several different types of circuits are used in the substation:

- Analog (current and voltage)
- Binary—protection and control signals
- Power supply—DC or AC

A typical conventional substation design is shown in Figure 22.8.

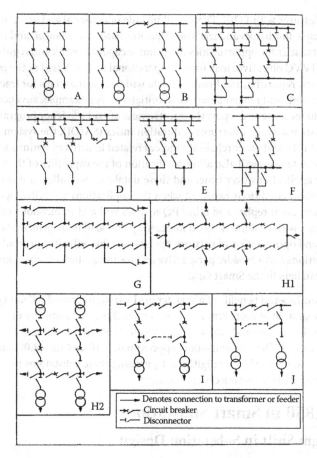

FIGURE 22.7 Typical traditional primary substation schemes. A–J refers to different substation primary plant topologies as they are used within the Cigre report that this figure comes from. (From Cigré Technical Brochure 069, General guidelines for the design of outdoor a.c. substations using facts controllers, p. 8. Copyright 2012 Cigré.)

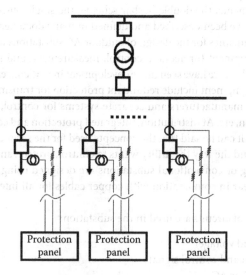

FIGURE 22.8 Typical conventional substation design. (From Apostolov, A. and Janssen, M., IEC 61850 impact on substation design, paper number 0633, IEEE PES. Copyright 2012 IEEE.)

Depending on the size of the substation, the location of the switchgear components, and the complexity of the protection and control system, there, very often, are a huge number of cables with different lengths and sizes that need to be designed, installed, commissioned, tested, and maintained.

A typical conventional substation has multiple instrument transformers and circuit breakers associated with the protection, control, monitoring, and other devices being connected from the switch yard to a control house or building with the individual equipment panels.

These cables are cut to a specific length and bundled, which makes any required future modification very labor intensive. This is especially true in the process of refurbishing old substations where the cable insulation is starting to fail.

The large amount of copper cables and the distances that they need to cover to provide the interface between the different devices expose them to the impact of electromagnetic transients and the possibility for damages as a result of equipment failure or other events.

The design of a conventional substation needs to take into consideration the resistance of the cables in the process of selecting instrument transformers and protection equipment, as well as their connection to the instrument transformers and between themselves. The issues of CT saturation are of special importance to the operation of protection relays under maximum fault conditions. Also, Ferro resonance in voltage transformers has to be considered with relation to the correct operation of the protection and control systems.

Failures in the cables in the substation may lead to misoperation of protection or other devices and can represent a safety issue. In addition, open CT circuits, especially when it occurs while the primary winding is energized, can cause severe safety issues as the induced secondary e.m.f. can be high enough to present a danger to people's life and equipment insulation.

The aforementioned list is definitely not a complete list of all the issues that need to be taken into consideration in the design of a conventional substation. It provides some examples that will help better understand the impact of IEC 61850 in the substation.

In order to take full advantage of any new technology, it is necessary to understand what it provides. The next section of the chapter gives a short summary of some of the key concepts of the standard that have the most significant impact on the substation design.

22.5.2 Interoperability and IEC 61850

IEC 61850 is a vendor-neutral, open systems standard for utility communications, significantly improving functionality while yielding substantial customer savings. The standard specifies protocol independent and standardized information models for various application domains in combination with abstract communication services, a standardized mapping to communication protocols, a supporting engineering process, and testing definitions. This standards allows standardized communication between IEDs located not only within electric utility facilities, such as power plants, substations, and feeders, but also outside these facilities such as wind farms, electric vehicles, storage systems, meters, etc. The standard also includes requirements for database configuration, object definition, file processing, and IED self-description methods. These requirements will make adding devices to a utility automation system as simple as adding new devices to a computer using "plug and play" capabilities. With IEC 61850, utilities will benefit from cost reductions in system design, substation wiring, redundant equipment, IED integration, configuration, testing, and commissioning. Additional cost savings will also be gained in training, MIS operations, and system maintenance.

IEC 61850 has been identified by the National Institute of Standards and Technology (NIST) as a cornerstone technology for field device communications and general device object data modeling. IEC 61850 Part 6, which defines the configuration language for systems based on the standard. Peer-to-peer communication mechanisms such as the Generic Object Oriented System Event (GOOSE) will minimize wiring between IEDs. The use of peer-to-peer communication in combination with the use of sampled values from sensors will minimize the use of copper wiring throughout the substation leading to significant benefits in cost savings, more compact substation designs, and advanced and more flexible automation systems, to name a few. With high-speed Ethernet, the IEC 61850-based communication system will be able to manage all of the data available at the process level as well as at the station level.

The IEC 61850 standard was originally designed to be a substation communication solution and was not designed to be used over the slower communication links typically used in distribution automation (DA). However, as wide area and wireless technologies (such as WiMAX) advance, IEC 61850 communications to devices in the distribution grid will become possible. It is therefore possible that IEC 61850 will eventually be used in all aspects of the utility enterprise. At this time, an IEC WG is in the process of defining new logical nodes (LNs) for distributed resources—including photovoltaic, fuel cells, reciprocating engines, and combined heat and power.

With the introduction of serial communication and digital systems, the way we look at secondary systems is fundamentally changing. Not only are these systems still meant to control, protect, and monitor the primary system but also we expect these systems to provide more information related to a realm of new functions. Examples of new functions include the monitoring of the behavior, the aging, and the dynamic capacity of the system. Many of the new functions introduced in substations are related to changing operating philosophies, the rise of distributed generation, and the introduction of renewable energy. For protection, new protection philosophies are being introduced that focus more on the dynamic adaption of protection functions to the actual network topology, wide-area protection and monitoring, the introduction of synchrophasors, and many more.

This tendency is not new. Ever since the introduction of the first substation automation systems and digital protection, we have been searching for ways to make better use of the technologies at hand. After many experiments and discussions, this has led to the development of IEC 61850, originally called "communication networks and systems in substations." It has now evolved into a worldwide standard called "communication networks and systems for power utility automation" providing solutions for many different domains within the power industry.

The concepts and solutions provided by IEC 61850 are based on three cornerstones:

- *Interoperability*: The ability of IED from one or several manufacturers to exchange information and use that information for their own functions.
- *Free configuration*: The standard shall support different philosophies and allow a free allocation of functions, for example, it will work equally well for centralized (RTU like) or decentralized (SCS like) systems.
- *Long-term stability*: The standard shall be future proof, that is, it must be able to follow the progress in communication technology as well as evolving system requirements.

This is achieved by defining a level of abstraction that allows for the development of basically any solution using any configuration that is interoperable and stable in the long run. The standard defines different logical interfaces within a substation that can be used by functions in that substation to exchange information between them. This is shown in Figure 22.9.

IEC 61850 does not predefine or prescribe communication architectures. The interfaces shown in Figure 22.9 are logical interfaces. IEC 61850 allows in principle any mapping of these interfaces on communication networks. A typical example could be to map interfaces 1, 3, and 6 on what we call a station bus. This bus is a communication network focused on the functions at bay and station level. We also could map interfaces 4 and 5 on a process bus, a communication network focused on the process and bay level of a substation. The process bus may in such a case be restricted to one bay, while the station bus might connect functions located throughout the substation. However, it may be possible as well to map interface 4 on a point-to-point link connecting a process-related sensor to the bay protection.

IEC 61850 is in principle restricted to digital communication interfaces. However, IEC 61850 specifies more than the communication interfaces. It includes domain-specific information models. In case of substation, a suite of substation functions have been modeled providing a virtual representation of the substation equipment. The standard, however, also includes the specification of a configuration language. This language

Role of Substations in Smart Grids   **22**-19

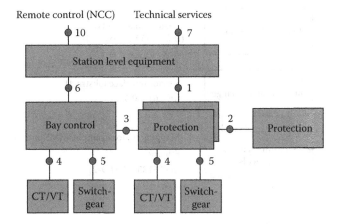

FIGURE 22.9 Interfaces within a substation automation system. (Courtesy of Marco Janssen. Copyright 2012.)

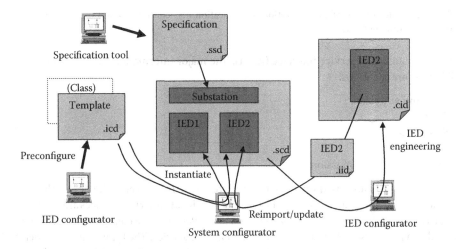

FIGURE 22.10 The engineering approach in IEC 61850. (Courtesy of Marco Janssen. Copyright 2012.)

defines a suite of standardized, XML-based, files that can be used to define in a standardized way the specification of the system, the configuration of the system, and the configuration of the individual IEDs within a system. The files are defined such that they can be used to exchange configuration information between tools from different manufacturers of substation automation equipment. This is shown in Figure 22.10.

The definitions in IEC 61850 are based on a layered approach. In this approach, the domain-specific information models, abstract communication services, and the actual communication protocol are defined independently. This basic concept is shown in Figure 22.11.

IEC 61850 is divided in parts, and in parts 7-3 and 7-4xx, the information model of the substation equipment is specified. These information models include models for primary devices, such as circuit breakers, and instrument transformers, such as CTs and VTs. They also include the models for secondary functions such as protection, control, measurement, metering, monitoring, and synchrophasors.

In order to have access to the information contained in the information models, the standard defines protocol-independent, abstract communication services. These are described in part 7-2, such that the information models are coupled with communication services suited to the functionality making use of the models. This definition is independent from any communication protocol and is called the

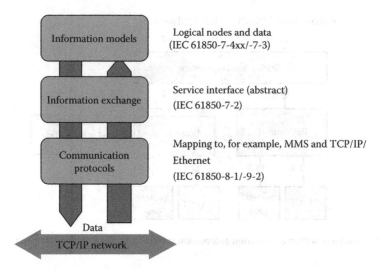

FIGURE 22.11 The concept of the separation of application and communication in IEC 61850. (Courtesy of Marco Janssen. Copyright 2012.)

abstract communication service interface (ACSI). The major information exchange models defined in IEC 61850-7-2 are as follows:

- Read and write data
- Control
- Reporting
- GOOSE
- Sampled value transmission

The first three models are based on a client–server relation. The server is the device that contains the information, while the client is accessing the information. Read and write services are used to access data or data attributes. These services are typically used to read and change configuration attributes. Control model and services are somehow a specialization of a write service. The typical use is to operate disconnector, earthing switches, and circuit breakers. The reporting model is used for event-driven information exchange. The information is spontaneously transmitted, when the value of the data changed.

The last two models are based on a publisher/subscriber concept. In IEC 61850 for this the term peer-to-peer communication is introduced to stress that publisher subscriber/communication involves mainly horizontal communication among peers. These communication models are used for the exchange of time critical information. The device being the source of the information is publishing the information. Any other device that needs the information can receive it. These models are using multicast communication (the information is not directed to one single receiver).

The GOOSE concept is a model to transmit event information in a fast way to multiple devices. Instead of using a confirmed communication service, the information exchange is repeated regularly. Application of GOOSE services are the exchange of position information from switches for the purpose of interlocking or the transmission of a digital trip signal for protection-related functions.

The model for the transmission of sampled values is used when a waveform needs to be transmitted using digital communication. In the source device, the waveform is sampled with a fixed sampling frequency. Each sample is tagged with a counter representing the sampling time and transmitted over the communication network. The model assumes synchronized sampling, that is, different devices are sampling the waveform at exactly the same time. The counter is used to correlate samples from different sources. That approach creates no requirements regarding variations of the transmission time.

While IEC 61850-8-x specifies the mapping of all models from 7-2 with the exception of the transmission of sampled values, IEC 61850-9-x are restricted to the mapping of the transmission of sampled value model. While IEC 61850-9-2 is mapping the complete model, IEC 61850-9-1 is restricted to a small subset using a point-to-point link providing little flexibility. Both mappings are using Ethernet as communication protocol.

Of course in order to create real implementations, we need communication protocols. These protocols are defined in parts 8-x and 9-x. In these parts is explained how real communication protocols are used to transmit the information in the models specified in IEC 61850-7-3 and -7-4xx using the abstract communication services of IEC 61850-7-2. In the terminology of IEC 61850, this is called "specific communication service mapping" (SCSM).

Through this approach, an evolution in communication technologies is supported since the application and its information models and the information exchange models are decoupled from the protocol used, allowing for upgrading the communication technology without affecting the applications.

The core element of the information model is the logical node. A logical node is defined as the smallest reusable piece of a function. It as such can be considered as a container for function-related data. Logical nodes contain data and these data and the associated data attributes represent the information contained in the, part of the, function. The name of a logical node class is standardized and comprises always four characters. Basically, we can differentiate between two kinds of logical nodes:

- Logical nodes representing information of the primary equipment (e.g., circuit breaker—XCBR or current transformer—TCTR). These logical nodes implement the interface between the switchgear and the substation automation system.
- Logical nodes representing the secondary equipment including all substation automation functions. Examples are protection functions, for example, distance protection—PDIS or the measurement unit—MMXU.

The standard contains a comprehensive set of logical nodes, allowing modeling many of, not all, substation functions. In case a function does not exist in the standard extension rules for logical nodes, data and data attributes have been defined, allowing for structured and standardized extensions of the standard information models.

The mappings currently defined in IEC 61850 (parts 8-x and 9-x) are using the same communication protocols. They differentiate between the client/server services and the publisher/subscriber services. While the client/server services use the full seven-layer communication stack using MMS and TCP/IP, the publisher/subscriber services are mapped on a reduced stack, basically directly accessing the Ethernet link layer.

For the transmission of the sampled values, IEC 61850-9-2 is using the following communication protocols:

- *Presentation layer*: ASN.1 using basic encoding rules (BER) (ISO/IEC 8824-1 and ISO/IEC 8825)
- *Data link layer*: Priority tagging/VLAN and CSMA/CD (IEEE 802.1Q and ISO/IEC 8802-3)
- *Physical layer*: Fiber optic transmission system 100-FX recommended (ISO/IEC 8802-3)

Ethernet is basically a nondeterministic communication solution. However, with the use of switched Ethernet and priority tagging, a deterministic behavior can be achieved. Using full duplex switches, collisions are avoided. Tagging the transmission of sampled values—which requires, due to the cyclic behavior, a constant bandwidth—with a higher priority than the nondeterministic traffic used for, for example, reporting of events ensures that the sampled values always get through.

The model for the transmission of sampled values as specified in IEC 61850-7-2 is rather flexible. The configuration of the message being transmitted is done using a sampled value control block. Configuration options include the reference to the dataset that defines the information contained in one message, the number of individual samples that are packed within one message, and the sampling rate.

While the flexibility makes the concept future proof, it adds configuration complexity. That is why the utility communication architecture (UCA) users' group has prepared the "Implementation guideline for digital interface to instrument transformers using IEC 61850-9-2." This implementation guideline is an agreement of the vendors participating in the UCA users' group, how the first implementations of digital interfaces to instrument transformers will be. Basically, the implementation guideline is defining the following items:

- A dataset comprising the voltage and current information for the three phases and for neutral. That dataset corresponds to the concept of a merging unit (MU) as defined in IEC 60044-8.
- Two sampled value control blocks. A first one for a sample rate of 80 samples per period, where for each set of samples an individual message is sent. A second one for 256 samples per period, where eight consecutive set of samples are transmitted in one message.
- The use of scaled integer values to represent the information including the specification of the scale factors for current and for voltage.

22.5.3 Impact of IEC 61850 in the Substation

In a Smart Grid environment, availability of, and access to, information is key. Standards like IEC 61850 allow the definition of the available information and access to that information in a standardized way.

The IEC 61850 standard Communication Networks and Systems for Utility Automation allows the introduction of new designs for various functions, including protection inside and outside substations. The levels of functional integration and flexibility of communication-based solutions bring significant advantages in costs at various levels of the power system. This integration affects not only the design of the substation but almost every component and/or system in it such as protection, monitoring, and control by replacing the hardwired interfaces with communication links. Furthermore, the design of the high-voltage installations and networks can be reconsidered regarding the number and the location of switchgear components necessary to perform the primary function of a substation in a high-voltage network. The use of high-speed peer-to-peer communications using GOOSE messages and sampled values from MUs allows for the introduction of distributed and wide-area applications. In addition, the use of optical LANs leads in the direction of copper-less substations.

The development of different solutions in the substation protection and control system is possible only when there is good understanding of both the problem domain and the IEC 61850 standard. The modeling approach of IEC 61850 supports different solutions from centralized to distributed functions. The latter is one of the key elements of the standard that allows for utilities to rethink and optimize their substation designs.

A function in an IEC 61850-based integrated protection and control system can be local to a specific primary device (distribution feeder, transformer, etc.) or distributed and based on communications between two or more IEDs over the substation LAN.

Considering the requirements for the reliability, availability, and maintainability of functions, it is clear that in conventional systems numerous primary and backup devices need to be installed and wired to the substation. The equipment as well as the equipment that they interface with must then be tested and maintained.

The interface requirements of many of these devices differ. As a result, specific multicore instrument transformers were developed that allow for accurate metering of the energy or other system parameters, on the one hand, and provide a high dynamic range used by, for example, protection devices.

With the introduction of IEC 61850, different interfaces have been defined that can be used by substation applications using dedicated or shared physical connections—the communication links between the physical devices. The allocation of functions between different physical devices defines the requirements for the physical interfaces and, in some cases, may be implemented in more than one physical LAN or by applying multiple virtual network on a physical infrastructure.

Role of Substations in Smart Grids

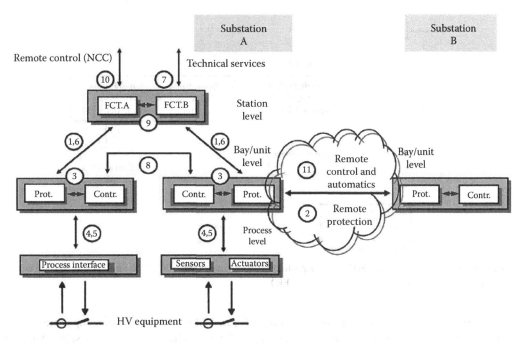

FIGURE 22.12 Logical interfaces in IEC 61850. (From IEC TR61850-1. Copyright 2012 IEC.)

The functions in the substation can be distributed between IEDs on the same or on different levels of the substation functional hierarchy—station, bay, or process as shown in Figure 22.12.

A significant improvement in functionality and reduction of the cost of integrated substation protection and control systems can be achieved based on the IEC 61850-based communications as described in the following.

One example where a major change in substation is expected is at the process level of the substation. The use of nonconventional and/or conventional instrument transformers with digital interface based on IEC 61850-9-2 or the implementation guideline IEC 61850-9-2 LE result in improvements and can help eliminate issues related to the conflicting requirements of protection and metering IEDs as well as alleviate some of the safety risks associated with current and voltage transformers.

The interface of the instrument transformers (both conventional and nonconventional) with different types of substation protection, control, monitoring, and recording equipment as defined in IEC 61850 is through a device called an MU. The definition of an MU in IEC 61850 is as follows:

> Merging unit: interface unit that accepts multiple analogue CT/VT and binary inputs and produces multiple time synchronized serial unidirectional multi-drop digital point to point outputs to provide data communication via the logical interfaces 4 and 5.

MUs can have the following functionality:

- Signal processing of all sensors—conventional or nonconventional
- Synchronization of all measurements—three currents and three voltages
- Analogue interface—high- and low-level signals
- Digital interface—IEC 60044-8 or IEC 61850-9-2

It is important to be able to interface with both conventional and nonconventional sensors in order to allow the implementation of the system in existing or new substations.

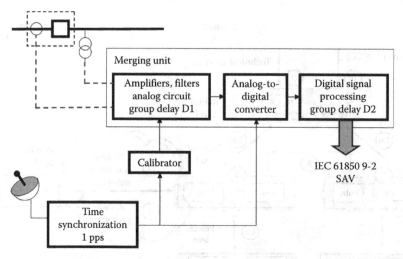

FIGURE 22.13 Concept of the MU. (From Apostolov, A. and Janssen, M., IEC 61850 impact on substation design, paper number 0633, IEEE PES. Copyright 2012 IEEE.)

The MU has similar elements as can be seen from Figure 22.13 as a typical analog input module of a conventional protection or multifunctional IED. The difference is that in this case the substation LAN performs as the digital data bus between the input module and the protection or functions in the device. They are located in different devices, just representing the typical IEC 61850 distributed functionality.

Depending on the specific requirements of the substation, different communication architectures can be chosen as described hereafter.

IEC 61850 is being implemented gradually by starting with adaptation of existing IEDs to support the new communication standard over the station bus and at the same time introducing some first process bus–based solutions.

22.5.4 Station Bus–Based Architecture

The functional hierarchy of station bus–based architectures is shown in Figure 22.14. It represents a partial implementation of IEC 61850 in combination with conventional techniques and designs and brings some of the benefits that the IEC 61850 standard offers.

The current and voltage inputs of the IEDs (protection, control, monitoring, or recording) at the bottom of the functional hierarchy are conventional and wired to the secondary side of the substation instrument transformers using copper cables.

The aforementioned architecture, however, does offer significant advantages compared to conventional hardwired systems. It allows for the design and implementation of different protection schemes that in a conventional system require significant number of cross-wired binary inputs and outputs. This is especially important in large substations with multiple distribution feeders connected to the same medium voltage bus where the number of available relay inputs and outputs in the protection IEDs might be the limiting factor in a protection scheme application. Some examples of such schemes are a distribution bus protection based on the overcurrent blocking principle, breaker failure protection, trip acceleration schemes, or a sympathetic trip protection.

The relay that detects the feeder fault sends a GOOSE message over the station bus to all other relays connected to the distribution bus indicating that it has issued a trip signal to clear the fault. This can be considered as a blocking signal for all other relays on the bus. The only requirement for the scheme implementation is that the relays connected to feeders on the same distribution bus have to subscribe to receive the GOOSE messages from all other IEDs connected to the same distribution bus.

Role of Substations in Smart Grids

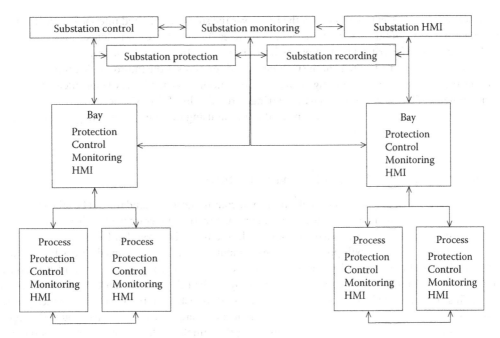

FIGURE 22.14 Station bus functional architecture. (Courtesy of Marco Janssen. Copyright 2012.)

The reliability of GOOSE-based schemes is achieved through the repetition of the messages with increased time intervals until a user defined time is reached. The latest state is then repeated until a new change of state results in sending of a new GOOSE message. This is shown in Figure 22.15.

The repetition mechanism does not only limit the risk that the signal is going to be missed by a subscribing relay. It also provides means for the continuous monitoring of the virtual wiring between the different relays participating in a distributed protection application. Any problem in a device or in the communications will immediately, within the limits of the maximum repetition time interval, be detected and an alarm will be generated and/or an action will be initiated to resolve the problem. This is not possible in conventional hardwired schemes where problems in the wiring or in relay inputs and outputs can only be detected through scheduled maintenance.

One of the key requirements for the application of distributed functions using GOOSE messages is that the total scheme operating time is similar to or better than the time of a hardwired conventional scheme. If the different factors that determine the operating time of a critical protection scheme such as breaker failure protection are analyzed, it is clear that it requires a relay to initiate the breaker failure protection through a relay output wired into an input. The relay output typically has an operating time of 3–4 ms and it is not unusual that the input may include some filtering in order to prevent an undesired initiation of this critical function.

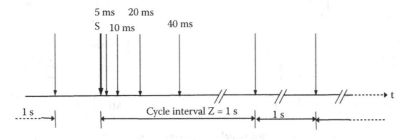

FIGURE 22.15 GOOSE message repetition mechanism. (From IEC TR 61850. Copyright 2012 IEC.)

As a result in a conventional scheme, the time over the simple hardwired interface, being the transmission time between the two functions, will be between 0.5 and 0.75 cycles—longer than the required 0.25 cycles defined for critical protection applications in IEC 61850-based systems.

Another significant advantage of the GOOSE-based solutions is the improved flexibility of the protection and control schemes. Making changes to conventional wiring is very labor intensive and time consuming, while changes of the "virtual wiring" provided by IEC 61850 peer-to-peer communications requires only changes in the system configuration using the substation configuration language (SCL)–based engineering tools.

22.5.5 Station and Process Bus Architecture

Full advantage of all the features available in the new communication standard can be taken if both the station and process bus are used. Figure 22.16 shows the functional hierarchy of such a system.

IEC 61850 communication-based distributed applications involve several different devices connected to a substation LAN. MUs will process the sensor inputs, generate the sampled values for the three-phase and neutral currents and voltages, format a communication message, and multicast it on the substation LAN so that it can be received and used by all the IEDs that need it to perform their functions. This "one-to-many" principle similar to that used to distribute the GOOSE messages provides significant advantages as it not only eliminates current and voltage transformer wiring but also supports the addition of new ideas and/or applications using the sampled values in a later stage as these can simply subscribe to receive the same sample stream.

Another device, the IO unit (IOU) will process the status inputs, generate status data, format a communication message, and multicast it on the substation LAN using GOOSE messages.

All multifunctional IEDs will receive the sampled value messages as well as the binary status messages. The ones that have subscribed to these data then process the data, make a decision, and operate by sending another GOOSE message to trip the breaker or perform any other required action.

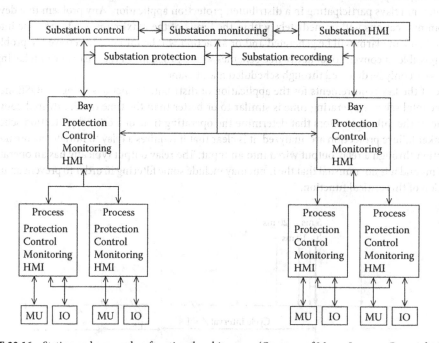

FIGURE 22.16 Station and process bus functional architecture. (Courtesy of Marco Janssen. Copyright 2012.)

Role of Substations in Smart Grids

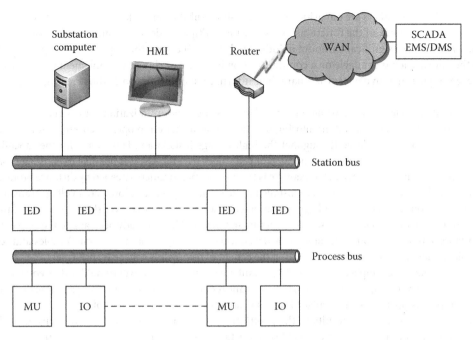

FIGURE 22.17 Communication architecture for process and station bus. (Courtesy of Marco Janssen. Copyright 2012.)

Figure 22.17 shows the simplified communication architecture of the complete implementation of IEC 61850. The number of switches for both the process and substation busses can be more than one depending on the size of the substation and the requirements for reliability, availability, and maintainability.

Figure 22.18 is an illustration of how the substation design changes when the full implementation of IEC 61850 takes place. All copper cables used for analog and binary signals exchange between devices are replaced by communication messages over fiber. If the DC circuits between the substation battery

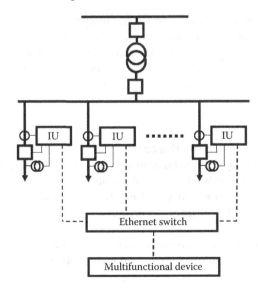

FIGURE 22.18 Alternative substation design. (From Apostolov, A. and Janssen, M., IEC 61850 impact on substation design, paper number 0633, IEEE PES. Copyright 2012 IEEE.)

and the IEDs or breakers are put aside, the "copper-less" substation is a fact. We can then even go a step further and combine all the functions necessary for multiple feeders into one multifunctional device, thus eliminating a significant amount of individual IEDs. Of course the opposite is also possible. Since all the information is available on a communication bus, we can choose to implement relatively simple or even single function devices that share their information on the network, thus creating a distributed function.

The next possible step when using station and process bus is the optimization of the switchgear. In order for the protection, control, and monitoring functions in a substation to operate correctly, several instrument transformers are placed throughout the high-voltage installation. However, with the capability to send voltage and current measurements as sampled values over a LAN, it is possible to eliminate some of these instrument transformers. One example is the voltage measurements needed by distance protections. Traditionally, voltage transformers are installed in each outgoing feeder. However, if voltage transformers are installed on the busbar, the voltage measurements can be transmitted over the LAN to each function requiring these measurements. These concepts are not new and have already been applied in conventional substations. In conventional substations, however, it requires large amounts of (long) cables and several auxiliary relays limiting or even eliminating the benefit of having less voltage transformers.

Process bus–based applications offer important advantages over conventional hardwired analog circuits. The first very important one is the significant reduction in the cost of the system due to the fact that multiple copper cables are replaced with a small number of fiber-optic cables.

Using a process bus also results in the practical elimination of CT saturation of conventional CTs because of the elimination of the current leads resistance. As the impedance of the MU current inputs is very small, this results in the significant reduction in the possibility for CT saturation and all associated with its protection issues. If nonconventional instrument transformers can be used in combination with the MUs and process bus, the issue of CT saturation will be eliminated completely as these nonconventional CTs do not use inductive circuits to transduce the current.

Process bus–based solutions also improve the safety of the substation by eliminating one of the main safety related problems—an open current circuit condition. Since the only current circuit is between the secondary of a current transformer (CT) and the input of the MU is located right next to it, the probability for an open current circuit condition is very small. It becomes nonexistent if optical current sensors are used.

Last, but not least, the process bus improves the flexibility of the protection, monitoring, and control systems. Since current circuits cannot be easily switched due to open circuit concerns, the application of bus differential protection, as well as some backup protection schemes, becomes more complicated. This is not an issue with process bus because any changes will only require modifications in the subscription of the protection IEDs receiving the sampled analog values over IEC 61850 9-2.

22.5.6 Summary

IEC 61850 is a communication standard that allows the development of new approaches for the design and refurbishment of substations. A new range of protection and control applications results in significant benefits compared to conventional hardwired solutions.

It supports interoperability between devices from different manufacturers in the substation that is required in order to improve the efficiency of microprocessor-based relay applications and implement new distributed functions.

High-speed peer-to-peer communications between IEDs connected to the substation LAN based on exchange of GOOSE messages can successfully be used to replace hardwiring for different protection and control applications.

Sampled analog values communicated from MUs to different protection devices connected to the communication network replace the copper wiring between the instrument transformers in the substation yard and the IEDs.

Such systems provide some significant advantages over conventional protection and control systems used to perform the same functions in the substations:

- Reduced wiring, installation, maintenance, and commissioning costs
- Optimization possibilities in the design of the high-voltage system in a substation
- Easy adaptation to changing configurations in the substation
- Practical elimination of CT saturation and open circuits
- Easier implementation of complex schemes and solutions as well as easier integration of new applications and IEDs by using GOOSE messages and sampled values that are multicasted on the communication network and that the applications and IEDs can simply subscribe to

22.6 Smart Grid, Where Do We Go from Here?

The integration of relatively large scale of new generation and active load technologies into the electric grid introduces real-time system control and operational challenges around reliability and security of the power supply. These challenges, if not addressed properly, will result in degradation of service, diminished asset service life, and unexpected grid failures, which will impact the financial performance of the utility's business operations and public relationship image. If these challenges are met effectively, optimal solutions can be realized by the utility to maximize return on investments in advanced technologies. To meet these needs, a number of challenges must be addressed:

- Very high numbers of operating contingencies different from "system as design" expectations
- High penetration of intermittent renewable and distributed energy resources, with their (current) characteristic of limited controllability and dispatchability
- Power quality issues (voltage and frequency variation) that cannot be readily addressed by conventional solutions
- Highly distributed, advanced control and operation logic
- Slow response during quickly developing disturbances
- Volatility of generation and demand patterns and wholesale market demand elasticity
- Adaptability of advanced protection schemes to rapidly changing operational behavior due to the intermittent nature of renewable and distributed generation (DG) resources

In addition, with wide deployment of Smart Grid, there will be an abundance of new operational and nonoperational devices and technologies connected to the wide-area grid. The wide range of devices will include smart meters; advanced monitoring, protection, control, and automation; EV chargers; dispatchable and nondispatchable DG resources; energy storage; etc. Effective and real-time management and support of these devices will introduce enormous challenges for grid operations and maintenance. To effectively address all these challenges, it is necessary to engineer, design, and operate the electric grid with an overarching solution in mind, enabling overall system stability and integrity. A Smart Grid solution, from field devices to the utility's control room, utilizing intelligent sensors and monitoring, advanced grid analytical and operational and nonoperational applications, comparative analysis and visualization, will enable wide-area and real-time operational anomaly detection and system "health" predictability. These will allow for improved decision-making capabilities, PQ, and reliability. An integrated approach will also help to improve situational awareness, marginal stress evaluation, and congestion management and recommend corrective action to effectively manage high penetration of new alternative generation resources and maximize overall grid stability.

It is without a doubt that the Smart Grid hype has gained much interest and backing worldwide, driven by regulatory initiatives in advanced metering and from funding opportunities, not only in the United States but also in other regions of the world. Smart Grid has helped bring more focus on cleaner, more efficient, and greener solutions to electric energy usage. The discussion about Smart Grids in combination with the political awareness regarding carbon emissions and the finiteness of fossil fuels has helped heighten the interest, awareness, and commitment from consumers in energy reduction and savings. While advanced technologies have been applied to the electric utility industry for many years, the Smart Grid brings promises of synergies and advancements in integrating technologies and focusing on both the consumer and the utility. Changes in the utility industry have historically been very slow and while there has been a tremendous momentum in the Smart Grid evaluation and foundation work of standards, interoperability and security, it will take a few more years to determine if the efforts are just a dot.com fad that will pass, or if the support and excitement will certainly drive the industry into a new era where hopefully Thomas Edison would have a hard time recognizing the utility if he were alive today.

References

1. Available at http://www.smartgrids.eu
2. GRID 2030: A National Vision for Electricity's Second 100 years, United States Department of Energy, Office of Electric Transmission and Distribution, 2003.
3. Report to NIST on the Smart Grids Interoperability Standards Roadmap, EPRI, 2009.
4. Available at GE website: http://ge.ecomagination.com/smartgrid, 2009.
5. Cigré Technical Brochure 069. General guidelines for the design of outdoor a.c. substations using facts controllers. p. 8.
6. A. Apostolov and M. Janssen. IEC 61850 impact on substation design, paper number 0633, IEEE PES, 2008.

Index

A

AAA servers, *see* Authentication, Authorization, and Accounting (AAA) servers
Aboveground GIL installation method, 18-6, 18-8
AC and DC yard, 5-29
Accidental ground circuit
    conditions, 11-2–11-5
    high-speed fault clearing, 11-6
    permissible body current limits, 11-5–11-6
    tolerable voltages, 11-7–11-9
Advanced mobile phone system (AMPS), 15-19
Advanced radio data information service (ARDIS), 15-18–15-19
Aging infrastructure management
    description, 19-10
    management concept
        inexorable trend, 19-11
        master lifetime management plan, 19-12
        practical issues, lifetime management, 19-12–19-14
    operating policy, 19-11
    sustainable point, 19-10
Air connection, 2-9–2-10
Air-insulated substations (AIS), 5-1, 9-6
All dielectric self-supporting (ADSS) cables, 15-20
Aluminum
    corrosion protection, 18-31
    GIS, 2-4
    uses, 11-20
AMPS, *see* Advanced mobile phone system (AMPS)
Analog-to-digital (A/D) conversion, 6-7–6-9, 6-15–6-16
Animal deterrents, 10-1–10-2
Animal types
    birds, 10-3
    clearance requirement, 10-2
    raccoons, 10-3
    snakes, 10-3
    squirrels, 10-2–10-3
ARDIS, *see* Advanced radio data information service (ARDIS)

Asset management
    business-driven approach
        coordinated cross-functional decision making, 19-5–19-6
        data-driven process, 19-2
        description, 19-2–19-3
        framework, 19-3–19-5
        one utility, KPIs, 19-3
    description, 19-1–19-2, 19-36–19-37
    electric utility
        efficient frontier, 19-15–19-17
        Pareto curves, 19-15–19-17
        risk-based asset management methods, 19-17–19-19
        standards-driven to business-based management shift, 19-14–19-15
    functional elements
        aging infrastructure management, 19-10–19-14
        multi-KPI contributions evaluation, 19-7
        portfolio management, 19-6–19-7
        probabilistic risk management, 19-8–19-10
        total KPI contribution, prioritization based, 19-7–19-9
    half-measure approaches, 19-25
    must-do projects limitations, 19-25–19-26
    philosophy and approach changes
        business-case approach, 19-26
        comprehensive and balanced integration, 19-26
        lower standards, 19-27–19-28
        perspective and organizational culture, 19-28
        reduce spending, 19-28
        standardization, 19-27
    project evaluation and portfolio selection implementation
        conditional approval, 19-25
        constraints, 19-22–19-24
        cultural changes, 19-21
        decision-making process, 19-21–19-22
        portfolio optimization, 19-21, 19-23

substation
- advanced technology, 19-33
- BBOM, 19-32
- condition-based maintenance, 19-31–19-32
- Eastside substation, 19-20–19-21
- equipment stewardship, 19-33–19-36
- life cycle management, 19-29–19-31
- priorities setting, 19-32
- utility's guideline, 19-19
- utilization, 19-29–19-31

Asymmetrical currents, 11-21
Asynchronous transfer mode (ATM), 15-27
Authentication, Authorization, and Accounting (AAA) servers, 17-4–17-5
Average loss of expected life (ALEL), 19-13–19-14
A-weighted sound level, 9-14

B

Bald and Golden Eagle Protection Act, 10-2
Big bad outage measure (BBOM), 19-32
BPS, *see* Bulk-power system (BPS)
Breaker control interface, 6-19
Breaker-and-a-half arrangement, 3-5–3-7
Bulk-power system (BPS), 17-5
Business-driven approach, asset management
- coordinated cross-functional decision making, 19-5–19-6
- data-driven process, 19-2
- description, 19-2–19-3
- framework
  - business driven, 19-3
  - multiple objectives, 19-3–19-4
  - risk management, 19-4–19-5
- one utility, KPIs, 19-3

C

Cable bus, 9-6
CES, *see* Community energy storage (CES)
Chain-link fences, 9-4
CIGRE, 17-5
Circuit breakers
- GIS, 2-5
- SF 6 gas dead tank, 4-22, 4-24
- single-tank bulk oil, 4-22–4-23, 4-25
- three-tank bulk oil, 4-22–4-23, 4-25
- types, 4-22
- vacuum, 4-22

Circuit switchers
- horizontal interrupter, 4-20–4-21
- integral horizontal interrupter, 4-20–4-21
- multiple interrupter gap per phase, 4-19–4-20
- vertical interrupter, 4-20

Commercial zones, 9-14

Community acceptance
- aesthetics
  - attenuation of noise, 9-7
  - bus design, 9-6
  - color, 9-4–9-5
  - enclosures, 9-5
  - equipment noise levels, 9-7
  - governmental regulations, 9-7–9-8
  - landscaping and topography, 9-4
  - lighting, 9-5
  - noise abatement methods, 9-8–9-9
  - noise sources, 9-6–9-7
  - structures, 9-5
  - visual simulation, 9-3–9-4
- construction
  - hazardous material, 9-12
  - noise, 9-12
  - safety and security, 9-12
  - site housekeeping, 9-12
  - site preparation, 9-11–9-12
- electric and magnetic fields, 9-9–9-10
- operations
  - fire protection, 9-13
  - hazardous material, 9-13
  - site housekeeping, 9-12–9-13
- permitting process, 9-11
- safety and security, 9-10–9-11
- site location and selection, and preparation
  - factors, 9-2
  - potable water and sewage, 9-3
  - site contamination, 9-3
  - wetlands, 9-2

Community energy storage (CES)
- benefits, 21-8–21-9
- community energy devices, 21-8
- description, 21-7
- power factor correction, 21-8
- specification, 21-7

Conductors, selection of
- asymmetrical currents, 11-21
- materials, 11-19–11-20
- sizing factors, 11-20
- symmetrical currents, 11-20–11-21

Connections, selection of, 11-21–11-22

Containment systems
- collecting ponds with traps, 8-6
- fire-quenching, 8-7
- oil-containment equipment pits, 8-6–8-7
- substation ditching, 8-5
- typical equipment containment solutions, 8-8–8-11
- volume requirements, 8-7–8-8
- yard surfacing and underlying soil, 8-5

Continuous audible sources, 9-6
Continuous radio frequency (RF) sources, 9-7
Control circuit designs, 6-20–6-21
Coordinated cross-functional decision making, 19-5–19-6

Index

Copper clad steel, 11-19–11-20
Corrosion protection, GIL system
    active, 18-33
    passive, 18-31–18-32
Covered trench method, 18-10
Current transformers (CTs)
    GIS, 2-5–2-6
    instrument transformers, 6-11–6-13
    Smart Grids, 22-28
Cyber intrusion, SA system
    corporate network
        AAA server, 17-16
        authentication, 17-18–17-19
        dial-up lines elimination, 17-21
        dial-up lines to IEDs, 17-18
        encrypting communications, 17-19–17-21
        measurements, 17-15
        multifactor authentications, 17-16
        password policies, 17-15–17-16
        RBAC, 17-16
        SCADA communication lines, 17-17–17-18
    detection, 17-21–17-23
    recording, reporting, and restoring, 17-23–17-24
Cyber security
    authentication and encryption, 17-26
    contemporary threats, 17-4
    definitions and terminology
        AAA server, 17-4–17-5
        BPS, 17-5
        CIGRE, 17-5
        DNP3, 17-5
        firewall, 17-5
        IDS, 17-5–17-6
        IEC, 17-6
        IED, 17-6
        IPS, 17-6
        NAC, 17-6
        password, 17-5
        port, 17-6
        protocol, 17-6
        remote access, 17-6–17-7
        RTU, 17-7
        security, 17-7
        SEM, 17-7
    detection systems and firewalls, 17-27
    functional testing and certification, 17-26
    incident reporting sites, 17-26–17-27
    intrusion prevention, 17-27
    SA system (*see* Substation automation (SA) system)
    secure real-time operating systems, 17-26
    secure recovery, 17-27
    security program, 17-24–17-25
    Smart Grid paradigm changing technologies, 17-2
    Stuxnet, 17-1–17-3
    technical standards and guidelines, 17-27–17-28
    test beds, 17-26
    traditional threats, 17-4

D

Data concentrators, 7-4, 7-11
DC and AC transmission, power capacities, 5-35
Digital microwave systems, 15-19–15-20
Digital subscriber loop (DSL), 15-25–15-26
Direct lightning stroke shielding system
    active lightning terminals, 12-19
    classical design methods
        empirical curves, 12-6–12-9
        fixed-angle, 12-5–12-6
    EGM
        development, 12-7
        discrepancy, 12-9
        minimum stroke current, 12-17
        Monte Carlo simulation, 12-9
        multiple shielding electrodes, 12-15–12-17
        revised, 12-9–12-11, 12-17–12-18
        rolling sphere method, 12-11–12-15
        voltage level changes, 12-16
        Whitehead's model, 12-9
    ESM lightning rods, 12-19
    failure probability, 12-19
    lightning parameters
        detection networks, 12-5
        GFD, 12-5
        Keraunic level, 12-4
        strike distance, 12-2–12-3
        stroke current magnitude, 12-3–12-4
    lightning rods, radioactive tips, 12-19
    point discharge phenomenon, 12-19
    protection design, 12-2
    transients prevention, 12-1
Direct sequence spread spectrum (DSSS), 15-24
Direct transformer connections, 2-10–2-12
Discharge control systems
    flow blocking systems, 8-12–8-15
    oil–water separator systems, 8-11–8-12
Disconnect switches, 2-6–2-7, 4-2–4-13
Distributed network protocol (DNP) 3.0, 7-16–7-17, 17-5
Double bus–double breaker arrangement, 3-2–3-3
Double bus–single breaker arrangement, 3-4–3-5
DSL, *see* Digital subscriber loop (DSL)
DSSS, *see* Direct sequence spread spectrum (DSSS)

E

Early Streamer Emission (ESM) lightning rods, 12-19
Earthquakes, 13-1–13-4
Efficient frontier
    definition, 19-16
    description, 19-15
EGM, *see* Electrogeometric model (EGM)
Electric utility
    efficient frontier, 19-15–19-17
    Pareto curves, 19-15–19-17

risk-based asset management methods, 19-17–19-19
standards-driven to business-based management shift, 19-14–19-15
Electrogeometric model (EGM)
- development, 12-7
- discrepancy, 12-9
- minimum stroke current, 12-17
- Monte Carlo simulation, 12-9
- multiple shielding electrodes, 12-15–12-17
- revised EGM
  - application, 12-11–12-15
  - BIL, 12-10–12-11
  - bus insulators, 12-10
  - description, 12-10
  - Mousa and Srivastava method, 12-17–12-18
  - negative polarity impulse critical flashover, 12-10–12-11
  - *vs.* Whitehead's model, 12-10
  - withstand voltage, insulator strings, 12-11
- rolling sphere method
  - principle, 12-12
  - shield mast protection, stroke current, 12-12–12-15
  - use of, 12-12
- voltage level changes, 12-16
- Whitehead's model, 12-9

Empirical curves design method, 12-6–12-7
Encryption, 15-16–15-17
Endangered Species Act, 10-2
Energy storage
- benefit and maximum market potential estimation, 21-2–21-3
- CES
  - benefits, 21-8–21-9
  - community energy devices, 21-8
  - description, 21-7
  - power factor correction, 21-8
  - specification, 21-7
- characteristics, 21-2
- comparing technologies
  - capital costs, 21-20–21-21
  - system ratings, 21-20
- description, 21-1–21-2
- development technologies, 21-19–21-20
- financial impacts, benefits, 21-11
- first power system, 21-2
- flywheels, 21-18
- lead-acid batteries, 21-12
- lithium-ion batteries, 21-14
- load shifting, 21-4–21-5
- NaS batteries, 21-14–21-15
- NiCd battery, 21-12–21-13
- NiMH batteries, 21-12–21-13
- radial system reliability improvement, 21-7
- solar-storage applications, 21-10–21-11
- stability-related applications, 21-5

- storage and distributed solar, 21-10
- storage technologies, 21-11–21-12
- system peak *vs.* solar peak output comparison, 21-10–21-11
- transmission and distribution
  - applications, 21-4
  - capital deferral, 21-9–21-10
  - deferral, 21-6
- transmission time, 21-4–21-5
- utilities and merchant operators, 21-2
- vanadium redox batteries
  - battery technology reactions, 21-17
  - diagram of, 21-17
  - redox flow systems, 21-16
  - system, 21-16
- voltage support, 21-4
- wholesale energy applications
  - congestion relief, 21-4
  - description, 21-3
  - regulation services, 21-3–21-4
  - renewable firming, 21-4
  - time arbitrage of energy, 21-4
- zinc-bromine batteries, 21-15–21-16

Environmental Liberation Front (ELF), 16-6
Equipment stewardship
- comprehensive asset management approach, 19-34, 19-36
- perspective and culture change, 19-33
- single-planning group approach, 19-35
- traditional standards-driven utility paradigm, 19-34

ESM lightning rods, *see* Early Streamer Emission (ESM) lightning rods

F

Factory acceptance tests (FATs), 1-8
Federal Communications Commission (FCC), 15-20–15-21, 15-23
FHSS, *see* Frequency hopping spread spectrum (FHSS)
Fiber-optic (F/O) technology, 6-3, 6-24–6-26, 15-20
Firewall, 17-5
Fixed-angle design method, 12-5–12-6
Flexible AC transmission systems (FACTS)
- configurations and applications, 5-18
- fixed series compensation (FSC), 5-20, 5-22
- static VAr compensators (SVCs), 5-18–5-19
- SVC "Classic" controls, 5-31
- SVC PLUS *vs.* SVC Classic, 5-22–5-23
- thyristor-controlled series compensation (TCSC), 5-20

Flywheels, 21-18
Four-pin method, 11-14
Frame relay, 15-26–15-27
Frequency hopping spread spectrum (FHSS), 15-24

G

Gas-insulated substation (GIS)
  air connection, 2-9–2-10
  bus designs, 9-6
  circuit breaker, 2-5
  control system, 2-13–2-14
  current transformers, 2-5–2-6
  direct transformer connections, 2-10–2-12
  disconnect switches, 2-6–2-7
  economics, 2-18
  electrical and physical arrangement, 2-15–2-16
  gas compartments and zones, 2-14–2-15
  gas monitor system, 2-14
  ground switches, 2-7–2-8
  grounding, 2-16–2-17
  installation, 2-17
  interconnecting bus, 2-8–2-9
  maintenance, 2-18
  operation and interlocks, 2-17
  power cable connections, 2-9–2-11
  power electronic substations, 5-1
  single-phase enclosure, 2-3
  sulfur hexafluoride (SF_6), 2-1–2-3
  surge arrester, 2-11–2-12
  testing, 2-17
  three-phase enclosure, 2-3–2-4
  voltage transformers, 2-6
Gas-insulated transmission line (GIL) system
  advantages of
    electrical losses, 18-24
    magnetic fields, 18-25–18-27
    safety and gas handling, 18-24–18-25
  corrosion protection
    active, 18-33
    passive, 18-31–18-32
  description, 18-1–18-2
  design tests, 18-35
  developments, 18-10–18-11
  diagnostic tools, 18-30–18-31
  gas mixture, 18-11
  high-power interconnections
    metropolitan areas, 18-35–18-37
    traffic tunnels usage, 18-37–18-38
  history of, 18-2–18-3
  long-duration tests
    calculation model, 18-20–18-21
    comparison of calculations, 18-21–18-22
    on directly buried GIL, 18-17–18-20
    load cycles and intermediate test, 18-14–18-15
    mechanical aspects, 18-23
    parameters test, 18-13, 18-15
    test cycle, 18-14–18-15
    thermal aspects, 18-20
    on tunnel-laid GIL, 18-16–18-18
  on-site tests, 18-35
  overhead lines, 18-1–18-2
  quality control, 18-30–18-31
  second-generation application
    directly buried, 18-29–18-30
    tunnel-laid, 18-26–18-29
  system design
    laying methods, 18-6–18-10
    standard units, 18-4–18-6
    technical data, 18-4
  transmission net
    description, 18-38
    integrating renewable energy, 18-41
    requirements, 18-39–18-40
    technical solutions, 18-40–18-41
    transmission losses, 18-41
  type tests
    dielectric tests, 18-13–18-14
    internal-arc test, 18-12–18-13
    short-circuit withstand tests, 18-12
  voltage stress, electric power net
    calculations results, 18-34–18-35
    external application, 18-34
    insulation coordination, 18-35
    integrated surge arresters, 18-34
    lightning strokes, 18-33
    modes of operation, 18-34
    overvoltage stresses, 18-33
Gas monitor system, 2-14
Generic Object Oriented System Event (GOOSE) concept
  IEC 61850, 7-18–7-19
  station bus-based architecture, 22-20–22-21
GFD, *see* Ground flash density (GFD)
Green Field *vs.* Brown Field, 7-7
Grid current, 11-17–11-19
Grid resistance, 11-15–11-16
Grid transformation
  distribution, 22-7
  engineering and design, 22-8
  enterprise integration, 22-10
  heavy metal electric delivery system, 22-7
  information infrastructure, 22-9
  operation and maintenance, 22-9–22-10
  substations, 22-8
  testing and commissioning, 22-10
Ground flash density (GFD), 12-5
Ground potential rise (GPR), 11-7
Ground switches, 2-7–2-8

H

Hazardous material, 9-14
High-power interconnections
  metropolitan areas
    in 1970, 18-36
    in 2000, 18-36–18-37
    in 2010, 18-37
    description, 18-35
  traffic tunnels usage, 18-37–18-38
High-speed fault clearing, 11-6

High-speed grounding switches, 4-17
High-voltage direct current (HVDC) transmission systems
    AC systems, 5-4
    back-to-back converter station, 5-6
    "Classic"–basic control functions, 5-30
    "Classic" control hardware, 5-30
    configuration possibilities, 5-2
    DC breakers, 5-5
    DC energy bridge, 5-8
    ground return transfer switches (GRTS), 5-5
    Hudson transmission project, 5-7
    HVDC PLUS (VSC) technology, 5-17–5-18
    hydro-power-based electrical energy, 5-9
    long-distance transmission station, 5-3
    metallic return transfer breakers (MRTB), 5-5
    Ningdong–Shandong DC system, 5-5
    RES, 5-17
    transformer bushings, 5-12
    UHV DC station equipment, 5-14–5-15
    Yunnan–Guangdong UHV DC system, 5-10–5-11, 5-13
High-voltage power electronic substations
    civil works, 5-32
    control and protection system, 5-28–5-31
    FACTS controllers (*see* Flexible AC transmission systems (FACTS))
    HVDC converters (*see* High-voltage direct current (HVDC) transmission systems)
    losses and cooling, 5-31–5-32
    reliability and availability, 5-32–5-33
    smart power and grid access, 5-23–5-28
High-voltage switching equipment
    ambient conditions, 4-1
    circuit breakers
        SF 6 gas dead tank, 4-22, 4-24
        single-tank bulk oil, 4-22–4-23, 4-25
        three-tank bulk oil, 4-22–4-23, 4-25
        types, 4-22
        vacuum, 4-22
    disconnect switches
        center break, 4-6, 4-9
        contact design, 4-11
        contact design vertical break, 4-7, 4-11
        devices, 4-2
        double end break, 4-6, 4-8
        double end break open position, 4-7, 4-11
        double end break two column structure, 4-9, 4-12
        gear crank operator, 4-4
        grounding, 4-6, 4-10
        grounding switch integrally attachment, 4-13
        hook stick operation, 4-6, 4-11
        horizontal mounted, 4-6
        key type interlock, 4-2–4-3
        mechanical cam-action type interlock, 4-2–4-3
        motor operator, 4-5
        pantograph, 4-6, 4-10
        single side break, 4-6, 4-10
        solenoid type interlock, 4-2–4-3
        stored energy motor operator, 4-5
        swing handle operator, 4-4
        under hung mounted, 4-6–4-7
        vertical break, 4-6–4-7
        vertical mounted, 4-6
        vertically mounted center break, 4-12
    high-speed grounding switches, 4-17
    load break switches
        arcing horns vertical break, 4-14
        factor influence, 4-16–4-17
        high-speed arcing horns, 4-14
        multibottle vacuum interrupters., 4-15–4-16
        SF 6 interrupter's pressure indicator, 4-15–4-16
        SF 6 interrupters, 4-15
    power fuses
        horizontal interrupter, 4-20–4-21
        incoming line structure mounted, 4-18–4-19
        integral horizontal interrupter, 4-20–4-21
        multiple interrupter, 4-19–4-20
        power transformer protection, 4-17–4-18
        vertical interrupter, 4-20
        vertically mounted, 4-18–4-19
Human machine interface (HMI), 6-3

I

IDS, *see* Intrusion detection system (IDS)
IEC, *see* International Electrotechnical Commission (IEC)
IEC 60870, 7-17
IEC 61850, smart substations
    communication standard, 22-28–22-29
    impact of
        communication network, 22-22
        interface requirements, 22-22
        logical interfaces, 22-23
        MU, 22-23–22-24
        systems for utility automation, 22-22
    interoperability
        application separation and communication concept, 22-19–22-20
        communication protocols, 22-21
        cornerstone technology, 22-17–22-18
        engineering approach, 22-19
        Ethernet, 22-21
        GOOSE concept, 22-20–22-21
        IED, 22-17
        implementation guidelines, 22-22
        information exchange models, 22-20
        logical nodes, 22-21
        SCSM, 22-21
        substation automation system interfaces, 22-18–22-19
        substation communication solution, 22-18
        transmission model, 22-21–22-22

paradigm shift
  circuit types, 22-15
  conventional substation design, 22-15–22-17
  traditional primary substation schemes, 22-15–22-16
station and process bus architecture
  alternative substation design, 22-27
  communication architecture, 22-27
  CT, 22-28
  functional architecture, 22-26
  process bus-based applications, 22-28
station bus-based architecture
  application requirements, 22-25–22-26
  functional architecture, 22-24–22-25
  GOOSE message, 22-24–22-25
  GOOSE-based schemes, 22-25
IED, *see* Intelligent electronic devices (IEDs)
IEEE 693 and complementary standards, 13-5–13-6
IEEE Standard 1402-2000, 16-4–16-5
Impulse sources, 9-7
Industrial zone, 9-14
INELFE project, 5-27–5-28
Integrated services digital network (ISDN), 15-25
Intelligent electronic devices (IEDs); *see also* Substation automation (SA) system
  control functions, 6-2–6-5, 6-15
  cyber security, 17-6
Interconnecting bus, 2-8–2-9
Intercontrol center communications, 7-17
International Electrotechnical Commission (IEC), 17-6
Intrusion detection system (IDS), 17-5–17-6
Intrusion prevention systems (IPSs), 17-6, 17-8
ISDN, *see* Integrated services digital network (ISDN)
Isolation devices, 10-5

K

Keraunic level, 12-4
Key performance indicators (KPIs)
  contributions
    prioritization based, 19-7–19-9
    projects based evaluation, 19-7

L

Latching devices, 6-21
Lead-acid batteries, 21-12
Legacy real-time operating systems, 17-12
Lifetime management plan
  practical issues
    age, 19-12
    age distribution, 19-13
    aging rate, 19-13
    ALEL, 19-13–19-14
    average age, 19-13
    basic metrics, 19-13
    breakdown age and rate, 19-14
    failure aging rate, 19-14
    failure and breakdown, 19-13
    long-term data, 19-12
    reliability, 19-13
    system equivalent failure age, 19-14
  value, 19-12
Lithium-ion batteries, 21-14
Load shifting, 21-4–21-5
Long-duration tests
  calculation model, 18-20–18-21
  comparison of calculations, 18-21–18-22
  on directly buried, 18-17–18-20
  load cycles and intermediate test, 18-14–18-15
  mechanical aspects, 18-23
  parameters test, 18-13, 18-15
  test cycle, 18-14–18-15
  thermal aspects, 18-20
  on tunnel-laid, 18-16–18-18

M

Magnetic fields
  GIL system
    devices, 18-24
    measurements at PALEXPO, Geneva, 18-25–18-27
    solid grounded earthing system, 18-25
  substations, 9-9–9-10
MAS radio, *see* Multiple address (MAS) radio
Merging unit (MU), 22-23–22-24
Mesh voltage, 11-7, 11-9–11-12
Migratory Bird Treaty Act, 10-2
Mitigation methods, animals
  barriers, 10-4
  deterrents, 10-4
  insulation, 10-4–10-5
  isolation devices, 10-5
Mobitex packet radio, 15-21–15-22
Monte Carlo simulation, EGM, 12-9
Mousa and Srivastava method, 12-17–12-18
MPLS, *see* Multiprotocol label switching (MPLS)
Multiple address (MAS) radio, 15-21
Multiple shielding electrodes, EGM, 12-15–12-17
Multiprotocol label switching (MPLS), 15-27

N

NAC, *see* Network access control (NAC)
NaS batteries, *see* Sodium-Sulfur (NaS) batteries
Network access control (NAC), 17-6
Nickel-cadmium (NiCd) battery, 21-12–21-13
Nickel-metal hydride (NiMH) batteries, 21-12–21-13
Ningdong–Shandong DC system, 5-5
Noise sources
  continuous radio frequency (RF) sources, 9-7
  impulse sources, 9-7
  site housekeeping, 9-14

O

Oil containment
  conditions, 8-1–8-2
  oil-filled equipment
    cables, 8-2
    large oil-filled equipment, 8-2
    mobile equipment, 8-2
    oil-handling equipment, 8-2–8-3
    oil storage tanks, 8-3
    other sources, 8-3
    spill risk assessment, 8-3–8-4
  selection criteria, 8-4–8-5
  spill prevention control and countermeasure (SPCC) plan, 8-1
  spill prevention techniques
    containment systems, 8-5–8-10
    discharge control systems, 8-10–8-14
    warning alarms and monitoring, 8-14–8-15
Oil-filled equipment
  cables, 8-2
  large oil-filled equipment, 8-2
  mobile equipment, 8-2
  oil-handling equipment, 8-2–8-3
  oil storage tanks, 8-3
  other sources, 8-3
  spill risk assessment, 8-3–8-4
Oil storage tanks, 8-3
Optical fiber systems
  Ethernet over fiber, 6-25
  fiber loops, 6-24–6-25
  fiber stars, 6-25
  message limitations, 6-25
  SCADA and automation, 6-23–6-24
Optical power ground wire (OPGW) cables, 15-20
Orthogonal frequency division multiplexing (OFDM), 15-24

P

Paging systems, 15-22
Pareto curves
  characteristics
    bumps and small shifts in slop, 19-17
    decreasing marginal return, 19-17
    must-do region, 19-16–19-17
    portfolios, 19-17
  description, 19-15
Peer-to-peer networks, 6-23
PLC systems, *see* Power line carrier (PLC) systems
Point-to-multipoint networks, 6-22
Point-to-point networks, 6-22
Portfolio management, 19-6–19-7
Power fuses
  horizontal interrupter, 4-20–4-21
  incoming line structure mounted, 4-18–4-19
  integral horizontal interrupter, 4-20–4-21
  multiple interrupter, 4-19–4-20
  power transformer protection, 4-17–4-18
  vertical interrupter, 4-20
  vertically mounted, 4-18–4-19
Power line carrier (PLC) systems, 15-22–15-23
Probabilistic risk management, 19-8–19-10
Product verification plan (PVP), 1-8
Programmable logic controllers (PLCs), 6-2–6-3, 6-21–6-22
Project closeout
  accounting, 20-10
  archive records, 20-9
  closure of, 20-8–20-9
  construction drawings, 20-5–20-6
  equipment manuals, 20-6
  final governing agency inspections, 20-8–20-9
  final walk-through/inspection, 20-4
  invoices, 20-6–20-7
  lessons learned opportunities, 20-11
  operations instructions, 20-6
  permits/sureties required, 20-8
  punch list items, 20-4–20-5
  stakeholder notification, 20-10–20-11
  unit cost development, 20-9–20-10
Pulse accumulators (PAs), 6-16

Q

Quality of service (QoS), 15-7

R

Relay systems, SCADA
  performance factors, 15-5
  phasor measurements, 15-7
  quality of service (QoS), 15-7
  Smart Grid characterization, 15-6–15-7
  time distribution, 15-7
Remote terminal unit (RTU), 17-7
Renewable energy source (RES)
  GIL system, 18-41
  HVDC transmission systems, 5-17
Residential zone, 9-14
Revised electrogeometric model
  application, 12-11–12-15
  BIL, 12-10–12-11
  bus insulators, 12-10
  description, 12-10
  Mousa and Srivastava method, 12-17–12-18
  negative polarity impulse critical flashover, 12-10–12-11
  vs. Whitehead's model, 12-10
  withstand voltage, insulator strings, 12-11
Ring bus arrangement, 3-5
Risk-based asset management methods
  analysis performance, 19-18–19-19
  description, 19-17

SAIDI, 19-17–19-18
utility asset management, 19-18
Rolling sphere method, EGM
  principle, 12-12
  shield mast protection, stroke current, 12-12–12-15
  use of, 12-12
RS-232/EIA-232, 7-7–7-8
RS-422, 7-8
RS-485/EIA-485/TIA-485, 7-8
RTU, *see* Remote terminal unit (RTU)

S

SA system, *see* Substation automation (SA) system
Safety
  fences and walls, 9-10
  fire protection, 9-11
  GIL system, 18-24–18-25
  grounding, 9-11
  lighting, 9-10
Satellite systems, 15-23
SCADA systems, *see* Supervisory control and data acquisition (SCADA) systems
Secondary oil-containment, 8-3–8-4
Security event manager (SEM), 17-7
Seismic considerations
  ASD, 13-11
  design process, 13-6
  design standards
    benefits of, 13-4–13-5
    IEEE 693 and complementary standards, 13-5–13-6
  earthquakes and substations, 13-1–13-4
  installation, 13-14–13-15
  qualification
    acceptance criteria, 13-11–13-12
    analytical methods, 13-10–13-11
    earthquake hazard method, 13-8
    equipment supports, 13-12–13-13
    level and design earthquake, 13-7
    level selection, 13-8–13-10
    methods, 13-10–13-12
    performance level, 13-13–13-14
    projected performance level, 13-13–13-14
    test methods, 13-10–13-11
SEM, *see* Security event manager (SEM)
Short message system (SMS), 15-23–15-24
Single bus arrangement, 3-1–3-2
Site acceptance tests (SATs), 1-8
Site delivery acceptance test (SDAT), 1-8
Site housekeeping, 9-12–9-13
Site preparation
  clearing, grubbing, excavation, and grading, 9-11
  site access roads, 9-11
  water drainage, 9-12
Smart feeder applications platform, 22-14–22-15

Smart Grids
  benefits
    broader deployment, 22-5–22-6
    components, 22-4–22-5
    enterprise-wide solution, 22-6
    importance of, 22-5
    operational efficiency, 22-5
    reliability, 22-5
  challenges, 22-29
  description, 22-1–22-3
  drivers, 22-3–22-4
  electric grid, 22-2
  grid transformation
    distribution, 22-7
    engineering and design, 22-8
    enterprise integration, 22-10
    heavy metal electric delivery system, 22-7
    information infrastructure, 22-9
    operation and maintenance, 22-9–22-10
    substations, 22-8
    testing and commissioning, 22-10
  IEC 61850, smart substations
    communication standard, 22-28–22-29
    impact of, 22-22–22-24
    interoperability, 22-17–22-22
    paradigm shift, 22-15–22-17
    station and process bus architecture, 22-26–22-28
    station bus-based architecture, 22-24–22-26
  real-time system control, 22-29
  SCADA systems
    data flow, 22-12, 22-13
    IEDs, 22-12–22-14
    network-based architecture, 22-11
    protocol architecture, 22-11
    RTU, 22-10–22-11
    security architecture, 22-12–22-13
    server-based control system architecture, 22-12
    substation migration, 22-12, 22-14
    utility communication, 22-12–22-13
  smart feeder applications platform, 22-14–22-15
  technology framework, 22-6–22-7
  usages, 22-29–22-30
  wide deployment, 22-29
Smart substations, IEC 61850
  communication standard, 22-28–22-29
  impact of
    communication network, 22-22
    interface requirements, 22-22
    logical interfaces, 22-23
    MU, 22-23–22-24
    systems for utility automation, 22-22
  interoperability
    application separation and communication concept, 22-19–22-20
    communication protocols, 22-21
    cornerstone technology, 22-17–22-18

engineering approach, 22-19
Ethernet, 22-21
GOOSE concept, 22-20–22-21
IED, 22-17
implementation guidelines, 22-22
information exchange models, 22-20
logical nodes, 22-21
SCSM, 22-21
substation automation system interfaces, 22-18–22-19
substation communication solution, 22-18
transmission model, 22-21–22-22
paradigm shift
circuit types, 22-15
conventional substation design, 22-15–22-17
traditional primary substation schemes, 22-15–22-16
station and process bus architecture
alternative substation design, 22-27
communication architecture, 22-27
CT, 22-28
functional architecture, 22-26
process bus-based applications, 22-28
station bus-based architecture
application requirements, 22-25–22-26
functional architecture, 22-24–22-25
GOOSE message, 22-24–22-25
GOOSE-based schemes, 22-25
Sodium-Sulfur (NaS) batteries, 21-14–21-15
Soil resistivity, 11-14–11-15
SONET, *see* Synchronous optical networking (SONET)
Station commissioning
coordination, 20-2–20-3
legal notification, 20-3
project closeout
accounting, 20-10
archive records, 20-9
closure of, 20-8–20-9
construction drawings, 20-5–20-6
equipment manuals, 20-6
final governing agency inspections, 20-8–20-9
final walk-through/inspection, 20-4
invoices, 20-6–20-7
lessons learned opportunities, 20-11
operations instructions, 20-6
permit closure process, 20-8–20-9
permits/sureties required, 20-8
punch list items, 20-4–20-5
stakeholder notification, 20-10–20-11
unit cost development, 20-9–20-10
site issues, 20-3
testing, 20-1–20-2
Step voltage
calculation, 11-13
circuit, 11-2, 11-4
definition, 11-7

exposure to, 11-2, 11-3
geometrical factor, 11-13–11-14
Thevenin equivalent impedance, 11-4
Storage technologies, 21-11–21-12
Stress
electrical, 2-5
voltage, electric power net
calculations results, 18-34, 18-35
external application, 18-34
insulation coordination, 18-35
integrated surge arresters, 18-34
lightning strokes, 18-33
modes of operation, 18-34
overvoltage stresses, 18-33
Substation asset management; *see also* Asset management
advanced technology, 19-33
BBOM, 19-32
condition-based maintenance, 19-31–19-32
equipment stewardship
comprehensive asset management approach, 19-34, 19-36
perspective and culture change, 19-33
single-planning group approach, 19-35
traditional standards-driven utility paradigm, 19-34
life cycle management, 19-29–19-31
priorities setting, 19-32
utilization, 19-29–19-31
Substation automation (SA) system
air core current transformers, 6-13
communication networks
assessing channel capacity, 6-26
network reliability, 6-26
optical fiber systems, 6-23–6-25
peer-to-peer networks, 6-23
point-to-multipoint networks, 6-22
point-to-point networks, 6-22
control functions
control circuit designs, 6-20–6-21
intelligent electronic devices, 6-21–6-22
interposing relays, 6-19–6-20
latching devices, 6-21
cyber security enhance measurement
cyber intrusion, 17-15–17-24
principles, 17-15
electrical measuring interface, 6-6
hall effect sensor, 6-13
instrument transformers
current transformers, 6-11–6-12
voltage sources, 6-13
integrated energy measurements, 6-16
magneto-optic sensors, 6-13
measurements
digitized measurements, 6-8–6-11
IED measurements, 6-5–6-6
performance requirements, 6-6–6-8

measuring devices
    intelligent electronic devices, 6-15
    scaling measured values, 6-15–6-16
    transducers, 6-14–6-15
physical challenges
    components, 6-1–6-2
    electrical environment, 6-4–6-5
    environment, 6-4
    locating interfaces, 6-2–6-3
resistor divider, 6-13
Rogowsky coils, 6-13
SCADA and automation communications, 6-25–6-26
security challenges
    authentication lack, 17-13
    centralized system administration lack, 17-14
    description, 17-11
    insecure communication media, 17-12
    legacy real-time operating systems, 17-12
    open protocols, 17-12–17-13
    organizational issues, 17-13–17-14
    remote devices, 17-14
    stringent real-time constraints, 17-11–17-12
    substation diagnostic systems, 17-14
security misconceptions
    electronic security perimeters, 17-7–17-8
    ICS/SA devices, 17-9
    IPSs, 17-8
    network, 17-8–17-9
    security technologies, 17-9–17-10
    VPN, 17-8
status monitoring
    ambiguity, 6-17–6-18
    contact performance, 6-17
    wetting sources, 6-18
    wiring practices, 6-19
substation wiring practices, 6-14
testing automation systems
    control, 6-27
    in-service testing, 6-28–6-29
    measurements, 6-27–6-28
    programmed logic obtains data, 6-28
    status points, 6-27
    test plan, 6-28
threat actors, 17-10–17-11
Substation diagnostic systems, 17-14
Substation fire protection
    automatic, 14-4–14-5
    building
        active measures, 14-12
        assessment, 14-22–14-23
        exit facilities, 14-11
        hazard assessment, 14-23
        life safety assessment, 14-22
        manual measures, 14-13
        passive measures, 14-12
        risk assessment, 14-22
        safety measures, 14-10–14-11
    economic risk analysis, 14-18–14-19
    fire recovery, 14-5
    hazards
        building, 14-7–14-8
        fuel control, 14-6
        high-voltage DC valves, 14-7
        ignition sources, 14-6
        oil-insulated equipment, 14-7
        switchyard, 14-8–14-9
    incident management, 14-5
    management process
        environmental preparedness, 14-21
        fire safety plans, 14-19–14-20
        fire training, 14-21
        operations plan, 14-20–14-21
        recovery, 14-21–14-22
    manual fire suppression, 14-5
    objectives
        compliance, 14-3–14-4
        electrical supply reliability, 14-2
        operational safety, 14-2–14-3
        revenue and asset preservation, 14-3
        risk management, 14-4
    preparedness philosophy, 14-5
    prevention/safety, 14-4
    selection criterion, 14-18
    site-related considerations, 14-9
    switchyard mitigation measures
        active systems, 14-16–14-17
        fire spread assessment, 14-24
        manual systems, 14-17
        passive measures, 14-13–14-16
        radiant exposure assessment, 14-23–14-24
        risk assessment, 14-23
Substation grounding system
    accidental ground circuit
        conditions, 11-2–11-5
        high-speed fault clearing, 11-6
        permissible body current limits, 11-5–11-6
        tolerable voltages, 11-7–11-9
    description, 11-1
    design criteria
        computer programs, 11-19
        conductors, selection of, 11-19–11-21
        connections, selection of, 11-21–11-22
        fence, 11-22
        grid current, 11-17–11-19
        grid resistance, 11-15–11-16
        mesh voltage, 11-9–11-12
        soil resistivity, 11-14–11-15
        step voltage, 11-12–11-14
    safety requirements, 11-1–11-2
Substation integration
    asset management, 7-4–7-5
    automation applications, 7-14–7-15
    components

    Bay controllers, 7-12
    data concentrators, 7-11
    Ethernet switches, 7-13–7-14
    gateways, 7-11
    human machine interface (HMI), 7-12–7-13
    logic processors, 7-12
    protocol convertors, 7-11
    remote input/output devices, 7-12
    remote terminal unit (RTU), 7-10–7-11
    routers and layer 3 switches, 7-14
  configuration data, 7-3
  cyber security, 7-14
  data flow
    data concentrator, 7-4
    field devices, 7-4
    methods, 7-3
    SCADA and data warehouse, 7-4
    segregation, 7-3
    substation communications, 7-4
  DNP 3.0, 7-16–7-17
  factory acceptance test, 7-10
  highly available networks, 7-9–7-10
  IEC 60870, 7-17
  IEC 61850
    configuration paradigm, 7-18
    GOOSE, 7-18–7-19
    station bus and process bus, 7-19
  modbus protocol, 7-17
  nonoperational data, 7-2–7-3
  open systems, 7-2
  operational data, 7-2
  OSI communications model
    application layer, 7-15
    data-link layer, 7-16
    network layer, 7-16
    physical layer, 7-16
    presentation layer, 7-15
    session layer, 7-15–7-16
    transmission layer, 7-16
  proprietary protocols, 7-17
  protocol considerations, 7-6
  redundancy, 7-5
  serial communications
    fiber optics, 7-8
    RS-232/EIA-232, 7-7–7-8
    RS-422, 7-8
    RS-485/EIA-485/TIA-485, 7-8
  synchrophasors
    C37.118, 7-19
    phasor data concentrator (PDC), 7-20–7-21
    phasor measurement units (PMUs), 7-20
    wide area situational awareness, 7-20
  system architecture
    design, 7-7
    documentation, 7-6–7-7
Substations
  budgeting process, 1-2–1-3

  commissioning, 1-8
  construction, 1-8
  customer requirements, 1-2
  design, 1-5–1-8
  direct lightning stroke shielding (*see* Direct lightning stroke shielding system)
  engineering, 1-3–1-4
  equipment type, 1-2
  financing, 1-3
  site selection and acquisition, 1-4–1-5
  switchyard, generating station, 1-1
  system requirements, 1-2
  traditional and innovative substation design, 1-3–1-4
  types, 1-1
Substations, physical security
  comprehensive system analysis
    criticality assessment, 16-8–16-9
    risk assessment, 16-9–16-10
    vulnerability assessment, 16-9
  contractual methods, 16-19
  electric system, 16-3–16-5
  intrusion
    definition, 16-6
    disgruntled employees, 16-7
    general public, 16-6
    resources, 16-8
    terrorists, 16-7–16-8
    thieves, 16-6–16-7
    vandals, 16-7
  management/organizational methods, 16-19–16-20
  physical methods
    barriers, 16-14
    clear areas and safety zones, 16-15–16-16
    control building design, 16-15
    drawings and information books, 16-17
    fences and walls, 16-13–16-14
    gates and locks, 16-14
    ground mats, 16-14–16-15
    grounding, 16-14–16-15
    intrusion detection systems, 16-16
    landscaping, 16-14
    lighting, 16-15
    personnel access, 16-17
    relay and control equipment, 16-18
    SCADA/communication equipment, 16-17–16-18
    security patrols, 16-15
    signs, 16-15
    site maintenance, 16-16
    substation service, 16-17
    video motion detection systems, 16-17
  presence/capability/intent, 16-5–16-6
  responsibility for
    federal government, 16-11–16-12
    NERC security guidelines, 16-12
    owner/operator, 16-11
  risk management, 16-10–16-11

system methods, 16-18–16-19
threat assessment, 16-5
Sulfur hexafluoride (SF$_6$), 2-1–2-3
Sump pump water discharge, 8-14–8-15
Supervisory control and data acquisition (SCADA) systems
  AMPS, 15-19
  ARDIS, 15-18–15-19
  CDPD, 15-19
  common-carrier digital systems, 15-19
  communication protocol
    description, 15-9
    distributed network protocol, 15-13
    end-to-end messaging, 15-10–15-11
    ICCP, 15-14
    IEC 60870-5, 15-13
    IEC 61850, 15-14
    IEC TC57 Working Group, 15-14–15-15
    Internet protocol (IP), 15-14
    layered message structure, 15-10–15-11
    OSI reference model, 15-9–15-10
    standard process, 15-12
    TCP, 15-14
    UCA 1.0, 15-13
    UCA 2.0, 15-14
    work of, 15-9
  communication requirements, 15-5
  data flow, 22-12–22-13
  digital microwave systems, 15-19–15-20
  DSSS, 15-24
  electromagnetic environment, 15-17–15-18
  FHSS, 15-24
  fiber optics, 15-20
  F/O technology, 6-25, 6-26
  history of, 15-2–15-4
  hybrid fiber/coax, 15-20–15-21
  IEDs, 22-12–22-14
  MAS radio, 15-21
  mobile computing infrastructure, 15-21
  mobile radio systems, 15-21
  Mobitex packet radio, 15-21–15-22
  modern, 15-8
  network-based architecture, 22-11
  networked communications, 15-8–15-9
  OFDM, 15-24
  paging systems, 15-22
  PLC systems, 15-22–15-23
  power system relay communication references, 15-28–15-29
  protocol architecture, 22-11
  relay systems
    performance factors, 15-5
    phasor measurements, 15-7
    QoS, 15-7
    Smart Grid characterization, 15-6–15-7
    time distribution, 15-7
  RTU, 6-2–6-3, 22-10–22-11
  satellite systems, 15-23
  security
    authorization violation, 15-15
    by obscurity, 15-16
    denial-of-service attack, 15-16–15-17
    eavesdropping, 15-15
    encryption, 15-16–15-17
    information leakage, 15-15
    intercept/alter, 15-15
    ISO 7498-2 definitions, 15-15
    masquerade, 15-16
    message data integrity checking, 15-16
    replay, 15-16
  security architecture, 22-12–22-13
  server-based control system architecture, 22-12
  SMS, 15-23–15-24
  spread spectrum technology, 15-24
  Standards
    DNP3 Specifications for Device Communication, 15-30
    IEC 60870 Standards for Telecommunication, 15-29–15-30
    IEC 60870-6 TASE.2 (UCA/ICCP) Control System Communications, 15-30–15-31
    IEC 61850/UCA Standards for Substation Systems, 15-31–15-32
    IEC 61968 Standards for Distribution Application Integration, 15-32–15-33
    IEC 61970 Standards for Energy Management System Integration, 15-33
    IEEE 802.x Networking Standards, 15-29
    IEEE Electromagnetic Interference Standards, 15-29
    ISO Reference Models, 15-34
  substation migration, 22-12, 22-14
  telephone-based systems
    ATM, 15-27
    dial-up telephone lines, 15-24–15-25
    DSL, 15-25–15-26
    frame relay, 15-26–15-27
    ISDN, 15-25
    MPLS, 15-27
    SONET, 15-27
    T1 and fractional T1 technology, 15-26
  traditional, 15-8
  utility communication, 22-12–22-13
  websites, 15-28
  wireless LANs, 15-24
Switchgear, 9-6
Switching devices and buses
  breaker-and-a-half arrangement, 3-5–3-7
  comparison, 3-6–3-7
  double bus–double breaker arrangement, 3-2–3-3
  double bus–single breaker arrangement, 3-4–3-5
  main and transfer bus arrangement, 3-3–3-4
  ring bus arrangement, 3-5
  single bus arrangement, 3-1–3-2

Switchyard mitigation measures
   active systems
      automatic fire protection systems, 14-16–14-17
      explosion suppression, 14-17
   dry chemical extinguishers, 14-17
   fire spread assessment, 14-24
   passive measures
      cable systems, 14-16
      fire barriers, 14-15–14-16
      fire spill containment, 14-15
      ground cover, 14-15
      spatial separation, 14-13–14-15
   portable fire extinguishers, 14-17
   radiant exposure assessment, 14-23–14-24
   risk assessment, 14-23
   water supplies, 14-17
Symmetrical currents, 11-20–11-21
Synchronous optical networking (SONET), 15-27
Synchrophasors
   C37.118, 7-19
   phasor data concentrator (PDC), 7-20–7-21
   phasor measurement units (PMUs), 7-20
   wide area situational awareness, 7-20

T

Thevenin equivalent impedance, 11-4
Tolerable voltages, 11-7–11-9
Touch voltage
   circuit, 11-2–11-3
   definition, 11-8
   exposure to, 11-2–11-3
   metal-to-metal, 11-7
   Thevenin equivalent impedance, 11-4
Transmission net, GIL system
   description, 18-38
   integrating renewable energy, 18-41
   requirements, 18-39–18-40
   technical solutions, 18-40–18-41
   transmission losses, 18-41
Tunnel laying method, 18-8–18-9

U

Unified power flow controller (UPFC), 5-27
Uniform soil resistivity, 11-15

V

Vanadium redox batteries
   battery technology reactions, 21-17
   diagram of, 21-17
   redox flow systems, 21-16
   system, 21-16
Ventricular fibrillation, 11-5
Virtual private network (VPN), 17-8
Voltage
   conversion, 1-1
   stress, electric power net
      calculations results, 18-34, 18-35
      external application, 18-34
      insulation coordination, 18-35
      integrated surge arresters, 18-34
      lightning strokes, 18-33
      modes of operation, 18-34
      overvoltage stresses, 18-33
   touch
      circuit, 11-2–11-3
      definition, 11-8
      exposure to, 11-2–11-3
      metal-to-metal, 11-7
      Thevenin equivalent impedance, 11-4
   transformers, 2-6
   VT fuses, 6-13
VPN, *see* Virtual private network (VPN)

W

Wenner method, 11-14
Wetlands, 9-14
Whitehead's model, EGM, 12-9
Wood fences, 9-4

Y

Yunnan–Guangdong UHV DC system, 5-10–5-11, 5-13

Z

Zinc-bromine batteries, 21-15–21-16